Planet Earth

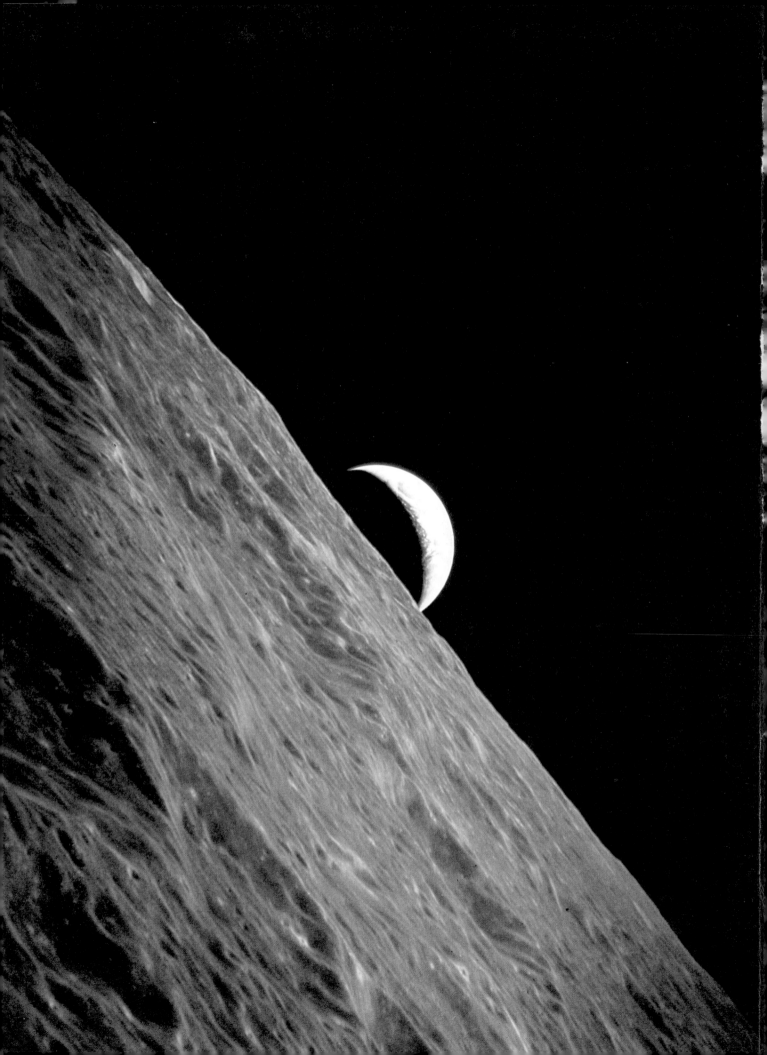

An Encyclopedia of Geology

Planet Earth

Consultant Editor
A. Hallam PhD
*Lapworth Professor of
Geology
University of Birmingham*

ELSEVIER PHAIDON

Frontispiece : Earthrise, seen from lunar orbit

Editors :
Peter Hutchinson PhD
Paul Barnett

Managing Editor :
Giles Lewis

Picture Editors :
Hilary Kay
Andrew Lawson

Design :
Adrian Hodgkins

Production :
Andrew Ivett

Visual Aids :
Roger Gorringe, Bryon Harvey, John
Fuller, Tony Morris, Oxford
Illustrators Ltd., Lovell Johns Ltd.,
Diagram Visual Information Ltd.,
Allard Graphics Ltd.

Maps :
Lovell Johns Ltd.

ISBN : 0 7290 0055 9

Elsevier-Phaidon, an imprint of
Phaidon Press Ltd, Littlegate House, St Ebbe's
Street, Oxford

Planned and produced by
Elsevier International Projects Ltd., Oxford
Copyright © 1977
Elsevier Publishing Projects SA, Lausanne

Filmset by Keyspools Ltd., Golborne,
Lancashire, Great Britain
Origination by Art Color Offset, Rome, Italy
and City Engraving Ltd., Hull, Great Britain
Printed by Jolly and Barber Ltd., Rugby,
Great Britain
Bound by Webb Son and Co. Ltd., Ferndale,
Great Britain

Contents

The Earth and Its Neighbors
Planet Earth 9
The Moon 14
Mercury 18
Venus 19
Mars 20
Meteorites 24
Tektites 27
Geophysics 28
Geochemistry 33
Climatic Zones 36

Processes That Shape the Earth
Continental Drift 39
Plate Tectonics 41
Weathering 46
Erosion 48
Deposition 52
Glaciation 56
Diagenesis 58
Faulting 60
Earthquakes 62
Folding 66
Igneous Intrusion 68
Vulcanicity 70
Geosynclines 72

Landscapes
Mountain Chains 73
River Valleys 75
Rift Valleys 78
Plateaux 79
Karst Landscapes 81
Arid Landscapes 83
Glacial Landscapes 86
Tundra Landscapes 88
Coastal Plains 89
Deltas 90
Coastlines 93
Volcanic Landscapes 95

The Ocean Floor
The Ocean Floor and Sea-Floor
  Spreading 97

Economic Geology
Engineering Geology 103
Hydrogeology 107
Mining Geology 109
Ore Deposits 112
Crystals 113
Gemstones 118
Minerals 119
Elements 139
Petroleum 149
Natural Gas 152
Coal 153

Salt 157
Building Stone 159

The Rocks of the Earth
Igneous Rocks 160
Sedimentary Rocks 167
Metamorphic Rocks 174
Soils 178

The Geological History of the Earth
The Age of the Earth 180
Principles of Stratigraphy 184
Precambrian 187
Paleozoic 190
  Cambrian 190
  Ordovician 194
  Silurian 197
  Devonian 201
  Carboniferous (Mississippian
    and Pennsylvanian) 204
  Permian 209
Mesozoic 213
  Triassic 213
  Jurassic 217
  Cretaceous 221
Cenozoic 225
  Tertiary 225
    Paleocene 225
    Eocene 227
    Oligocene 229
    Miocene 229
    Pliocene 231
  Quaternary 232

The History of Life on Earth
The Origin of Life 235
Evolution 237
The Fossil Record 240
Paleoecology 245
Plants and Animals 248
Invertebrate Animals 253
Vertebrate Animals 264
  Jawless Fishes 264
  Jawed Fishes 265
  Amphibians 269
  Reptiles 270
  Birds 278
  Mammals 279
  Hominids 284

The Making of Geology
Man and the Earth 288
Great Geologists 297

Glossary 310
Index 315

List of Contributors

DVA   Professor D. V. Ager DSc PhD
Department of Geology and
Oceanography
University College of Swansea

FBA   F. B. Atkins MA DPhil
Department of Geology and
Mineralogy
University of Oxford

JA   J. Anderson BSc
Department of Geology
University of Witwatersrand

MCA-C   M. C. Audley-Charles PhD BSc
Department of Geology
Imperial College
University of London

GMB   Professor G. M. Brown DSc DPhil FRS
Department of Geological Sciences
University of Durham

GSB   G. S. Boulton BSc PhD FGS
School of Environmental Sciences
University of East Anglia

JDB   J. D. Bell MA DPhil
Department of Geology and
Mineralogy
University of Oxford

MDB   M. D. Brasier BSc PhD
Department of Geology
University of Hull

PCSB   Professor P. C. Sylvester Bradley
Department of Geology
University of Leicester

R & DB   R. P. Beckinsale MA DPhil
School of Geography
University of Oxford
and
D. M. Beckinsale BA DipEd
late of Department of Geography
University of Reading

WAB   W. A. Berggren DSc
Woods Hole Oceanographic Institution
Massachusetts

CBC   Professor C. B. Cox MA PhD DSc
Department of Zoology
Kings College
University of London

CDC   C. D. Curtis BSc PhD MIMM CEng
Sorby Laboratory
Department of Geology
University of Sheffield

JDC   J. D. Collinson MA DPhil
Department of Geology
University of Keele

KGC   K. G. Cox PhD
Department of Geology and
Mineralogy
University of Oxford

RD   Professor R. Davies CEng FIMM FGS
Department of Geology
Imperial College
University of London

DE   D. Edwards MA PhD
Department of Botany
University College of Cardiff

DGF   D. G. Fraser DPhil BSc
Department of Geology and
Mineralogy
University of Oxford

GF   G. Fielder BSc PhD
Lunar and Planetary Unit
University of Lancaster

PF   P. Forey PhD
Department of Paleontology
British Museum (Natural History)

AG   A. S. Goudie MA PhD
School of Geography
University of Oxford

FG   F. H. W. Green MSc
Department of Agricultural Science
University of Oxford

PCG   P. C. Gardner MA
late of Jesus and Wolfson Colleges
Oxford

RG   R. Goldring PhD
Department of Geology
University of Reading

AH   Professor A. Hallam
Lapworth Professor of
Geology
University of Birmingham

CH   C. Harrison PhD
Sub-Department of Ornithology
British Museum (Natural History)

LBH   L. B. Halstead PhD DSc
Departments of Geology and Zoology
University of Reading

AI   additional information supplied by
the editorial staff of Elsevier
International Projects Ltd

GALJ   G. A. L. Johnson DSc PhD
Department of Geology
University of Durham

GPJ   G. P. Jones
Department of Geology
University College
University of London

CAMK   Professor C. A. M. King MA PhD ScD
Department of Geography
University of Nottingham

JLK   Professor J. L. Knill PhD MICE
Department of Geology
Imperial College
University of London

WJK   W. J. Kennedy PhD
Department of Geology and Mineralogy
University of Oxford

DGM   D. G. Murchison BSc PhD FGS FRSE
Department of Geology
University of Newcastle-upon-Tyne

RBM   R. B. McConnell DPhil DèsSc
late of Overseas Geological Survey

SM   S. Moorbath MA DPhil DSc
Department of Geology and
Mineralogy
University of Oxford

SCM   S. Conway-Morris BSc PhD
Department of Geology
Sedgwick Museum
Cambridge

TM   Professor T. Mutch
Department of Geological Sciences
Brown University
Providence, RI

WSM   W. S. McKerrow DPhil
Department of Geology and
Mineralogy
University of Oxford

ERO   E. R. Oxburgh MA PhD
Department of Geology and
Mineralogy
University of Oxford

EMP   E. M. Parmentier PhD
Department of Geology and
Mineralogy
University of Oxford

NJP   N. J. Price DSc PhD FGS
Department of Geology
Imperial College
University of London

RP   R. Porter MA PhD
Churchill College
Cambridge

SWR   S. W. Richardson MA DPhil
Department of Geology and Mineralogy
University of Oxford

DS   Professor D. Skevington PhD DSc
Department of Geology
University College
Galway

ETCS   E. T. C. Spooner MA
Department of Geology and Mineralogy
University of Oxford

MMS   M. M. Sweeting MA PhD
School of Geography
University of Oxford

RS   R. C. Selley PhD DIC BSc
Department of Geology
Imperial College
University of London

RCS   R. C. Searle PhD
Institute of Oceanographic Sciences
Surrey

RJGS   R. J. G. Savage PhD
Department of Geology
University of Bristol

HST   H. S. Torrens PhD
Department of Geology
University of Keele

JT   J. Tarney PhD
Department of Geological Sciences
University of Birmingham

AW   A. Wesley BSc ARCS FGS FLS
Department of Plant Sciences
University of Leeds

MHW   M. H. Worthington PhD
Department of Geology and Mineralogy
University of Oxford

RCLW   R. C. L. Wilson PhD
Department of Earth Sciences
The Open University

D. H. Tarling PhD of the Department of
Geophysics and Planetary Physics, University of
Newcastle-upon-Tyne, compiled the briefs from
which the paleographical maps were prepared.

Preface

Everyone has looked from the window of a car or train and seen a mountainside or the meandering course of a river and wondered how it came to be there. Everyone has encountered in the course of a country walk an unusual rock or a cave and wondered about its past history. Some have indulged their wonder further than this by spending an occasional weekend searching for fossils or reading a book on the evolution of the landscape. Some have gone still further to become professional geologists.

Geology has long been a subject of interest to the layman: indeed, contributions made by amateurs to our understanding of the Earth's history have been of great significance during the last two centuries, and there are still many fields of geology in which the amateur can play a major part. And the science of geology itself is currently enjoying a renaissance as new ideas are coming to the fore and old ideas are being discarded.

There are many features of *Planet Earth* that would have been drastically different – perhaps even excluded – had the encyclopedia been published only a few years ago. This is the result of two "revolutions". The first is the elucidation of the process known as plate tectonics. Perhaps as much as fifty per cent of this book relates in some way to this all-important concept, which provides answers to problems posed by paleontologists, stratigraphers and geophysicists for decades. A single example is provided by the unique series of maps towards the end of this book showing the distribution of the Earth's landmasses at various stages during the geological past. The second of these "revolutions" has come about through our exploration of the nearby planets: space probes have recently given us our first fascinating clues to the geologies of the Moon, Mercury, Venus and Mars.

Because an understanding of the Earth depends on an appreciation of the fruits of recent research, the authors of the articles in this encyclopedia have been selected from those actively pursuing such research. They have been conscious that technical terminology is unavoidable in the literature of any scientific discipline, but have ensured that specialist terms are thoroughly explained in everyday language, either *in situ* or, where a term occurs frequently, in the glossary. For both these reasons, the editors have been able to include in *Planet Earth* much that has appeared before only in learned journals. Chemical formulae and, occasionally, mathematical notation have been used but in all cases the less scientifically oriented reader will find no difficulty in following the argument simply by reading the surrounding text.

Although each article is self-contained, in cases where a topic mentioned is dealt with more fully in another article this is indicated by a star (*) immediately before the name of that article. Such cross-references, together with the many shorter entries and the comprehensive index, permit rapid access to the information contained within this book.

Nearly half of the available space in *Planet Earth* is devoted to photographs, maps and diagrams. Photographs have been chosen not only for their beauty but also, more particularly, to amplify or extend points made in the text. Maps and diagrams have been specially commissioned with the assistance of the authors, in particular, the section *The History of Life on Earth* is illustrated with a unique series of some seventy panels that show the structure and geological range of every plant and animal group that has been preserved in the fossil record. These illustrations, of a characteristic member of each group, will enable the amateur collector to identify the important features of any specimen he may find in the field or see in any museum collection.

The Earth and its Neighbors

Planet Earth

Over 4.6 billion years ago, a tenuous, rotating cloud of dust and gas drifting through interstellar space began to contract. The contraction was caused by gravitational attraction between different parts of the cloud with – perhaps – a little push from the explosion of a supernova or the light of surrounding stars. As the cloud contracted, the temperature and pressure near its center increased. A nuclear reaction, fusing hydrogen atoms into helium, was triggered by the high temperature and pressure, and so, close to the center of the cloud, the Sun began to shine. Meanwhile the dust and gas further from the center of the cloud were coalescing to form the planets – among which, of course, was the planet Earth.

This is the nebular hypothesis for the origin of the Solar System, the contracting cloud of dust and gas being termed the solar nebula. In its most general form, the nebular hypothesis is several centuries old, but only very recently have the processes which could have occurred within the nebula been studied in sufficient detail to tell us something of the origin of the Earth.

There are several important features of the Solar System which must be explained by any hypothesis of its origin. First is the regularity of the motion of the planets about the Sun. All the planets, with the exception of Pluto, revolve around the Sun in nearly circular orbits which lie close to a common plane. They also revolve around the Sun in the same direction as the Sun itself rotates. Furthermore, nearly all of the moons of the various planets revolve around the planets in the same direction that the planets rotate – and *that* is in the same direction as the Sun rotates. Although the Sun contains 99.9% of the mass of the Solar System, it possesses only 2% of the rotational energy, or angular momentum. Distinctive chemical differences also exist among the planets of the Solar System. The innermost planets – Mercury, Venus, Earth and Mars – are composed primarily of compounds of iron, magnesium, and silicon: these planets, known as the terrestrial planets, are small and of high density. The outer or Jovian

planets – Jupiter, Saturn, Uranus, and Neptune – are large and of low density. Jupiter and Saturn are composed primarily of hydrogen and helium and have an overall chemical composition close to that of the Sun. Uranus and Neptune have a higher density than Jupiter and Saturn and probably consist of carbon, nitrogen and oxygen combined with hydrogen to form solid methane, ammonia, and ice. Pluto's nature is little known.

The nebular hypothesis provides an acceptable explanation for many of the regularities of motion noted above. A rotating cloud of dust and gas will eventually take the form of a disc whose plane is perpendicular to its axis of rotation. Hence, as planetary bodies coalesce from the nebula, they will revolve about the Sun in the same direction and in coplanar, approximately circular orbits. Much more difficult to explain is the observed distribution of rotational energy: unless significant forces act between different parts of the nebula, the rotational energy of each part will be preserved as the nebula contracts, and so most of the rotational energy should reside with the Sun.

Since early theories could offer no explanation of the forces required to redistribute rotational energy, the nebular hypothesis was abandoned in favor of catastrophic theories. These proposed that the close approach to the Sun of another star or perhaps a comet pulled a stream of hot gas from the Sun which on cooling condensed to form the planets. It is now recognized that such a theory provides no better an explanation than does the nebular hypothesis. (The catastrophic theories have the additional difficulty that the hot gases pulled from the Sun would disperse into space before cooling sufficiently to condense.)

The nebular hypothesis regained credibility when several mechanisms were suggested to redistribute angular momentum within the contracting nebula. It has been proposed that large turbulent eddies would form in the nebula causing viscous forces that could be responsible for the redistribution of angular momentum. Another possibility is the interaction of ionized gas in the nebula with the magnetic field generated within the Sun: differential rotation between the Sun's magnetic field and the ionized gas would introduce forces that would tend to make the gas rotate more rapidly.

Astronomers now believe that stars like

our Sun can be observed forming today within our own galaxy. In addition, studies of certain of the closer stars have revealed perturbations in their motion that seem almost certainly to be due to the gravitational effects of planets that are orbiting them: indeed, if the nebular hypothesis is correct, as we believe, then planetary systems should be a very common thing. On the other hand, if the Solar System were formed by a catastrophic event such as the near collision of two stars, this would be an exceedingly rare event, and so planetary systems like our own would be very limited in number. Therefore, understanding how the Solar System formed is important not only to our understanding of the origin of the Earth but also to our understanding of our own place within the universe.

In addition to explaining regularities of motion of the Solar System, the nebular hypothesis must also account for the differences in chemical composition between the terrestrial and the Jovian planets – and also differences occurring within these two groups. It is generally accepted that the composition of the nebula was the same as that of the Sun. It is then necessary to explain how the terrestrial planets have been almost completely depleted in hydrogen and helium, which together make up 99% of the Sun. A reasonable explanation is that temperature varied with position in the cooling nebula. The center, near the Sun, was hot, the edges cooler, perhaps only a few degrees above absolute zero. The chemical compounds that condense as solids from a gas depend on temperature. In that region of the nebula occupied by the terrestrial planets, the temperature was never low enough for hydrogen, helium, and other inert gases to condense, but in the outer regions complete condensation occurred so that Jupiter and Saturn reflect the composition of the nebula.

Gases in the vicinity of the terrestrial planets must have been dispersed from the Solar System, and several mechanisms by which this could have occurred have been suggested. The same magnetic forces on ionized gas in the nebula as were suggested to explain the redistribution of angular momentum could also have pushed ionized gases out of the inner region of the nebula. Again, there is a possibility that, as nuclear reactions began within the Sun, large amounts of mass were lost: the wind set up by this mass loss could have swept the inner region of the nebula clear of uncondensed gases.

The nebula M42 in the sword of Orion, one of the largest in the Galaxy. In this nebula, discovered in 1610, astronomers believe that new stars are forming. The cloud of interstellar dust and gas glows because of the cluster of very young, hot stars embedded in its center. The blue color is due mostly to light emitted by ionized oxygen, the red to the presence of hydrogen.

The planets formed from the condensed matter of the nebula. Dust grains forming within the cooling nebula collided and adhered to each other, gradually accumulating to form small planetary bodies which themselves collided to form larger bodies. The present planets of the Solar System and the moons which orbit them represent the end stage of this accumulation process. An alternative theory is that the planets could have formed by the direct condensation of gases from the nebula onto the surfaces of growing protoplanets.

These two processes would have each implied a very different form for the early planets: the physical accumulation of dust and planetesimals would result in chemically homogeneous planets, while direct condensation of gases onto planetary cores would give rise to an onion-like layer structure, the composition of successive layers being determined by the sequence in which various solids condensed as the nebula cooled. Refractory oxides of calcium and aluminum would have condensed first, followed by metallic iron-nickel, then the magnesium silicate mineral enstatite and, at still lower temperature, water would condense in the form of ice. As each compound condensed it would be buried in the forming protoplanet and would have no further chance to react with the gases remaining in the nebula. If the solids condensed as small grains of dust, chemical equilibrium could be maintained and continued reaction between the dust and gases would be possible as the nebula cooled.

One line of evidence to distinguish between these two processes comes from *meteorites. Meteorites are fragments from the breakup of earlier planetary bodies in the Solar System – perhaps from the asteroid belt between Mars and Jupiter. Various types of meteorites have been identified. The major division is into "irons", composed primarily of an alloy of iron-nickel, and "stones", composed primarily of silicate minerals such as the enstatite we have already mentioned.

Meteorites show differing degrees of previous involvement with planetary bodies. One type, the carbonaceous chondrites, are rich in volatile materials and highly homogeneous, suggesting that they have at no time been a part of a planetary body: these meteorites also have elemental abundances very similar to that of the Sun. It is therefore natural to regard the carbonaceous chondrites as samples of the primitive nebular condensate. Since they contain abundant serpentine and troilite, they lend strong support to the equilibrium-condensation hypothesis.

Using an equilibrium-condensation model, reasonable temperatures can be ascribed to the formation of the various planetary bodies in the Solar System according to their distance from the Sun; but it is still hard to explain the loss of hydrogen and helium from Uranus and Neptune. It is important to realize that any model for the formation of the planets from a condensing

nebula is based on limited data and that our ideas on such matters are likely to change as exploration of the Solar System continues.

More specifically, let us consider the conditions under which the Earth would have formed as predicted by such models. In addition to hydrogen, helium and other inert gases, it seems likely that water did not directly condense out of the nebula to form a part of the primitive Earth. This means that the water presently at the surface of the Earth was originally contained in hydrous silicate minerals, like serpentine, amphibole and mica. These minerals lose their water above a temperature of about 230°C (450°F) so that a relatively cool origin for the Earth is indicated. Since the gravitational energy released by infalling planetesimals is large, enough to raise the temperature of the whole Earth by over 10,000 C° (18,000 F°), accumulation must have been slow enough to allow most of this energy to be radiated into space.

As the Earth grew in size by the further accumulation of planetesimals, its interior was heated by the decay of radioactive isotopes. The long-lived isotopes of uranium, thorium and potassium continue to be an important heat source within the Earth today, but other short-lived isotopes could have been important early in its history.

Chemical differentiation of the Earth also occurred. Metallic iron compounds, being heavier than silicates, sank towards the center of the Earth to form what we now recognize as the core: the silicates formed the outer shell, or mantle, of the Earth. As heating continued, some melting of the silicate mantle probably occurred and the less dense liquid fraction rose to the surface to form a primitive crust. As this occurred, water would also have been liberated from the constituents of the mantle to form the oceans.

In the geological record, nothing remains of this early period of the Earth's evolution. The oldest known surface rocks are about 3.7 billion years old, nearly a billion years younger than the formation of the Earth. Although we have no evidence, it seems likely that at this time large-scale convection currents were established in the Earth's solid mantle. These currents carried heat from the interior to the surface and aided in the chemical differentiation described above: such convection currents are thought to be important to many of the geological processes occurring at the surface of the Earth today, and have played a dominant role in establishing the structure of the present-day Earth. They provide the genesis of the process known as *plate tectonics.

Structure of the Earth. The radius of the Earth is determined by astronomical measurements to be 6371km (3950mi), and the mass can be determined, from the value of gravity measured at the surface, as 5.976×10^{27}g (about 60×10^{20}tons). These values combine to give a mean density for the Earth of 5.2g/cm^3 (356lb/ft^3).

To study the internal structure of the

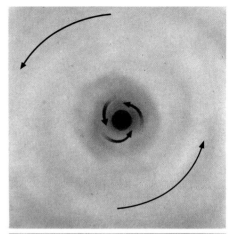
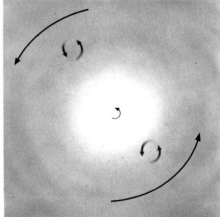
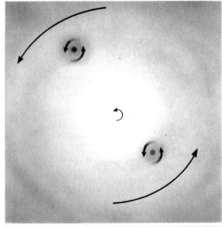
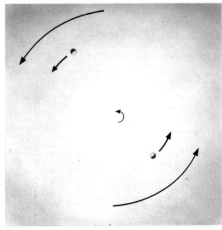

The nebular hypothesis for the origin of the Solar System. As a swirling mass of interstellar gas contracts, the Sun begins to shine: eddies in the outer regions give rise to planets.

Earth in detail, indirect methods must be used since only the upper few kilometres are accessible to direct examination. The Earth's deep interior is studied primarily by seismic methods (see *geophysics).

By studying in the laboratory the velocity of propagation of seismic waves through various representative materials, the relationship between density and seismic velocity can be determined. Observed seismic velocities then allow the variation of density with depth in the Earth to be determined. On the basis of density, the Earth is divided into three major zones: the crust, the mantle and the core. In addition to determining density, seismic-wave velocity places important constraints on the chemical composition of each zone.

The Earth's Crust. The crust is the thin, low-density outermost layer of the Earth. It is separated from the underlying mantle by an abrupt increase in seismic velocity and density termed the Mohorovičić Discontinuity, or Moho. Across the Moho seismic velocity increases from about 6.5km/s (14,500mph) in the crust to 8km/s (17,850mph) in the mantle, and this indicates a density change from 3g/cm^3 (190lb/ft^3) to 3.3g/cm^3 (210lb/ft^3). Defined in this way, the crust represents less than 1% of the Earth by volume and less than 0.5% by mass.

Two distinctly different types of crust are recognized: oceanic and continental.

The crust beneath the oceans, comprising 70% of the surface of the Earth, is by far the simpler. It has a fairly uniform thickness of around 5km (3mi) and is composed of *gabbro and *basalt, rocks made up primarily of feldspar, quartz and pyroxene and produced by partial melting of the underlying mantle in the process called seafloor spreading (see *ocean floor; *plate tectonics). All the present oceanic crust has been created within the last two hundred million years. Beneath a superficial layer, called layer 1, of sediments that collect on the ocean bottom, the crust has two distinct layers, layers 2 and 3. The 5km/s (11,000mph) seismic velocity of layer 2 is consistent with it being either highly compacted sediments or a mixture of sediments and basalt: results of deep-sea drilling suggest that it is the latter. Layer 3, with a seismic velocity of 6.5km/s (14,500mph), is probably composed of gabbro, the deepseated equivalent of basalt.

The continental crust is more complicated in having a more variable thickness and a less well defined structure. In areas which have not undergone recent deformation or mountain-building, continental crust has a thickness of about 35km (22mi), while beneath mountain belts it can be as much as 50km (30mi). In some areas an upper and lower crust can be identified on the basis of a change in seismic velocity and density: this is called the Conrad Discontinuity and occurs at depths of 15–20km (9–12mi). In other areas, the Conrad Discontinuity cannot be observed and density increases more uniformly through the crust.

The upper part of the continental crust is accessible for detailed study, drill holes having penetrated to a depth of 8km (5mi) or so. It has an average density of 2.8g/cm^3 (180lb/ft^3) and, beneath a superficial layer of sediments, is composed of a granite-like rock type called *granodiorite whose most common minerals are feldspar, quartz, hornblende, and pyroxene. It differs from the gabbro of oceanic crust in being much richer in silicon and poorer in iron and magnesium. The continental crust also contains relatively high concentrations of the incompatible elements, so named because they do not fit into the structure of minerals making up the mantle: the heat-producing elements, *uranium, *thorium and *potassium, are among these incompatible elements, so that the upper continental crust has a high heat productivity compared with other parts of the Earth. This is of very great importance for understanding the flow of heat from the Earth's interior.

The lower continental crust is of higher density, about 3g/cm^3 (190lb/ft^3), and its composition is a matter of controversy. The similarity of density and seismic velocity suggests that it may have a composition much like oceanic crust; but density and seismic velocity depend not only on chemical compositions but also on the minerals which make up the rock. At high pressures, a phase transformation can occur in which the minerals feldspar and pyroxene react to form garnet and quartz, and this results in the transformation of gabbro into a rock type called *eclogite, which has a very high density (3.4g/cm^3 (220lb/ft^3)) and seismic velocity primarily due to the presence of *garnet. Laboratory studies suggest that gabbro, although stable in the oceanic crust, would transform to eclogite in the higher-pressure environment of the lower continental crust; and this would exclude the possibility that lower continental crust is similar in composition to oceanic crust. The lower continental crust must be richer in silicon – that is, more like the rock of the upper continental crust. Hopefully, further laboratory study and better seismic data can resolve this uncertainty.

Although oceanic crust is produced by partial melting of the mantle, the origin of continental crust is much more complex. The older view is that most continental crust was produced early in the Earth's history by differentiation of the primitive Earth. We now believe that continental crust may be produced more uniformly with time and that the area of the continents is still growing at the present day. Continental crust is probably not produced by direct partial melting of the mantle, but by remelting oceanic crust.

The Mantle. Beneath the Moho, the mantle extends to a depth of 2900km (1800mi). It is characterized by seismic velocities and densities which generally increase with depth. Based on its density distribution, the mantle has been divided into three parts: the upper mantle, which

extends to a depth of 400km (250mi); the transition zone, which extends from 400 to about 700km (250–430mi); and the lower mantle. Most of the volume and mass of the Earth – approximately 83% and 67% respectively – are contained within the mantle. It is not simply because it makes up so large a part of our planet that study of the mantle is important: the processes that operate within it are responsible for crustal plate movements (see * plate tectonics).

The upper mantle has a density of about 3.3g/cm^3 (210lb/ft^3) and a P-wave (seismic pressure-wave) velocity of about 8km/s (17,850mph). This density is consistent with it having the composition of *peridotite, a rock composed of the minerals olivine and pyroxene with small amounts of garnet. These minerals are all magnesium-iron silicates, with the exception of garnet which also contains aluminum. Our knowledge of the composition of the upper mantle is supplemented by information from other sources. Basalt, mentioned in connection with oceanic crust, is a very abundant volcanic rock type in the crust – indeed, it is so abundant and widespread that it must be produced by partial melting of the upper mantle. Therefore, by comparing the composition of rocks produced by partial melting of representative source materials to natural basalts, additional constraints on upper mantle composition have been derived. Volcanic eruptions also transport to the surface of the Earth fragments of rock that appear to have been unaltered by their transport in volcanic liquids. These fragments, called xenoliths, come from depths as great as 200km (125mi). Although the details remain controversial, all of these sources suggest an upper mantle composed of peridotite with a ratio of iron to magnesium of about one to ten.

In detail, the seismic-velocity structure of the upper mantle is complex. Under oceanic and some continental areas, the velocity of seismic waves decreases with depth in the uppermost 100km (60mi). This variation gives rise to a zone of low velocity about 100km thick at the base of which seismic velocity again increases with depth. In the low-velocity zone, seismic waves are also more strongly attenuated than in the rest of the mantle. Lateral inhomogeneity also exists in the upper mantle since a low-velocity zone is not detected under some continental areas.

In the transition zone several abrupt increases of seismic velocity with depth are observed. The first of these occurs at a depth of about 400km (250mi) and corresponds to a phase transition in which the molecular structure of *olivine, under increased pressure, changes to a denser form called *spinel: unlike the case of the velocity discontinuity at the crust/mantle boundary, such a phase transition does not involve differences in chemical composition but only a spatial rearrangement of the atoms in the silicate structure. A further distinct velocity increase, again probably

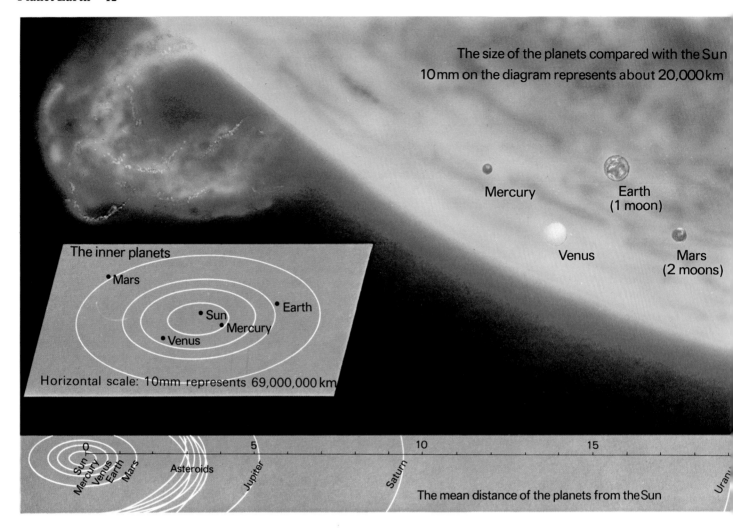

The size of the planets compared with the Sun
10 mm on the diagram represents about 20,000 km

Mercury

Earth
(1 moon)

Venus

Mars
(2 moons)

The inner planets

•Mars

•Sun

•Mercury

•Venus

•Earth

Horizontal scale: 10mm represents 69,000,000 km

0   5   10   15

Sun
Mercury
Venus
Earth
Mars
Asteroids
Jupiter
Saturn
Uranus

The mean distance of the planets from the Sun

The scale of the Solar System. The nearest star, α Centauri, is about 41,040 thousand million km (25,501 thousand million mi) from the Sun: to the scale of the lower part of this diagram it would be some 3km (1.9mi) distant from the Sun.

caused by a phase transition, occurs at a depth of about 650km (400mi). This could correspond to the breakdown of silicate minerals into their component oxides. For example, olivine (Mg_2SiO_4) breaks down into periclase (MgO) and stishovite (SiO_2) with about a 10% increase in density. Such a phase transition would seem to account for the density increase in the lower transition zone. However, density is also sensitive to the relative proportions of iron and magnesium, so that part of the density increase could be due to an increase in iron relative to magnesium, although this does not seem to be required by the available seismic data. As in the upper mantle, there is evidence of lateral inhomogeneity in that the 650km velocity discontinuity is not observed under all areas of the Earth's surface. The reasons for this are not clear and may be related only to the difficulty of observing the velocity discontinuity in seismic data which are influenced by shallower crustal inhomogeneities.

In the lower mantle, seismic velocity and density increase uniformly with depth all the way to the core-mantle boundary. The data are consistent with a composition of mixed oxides having the same bulk composition as the upper mantle, density increasing primarily due to the compression caused by increasing pressure with depth. Again, some compositional variations could occur, but this has not yet been shown.

The class of *meteorites called chondrites has also influenced our thinking on the overall composition of the mantle. These meteorites, comprising 90% of the stony meteorites, are thought to have evolved from primordial condensate in the interior of small protoplanets perhaps broken up by collisions early in the evolution of the Solar System. The mineralogy of the chondrites – 46% olivine, 25% pyroxene, 11% feldspar, and 12% iron-nickel – is consistent with what we know of the mantle from other sources, the only difference being the metallic iron-nickel, which segregated from the mantle to form the core.

The Core of the Earth. The core extends from the base of the mantle to the center of the Earth, and contains 16% of the volume and 32% of the mass of the Earth. It is metallic.

The core is in two parts. The outer core does not transmit shear waves and so must be liquid; the inner core, having a radius of 1220km (755mi), is solid and is thought to be composed of an iron-nickel alloy similar to that found in meteorites. It has a density in the range of $8-12g/cm^3$ ($510-770lb/ft^3$), which is consistent with the measured density of iron-nickel extrapolated to the pressures of over 4 million atmospheres found in the core: when combined with crust- and mantle-densities, our figures for the core give the correct average density for the Earth as a whole.

The composition of the outer core is controversial. It has a density too low to be a pure iron-nickel melt, so some lighter element must be mixed in with the iron-nickel. The element silicon has been considered most likely, though recently sulfur has been suggested as another possibility. This difference could be significant since, if sulfur is the light element in the outer core, the formation of compounds of potassium and sulfur could mean that a significant fraction of the Earth's potassium is in the core. Since radioactive potassium is an important heat-producing element, this could seriously influence our ideas about temperatures within the Earth.

Internal Temperatures. We know the Earth's internal temperatures only approximately. Once again, the most direct information is available for shallow depths – the crust and uppermost mantle. Many measurements of heat-flow from the interior have been made at the Earth's surface, the temperature within the upper kilometre or so being measured in drill holes. If the physical properties and heat

Uranus
(5 moons)

Neptune
(2 moons)

Pluto

Jupiter
(13 moons)

Saturn
(10 moons)

25   30   35   40AU

The scale is marked in Astronomical Units (AU)   1 AU = 149,600,000 km

productivity of crustal rock can be measured or inferred, the observed temperature can be extrapolated to greater depth. As measured in this way, temperature in continental areas rises by about 30C° for each kilometre of depth (87F°/mi). This is an average, since measured values range from 10C°/km (29F°/mi) in some areas to over 100C°/km (290F°/mi) in others.

In interpreting the observed heat-flow, it is helpful to recognize several sources of heat. One source, already mentioned, is the high concentration of heat-producing isotopes in the crust. A second source is the heat of formation of the crust: since crustal rock is produced by melting it forms at a high temperature and cools with time. A third source is the heat supplied to the crust and upper mantle from deeper within the Earth. The relative importances of these different sources of heat differ between oceanic and continental areas. Oceanic crust is younger, thinner, and has lower concentrations of heat-producing elements than has continental crust. In oceanic areas almost all the heat-flow can be attributed to cooling of the crust and underlying mantle. In continental areas heat production by radioactive decay and heat supplied from the interior of the Earth are of approximately equal magnitude: in older continental crust, the remaining heat of formation is small. In both oceanic and continental

areas, the rate of increase in temperature with depth decreases significantly with increasing depth.

The xenoliths carried to the surface of the Earth in volcanic liquids provide information on temperatures to depths of about 200km (125mi). From laboratory studies of the composition of various coexisting minerals over a range of temperatures and pressures, the pressure and temperature conditions under which the minerals composing the xenoliths formed can be determined. Temperatures determined in this way agree with extrapolations of surface heat-flow measurements in predicting a temperature of 1000°C (1800°F) at a depth of 150km (95mi).

In the deeper mantle, temperature can be estimated only by indirect methods. The electrical conductivity of mantle rock is sensitively related to temperature: using measurements of the temporal variation of natural magnetic fields, conductivity can be determined as a function of depth, and temperature can then be inferred from conductivity. Unfortunately, conductivity also depends on other variables such as composition, and so this method does not give very accurate results. The phase transitions discussed earlier, particularly that of olivine transforming to a spinel structure, occur at a particular temperature which depends on pressure. Therefore, the depth

at which this transition occurs could be used to measure temperature at that depth. Again uncertainty arises because the depth at which the phase transition occurs is not known accurately and because the temperature at which the phase transition occurs depends on composition.

Seismic data indicate that the mantle is solid, so that the melting temperature is nowhere exceeded: melting temperature then provides an upper limit on temperatures in the mantle. Extrapolating measured melting temperatures to the high-pressure conditions of the deep mantle indicates an upper limit on melting temperature of about 5000°C (9000°F) at the core/mantle boundary. The complex structure of the upper mantle may be related to incipient melting. One explanation of the low-velocity zone is the partial melting of silicate minerals in the presence of small amounts of water and carbon dioxide. If this is correct, then temperatures within the low-velocity zone must be about 1100°C (2000°F).

Thus our knowledge of temperatures within the Earth, particularly in the deep interior, is very meager. It is an area of study in which much more information is needed if we are to fully understand the present structure of the Earth and how this structure may have evolved through time.

EMP

The Moon

On July 20, 1969, men from Earth stood for the first time on the surface of an almost unknown planetary body, the Moon.

The successful landing of Apollo 11 at Tranquillity Base marked the culmination of over ten years of unmanned lunar exploration by remote-controlled orbiting, hard-landing and soft-landing spacecraft from the USA and the USSR. It also signalled the beginning of an extensive scientific study of the rocks and minerals, the interior, and the surface environment of the Moon; the new science of lunar geology.

About a hundred and fifty scientists from laboratories throughout the world were selected to conduct a wide range of geological, chemical, physical and biological tests on the rocks returned to Earth, and to design measuring instruments for the astronauts to place on critical areas of the Moon's surface. These scientific studies have led to a host of new discoveries, not only about the Moon but also about its relationship to the Earth and to other planets. Most dramatic of all was the discovery that the Moon was born at the same time as the Earth, about 4600 million years ago.

Such studies will continue for many more years, as each discovery poses new problems to be solved. For example, only recently has it been discovered that lunar crystals contain a fission-track record of particles from the Sun that have bombar-

Apollo-17 astronaut Schmitt collects rocks from the moon's boulder-strewn, dusty surface.

ded the Moon for thousands of millions of years. The Earth's atmosphere has prevented such particles, cosmic rays, from leaving such a record here of variations in the Sun's activity through time. Another recent discovery is that the surface of *Mercury seems much like that of the Moon, and that Moon-like craters are abundant also on *Mars. It will be decades before these planets can be visited by Man, but by then the Moon should be well understood.

The first astronauts to land on the Moon, Armstrong and Aldrin, collected 22kg (48lb) of rocks. Since then, another five

Apollo lunar missions have taken place, the last of the series being the Apollo 17 landing on December 11 1972. The total rock collected amounts to 382kg (845lb) plus 0.13kg (4.6oz) from the two unmanned Russian Luna missions. At the Lunar Receiving Laboratory in Houston, USA, are catalogued 35,600 small pieces of the Moon.

The first aim of the researchers has been to discover whether any forms of life exist or have existed on the Moon. The most sophisticated methods of biological analysis currently possible have failed to reveal even the most primitive life-forms or their molecular precursors. Secondly, the composition of the Moon and its evolutionary history needed to be known, and particularly the nature of the processes that have pocked its surface with the huge craters visible through terrestrial telescopes. Thirdly, we needed to know more about how the Moon originated, since it is the satellite of our planet and since it is the closest to us of the 32 known moons in the Solar System. Some people thought that we also needed to know whether precious minerals existed on the Moon, but geologists already knew that the processes that operate on Earth to concentrate minerals of economic importance were unlikely to have operated on the Moon; also, economics tells us that even diamonds would be barely worth collecting in any abundance from a world 385,000km (240,000mi) from Earth. After all, the short Apollo program alone cost around $25,000 million ($14,000 million).

General Physical Properties. Our satellite is about a quarter of the Earth's size and one eightieth of its mass. Hence its gravity is only a sixth of Earth's; and so the Moon was unable to retain its original gases, such as water vapor and oxygen, to form an atmosphere and oceans.

The average density of the Moon is only $3.34g/cm^3$ ($215lb/ft^3$) compared with Earth's $5.52g/cm^3$ ($356lb/ft^3$), so it cannot have a dense metallic core of any appreciable size. In fact, its density is similar to that of the Earth's mantle and higher than that of the Earth's crust. The pressure at the center is about 50kb and the temperature about $1000°C$ ($1800°F$), compared with 300kb and $4500°C$ ($8100°F$) for the Earth.

The Moon spins on its axis once a month (Earth time), so most parts of its surface are heated and illuminated by the Sun's rays for around 15 days, and then are in cold darkness for the next 15 days or so. The temperature changes between lunar day and night are extreme by Earth standards, because the Moon's atmosphere, being virtually a vacuum, provides little protection from the direct heat of the Sun, and little insulation from the ultracold of space. Temperature ranges from $110°C$ ($230°F$) at the height of the lunar day to $-170°C$ ($-275°F$) in the lunar night are typical, and result in *erosion of rock to dust through cracking by thermal expansion and contraction. The only other erosional agent is the solar wind, the constant stream of

protons from the Sun, which has given melted, glassy skins to the rocks.

The Moon moves round the Earth in an elliptical path, so that it is illuminated to varying degrees by the Sun as viewed from Earth. This path, like the axial spin, also takes a month, and from our point of view the Moon varies within the month from being fully illuminated (full moon), through partial (crescent moon) to non-illumination (new moon). A feature arising from the mechanics of this coupled Earth-Moon system, where each body exerts a pull on the other, is the tidal effect. As noted above, the Moon spins round its axis at the same rate as it rotates round the Earth. The result is that the Moon always presents approximately the same face to the Earth.

Before leaving the subject of the Moon's illumination it may be mentioned that "moonlight" is simply a reflection of sunlight from the Moon's surface. But the degree of reflection, the albedo, varies across the face of the Moon so that some areas appear light, and other areas dark. The early astronomers, starting with Galileo who, in 1610, first used a telescope to observe lunar surface features, called the dark areas "maria" (singular *mare*), because they looked like seas, and the light areas "terrae", or lands.

We now know that the "land" areas (or highlands) are the primitive Moon's crust, and that the maria are meteorite-excavated basins filled with dark-colored volcanic lavas. But, more than that, we now know the chemical compositions, ages, and evolutionary history of those huge rock masses.

Surface Features. The near face of the Moon, as observed by the naked eye or through a telescope, consists of whitish and blackish areas nowadays referred to as the highlands and the maria, respectively. In contrast, the far side is composed almost entirely of highlands with only a few small maria. This asymmetry is also a characteristic of Mars and Mercury and the reasons for it are not yet understood.

The highlands once seemed to resemble terrestrial mountain belts and were named accordingly. The topography is not rugged by terrestrial standards but the Apennines range, for example, is 650km (400mi) long and includes about 3000 prominent peaks with gentle slopes, the highest being Leibnitz at 6km (20,000ft) above base level. It is now known that these mountain ranges were not produced by the mountain-building processes we know on Earth, but are simply huge piles of rock debris encircling the craters and maria basins. The debris consists of jumbled boulders and rock dust thrown from craters as they were formed by the impact of giant meteorites. The heat generated by the impacts resulted in some rock melting, so that the debris is usually welded into *breccias.

The maria prominent on the near face are really huge craters now filled with volcanic *lavas. The largest, Mare Imbrium, is 1250km (775mi) in diameter.

There must have been a deluge of huge

meteorites early in the Moon's history, Mare Tranquillitatis being excavated early on. It was followed by 16 more until the culmination of Mare Imbrium and Mare Orientale about 3900 million years ago. During the next 700 million years, these basins were filled with volcanic lava. Meteorites have continued to bombard the Moon's surface, so that the older lava-filled basins are more damaged than the younger ones: a glance at a photograph of the Moon shows clearly the younger darker areas as being less scarred than are the older. Close-up photographs of any part of the lunar surface show an abundance of small pits, indicating meteoritic and micro-meteoritic bombardment of virtually every part of the surface, including the lava flows. As a result, the surface sampled by the astronauts is a boulder- and dust-strewn terrain. The debris blanket is given the name "regolith" and is, on average, 15m (50ft) thick.

Outside the 17 lava-filled basins, over 300,000 craters are significant enough to have been named. Two prominent examples are Tycho and Copernicus, each about 90km (55mi) in diameter. They are prominent because each is surrounded by a ray system consisting of streaks of light-colored rock radiating from the crater. The streaks are of highland-rock debris, excavated so recently that the rays have not yet been darkened either by the glassy skins resulting from long periods of cosmic-ray bombardment or by a covering of volcanic lava. These craters are about 900 millions years old, which is young by Moon-activity standards.

Most craters possess central peaks and this feature, together with measurements of depth/diameter ratios, indicates that they were produced by impacts rather than by volcanic processes. However, some volcanic features on the Moon have a few associated craters of a non-impact origin. These features, known as rilles, are shallow, often sinuous, valleys that cut across the maria. They are commonly more than 5km (3mi) wide and may be hundreds of kilometres long. Once they were thought to be ancient river channels but such water bodies are now discounted and the rilles are more closely comparable with lava channels such as those in Idaho and Iceland. There, where the lava has flowed through subterranean tubes, later collapse of the upper skin has given rise to rille features. The Hyginus rille on the Moon is peppered along its length with small craters, which must be due to circular collapse-depressions of volcanic origin.

The surface features of the Moon can thus be attributed to only two major processes: impacts of meteorites from outside the Moon (exogenic), and *vulcanicity induced by melting of the interior of the Moon (endogenic). We should therefore expect to find only two rock groups, meteorite debris and crystallized Moon lavas. Although this is broadly true, the samples returned to Earth are much more complex.

Meteorite debris amounts to only about 2% of the sampled regolith, which suggests that the huge crater-producing meteorites were mostly vaporized on impact. Volcanic lavas of various ages are abundant amongst the rock samples, but so also are pale-colored rocks from the highlands that yield evidence of such ancient ages that they have been recognized as broken-up primitive lunar crust. The most complex of all are the breccias, made up of rock fragments representative of numerous cratering events throughout the Moon's history.

Moon Rocks and Minerals. From each landing site the Apollo astronauts hammered fragments from large boulders, collected a host of smaller rocks and scooped samples of the regolith gravel and dust. Yet it could be argued that the sampled population is too small to provide sure knowledge of the composition of the Moon in its entirety. This criticism would be valid also for Earth. But, firstly, the Moon is fairly simple in its variety of rock units; secondly, continued meteorite impacting has redistributed rock debris around the Moon's surface with great ease due to the low force

Crater Schmidt on the western edge of the Mare Tranquillitatis is some 4.5km (7mi) across.

Looking into Tsiolkovsky crater on the far side of the Moon. A lava pool has flooded the adjacent valleys and encircled the central mountain peak. This crater-strewn terrain contrasts strongly with the comparative evenness of the lunar maria.

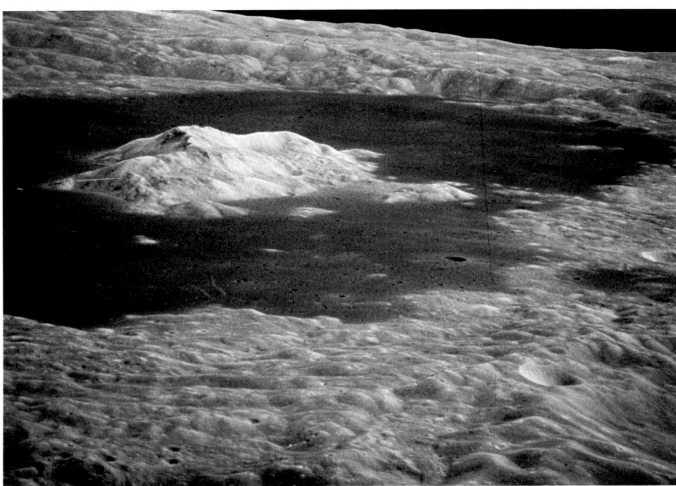

of gravity; thirdly, analyzing instruments operated by the third astronaut of each mission during numerous orbits of the Moon's surface were carried by the later Apollo Command Modules: the measurements were crude but could be refined by comparison with the properties of the collected rocks.

A typical example of a scooped sample is not a dust or soil, but consists of variously sized fragments of rock. Such samples were separated into hundreds of individual fragments, and each fragment analyzed for its mineral and chemical composition. Such properties as isotopic ages, cosmic-ray damage, magnetism or density, were also measured; while a few samples were treated at high temperatures and pressures in order to establish the conditions under which they were formed. Although such fragment samples contained a host of different rock types redistributed from distant parts of the Moon, every set of landing-site samples had a characteristic unique to that site, indicating that the local bedrock had contributed most of the material of which they are composed. This was important because, for example, it permitted the dating of each visited maria lava site. The fragment samples were supplemented by larger pieces of rock, 10cm (2.5in) or so across, from which pieces could be separated for various analytical programmes, and the whole data then correlated. This large-rock sampling was unique to the manned Apollo missions, and could not be achieved on the unmanned Luna missions.

The volcanic lavas were soon shown to be of basalts similar to those erupted from Earth volcanoes: they were molten around 1200°C (2200°F). They vary from an older group, rich in *titanium-bearing minerals, to a younger group low in these minerals. All consist predominantly of the minerals *plagioclase, *pyroxene, and variable amounts of *ilmenite. Other common minerals are *olivine and *cristobalite. Important accessory minerals include metallic iron-nickel, which indicates crystallization under strongly reducing conditions in the absence of an oxygen atmosphere. Earth minerals notably absent are those, such as *hematite or *mica, containing ferric iron or hydroxyl (OH) ions that require the presence of free oxygen or water. In all, about forty mineral species have been recognized and analyzed from various types of basalt. These include minerals not found on Earth, occurring as tiny crystals that can be analyzed only by use of a special instrument, the electron microprobe. One such mineral, named armalcolite (for Armstrong, Aldrin and Collins), is a titanium-rich iron oxide. Another, named tranquillityite, is rich in zirconium, uranium and rare earth elements. At least five other minerals related to tranquillityite have since been discovered. Patches of granitic glass occur alongside these rare minerals, as the last material to solidify within the basalts.

The light-colored rocks from the high-

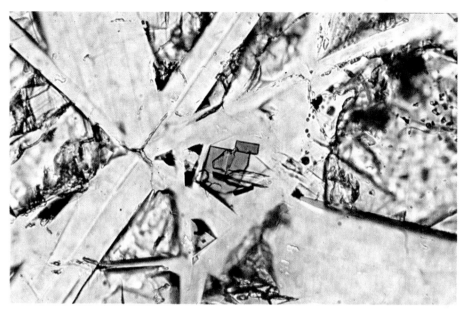

A color photomicrograph of tranquillityite, a mineral discovered only on the Moon.

lands are very different from the dark maria basalts. They are rich in plagioclase feldspar, and their average density (2.9) is much lower than that of the basalts (3.4). Several fragments are of pure feldspar (anorthosite), but the bulk composition of the highlands is believed to be that of a pyroxene-bearing, gabbroic anorthosite. These rocks are older than the maria basalts, and have been deformed by meteorite impacting. Because the feldspar is white, they impart the pale color to the highlands as seen from Earth.

Most of the highlands rocks forming the mountain belts that encircle the basalt basins are layers of breccia comprising welded rock fragments. Rock fragments in the breccia, and loose in the regolith, have sometimes been converted to glass beads by impact melting. Although the fragments are chiefly of gabbroic anorthosite and anorthosite, some are of feldspathic basalt, of plutonic rocks rich in olivine and pyroxene (dunite and norite), and more rarely of granitic rocks. The story here is a complex but intriguing one. The breccias are revealing to us the relics of an original lunar crust that formed and was reconstituted several times during violent meteoritic bombardment in the period 4600–3900 million years ago.

Isotopic Ages and Chemical Properties. All the known *elements have been sought in the Moon rocks, and their relative abundances used both to trace the Moon's evolution and for comparison with the Earth's geochemical processes.

The measurements of certain isotopic ratios provide an accurate means of determining the ages of rocks. In particular, the ratio of strontium-87 (produced by radioactive decay of rubidium-87) to common strontium-86 is of value in lunar chronology. The age of the Moon has thus been shown to be about 4600 million years (expressed as 4.6AE, where 1 aeon = 10^9 years). This is the age attributed to the Earth and other planets. The ages of major

impact basins range from 4.6 to 3.9AE. The volcanic lavas were erupted from about 3.8AE to 3.15AE. Since then, the Moon has been a "dead" planetary body, except for continued meteorite cratering, including the Copernicus rayed crater at about 0.9AE.

The Moon's basalt lavas originated by partial melting of the Moon's upper mantle of *peridotite, through heat produced by radioactive decay of elements such as uranium, thorium and potassium. Such heating took about 700 million years to develop to the point of rock melting, followed by a further 700 million years during which the heat-producing elements were transported in the lavas to the Moon's surface. The lava compositions thus reflect the composition of the upper mantle from which they were derived by partial melting.

The surprise in the basalt compositions is that when they are compared with Earth basalts, or with average Solar-System compositions as derived from meteorite analyses, they are found to be depleted in certain chemical elements – broadly, the volatile elements such as sodium and chlorine, and those elements that accompany iron and sulfur, such as nickel, cobalt, platinum and zinc. They are enriched in refractory elements such as uranium, titanium and the rare earths.

The low oxidation states result in some low-valency elements such as divalent europium, which can then enter feldspar minerals in abundance. Using this feature, geochemists have shown that the Moon's feldspathic crust (rich in europium) formed as a low-density, complementary crystal fraction to the underlying basalt source layers (low in europium).

Physical Measurements. One of the Apollo Command Module instruments measured gravity profiles and explained the mass concentrations in the maria basins as due to deep, solidified lava lakes in old crater basins. The existence of such mass anomalies for over 3000 million years im-

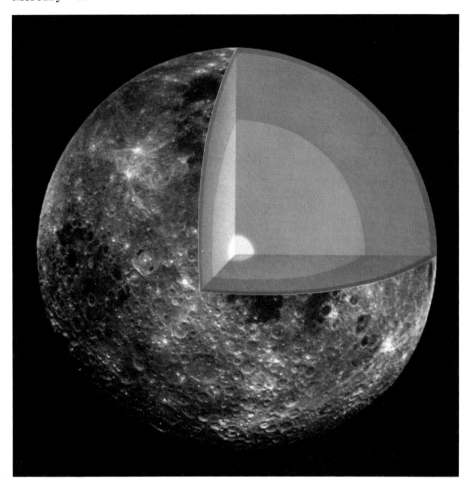

A section through the moon. Central is an irregularly shaped asthenosphere, with a radius of about 1100km (about 700mi), which may contain a core. Above this is the mantle, with a radius around 1680km (about 1000mi), and the crust, which is some 60km (37mi) thick. The surface crust is of two types: primitive anorthositic lunar crust and more recent basalt lavas.

plies a thick, rigid lithosphere. The X-ray spectrometer mapped aluminum:silicon ratios and delineated the extent of the aluminous highlands. The gamma-ray spectrometer mapped the distribution of radioactive, heat-producing elements.

Instruments placed on the surface at each landing site measured moonquakes, heat-flow from the interior, cosmic-ray compositions and magnetic field. The strength of the magnetic field is now about one-thousandth of that of the Earth, but the lavas show that 3500 million years ago they crystallized in a stronger magnetic field, around 1/20th of the Earth's. One reason could be that the Moon's metallic core was liquid, so working rather like a dynamo, whereas now it is solid.

The seismometers measured the travel speeds of natural and artificially induced moonquake vibrations, and thus the internal structure of the Moon. There is a distinct boundary between the crust and the upper mantle, and a deeper, semi-liquid "asthenosphere" that incorporates a zone that is possibly a metallic core. The main difference from Earth is that the Moon's rigid "lithosphere" is now very thick, (about 600km (370mi)), which explains the absence of orogeny and young volcanism.

Origin and Evolution of the Moon. About 4600 million years ago, the Earth and Moon formed as planetary bodies. The Moon is too ancient to have been drawn from the Earth, except while both were forming; and even this is unlikely because of the bulk chemical differences. Alternatively, the Moon could have been captured by the Earth, but if so it should have a primitive composition related to Solar-System compositions, whereas it is in fact strongly differentiated in terms of several chemical elements.

Most probably, the Moon formed from a dense atmosphere, generated by the high temperatures of solid-particle accretion at the surface of the proto-Earth. This atmosphere then condensed to give a ring of small, solid objects around the Earth, like Saturn's rings. As the objects (planetesimals) collided and grew, so the Moon formed. It lost the volatile elements to outer space, and probably the metal-associated elements to its interior.

Shortly after formation, the outer 600km (370mi) or so of the Moon melted through accretional energy. Feldspar crystals of low density floated to give a light-colored, aluminous crust about 60km (37.5mi) thick. Denser minerals such as olivine and pyroxene sank from the molten liquid. Between 4.6AE and 3.9AE, the crust was pulverized by meteorites, and feldspathic lavas were generated as impact melts. Large basins such as Mare Imbrium developed, probably due to impacting by those large planetesimals that were still circling the Earth. Around 3.8AE ago, radioactive heating of the denser mineral layers underlying the crust gave rise to partial-melt liquids which flooded the maria as basalt lava flows. About 3.2AE ago, volcanic activity ceased but, because of the absence of an atmosphere, meteorites continued to bombard the Moon's surface – as they still do.

Conclusions. We now know a great deal about a part of our Solar System outside the Earth. On Earth, it is not possible to examine parts of the primitive crust, because Precambrian rocks older than about 3000 million years occur only as isolated patches of one-time sedimentary and igneous rocks that have suffered several stages of metamorphic deformation and recrystallization. In contrast, the Moon's surface is covered by the feldspathic crust that crystallized at the very beginning of the Moon's formation, together with still-fresh lavas that formed as long ago as 3800 million years. When we look at the Moon, we are looking at a sort of cinematic "still", preserved throughout the ages, of a planet in its early stages of evolution. Because it was small by Earth standards, and therefore could not retain water and other volatile components, and because it was initially richer in elements such as aluminum, it has a feldspathic crust that is probably not a safe guide to what the Earth's primitive crust was like – but the crust of Mercury, for example, may turn out to be more like that of the Moon.

A small accident of size prevented the Moon from retaining water and oxygen, and therefore from providing an environment in which life could develop and evolve. For 4600 million years, our nearest neighbor and relative has moved with us through space but has been unaffected by the dramatic developments of Earth history.

At least, that was so until 1969.

That was the year in which Man had evolved to the level at which he could move from Earth to another planetary body for the first time. He had opened a new gate of knowledge, not only by walking upon the Moon, but also by beginning to find answers to its ancient mysteries.　GMB

Mercury

Almost all of our knowledge regarding the geology of Mercury was acquired in a matter of hours, on March 29 1974, when the Mariner 10 spacecraft swept by the innermost planet of our Solar System. The photographs it took revealed a cratered surface very similar to that of the *Moon. Additional information, acquired as Mariner 10 circled the Sun and twice more passed close to Mercury, further confirmed the Moon-like appearance.

Mercury's average distance from the Sun is 57.9 million km (35.9 million mi). At closest approach to the Earth, Mercury is

Mariner-10 photograph of Mercury, taken on September 21 1974, showing typical craters formed by the impact of meteorites, and a compressional ridge extending from upper left to lower right which is more than 300km (185mi) long.

about 80 million km (50 million mi) distant from us. Its diameter is 4880km (3025mi), slightly less than half that of the Earth. Its average density is 5.4g/cm^3 (348lb/ft^3), approximately the same as the Earth's.

Mercury rotates relatively slowly, in about 58.7 days, and its period of revolution about the Sun is 88 days. These periods couple in such a way that certain longitudes are more frequently directly beneath the Sun when the planet is at its closest approach (perihelion). For this reason, even though the spin axis is approximately vertical to the orbital plane, places on Mercury experience seasonal changes in the amount of light and heat they receive.

Approximately one hemisphere of the planet was photographed by Mariner 10. The dominant features are meteoritic craters which, in the absence of atmospheric *erosion, have retained many of their original impact features. Deposits of material ejected from the craters, secondary craters

and bright rays and haloes are well preserved. Because the Mercurian gravitational field is stronger than that of the Moon, secondary craters and ejecta deposits tend to occur closer to the primary crater. Wall slumping owing to the greater gravity has led to shallower crater depths.

The large craters display central peaks and rings, as do lunar craters. At still larger diameters, several basins have been observed. Most prominent is the Caloris basin, 1300km (800mi) across: associated ejecta deposits are widespread over an entire hemisphere.

There are on Mercury smooth plains materials which occur both between craters and superimposed over subjacent cratered terrain. These may record widespread volcanic flooding, perhaps at the same time as the flooding of the lunar maria. However, in the absence of clear morphological evidence for volcanic activity, this interpretation can be no more than speculative. Plains materials might also form by emplacement of partially molten material ejected from large craters and basins.

Lobate scarps, ranging in length from 20 to 500km (12.5–300mi) and in height from a few hundred metres to 3km, are visible in

Mariner 10 pictures. They are generally interpreted as compressional ridges, perhaps formed during an early stage of crustal shortening.

Photometric measurements carried out with the Mariner 10 imaging system suggest that Mercury is covered with a dark, fine-grained soil similar to the lunar regolith. Where the terrain is heavily cratered it is somewhat brighter than are the lunar highlands. The dark smooth plains have color and albedo suggesting basaltic composition.   TAM

Venus

Venus has 0.815 times the mass of the Earth, and the solid part of the planet has a radius of 6070km (3770mi). It rotates once every 243 days, but in a direction opposite to that of the other terrestrial planets: the lack of a magnetic field may be a consequence of the slow rotation.

In spite of space research, the composition of Venus' bright clouds – found up to 70km (43 mi) above the surface – is unknown. The clouds preclude remote viewing of the surface, where the pressure

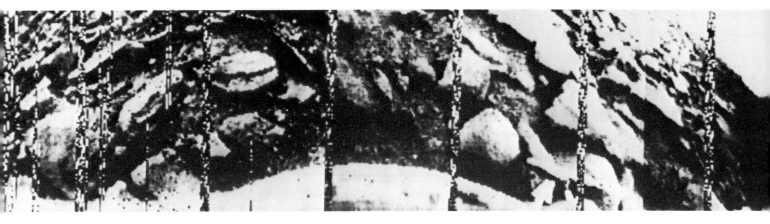

Panorama of the Venusian surface as seen on October 22 1975 from the Soviet soft-landing spacecraft Venera 9. Prior to the taking of this photograph it had been thought that Venus' surface would be sandy due to extreme erosion and weathering; but, as can be seen, angular rocks are in fact present. (The white arc at lower center is part of the spacecraft.)

can attain more than 90 atmospheres and the temperature is nearly everywhere around 500°C (900°F): this is too hot to support any known form of life. In addition, over 90% of the atmosphere is carbon dioxide; and the upper atmosphere also contains traces of the poisonous gas carbon monoxide and the corrosive agents hydrochloric acid and hydrofluoric acid. Perhaps only a few tenths of one per cent of the lower atmosphere (analyzed from the Soviet Veneras 4, 5 and 6) consists of molecular oxygen and water vapor.

It was widely conjectured (largely through interpretations of cloud motions observed using ultraviolet photography) that high winds wore down the rocks of Venus. However, using radar, US astronomers have demonstrated that Venus displays appreciable differences in surface altitude; and the surface panoramas transmitted back to Earth from the Soviet soft-landers Veneras 9 and 10 show a barren landscape that contains both angular and rounded rocks – proof that not all parts of the surface have been reduced by erosional processes to sand.   GF

Mars

Among the planets, Mars has always been singled out for speculation: throughout the more than three centuries that it has been observed telescopically, many laymen and scientists have considered the planet to be populated by an alien civilization.

These ideas were fueled by alleged sightings of a network of "man-made" canals, a misconception whose origins lie, oddly, in a mistranslation. The Italian astronomer Giovanni Schiaparelli made a detailed study of the Martian disc in the years 1877–81, and announced that certain of the surface markings could only be interpreted as a network of straight lines, which he named "channels" – in Italian, canali.

Canali was translated into English as "canals", and the excitement began.

In the early part of the 20th century the hypothesis was taken to its ultimate extent by a US astronomer, Percival Lowell, who wrote several best-selling volumes which described in detail how Martian citizens were living on a planet that was becoming progressively more desert-like. In order to husband the dwindling supply of water, canals carried water from the polar caps to the equatorial regions, where the natives lived underground.

Results from several spacecraft which passed by Mars – Mariner 4 in 1965, Mariners 6 and 7 in 1969 – and one spacecraft which orbited Mars – Mariner 9 in 1971 – have demolished such theories and radically changed and enlarged our impressions. The old ideas of Martian canals and civilizations have turned out to be complete fiction. However, Mariner pictures reveal a spectacular landscape, complete with new puzzles which stubbornly resist solution.

The most up-to-date information currently available is that from the US Viking probes 1 and 2, which arrived on Mars in the summer of 1976. After achieving orbital trajectories, the orbiters deployed landers which descended to the surface slowed by parachutes and retrorockets. Shortly afterward began a flood of transmitted data, visual and otherwise, concerning the nature of the red planet. Perhaps the most provocative investigations – in both intent and results – were those seeking microscopic lifeforms in the soil: the results were so positive that scientists on Earth could not believe that the reactions they observed were due to the activities of lifeforms! The question of the existence of life on Mars is one that still awaits an answer.

General Properties. Moving outward from the Sun, Mars is the next planet encountered after Earth. At the time of closest approach (opposition) the two planets are separated by about 60 million kilometres (38 million miles). The diameter of Mars is 6740km (4180mi), about half that of the Earth. The average density is $3.9g/cm^3$ ($251lb/ft^3$), compared with $5.52g/cm^3$ ($356lb/ft^3$) for Earth.

Mars completes one rotation in approximately 24h 37min compared with 24h for Earth. It takes 687 Earth days (670 Martian days) to complete one revolution about the Sun, so that a Martian year is approximately twice as long as a terrestrial year. Because the equator is at an angle of 25° to the orbital plane, different regions on the planet experience seasonal variations related to periodic changes in the amount of radiation they receive from the Sun. Polar caps of carbon dioxide (CO_2) ice grow and shrink each year.

The surface of Mars, viewed telescopically, shows irregular dark and bright markings of reddish hue. Photometric measurements suggest that the dark material is *basalt and that the reddish color is caused by *weathering of the basalt, resulting in the formation of hydrated iron-oxide minerals (e.g., *goethite). The configuration of dark and bright markings changes seasonally and annually, probably because of wind transport of bright, fine-grained sediment.

The atmosphere of Mars has one hundredth of the density of the Earth's. Carbon dioxide is the chief constituent, but there may be a reasonable amount of argon present.

Physiographic Provinces. Most of what is known regarding the geology of Mars is based on interpretations of pictures acquired via Mariner 9. The first thing that strikes one is the difference between the terrains of the northern and southern hemispheres: cratered terrain appears in the south; smooth, undivided plains in the north. Global elevations show a similar distribution, relatively high in the south and low in the north. The cratered terrain is relatively old, preserving the record of heavy bombardment that occurred in the first few hundred million years of the planet's history. The smooth plains probably represent regions of thin crust that have been fractured from beneath and flooded with volcanic lava and wind-deposited sediments.

Other terrains include the cratered plains. These surfaces have crater densities intermediate between cratered terrain and smooth plains. The materials of cratered plains resemble terrestrial *plateau basalts, and may record volcanic activity that occurred about 3000 million years ago. Hummocky terrain may have formed – in part, at least – by the melting and flowing away of subsurface ground-ice with attendant collapse and *erosion of surface materials.

Distinctive landforms – layered deposits and etched plains – are found in both polar regions. These indicate that the erosional-depositional cycle near the poles differs from that in the equatorial region. The succession of strata in the layered deposits has been attributed to periodic changes in climate, this in turn related to systematic variations in Mars' orbital configuration due to gravitational interaction with other bodies.

Craters. Almost all Martian craters have been formed by meteoritic impact. Crater morphology changes with size: many small craters (1km (0.62mi) in diameter and less) are bowl-shaped, while progressively larger craters show flat floors, central peaks, central-peak rings, and multiple structural rings. The same features are displayed by lunar and Mercurian impact craters.

Erosion has removed the traces of most of the ejecta deposits, the material scattered around the craters by the impacts. Bright haloes and rays, observed around youthful lunar craters, are almost totally absent. Hummocky ejecta deposits have typically been stripped or eroded back to form a narrow annulus close to the crater's upraised rim.

Ages of planetary surfaces can in suitable cases be estimated by measuring the number of craters per unit area. Assuming that all craters are formed by impact, adopting a model for erosional modification and obliteration, and employing an independently determined frequency of impacting bodies, it is possible to calculate an age for surface materials.

Absolute calibration of the meteoroid flux (i.e., the "population" of meteoroids) in the vicinity of Mars is more difficult. Early investigators assumed that, because Mars was close to the asteroid belt, impact rates might be ten or twenty times as great as for the Moon; but current investigations suggest that flux rates are and have been similar throughout the inner Solar System. Accordingly, Mars has probably experienced an impact history not unlike those of Earth, Moon and Mercury. For all these bodies, initially high impact rates were associated with the terminal stages of planetary accretion and the sweeping up of interplanetary debris. During the first few million years after this, impact rates must have dropped off rapidly, finally reaching a rate that has remained more or less constant to the present.

As we have already noted, then, the cratered and mountainous terrains probably indicate the appearance of the planet as it was fairly early on in Martian history, though erosion has played a modifying role. The lightly-cratered smooth plains are comparatively recent developments.

Volcanoes, Canyons and Channels. More than twenty unambiguous volcanic constructional landforms have been identified in Mariner 9 pictures. Most are

This color picture of the Martian landscape was taken on July 21 1976, the day after Viking 1 landed on the planet: it is roughly noon, Martian time. Orange-red surface materials cover most of the surface, apparently forming a thin veneer over darker bedrock. The surface materials are thought to be limonite: such weathering products form on Earth in the presence of water and an oxidizing atmosphere. The reddish cast of the sky is probably due to scattering and reflection from reddish sediment suspended in the lower atmosphere. The view is southeast from the Viking craft.

characterized by a conical shape, presence of summit collapse calderas, roughly textured slopes, lava channels and ridges radiating down-slope from the central caldera, and lava-flow fronts in adjacent plains materials.

Olympus Mons, the largest Martian volcano and almost certainly the largest in the Solar System, is approximately 600km (370mi) in diameter and rises to a height of 26km (16mi) above its base. In shape and surface morphology it resembles terrestrial shield volcanoes; however, it is more than twice the size of the largest terrestrial shield volcano complex, that underlying the island of Hawaii. Martian volcanism has probably

occurred episodically throughout the planet's history. The Olympus Mons eruptions are among the most recent, probably occurring only a couple of hundred million years ago.

A giant system of canyons, the Valles Marineris, extends from (Martian) latitudes 20°W to 100°W, a distance of 5000km (3100mi). These canyons were probably formed by erosional widening of crustal fractures that are part of a radiating set of extensional faults resulting from crustal doming in the vicinity of the Olympus Mons volcanic province.

Numerous irregularly sinuous channels are to be found on Mars, chiefly in the

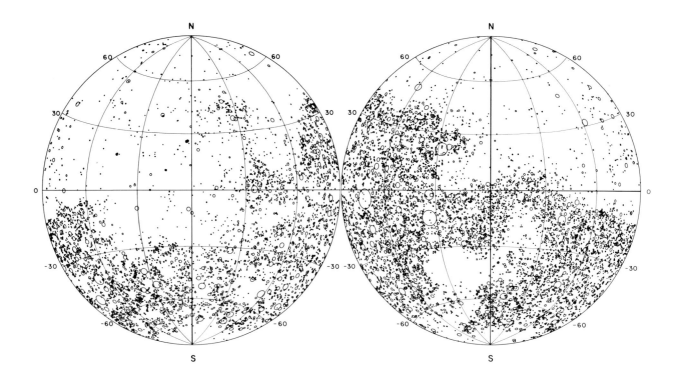

Polar Units

Permanent ice
Layered deposits
Etched plains

Ancient Units

Cratered terrain, undivided
Mountainous terrain

Volcanic Units

Volcanic constructs
Volcanic plains
Moderately cratered plains
Cratered plains

Modified Units

Hummocky terrain, chaotic
Hummocky terrain, fretted
Hummocky terrain, knobby

Channel deposits
Plains, undivided
Grooved terrain

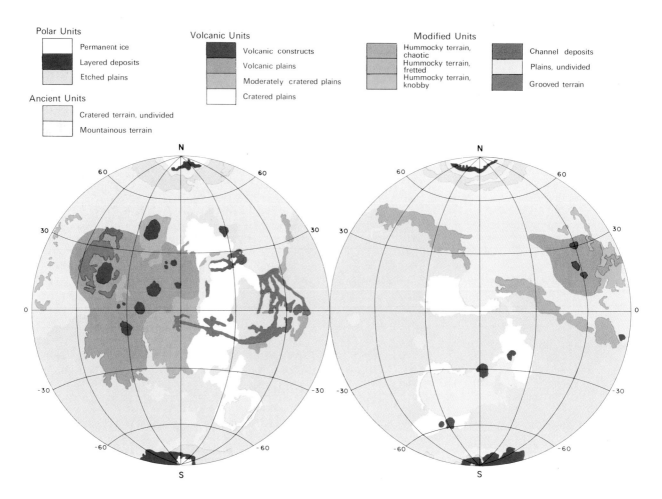

cratered terrain not far from the equator. The larger of these channels are several hundred kilometres long and tens of kilometres wide. Their origin is uncertain, but is generally believed to be related to surface flow of water during some former epoch – that is, they are the beds of Martian rivers of the past. (Under present tenuous atmospheric conditions liquid water would simply boil away.) Large amounts of water might have formed from a primitive atmosphere, have been stored in the soil as groundwater and ground-ice, and subsequently have been released through an artesian mechanism to form the observed channels.

Wind Deposits. Terrestrial observers have noted that, during some observing periods, the entire disc appears uniformly bright, with no dark surface markings visible. This condition has been attributed to global dust storms which occur most commonly during perihelion (closest approach to the Sun). The observed transport of sediment is somewhat unexpected, since wind velocities required to initiate movement are about six times as great as for Earth, due to the low Martian atmospheric density. The Mariner 9 spacecraft arrived at Mars during the height of one of these dust storms: during the craft's approach, only the highest (Olympus Mons) and brightest (polar caps) regions were visible.

Bright and dark streaks are visible in many Martian pictures, commonly occurring in association with small bowl-shaped craters. The orientation of the streaks is constant over large regions, but changes systematically with changing latitude. The streaks apparently form downwind of youthful craters, whose upraised rims perturb near-surface winds. High-speed winds cause turbulence in the lee of craters, and associated erosion leads to the exposure of dark bedrock. When wind speeds are lower, protected regions of deposition exist downwind of craters, and these are the bright streaks.

Some flat-floored craters have interior dark splotches, preferentially occurring on the downwind side (as determined by streak orientations). High-resolution pictures show that some of these dark splotches are transverse dune fields.

Some Martian landforms are due to wind erosion. These include irregularly-shaped, flat-floored deflation basins and cigar-shaped ridges of bedrock (yardangs).

Satellites. Mars has two small satellites, Phobos and Deimos. Both are densely cratered and irregularly shaped, lacking sufficient mass to yield a spherical "hydrostatic" shape. The dimensions of Phobos are roughly $13.5 \times 10.7 \times 9.5$km

Opposite, Above, a plot of all the craters of diameter greater than 15km (9.3mi). The uneven distribution of these is clear evidence both of erosion of surface features by wind and wind-transported particles and of tectonic activity. *Below,* the physiographic provinces of Mars, the different units being described in the accompanying legend.

Olympus Mons, Mars, the largest volcano in the Solar System. The diameter of its base is about 600km (370mi); its height is about 26km (16mi.).

100 km

A Mariner 9 photograph showing one of Mars' sinuous channels, thought to be possible valleys of vanished rivers.

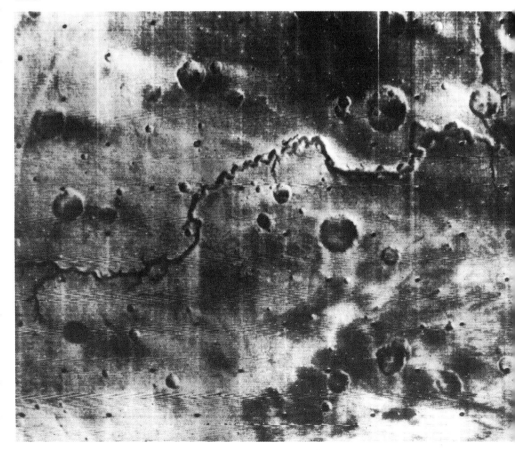

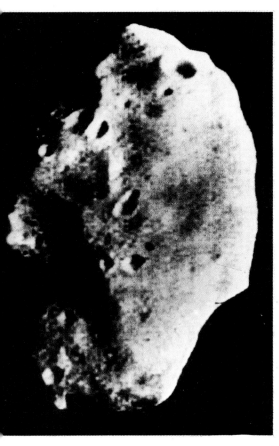

Phobos, the inner moon of Mars, from a distance of 5540km (3444mi). The profusion of craters suggests that Phobos is very old and of great structural strength.

(8.4 × 6.6 × 5.9mi), those of Deimos being 7.5 × 6.1 × 5.5 km (4.7 × 3.8 × 3.4mi). Both satellites have low albedo (reflectivity), and this dark appearance suggests a basaltic composition.

Phobos orbits Mars at a distance of 6100km (3780mi) every 7h 39min; while Deimos orbits at a distance of 20,000km (12,400mi) with a month of 30h 18min. Their orbits are very circular, and this has led some astronomers to suggest that the two moons, rather than being the captured asteroids that their appearance would imply, are the remains of a single, larger Martian satellite that was shattered by some unknown agent at some unknown point in the past.   TAM

Meteorites

On November 30 1954, Mrs E. H. Hodges of Sylacauga, Alabama, was resting on her sofa after lunch when a stone weighing about four kilograms crashed through the roof of the house, ricocheted off a radio, and struck her on the leg. This is the only authenticated instance of a person being hit by a meteorite, a rock from outer space captured by the Earth on its journey around the Sun.

Thousands, perhaps millions, of tonnes of extraterrestrial material enter the atmosphere each year. The fate of this material and the extent to which we are aware of it depends entirely upon particle size. Minute dust-like fragments make up most of the total, and these sink unnoticed to mix with terrestrial dust, so that they are extremely difficult to identify and recover. Somewhat larger particles, sand-sized grains, offer sufficient air resistance to be briefly heated to incandescence by friction before being entirely destroyed in the upper atmosphere. We are aware of these as transient streaks in the night sky – meteors, or shooting stars.

Other, much rarer fragments, weighing upwards of a few grams, are large enough to partially survive their passage through the atmosphere: they are reduced in size by ablation, and may be fragmented by the violence of their deceleration from cosmic velocities. Those remnants that reach the Earth's surface are called meteorites. It has been estimated that the total number of meteorites falling to Earth each year is about 500, of which perhaps 150 would be expected to fall on land. The world's population is, however, so irregularly distributed, with vast areas of land almost devoid of habitation, that an average of only four "falls" (defined as recovered stones whose arrival has been observed) are in fact recorded each year. "Finds" – those meteorites found on the surface, perhaps after plowing, excavation or some such, but whose date of fall is not known – also contribute to the growing list of authentic meteorites, presently totalling around 2000.

Meteorites vary in weight from a few tens of grammes to several tonnes, with a theoretical upper limit of about 100 tonnes. The largest known, the Hoba iron of Namibia (South West Africa), weighs about 60 tonnes but is much corroded. Calculations show that during its passage through the atmosphere a meteorite less than about 1 tonne in weight will lose all its cosmic velocity, which may have been as high as 70km/sec (over 150,000mph) relative to the Earth, and strike the ground with a velocity due only to gravitational attraction. Somewhat larger meteorites may retain their cosmic velocities; and those in excess of 100 tonnes will be hardly slowed by the atmosphere and so will impact at such high speeds as to be almost entirely vaporized in the resultant blast. For this reason giant meteorites are not found on the Earth, although some large craters are believed to be of impact origin.

The Arrival of a Meteorite. The fall of a meteorite of even quite modest dimensions is often accompanied by spectacular effects. In the absence of cloud cover, the meteorite itself is seen even during the day as a bright fireball, incandescent due to frictional heating, accompanied by a trail of ionized gases and dust which is brightly luminous by night and normally dark by day. Sometimes the meteorite is fragmented by the violence of its passage and a single fireball is observed to break into separate parts.

Sound effects which may be audible up to 80km (50mi) or more from the eventual impact point are variable and less easy to describe. Most witnesses hear one or more loud reports. Other sounds, including whistlings, cracklings, and noises like thunder or the tearing of cloth or the roaring of fire, have also been reported. An observer close to the point of landing may hear the thud of impact. Shock waves may be detected by seismographs.

Meteorites which have lost all their cosmic velocity during their journey through the atmosphere are often found lying on the surface of the ground, although they may penetrate soft earth to form craters up to a few metres deep. Stones from the 1869 Hessle fall in Sweden failed to break ice only a few centimetres thick.

Contrary to popular expectation, freshly fallen stones are usually quite cool to the touch. If a meteorite breaks up in the air, the resultant fragments are scattered on the ground within an elliptical area whose long axis is aligned with the flight direction.

The External Appearance of Meteorites. Ablation entirely destroys a high proportion (20–100%) of a meteoroid entering the atmosphere: the amount depends on such factors as original size, shape, cosmic velocity and angle of incidence to the atmosphere.

Freshly-fallen meteorites are usually covered with a thin black fusion crust, produced by melting of the leading surface as the meteorite decelerates through the atmosphere, molten material streaming backward to coat most or all of the specimen. If the meteorite has remained in a stable orientation during this stage, ablation tends to produce a conical or dome shape, sometimes bearing rather regularly disposed furrows and pits, so that the leading surface can be readily identified.

Classification. Meteorites are classified broadly into three groups, irons, stony-irons, and stones. Members of the first group consist almost entirely of iron and nickel alloyed together as the minerals kamacite (4–7% nickel) and taenite (30–60% nickel). Many irons reveal a striking structure when sawn, polished and etched with acid. This Widmanstätten pattern consists of bands of kamacite bordered by taenite and arranged parallel to the octahedral faces of an original homogeneous crystal of nickel-iron.

The stony-iron meteorites contain, as the name suggests, nickel-iron together with silicate minerals, the precise nature of which allows subdivision into four classes: the pallasites (metal plus *olivine); the mesosiderites (metal plus *pyroxene and *plagioclase); and two further classes each known from only a single example.

93% of all observed falls are of the third group, the stony meteorites. This figure probably approximates to their actual extraterrestrial proportion (the fact that irons predominate among *finds* is ascribed to their greater resistance to terrestrial weathering and to their more obviously unusual character). The stones consist chiefly of the minerals olivine, pyroxene, nickel-iron, plagioclase and *troilite (iron sulfide).

A view over Mauritania looking southeast toward the El Djouf desert. Below, in old sedimentary and igneous rocks, rising out of the desert are the famous Richat structures. The larger is about 50km (30mi) in diameter; the smaller, a little toward the south, about 8km (5mi) in diameter. It is now considered that these craters are almost certainly of impact origin, showing that the Earth, like its neighbors, has been subject to meteoritic bombardment in the past.

Minor amounts of other minerals, some unknown in terrestrial rocks, are also often present. Most stones are characterized by the presence of small, approximately spherical aggregates known as chondrules, and these meteorites (84% of the total falls) are called chondrites.

Some chondrites, especially a subgroup rich in carbon and called carbonaceous chondrites, have chemical, mineralogical and textural features which suggest that they are primitive, well preserved objects, less highly evolved than other meteorite types and terrestrial rocks. In particular, the carbonaceous chondrites approximate closely in composition to the non-volatile material of the Sun, as determined by spectroscopy. There are a number of other reasons for believing that these meteorites may represent primordial Solar System material, and that other meteorite types, the irons, stony-irons, and those stones lacking chondrules – the achondrites – may be derived from this primordial material by a variety of secondary processes and events.

The Origin of Meteorites. A considerable amount of evidence indicates that these extraterrestrial fragments are travelling in elliptical orbits around the Sun, and are thus genuinely part of the Solar System. Moreover, they appear to have originated in

Extraterrestrial material reaches the Earth in several different ways. Giant meteoroids (1), of mass greater than about 100 tonnes, are reduced in size during their passage through the atmosphere owing to ablation. They retain most of their cosmic velocity, and are almost entirely destroyed on impact with considerable release of heat and other energy. Their impacts produce large craters (e.g., that in Arizona) and may also be responsible for the formation of tektites. Meteoroids of mass between 1 gram and 100 tonnes (2 and 3) lose most or all of their cosmic velocity, and may fragment (3). They come to rest on the Earth's surface – or, perhaps, form shallow craters. Their fragmentation in the atmosphere may contribute to (4), meteors or shooting stars. These are primarily formed by meteoroid grains, which are totally destroyed by ablation, though their passage is seen as a trail of light (a meteor). Their destruction contributes to (5), meteoroid dust. These minute particles survive their trip through the atmosphere to the surface largely undetected, and mix with terrestrial dust. Known as micro-meteorites, they have been identified in deep-sea sediments and from the polar icecaps.

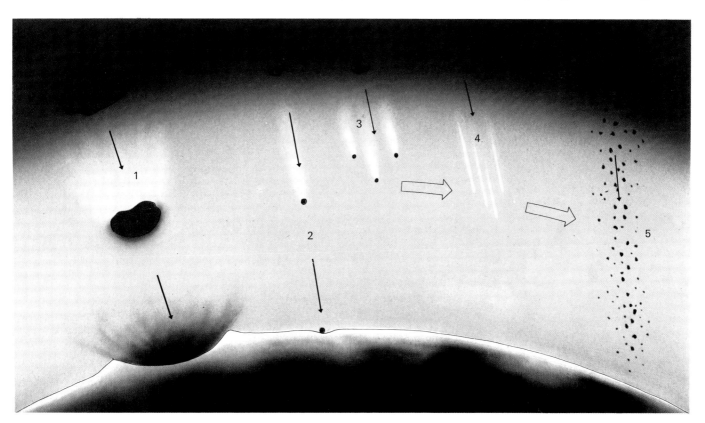

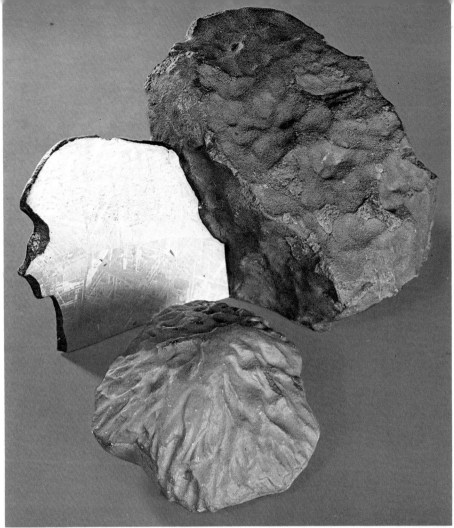

Meteorites. On the left is a polished and etched slice of the Amalia iron showing the Widmanstätten pattern. Behind it is a chondrite that fell in 1913; in the foreground is a cast of the Middlesbrough stone which fell in 1881.

the asteroid belt, the zone between the orbits of Mars and Jupiter which contains millions of small bodies that have failed to aggregate into a planet. Although most of the asteroids remain in more or less circular orbits beyond Mars, some travel in elliptical paths such that they cross the orbits of the inner planets. The origin of meteorites is thus linked with the nature and origin of the asteroids.

It is now generally held that the Solar System formed about 4.6×10^9 years ago from a rotating disc of dust and gas, and that the planets themselves were created by the progressive accumulation of small grains of solid matter. Initially, collisions were largely fortuitous but, as the bodies grew in size, the process was increasingly aided by the forces of gravitational attraction. Although smaller than the planets, the asteroid bodies formed in the same way, failing only to coalesce into masses more than a few hundred kilometres in diameter – Ceres, the largest, has a diameter of 755km (about 470mi).

Recent spectrographic studies show that the spectra of the light reflected by the asteroids fall into distinct classes, each class closely resembling the spectrum, as determined in the laboratory, of a particular meteorite group. Some 80% of the asteroids (generally the smaller measurable bodies) give spectra indicating close affinity with the carbonaceous chondrites, and as we have seen there is good reason for suppos-

ing that these meteorites are representative of the original material of the ancestral solar cloud. The chondrules themselves may outline primordial dust particles. The primitive mineralogy, chemistry and texture of the carbonaceous chondrites and a few of the common chondrites, less heated and metamorphosed than the other meteorites, survived only in the small asteroid bodies which were never large enough to become heated internally. Inside larger bodies, heating due to release of gravitational energy as well as radioactive decay processes initiated melting and the segregation of metal from silicate phases. The stony-iron meteorites represent an intermediate stage in this process, and probably originated in bodies 100–200km (60–125mi) in diameter. The iron meteorites and the achondrites probably represent the complementary core and outer shell regions of even larger, more fully differentiated bodies. The pallasites, for instance, appear to have formed in the core of bodies with a maximum diameter of about 600km (370mi). All these events are believed to have taken place within a relatively short time after the condensation of the Solar System. It is clear that, if the above sequence of events is broadly correct, then one or more of the larger parent bodies must have been fragmented in order to account for the arrival on Earth of small pieces of the more evolved meteorite types, including all the irons. The times of break-up of parent bodies have not been well

established, but many meteorites undoubtedly show features indicative of transient events of great violence.

The Age of Meteorites. Radiometric dating techniques, based on the decay of radioactive isotopes (see *age of the Earth), are potentially capable of dating five separate events in the history of a meteorite. The five events are:

(i) nucleosynthesis – i.e., the creation of the elements in the solar nebula which later accumulated to form the meteorites, asteroids and planets;

(ii) melting of the larger meteorite bodies when these had grown big enough to heat up spontaneously;

(iii) cooling of these parent bodies;

(iv) break-up of the parent bodies; and

(v) arrival on Earth.

The first event, nucleosynthesis, can in principle be dated by the measurement of an isotope of xenon produced by the decay of a short-lived and long-extinct isotope of iodine which is believed, for theoretical reasons, to have formed during nucleosynthesis. The melting and cooling ages of the parent bodies are determined, with rather more confidence, by comparing the amounts present of the long-lived radio-isotopes of uranium, potassium and rubidium with the amounts of their decay products – lead, argon and strontium respectively. These first three events all occurred early in the life of the Solar System, within the period $4.5–4.0 \times 10^9$ years ago, and such studies of meteorites have played an important part in establishing the age of the Solar System at about 4.6×10^9 years.

Later events, the break-up of parent bodies and the arrival of a meteorite on Earth, may be tentatively dated using a somewhat different principle. When a parent body fragments in space, the smaller bodies thus formed become exposed to cosmic rays which produce, in amounts proportional to the exposure time, certain new radioactive and stable isotopes. When a meteorite falls to Earth, it is shielded from further cosmic-ray irradiation, so that stable nuclides cease to be formed and their concentration becomes fixed, while radioactive nuclides formed during exposure begin to decay at known rates. Measurement of suitable nuclides thus permits the estimation of both exposure ages and terrestrial ages.

There are some theoretical difficulties in this method and uncertainties as to the correct interpretation of results. One rather puzzling result is a disparity between the exposure ages of stones (mostly in the range $20–30 \times 10^6$ years) and irons (mostly in the range $100–1000 \times 10^6$ years). Terrestrial ages so far obtained on finds are all geologically very young, less than 1 million years, a finding to be expected for objects susceptible to rapid weathering and erosion under terrestrial conditions.

Meteorite Craters. The heavily cratered surfaces of the *Moon, *Mars and *Mercury owe their topography chiefly to the impacts of vast numbers of meteoritic objects, acquired mainly in the early history of the Solar System. It is inconceivable that Earth in its early life was not likewise subjected to large-scale cosmic bombardment. The evidence for this, in the case of an active, mobile planet such as Earth, is now difficult to find. With its mobile crust, constantly being created and destroyed by plate movements (see *plate tectonics), and ceaselessly being modified by orogeny, erosion and sedimentation, there remain few areas of ancient terrain to be searched for impact structures. A few score of such structures have been tentatively identified, of which only a small number are undoubtedly of impact origin.

Conclusion. It is clear that meteorites are to Earth scientists much more than interesting but irrelevant curiosities. Apart from providing important evidence concerning the nature, origin and age of the Solar System – and the fact that, prior to the advent of space exploration, they were the only extraterrestrial material available for first-hand study – meteorites supply strong clues concerning the nature of the Earth itself. A whole range of geological and geophysical evidence suggests that, apart from the superficial crust, the Earth consists of a nickel-iron core (partly molten) and a silicate-rich mantle made up, at least in its outer parts, of olivine and pyroxene. It is believed that the iron and stony classes of meteorites are analogous, respectively, to these two regions of the Earth's interior. The stony-iron meteorites may broadly correspond to terrestrial rocks in the region of the core/mantle boundary at 2900km (1800mi) depth.

However, the analogy should not be taken too far. The size of the Earth is much greater than the sizes attained by even the largest of the meteorite parent bodies, and meteorites thus lack the high-pressure minerals which must be present at depth within the Earth. Nevertheless, among all natural objects available for study in the laboratory, meteorites are the closest relatives to the inaccessible matter that forms the bulk of the Earth.   FBA

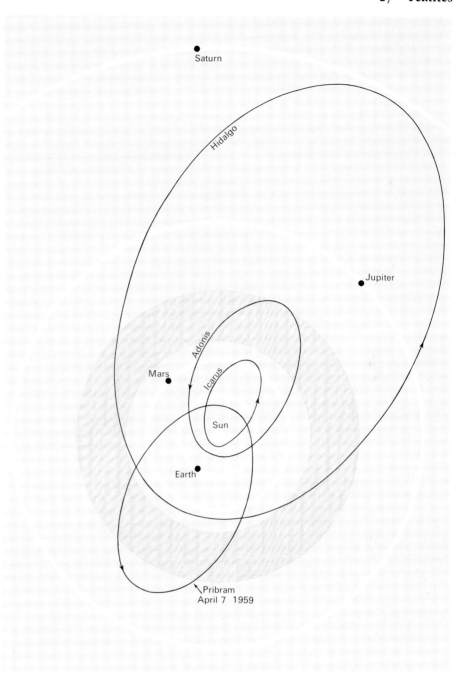

Most asteroids circle the Sun with orbits that lie within the asteroid belt, represented here as a stippled zone. Some, however, have much more elliptical orbits that intersect those of the inner planets: shown here are the orbits of Icarus and Adonis, the best known such asteroids. Hidalgo, the asteroid with the most eccentric orbit of all, approaches the orbit of Saturn at one extreme and that of Mars at the other. Also shown is the calculated orbit of the Pribram meteorite which fell to Earth on April 7 1959.

Tektites

A rare and problematical class of natural objects, tektites are much studied and little understood. They are small pieces of glass, varying in color from light green to black and having a chemical composition similar to that of acid *igneous rocks (about 75% SiO_2) but with a significant deficiency of alkalis. Many have distinct and characteristic shapes – spheres, buttons, dumbbells, etc. They have been found in only four areas of the Earth's surface, known as "strewn fields": southern Australia and southeast Asia, the Ivory Coast, Czechoslovakia, and Texas and Georgia, USA. Each group has its own distinctive age, but all are geologically very young, the Australian, Asian and Ivory Coast tektites being of *Pleistocene age (about 650,000 years and 1,300,000 years respectively), and the European and North American tektites having *Miocene (15,000,000 years) and *Oligocene (34,000,000 years) ages respectively. Older tektites are unknown.

Any theory which seeks to explain the origin of these mysterious objects must take account of their restriction in both space and time, and the fact that they bear no discoverable relation to their surroundings.

Other highly significant features which must be explained are the presence within some tektites of minute grains of the meteoritic minerals kamacite, troilite and schreibersite, and the recent discovery that contained gas bubbles are at pressures of only about one thousandth of an atmosphere.

In the last 200 years there has been no shortage of theories for the origin of tektites, although many must be dismissed as fanciful: modern evidence discounts, for example, an artificial, man-made origin; and Australian tektites have not been shaped in the gizzards of emus, as was proposed in 1911! Indeed some of their

A selection of tektites showing clearly the aero-dynamically molded shapes.

shapes have been shown by wind-tunnel experiments to have originated aero-dynamically, by ablation and molding during transit through the Earth's atmosphere.

Modern theories for the origin of tektites can be divided into three groups. The first claims an entirely extraterrestrial pro-venance, that tektites are quite simply *meteorites, albeit with novel composition. Candidates for the parent body include planets, asteroids, comets and the Moon, although the latter now seems an extremely improbable parent since no lunar rocks of remotely similar composition have been found *in situ*. The second group of theories invokes an entirely terrestrial origin, in-volving special and speculative types of volcanic or cryptovolcanic eruptions and processes.

There are formidable weaknesses in both these groups of theories, and the weight of evidence increasingly appears to favor an origin involving terrestrial material ejected from the ground in violent meteorite or comet impact. In its modern form, this hypothesis demands that the impact is sufficiently catastrophic to vaporize large amounts of surface and subsurface rock, the gases being ejected into suborbital trajec-tories. On cooling, the more refractory components condense and accrete into droplets that incorporate bubbles of gas at pressures appropriate to the atmosphere 30 to 40km (about 20–25mi) above the ground. Before returning to Earth in regions remote from, and perhaps even antipodal to, their source area the fragments are molded into aerodynamic shapes.

Spectacular and improbable though this hypothesis may appear, it is the best at present available, and it provides a model for future work on these enigmatic objects.

FBA

Geophysics

The classification of scientific research into neatly labelled compartments is a process traditionally alien to the geological sciences. Most geologists would consider that they were of necessity in close touch with many branches of chemistry, physics or zoology.

However, there is much scope within the Earth sciences for someone with a more specialized training in physics, and such a person is, quite logically, described as a geophysicist. In particular, various exper-imental techniques of a physical nature have been applied to studies of the Earth with spectacular success: the great re-volution in the Earth sciences brought about by the concept of *plate tectonics would never have occurred without the evidence obtained from geophysical meth-ods. The vast increase in the successful location of economic oil, gas and *ore deposits since the war has depended largely on the use of geophysical exploration tech-niques.

Seismology. Of all the methods available, seismic investigations are by far the most revealing in terms of structural contrast within the Earth's interior. An earthquake or explosion generates waves of elastic energy within the ground. The task of the seismologist is to record the portion of the wave motion that returns to the surface and study the recorded signals for information about the structures through which the waves have passed. The experimentalist is concerned with two basic types of waves, body waves and surface waves (see *earthquakes).

A body wave travels through the Earth as a simple pulse, whereas a surface wave

The distribution of tektites around the Earth: (1) Texan bediasites; (2) Georgian georgianites; (3) Ivory Coast tektites; (4) Czechoslovakian moldavites; (5) Indochinites; (6) Jawa and related tektites; (7) Philippinites; (8) Australites.

A seismic experiment in progress on the Moon. The leads trailing across the lunar surface are from individual sensors laid out at distance.

proceeds only along the surface of the Earth and its wave motion is spread out over a broad time interval. The reason for this dispersion of energy is that low-frequency portions of the surface wave travel through a thicker layer of the surface than do high-frequency portions, and hence in general there is a spread of velocities associated with the different frequencies of the surface waves in the initial "package".

This can be used to infer the average structure of the Earth between two recording stations. The inter-station velocity of the different frequency components of the surface wave can be measured directly from the seismograms made at the two stations. Since the different frequencies are sampling different depths within the Earth, these velocity/frequency data can be transformed into information about the variation of velocity with depth by a straightforward, though rather lengthy, computation. The method has been of great value in determining the Earth's gross structure.

A more recent development has been the detection of the natural vibrations of the Earth itself. These are excited to any measurable degree only by the largest earthquakes and are analogous to natural vibrations of a bell or of the air within an organ pipe. Models of the seismic velocity and density structure of the whole Earth are developed from these data in much the same way as surface waves are analyzed. However, very special instruments – normally called strain meters – are needed to detect these oscillations, which have periods of up to one hour. Strain meters consist essentially of two solid piers embedded in the ground some tens of metres apart. Apparatus which allows the very small relative displacements between them to be continuously measured, usually a laser generator and reflecting mirrors, is mounted on the two piers.

The interpretation of body-wave seismograms involves the correct identification as reflections or refractions from discontinuities within the Earth of different arrivals within the signals. For example, a seismic ray is refracted very sharply at the boundary between the core and the mantle. The result is that the core casts a large "seismic shadow" that is clearly observable if records from around the world are compared.

In the last few years, two abrupt increases in seismic velocity and density have been detected at depths of approximately 400km and 650km (about 250 and 400mi), and these are probably due to pressure-induced changes in the atomic packing of mantle material: a seismic body wave that has passed through these structures may consist of two or more interfering signals due to the reflections and refractions that occur at a boundary between materials that permit differing velocities of pressure waves.

Large spreads of seismometers have re-

cently been set up, one of the largest being in Norway where 198 instruments are distributed over an area of 40,000km^2 (about 15,500mi^2). The advantage of such a distribution is that, when the signals from all the seismometers are summed together, scattered energy tends to cancel out so that the coherent phases are more obviously apparent. In addition, it is possible to determine the direction and angle of approach of the waves from the small differences in their arrival times at the individual seismometers.

In the interpretation of body-wave seismograms, a wave that travels along the interface between two layers is often recorded. The thickness and seismic velocities of the layers can be deduced from the slopes of time-*vs*-distance (from source) graphs constructed from the records.

In the search for oil-bearing strata, geophysicists make use of the seismic-reflection technique to the virtual exclusion of all other exploration methods. This involves the deployment of a dense line of seismometers (normally referred to as geophones in this context) as close as possible to the shot point, the object being to detect only the reflections from directly beneath the shot/receiver location. The location is progressively shifted until a pattern of vertical reflections over the whole survey area is obtained. Finally, the seismograms from each geophone are plotted side by side, so that laterally varying structures can be clearly seen.

In practice, the interpretation of this kind of data is much more involved than one might expect. However, great advances have recently been made in improving the clarity of the records. We have already considered the spatial filtering that can be carried out using a pattern of receivers: equally, the signals can be frequency-filtered using standard electronic techniques and, in addition, the energy source can be enhanced by using a pattern of shots.

Gravity. The reader will be very familiar with the law of gravitational attraction between masses, if only because one is drawn so firmly towards the Earth when falling off a ladder. The precise value of the gravitational force is very slightly variable over the surface of the Earth and the changes reflect lateral inhomogenities of mass within our planet's interior.

In the gravimeter we have an instrument that allows us to measure the gravitational force to very high precision and thereby learn something about the Earth's internal structure. There is no problem in detecting variations in gravity due to dense ore bodies or basaltic intrusions. The difficulty comes in the correct interpretation of the data: any number of different shapes or density contrasts can give rise to any one set of gravity readings. The main value of the method comes when it is carried out in conjunction with other geophysical techniques.

The gravimeter does not make friends easily. It insists on being treated with extreme care at all times and has an in-

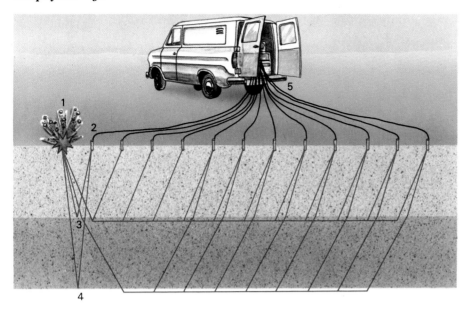

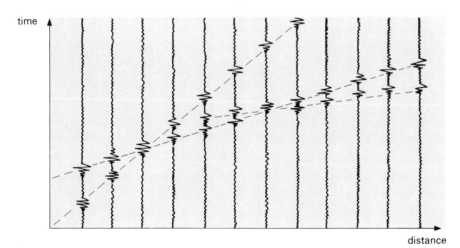

A seismic refraction experiment and an idealized trace of the results. By recording at different points the disturbance due to seismic waves generated by an explosion, a prospector can build up a picture of the underlying rock strata. This will indicate suitable sites for further exploration for oil or minerals. Shown in the diagram are (1) shot point; (2) first geophone; (3) and (4) direct reflections; and (5) equipment truck (not to scale).

fallible habit of refusing to settle to a steady reading towards the end of a long day when an icy wind is numbing the fingers. It is simply a small mass suspended on a very sensitive spring, though some form of mechanical amplification is also included. It is always necessary to protect the equipment from changes of temperature, either by including thermostatically controlled heating circuits, or by building the springs and levers out of various carefully chosen metals whose different degrees of expansion on heating interact in such a way as to cancel each other out: in effect the device itself makes the corrections for changes of temperature.

The gravity variations over the surface of the Earth are due to a combination of local anomalies, some of which we have already mentioned, and broad regional anomalies spread over hundreds or thousands of square kilometres which are probably due to mass anomalies in the deep interior. Our

knowledge of these features has been vastly increased in the last few years by observing the orbits of artificial satellites. The satellites precess, or progressively change their orbiting path relative to the Earth's axis, due to these broad variations in the gravity field, and it is possible to determine the location and amplitude of these anomalies from the rates of precession of a large number of satellite orbits. Future work will doubtless reveal more and more details of the gravity field.

Electrical and Electromagnetic Techniques. Many of the economically important ores are good conductors of electricity, particularly compared to the country rock in which they are embedded. Such ores include *graphite; *pyrite, which is an important ore of sulfur and is sometimes associated with gold; *chalcopyrite, the copper ore; *galena, the principal ore of lead; and the iron ore *magnetite. There are various methods of determining the electri-

cal conductivity of the ground over various depth ranges, and they have wider applications than to the location of mineral deposits: water content is a major factor in the conductivity of crustal rocks and consequently electrical methods are often used when an estimate of the depth of the water table is required.

From large-scale experiments that provide electrical conductivity values to depths in excess of 1000km (620mi) it has become possible to deduce (extremely approximate) temperature-depth curves for the Earth's interior. These estimates rest on the assumptions that the major constituents of the mantle are known, and that temperature is the dominant controlling factor on the electrical conductivity of mantle materials. For a variety of reasons, an improvement in our knowledge of the Earth's internal temperature would be of very great value, and there is currently in progress research aimed towards increasing our rather scanty knowledge of electrical conduction properties of Earth materials under high temperatures and pressures.

Electrical Prospecting Methods. The most straightforward approach to the measurement of Earth conductivity is the resistivity method. Two metal stakes are hammered into the ground and a direct or very low frequency alternating voltage is applied across them. The resulting distribution of current within the ground will depend on the variations in conductivity beneath the surface as well as on the separation of the two stakes, the separation being increased if a greater current penetration depth is required (there is a limit imposed by the power of the voltage source). The quantity that is measured is the drop in potential between two electrodes usually lying *between* the current electrodes. Generally speaking, a low drop in potential means that a region of high conductivity is being sampled. Correspondingly a large drop means low conductivity (high resistivity). Frequently, the interpretation of resistivity data is no more sophisticated than this: an anomalous region, once located, would be investigated further by other means. The association of particular distributions of conductivity with the shape of anomaly patterns in the readings taken at the surface is an extremely complex problem and an active area of current research.

Closely associated with the resistivity method is a technique known as induced polarization (IP). If, on completion of a resistivity reading, the current is turned off, the voltage measured across the potential electrodes may fall to zero virtually instantaneously or may do so gradually over as much as three minutes. The groundwater within the pores and cracks of the rock is an electrolyte – that is, it contains freely moving charged particles. These particles pile up at the interface of electrically conducting mineral grains when a current is flowing, and slowly diffuse back to an equilibrium state when the current is switched off. If ore minerals are spread out within the country

A contour map of the variations in the Earth's gravitational field obtained by observations from artificial satellites.

rock so that they have an effectively large total surface area, this effect will be enhanced and the measured voltage will take some time to die away.

In practice the interpretation of IP anomalies is much less clearcut than the above description suggests. Many other mechanisms have been proposed for the effect, and these may be applicable in particular circumstances.

Electromagnetic Prospecting and Deep-Sounding Methods. In general terms, electromagnetic (EM) prospecting equipment is like an oversized mine detector. An alternating current flowing in a large coil of wire gives rise to an alternating magnetic field in the surroundings. This field causes a current to flow in any conducting body in the vicinity, and a secondary magnetic field is then set up by the conducting body. All that is required is to detect the distortion due to the presence of the body of the primary magnetic field by the secondary. In the simplest type of EM detector, one measures the direction of the field by rotating a coil of wire attached to some earphones and listening for the minimum in the hissing sound produced by the induced current in the coil.

EM tests are frequently performed from the air as a preliminary search over a wide area with a view to locating regions that are worth studying in more detail on the ground. Surveys are usually flown at around 500ft.

Magnetics and Paleomagnetism. No single field of research has contributed more to the general acceptance of the concept of *plate tectonics than the study of rock magnetism.

We can consider the building-block of all matter, the atom, as behaving like a tiny bar magnet. Normally the directions of the myriads of tiny bar magnets making up a lump of matter are random. However, in a few special cases, there is an interaction between nearby atoms within solids that are composed of a particular ordered distribution of certain elements, so that magnetic directions of the atoms are aligned either all in the same direction or, alternatively, all in opposing directions: in rocks, the most common material of this type is the iron ore mineral *magnetite. The result is that the sample will have a spontaneous magnetization due to the resultant of the individual magnetic alignments of the atoms.

For all but the smallest samples, an equilibrium (low-energy) state results in which regions within the solid about 0.01mm wide are uniformly magnetized while adjacent regions, or domains, have quite different directions of magnetization. In the presence of an external magnetic field the domain boundaries are forced to move

slightly to give a net magnetization in the same direction as the field.

For very small samples the equilibrium state is when there is only one domain: the magnetic directions of these single-domain grains are extremely stable because, although an external magnetic field can provide the energy to shuffle grain boundaries about a little, much greater energies are required to actually change the direction of magnetization *within* a domain. However, one rather efficient way to destroy any magnetic symmetry is to heat the material until the thermal agitation of the atoms overcomes the magnetic ordering.

We can now consider what happens to the magnetite within a rock that solidifies from an initially molten state. Below a

particular temperature, the thermal energy is no longer sufficient to randomize the atomic magnets against their tendency to align themselves in the direction of the Earth's magnetic field. Hence the direction of the Earth's field at the time of formation is "frozen" into the rock. During later geological periods the geomagnetic direction will in general be quite different – due principally to the movement of the rock relative to the magnetic poles. By the time the geophysicist collects his sample, only the most magnetically stable single-domain grains still retain their initial magnetic direction.

In the immediate vicinity of a rock containing magnetic minerals, the Earth's magnetic field is distorted. The magnetic pros-

Horizontal-loop electromagnetic equipment in use. Note the connecting lead to the other operator in the distance.

pecting method involves measuring the Earth's magnetic field at regular intervals across the area of interest using an instrument called a magnetometer. The geophysicist is then faced with the – now familiar! – problem of interpreting the anomaly pattern in the readings. His task is one degree harder than with gravity techniques, for he is faced with the uncertainty of the resultant magnetic direction of the body (the sum of all the individual magnetic grains) in addition to its shape, depth and gravity or magnetic contrast. However, much information about structure can often be gained by simply plotting a contour map of the magnetic readings. Any systematic features or trends will then show up. A classic example of this is provided by the magnetic lineations on either side of the mid-ocean ridges.

If the *in situ* orientation of a rock sample is noted, it can then be removed to the laboratory to measure its magnetic direction. This is normally accomplished by rapidly rotating the sample on a platform, which has the effect of inducing a small alternating current in a set of pick-up coils. The direction of the sample's magnetism can then be determined by comparing the signals induced in the different coils. As we have seen, the resultant direction will be the combination of the directions of individual grains. However, by a process known as "magnetic cleaning", which normally involves successive heating and cooling of the sample in the absence of a magnetic field to demagnetize all but the most stable single-domain grains, the magnetic direction of the sample at the time of its formation can be extracted.

The data thus obtained have been used with great success to chart the motions of continents relative to the magnetic pole through geological time. One problem of this technique is that a continent could move only along a line of latitude, in which case the relative direction of the Earth's magnetic field would not vary: the result is that only changes of paleolatitude can be observed. However, the *relative* motions of different continents can often provide information on their past positions.

Conclusions. It has been pointed out more than once in this chapter that geophysical methods do not often give specific answers to the problems that they are supposed to solve, although the situation is normally improved by use of a combination of meth-

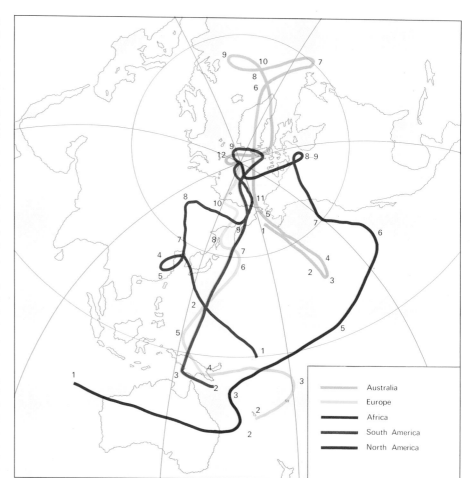

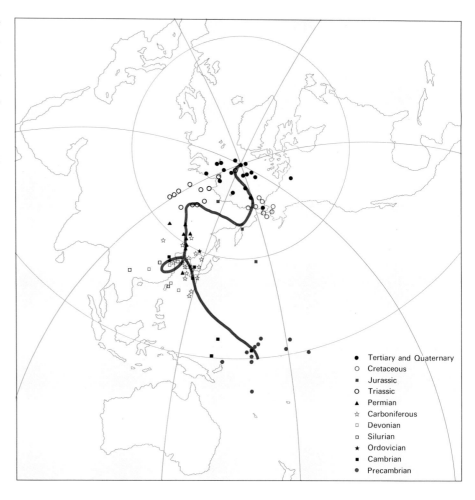

Above, polar wandering curves for the various continents through geological time, based on paleomagnetic data. The numbers refer to: 1, Precambrian; 2, Cambrian; 3 Ordovician; 4, Silurian; 5, Devonian; 6, Carboniferous; 7, Permian; 8, Triassic; 9, Jurassic; 10, Cretaceous; 11, Tertiary; 12, Quaternary. Since these curves, which describe the apparent positions of the north magnetic pole as determined from different continents, are not only not colinear but also show remarkable differences of form, the only apparent explanation of polar wandering would seem to be that the continents have moved relative to each other. *Below,* in more detail, the data and the resultant curve using paleomagnetic results from the North American continent alone.

ods. Our discussion has been concerned chiefly with techniques that are loosely combined under the heading "geophysics". However, these techniques, and their results, should never be considered in isolation from the other branches of geology, since they are often most effective when combined with geological constraints. MHW

Geochemistry

Nature is a very poor chemist – but an excellent experimental annealing oven. Man has learned to purify and concentrate small parts of the outermost skin of our planet so that chemists may work with and study the detailed behavior of pure substances. However, the rocks and fluids of Nature's experiments have not been subjected to the thousands of years of human effort which have produced pure substances for the chemical laboratory. Natural rocks and minerals, the oceans and atmosphere are in almost all cases complex mixtures of different chemical substances which are continually reacting in order to achieve a final condition of chemical equilibrium.

Geochemistry is the field of study dealing with the many diverse aspects of these reactions. It may be defined briefly as the study of the natural chemistry and evolution of the Earth and, more generally, of other cosmic bodies.

Formation of the Elements. The chemical compositions of the Sun and other stars, and of interstellar gas clouds, can be determined by observing their spectra. Such measurements show that, although the compositions of individual stars differ to some extent, all stars show similar general abundance patterns. From these data, together with analyses of meteorite compositions, relative "cosmic" abundances of elements in the Universe can be estimated.

Hydrogen (92.7%) and Helium (7.2%) together constitute 99.9% of all the matter to be found in the universe. The heavier elements, in particular oxygen, silicon, magnesium and iron, which make up the bulk of the Earth, together amount to a small fraction of 1% of the available matter.

The shape of the abundance pattern, where abundance is plotted against increasing atomic number (see *elements), carries several clues as to the manner in which these heavier elements were formed: the relative abundances decrease rapidly with increasing atomic number; the pattern has a characteristic "zig-zag" shape resulting from the greater stability of elements with even atomic numbers; the light elements lithium, beryllium and boron are present in anomalously low amounts; iron has an excessive abundance; and the abundances of elements with atomic number greater than 45 are approximately similar.

It is believed that the only mechanisms capable of supplying the vast amounts of energy radiated by a star are thermonuclear reactions. The most important of these is thought to involve the conversion of hy-

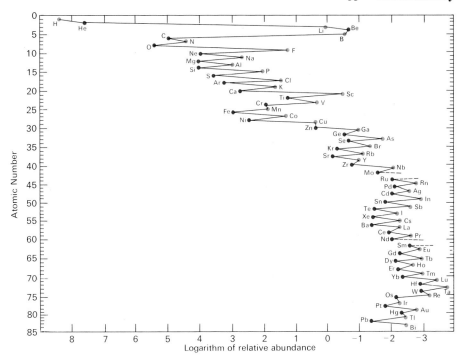

Relative cosmic abundances of the elements, referred to a value of 10^4 for silicon (Si: 4 on the logarithmic scale). Elements with even atomic numbers are plotted in blue, those with odd atomic numbers in red. Hydrogen (92.7%) and helium (7.2%) together constitute 99.9% of all the matter in the universe. The heavier elements, and in particular oxygen (O), silicon (Si), magnesium (Mg) and iron (Fe), which make up the bulk of the Earth, together amount to a fraction of 1% of the available matter.

drogen into helium. In this "hydrogen-burning" the combined masses of the products are slightly less than the mass of the reactants, the lost mass being converted into energy. At higher temperatures, approaching 100,000,000°C (180×10^{6}°F) or more, the helium produced in this way can react further to produce carbon. At still higher temperatures more fusion reactions may occur, to produce oxygen, neon, sodium, magnesium, aluminum, silicon, phosphorus and sulfur.

However, as the nuclear charges become higher, the repulsive forces between the nuclei also increase, so that impossibly high temperatures would be required to produce elements of higher atomic numbers by continued fusion alone. To produce such elements, different mechanisms (beyond the scope of this book) have been proposed.

Thus, starting with hydrogen, all the other elements can be built up in the interior of stars by a variety of nuclear reactions. The anomalously low abundances of lithium, beryllium and boron are thought to result from the rapid destruction of these nuclei in subsequent nuclear reactions. On the other hand, iron and silicon are unusually stable, as a result of their nuclear structures, and so are present in greater abundance than expected.

It is believed that Earth and the other planets formed by condensation of these various elements released after their formation in preexisting stellar reactors. The nature of the condensation process is still an active topic of research: however, it has been found that certain types of *meteorites have chemical compositions similar to the

cosmic abundances. These meteorites may represent relatively primitive material left over from the formation of the Solar System. The composition of the Earth, however, has changed since then as a result both of the loss of volatile elements and of chemical differentiation to produce the present distribution of elements on the planet.

Composition of the Earth. Direct observations can be made only of the compositions of the atmosphere and oceans, and of rocks near the Earth's surface. However, as a result of studies of nodules carried up in *kimberlite pipes from regions of high pressure, and analyses of seismic data (see *geophysics), the internal structure of the Earth is quite well understood.

The first task of the geochemist is to analyze the rocks and minerals which make up the Earth's crust. Once their compositions are known, experiments can be carried out over a wide range of temperatures and pressures to determine how the rocks may have been formed. Some of the techniques for doing so are discussed below.

Atomic Absorption and Emission Spectrometry. The rock to be analyzed is first ground to a fine powder and dissolved in hydrofluoric acid. After neutralizing or evaporating off any unreacted acid, the resulting rock solution is then sprayed into a hot flame. In the flame the high temperature causes some atoms to be excited into higher energy states. When electrons in the excited atoms drop back to their ground states, radiation of a characteristic color is emitted by each element. This is the basis of flame *emission* spectroscopy, which can itself be

X-ray diffraction photograph of finely powdered quartz. The film is curled into a cylinder, the sample mounted at its center, and a beam of X-rays passed in through a hole in the film. The beam is diffracted into a series of cones, represented on the film by pairs of curved lines whose positions and intensities provide a unique "fingerprint" of the mineral. (The marks on the right are later additions.)

used as a direct method of analysis. It is also used in the determination of the compositions of the atmospheres of stars.

Many atoms, however, are not excited in this way. If a lamp emitting the wavelengths of light characteristic of the element to be determined (e.g., red light for calcium, green for barium) is shone through the flame, the radiation can be absorbed by the atoms of that element present. The logarithm of the ratio between the intensities of the light before and after passing through the flame is found to be proportional to the concentration of the particular element in the rock solution, and hence in the rock itself. This is the basis of *atomic absorption* spectroscopy. This method of analysis is particularly useful for determining small or trace quantities of minor elements.

X-ray Fluorescence Spectroscopy. Another commonly used analytical technique involves the bombardment of a rock sample with X-rays. The rock may be finely powdered and pressed into a pellet, or fused into a glass bead using a low-melting-point flux. The pellet or glass bead is then exposed to radiation from an X-ray tube.

The atoms of the rock sample are again excited into high energy states. On decaying they emit fluorescent X-rays whose wavelengths, which are characteristic of the particular atoms present, can be measured.

The method of X-ray fluorescence spectroscopy has been further developed to allow a complete chemical analysis to be performed on a single mineral grain. Instead of exciting the specimen with a beam of X-rays, a focused beam of electrons is used. The electron beam, which can be focused into a spot less than a thousandth of a millimetre in diameter, again excites atoms into high energy states from which they fall back, emitting fluorescent X-rays. By collecting the X-rays from this tiny spot the chemical composition of the area hit by the electron-beam can be determined. This is the basis of the *electron microprobe microanalyzer*.

Chemical Compositions of Different Rock Types. Although *sedimentary rocks and *metamorphic rocks are frequently exposed at the Earth's surface, it is

estimated that 95% of the Earth's crust is composed of *igneous rocks.

The ocean basins which constitute around 70% of the Earth's surface are floored with *basalt*. Sometimes the igneous processes supplying basalt to spreading mid-oceanic ridges break through the surface to produce volcanic eruptions, as in the case of the island of Surtsey which was "born" in November 1963 (see *ocean floor; *vulcanicity). Large amounts of basalt are also found on the continents, together with other volcanic rocks and their intrusive slowly-cooled analogs.

Although analytical methods give data on the molar or atomic amounts of the different elements present, it has become standard practice for analysts to recast their determinations into the weight persentages of different *oxides* present. These analyses show that common igneous rock-types may vary from 48% to 72% silica (SiO_2) by weight, and an even larger relative change is shown for magnesium oxide (MgO) – from 8.1 to 0.5 wt%.

How can such variations in chemical composition be produced?

Once the chemical composition of a rock is known, laboratory experiments can be performed to determine how the minerals in the rock react under changing pressure and temperature conditions, and in particular how that rock composition melts. Most igneous rocks contain the minerals clinopyroxene (see *pyroxenes) and *plagioclase feldspar. In natural rocks neither is ever pure. Plagioclase feldspars are framework silicates ranging in composition from $NaAlSi_3O_8$ (Albite) to $CaAl_2Si_2O_8$ (Anorthite). Of clinopyroxenes the most important end-member composition is that of *diopside, $CaMgSi_2O_6$.

The system diopside-anorthite can therefore be considered as a (much-simplified) model basaltic rock composition. How does such a system behave at high temperatures?

Consider a rock consisting of 80% diopside crystals and 20% crystals of anorthite. Its temperature can be raised up to 1260°C (2300°F) without appreciable amounts of reaction between the crystals: however, at this temperature a melt starts to form. This melt has a composition richer in anorthite than the average composition of the rock, since more anorthite melts than diopside. This process continues at 1260°C until all the anorthite has been melted. Once this has happened, the temperature can continue to increase and the remaining diopside can dissolve in the melt, which becomes suc-

cessively richer in diopside with increasing temperature.

This simplified summary shows that, when a rock melts, the composition of the melt formed may be quite different from that of the original rock. When melting begins inside the Earth, the small quantities of liquid formed in this way may coalesce to form a body of molten magma. Since the magma is less dense than the surrounding rocks, the liquid may rise towards the Earth's surface, leaving behind a residue of minerals with higher melting points.

This process, partial melting, whereby melts having different compositions from the starting material may form and then migrate upwards has operated over the 4.6 thousand million years of the Earth's history, and is considered the primary mechanism of chemical differentiation of the Earth.

The Earth's Core and Mantle. Most geochemists now consider that the Earth formed by a gradual condensation and accretion of particles from the nebula which gave rise to the Solar System (see *planet Earth). In the earliest stages of the Earth's evolution, short-lived radioactive isotopes caused a rapid increase in temperature and so melting as described above. The core separated out as a dense metallic liquid rich in iron, nickel and possibly silicon and/or sulfur, and gravitional forces caused it to migrate to the center of the planet. At the same time, volatile elements were expelled as gases to form an early atmosphere.

The crust formed – and is probably still forming – from the less dense silicate melts, rich in magnesium, aluminum, silicon, sodium, potassium and calcium, which migrated upwards on partial melting.

The mantle represents the residual high-melting-point material left behind after partial melting. It is likely that the mantle consists dominantly of magnesian *olivine, $[Mg,Fe]_2SiO_4$, together with smaller amounts of *enstatite ($MgSiO_3$), diopside ($CaMgSi_2O_6$), pyrope *garnet ($Mg_3Al_2Si_3O_{12}$) and perhaps some phlogopite *mica ($KMg_3Si_3AlO_{10}(OH)_2$).

Evolution of the Atmosphere. It is likely that the proto-Earth had very little or no atmosphere at all. However, the heat liberated by gravitational collapse and by the decay of short-lived radioactive isotopes would soon result in melting of the Earth's interior to produce silicate melts and liquid iron. This heating would also liberate gases which would rise to the surface and escape. It is very likely that the entire atmosphere originated as volatile constituents expelled

as volcanic gases from the interior of the planet.

Three states in the evolution of the atmosphere are recognized. In the initial stage, the volcanic gases emitted would be in equilibrium with metallic iron in the mantle and would therefore have a much more reduced (less oxidized) character than those observed today. These volcanic gases would provide an initial atmosphere consisting largely of hydrogen, water vapor and carbon monoxide (H_2, H_2O and CO) with minor amounts of nitrogen, carbon dioxide and hydrogen sulfide (N_2, CO_2 and H_2S). The primeval atmosphere is therefore likely to have been quite different from that of today. As the gases cooled, H_2O condensed as water and the CO and CO_2 may have reacted with H_2 to form methane (CH_4), leaving an atmosphere of mainly H_2 and CH_4.

This stage came to an end following the separation of the metallic iron from the mantle to form the core. This allowed the oxidation state of the mantle to become much higher so that the volcanic gases in equilibrium with the mantle became more oxidized and were probably similar to those found today: H_2O, CO_2, CO, H_2, SO_2 (sulfur dioxide), N_2, Cl_2 (chlorine) and other minor constituents. In this second stage of evolution, the atmosphere probably contained mainly N_2, CO_2 and H_2O. The light gases, like H_2 and He (helium), continued to escape into space. Small amounts of free oxygen could have been produced in the upper atmosphere, but the oxygen would be rapidly consumed in the oxidation of the volcanic gases.

The third stage of evolution is marked by the appearance of free oxygen when oxygen production exceeded consumption. This is likely to be associated with the onset of green-plant photosynthesis – paleontological evidence suggests the existence of algae at least 2 thousand million years ago, and probably much earlier still. By 1.2 thousand million years ago, sufficient oxygen was present in the atmosphere to allow the formation of extensive sedimentary "redbeds" containing ferric iron.

The present composition of the Earth's atmosphere is 78% N_2, 21% O_2, 0.9% Ar (argon) and 0.03% CO_2, plus other gases such as Ne (neon), He, CH_4 and Kr (krypton) in smaller amounts (the amount of H_2O is variable, of course, depending on the degree of saturation). It is likely that the oceans play a crucial role in controlling the composition of the atmosphere by absorbing CO_2, which is later deposited as carbonates and organic carbon in *sedimentary rocks.

It is worth noting that the Venera and Mariner spacecraft sent to Venus found an atmosphere consisting of 97% CO_2 at temperatures around 460°C (860°F). It had been suggested that these conditions are the result of initially small build-ups of CO_2 in the Venusian atmosphere triggering a disastrously increasing greenhouse effect, caused by the retention of heat by CO_2 molecules. It appears possible that if the carbon dioxide content of the Earth's atmosphere were to exceed a critical level (perhaps as a result of the increased burning of fossil fuels), the Earth's surface temperature would rise, thereby releasing more CO_2 from the oceans and producing a similar effect.

A temperature of 460°C, as found on Venus, is clearly sufficient to vaporize all our rivers and oceans. Such a temperature would even release most of the CO_2 locked up in sedimentary carbonates. However, long before then, Earth would be, like Venus, a dead planet. It is therefore of obvious importance to increase our understanding of the controls on atmosphere composition and temperature on Earth.

Isotope Geochemistry. The number of protons (positively charged particles) in an atom's nucleus determines its atomic number, and hence its chemical properties. However, each *element may have different numbers of neutrons (non-charged particles) in the nucleus. Thus each element can have a number of different *isotopes* of different mass numbers (sum of protons and neutrons). Some of these isotopes may be radioactive, and can be used for radioactive age determinations (see *age of the Earth). Others are stable. Each element may therefore exist in nature as a mixture of different stable isotopes (as well as any radioactive isotopes that might remain).

Although the chemical properties of different isotopes of the same element are virtually identical, the differences in mass can give rise to small differences in both equilibrium and kinetic properties: when a cloud loses rainwater, the rain is enriched in heavy isotopes and the remaining vapor is richer in light isotopes. Isotopic measurements have made it possible to study the total circulation of water on the Earth.

Different isotopes of the same element can be separated using a *mass-spectrometer*. Modern mass spectrometers can measure variations in isotopic abundances to within about 1 part in 10,000. These measurements make it possible to measure isotopic variations for elements with atomic number up to about 20. Since the effects of relative differences in atomic mass are so small, they become much less important with increasing mass number, and variations in isotopic abundance cannot be detected at all for elements of atomic numbers above about 20. There is one group of important exceptions to this rule, isotopes of radioactive origin. Thus rather large variations in the isotopic composition of strontium, argon and lead are found in nature, due to the production of *strontium from radioactive *rubidium; argon from radioactive *potassium; and *lead from *uranium and *thorium.

The fractionation of light isotopes is most efficient in systems involving a gas phase when there is effectively a continuous fractional distillation. Thus, as mentioned above, the isotopic composition of water vapor in equilibrium with seawater indicates a depletion in heavy isotopes.

Current Topics of Geochemical Research. Considerable effort is being expended on efforts to place further geochemical constraints on possible mantle models and to elucidate the genesis of different igneous rocks. An understanding of the geochemistry of the igneous rocks erupted at spreading and destructive plate boundaries (see *plate tectonics) is expected to be of critical importance in investigating the mechanisms of plate movement.

To test models for the genesis of igneous rocks, detailed measurements are required not only of the phase chemistry of natural systems but also of the behavior of trace components. Experimental determinations of the distribution coefficients of trace components between minerals and melts are being carried out over wide ranges of temperature, pressure and melt composition. Recently improved techniques have extended the range of experimentation to 1.5 million atmospheres of pressure. This may make it possible to investigate the behavior of mantle materials at pressures near to the pressures experienced at the core-mantle boundary.

Experimental investigations of this type are carried out in conjunction with detailed theoretical treatments of the thermodynamic properties of the mineral and melt solutions. Thus studies in geochemistry can yield fundamental information to the fields of solid-state chemistry and thermodynamics – in addition to answering old questions and formulating new problems about the origin and chemical evolution of the Earth. DGF

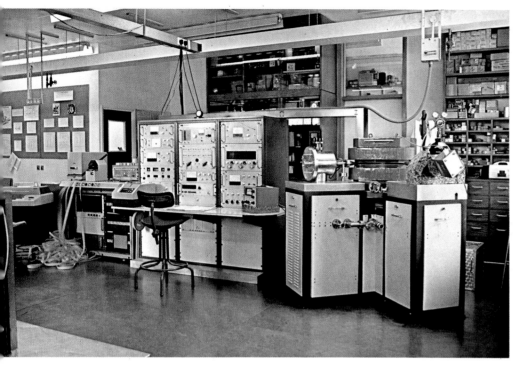

A mass spectrometer. On the left are the on-line computer, power supplies, vacuum gauges, electromagnet controls and visual display unit. Nearer, on the right, is the mass spectrometer itself: on its left is the sample chamber and ion source from which the ion beam passes through the electromagnet (blue, center). The beam is bent by the electromagnet, the lighter ions being deflected more than the heavier ones, to reach the collector on the right. Modern mass spectrometers can measure variations in isotopic abundances to within about 1 part in 10,000.

Climatic Zones

Any planet revolving round a sun inevitably has climatic zones. This arises from the simple fact that some parts of the planet receive a greater net amount of light and heat than others, and so there must be some sort of heat-flow from one part to another.

The variations which occur between the patterns of climatic zones of the different planets in our own Solar System depend not only on their possession or non-possession of a fluid mantle, but upon the period of rotation of the planet on its own axis, and the obliquity of this axis (i.e., the angle between its axis and the plane of its revolution around the Sun). The Earth's climatic zones have been subject to change through geological time, not only because its period of rotation and the obliquity of its axis have been subject to some variation, but because the fluid mantle has changed. Moreover, the changing distribution of land and sea, especially in relation to the position of the poles, greatly affects the heat-flow in the oceans and in the atmosphere. There have been changes in some other factors also, such as in the radiation output from the Sun, distance of the Earth from the Sun and from the Moon, and so on.

However, once the Earth had developed a solid core and a fluid mantle, it must have had climatic zones with a recognizable similarity to the present ones, corresponding to a circulation in the atmosphere resembling in general that of today. Given that the Earth has, during geological time, always been rotating daily on its axis with a period much shorter than its rate of revolution round the Sun (now just over 365 days), there will always have been a zone near to the equator where more radiation was absorbed than reflected, and polar areas where more was radiated outwards than received. The area of net absorption lies today between the Tropics of Cancer and Capricorn, each at an angular distance of 23.5° from the equator, corresponding to the 23.5° obliquity of the Earth's axis at the present time. The zone of maximum net absorption varies with the seasons of the year – i.e., the period of revolution of the Earth around the Sun. Correspondingly, the area of net loss of radiation varies through the year, the extreme variation being beyond the polar circles, whose latitudes are 66.5° (90° less 23.5°). The poles themselves experience continuous daylight for one half of the year, and continuous night for the other half. In between the Tropics and the polar circles are 43° of latitude in which the Sun is never vertically overhead, and in which there is never total darkness or total daylight for 24 consecutive hours at any time of the year.

Whether the Earth were rotating upon its axis or not, there would be a meridional (north-south) flow in the atmosphere to effect the necessary heat compensation between the tropics and the polar regions. There would be between the tropics, as today, rising air (upward convection), which necessarily spreads out at higher levels to flow northwards and southwards, to descend after travelling between a third and half-way to the poles, where there is in general a lack of upward convection. Some of this air then flows equatorwards close to the surface, to compensate for the rising air in the tropics. It follows that low pressure is usual near the equator, and high pressure in mid-latitudes. The vertical circulation here described is known as a Hadley Cell.

The basic situation near the poles, most marked in winter, is that the heavy, cold air – i.e., at high pressure – is always tending to spread equatorwards. It will be realized that the situation is inevitably somewhat complex where this air of polar origin encounters the descending air of mid-latitudes, and indeed this is where the Earth's most variable weather is experienced.

One now has to take into account the effect of the Earth's rotation, expressed as the Coriolis force, whose value is nil at the equator. The net effect of this force, combined with frictional drag, is to deflect the meridional flow more and more as one moves polewards. In the northern hemisphere deflection is to the right, so that, for example, the surface equatorwards flow (the Trade Winds) in lower latitudes becomes a northeast wind. In the southern hemisphere the corresponding winds are deflected to the left, becoming the southeast Trade Winds. The slowly outward-spreading air of the polar regions is similarly deflected. Some of the descending air just north of the Hadley Cell moves polewards, and the often strong surface winds are deflected to become (in the northern hemisphere) southwesterly or even westerly: the westerly wind belts are of considerable but varying width. In the southern hemisphere, the strong west winds of these latitudes have long caused them to be described as the Roaring Forties.

The vertical circulation is particularly complex in these latitudes. It is rendered all the more complex by the present distribution of oceans and continents, which in fact modifies the circulation everywhere. The oceans are more heat-conserving than the continents, and the effect of this is that the continents have much greater seasonal ranges of temperature, to the extent that winter in eastern Siberia is colder than at the poles, and this region experiences the highest atmospheric pressures known on Earth. Subsiding cold air tends to spread out then from these "cold poles". The converse is summer heating, leading to rising warm air and a compensatory flow of surface air in toward these areas: the Indian monsoon is the classic example of this.

It will be seen that there are areas which experience, at least seasonally, converging surface winds, of which the warmer is forced to rise over the cooler. When the differentiation between the two air masses is sharp there are distinct fronts, highly characteristic of the westerly wind belt. Rather less distinct fronts, occurring where there is strong convection, are found near the equator – the Intertropical Convergence Zone – between the Trade Winds of the two hemispheres.

The major climatic zones as they exist on the Earth today may be summarized as follows:

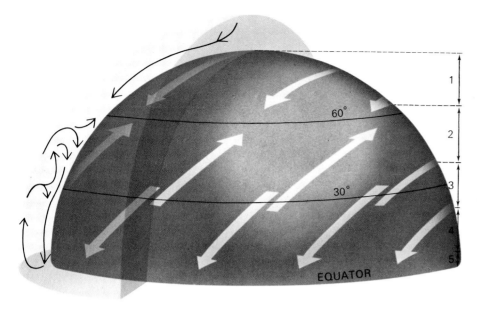

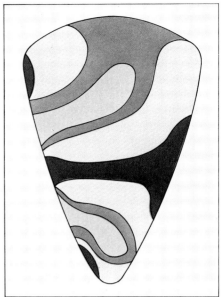

A cross-section of the northern hemisphere showing the climatic zones and a simplified representation of the atmospheric circulation patterns. Reading from the north pole toward the equator the zones are: (1) the polar zone; (2) the westerly wind belt; (3) the Mediterranean zone; (4) the Trade Wind belt; and (5) the equatorial region. Winds blow southward from the pole, and to both the northeast and southwest from roughly 30 latitude. On the left this circulation pattern is shown in its vertical aspect.

The pattern of precipitation that might develop on a hypothetical continent lying from 60° south to 80° north. Due to the rotation of the Earth, the pattern is not symmetrical about a north-south axis. Wet areas are shown in dark green, humid in pale green, sub-humid in brown and arid in pale brown.

(*i*) Equatorial regions, which are hot and wet throughout the year, as a result of continuously strong insolation (amount of solar heat received) and hence upward convection of moist air.

(*ii*) Tropical regions. Beyond about 5° latitude, north and south, the seasonal apparent shift of the Sun north and south begins to be reflected in the emergence of a dry season, which increases in length toward the Tropics of Cancer and Capricorn. The effect of the configuration of land and sea is that the length of the dry season increases more rapidly on the western sides of continents than on the eastern.

(*iii*) Sub-tropical deserts. The last mentioned effect is very noticeable in the Trade-Wind belt, for these deserts in no case extend to the eastern coasts of continents. The greatest extent of desert and semi-desert is, not unexpectedly, in the broadest land mass in the world, from the Saharan coast eastward *via* Arabia to inner Asia. It does of course border the Arabian Sea, but, although this brings it to the eastern coast of Africa, this is not in broad perspective the main eastern coast of the land mass known as the Old World.

(*iv*) The so-called Mediterranean and China climates embrace those areas where the effect of the north-south apparent movement of the Sun is a seasonal variation of the climatic zone between about 30° and 40° latitude (north and south) between the Trade-Wind belt in summer and the westerly wind belt in winter. In winter the on-shore westerlies bring rain to the western ("Mediterranean") coasts, while the easterly Trade Winds bring summer rain to the eastern ("China") coasts. Because of the barrier imposed by the north-south mountain chains in the American continents, the "Mediterranean" zones there are very narrow, and only in the European Mediterranean area has this zone a large east-west extent. The seasonal range of temperature is much greater in the "China" than in the "Mediterranean" type of climate.

(*v*) The westerly wind belt, which tends to have rain throughout the year. The seasonal range of temperature is relatively slight on the western margin of continents, as in the British Isles, but gradually increases eastwards as the total annual rainfall generally diminishes. But this is the zone of most complex meteorological activity and is characterized by travelling depressions and fronts, which, together with summer convectional activity, bring adequate rain in the growing season to large inland areas.

(*vi*) Polewards of the westerly wind belt is an area of less stormy conditions, but which becomes increasingly cold as one approaches the pole. It has a greater latitudinal extent on the eastward side of the northern landmasses – there is almost no land in the southern hemisphere until one reaches the ice-covered polar areas. In Siberia and northern America, there is sufficient warmth in summer for considerable vegetative growth, but not enough to thaw the soil at depth, and so the zone is characterized by permafrost (see *tundra landscapes).

There are of course various anomalous areas, such as the monsoon climate zones, but these may be regarded as major perturbations of the general zonation, and not as something entirely separate.   FG

Processes That Shape the Earth

Continental Drift

For those interested in the history of science, in the interplay of fact and hypothesis, of competing schools of thought and the pervasive influence of intellectual climates of opinion, there are few subjects more fascinating than the long-continued controversy over continental drift. Only within the last decade have the vast majority of geologists and geophysicists come to accept that the continents have moved with respect to each other, splitting up in some regions and colliding elsewhere. It is now customary to discuss such lateral mobility in terms of the comprehensive theory of plate tectonics, involving both continents and oceans, so that the older name, "continental drift", is falling into disuse.

Although a number of scholars had been impressed by the congruence of the (eastern) South American and West African coastlines, it was not until 1858, with the publication of *La Création et ses mystères dévoilés* by Antonio Snider-Pellegrini, that the first clear suggestion of a break-up and drifting apart of the Atlantic continents emerged. Snider's revolutionary views had no impact, however, upon the contemporary scientific community, mainly because of the fantastic nature of his old-fashioned catastrophist beliefs, for which the only support brought forward was the "jigsaw fit" of the two South Atlantic continents.

Although a few other, rather more respectable, figures speculated upon the possibility of lateral mobility of the continents, at the beginning of this century there was near-unanimity among geologists in favor of the alternative stabilist view.

This hypothesis stated that the continents had remained fixed in their relative positions since the time of their formation early in the history of our planet, and that therefore the ocean basins were permanent physiographic features.

The Earth was supposed to be still in the process of progressive solidification and contraction from a molten mass. Lighter rock materials had moved towards the surface to give rise to granitic-type rocks, and were underlain by denser rocks resembling, if not exactly matching, *basalt, *gabbro or *peridotite. Mountain ranges were produced by contraction in a manner somewhat analogous to the crinkles developed on a shrinking, drying apple. On a larger scale, an overall arching pressure caused certain sectors of the Earth's surface to collapse and subside, giving rise to the oceans, while the continents remained emergent.

Evidence of former land connections, or land-bridges, across what was now deep ocean was provided abundantly by the total or near identity of many fossil animals and plants found on different continents. Unless such transoceanic land-bridges had existed in the past these striking similarities were inexplicable in terms of Darwinian evolution, for genetic isolation should have given rise to morphological differences in the faunas of the different continents.

The first person to make a serious challenge to this long-standing orthodoxy was, strictly speaking, an outsider, because Alfred *Wegener was qualified professionally not in geology but in astronomy and meteorology. However, a man of wide interests and broad vision, he was able to perceive weaknesses in the conventional theory, and put forward an impressive array of evidence from several different research fields in support of his revolutionary theory, continental drift. This received its first public airing in a lecture in Frankfurt in 1912 and was followed later that year by two scientific papers. An enlarged version appeared in book form in 1915 but it was not until the translation into English of the third edition of his book, *The Origin of Continents and Oceans*, in 1924, that the theory began to attract widespread attention.

Wegener postulated a huge, primeval supercontinent, christened by him Pangaea (from the Greek, meaning "all land"), which had begun to split up late in the *Mesozoic era. Its various components had moved apart at different times: South America and Africa had started separating in the *Cretaceous, as had North America and Europe; the Indian Ocean had begun to open up in the *Jurassic but the principal drift movements had taken place later, during the Cretaceous and *Tertiary; Australia–New Guinea had split off from Antarctica in the *Eocene and moved northward, driving into the Indonesian archipelago in the late Tertiary. During the westward drift of the Americas, the western Cordilleran mountain ranges had been produced by compression at the leading edges; while a large area of land to the north of India had crumpled up in the path of the subcontinent during its northward movement, so forming the Himalayas. The Alpine ranges were likewise the consequence of north-south compression between Africa and Europe.

Supporting evidence for continental drift was brought forward from a variety of fields of study. The celebrated "jigsaw fit" of the Atlantic continents was, Wegener considered, no more than suggestive, because it could have been merely coincidental. Much more significant were the many indications from the geographic distribution of distinctive types of *fossil that there had been land connections between the southern continents in pre-Tertiary times.

To give just two examples: *Mesosaurus* is a small Permian reptile known only from South Africa and southern Brazil; and *Glossopteris* is a fossil leaf abundant in many deposits of the same age in South Africa, South America, Madagascar, India and Australia. The *Glossopteris* flora had indeed been known for some time and was the principal argument for the former existence of a southern supercontinent called Gondwanaland (the name is derived from that of a region in India).

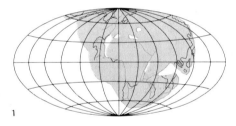

Wegener's reconstruction of the map of the world for (1) the upper Carboniferous (Pennsylvanian), (2) the Eocene and (3) the lower Quaternary.

Opposite: a satellite picture showing how separation of the African and Eurasian plates has led to the formation of the Red Sea and the Gulf of Aden.

Wegener argued that the traditional explanation of sunken land-bridges to account for such similarities of organisms between continents now separated by deep ocean was untenable on geophysical grounds. The principle of isostasy implies that lighter, continental crust cannot sink into denser, oceanic crust (the two types of crust were distinguished early this century on the basis of regional gravity measurements and differences in *igneous rocks). The only reasonable alternative explanation was that the various continents in question had drifted apart at some time in the geological past. Indeed the younger faunas, of the Tertiary and *Quaternary, tend to be quite different in each continent.

Another of Wegener's approaches was comparison of geological structures such as *mountain chains, or distinct groups of rocks, in the different continents. He pointed out a number of striking trans-Atlantic resemblances. For instance, the Cape fold mountains of South Africa appear to have a continuation in Buenos Aires Province in Argentina; and the late Paleozoic/early Mesozoic series of largely non-marine strata known in South Africa as the Karroo System are remarkably similar in many respects to what is known in Brazil as the Santa Catharina System. In Wegener's own (translated) words: "It is just as if we were to refit the torn pieces of a newspaper by matching their edges and then checking whether the lines of print run smoothly across. If they do, there is nothing left but to conclude that the pieces were in fact joined in this way."

A third line of argument was based on the reconstruction of ancient climates by investigating the distribution of certain types of sedimentary rocks. A number of distinctive boulder beds signifying deposition from an ice sheet (tillites) occur in late Paleozoic rock sequences in the southern, "Gondwana", continents. In South America, South Africa and India they are of late *Carboniferous age but in Australia there are important *Permian tillites as well. The distribution of these makes no kind of sense with the continents in their present relative positions but, if they are brought together as part of Pangaea, a position for the South Pole in the Carboniferous can be inferred just east of South Africa and within Antarctica.

Wegener suggested that the presence of Permian tillites in Australia and the disappearance of such deposits elsewhere at this time signified an eastward shift of the pole. Confirmatory evidence came from the Northern Hemisphere. Wegener found abundant evidence of a humid equatorial zone, in the form of thick beds of late *Carboniferous coal, extending from the eastern USA through Europe into China. These coals contain fossil plants of tropical type as indicated by, for instance, the lack of seasonal rings in the wood. In the overlying Permian deposits of Europe and the USA there are huge salt deposits, signifying a warm, arid climate. To Wegener this was

good evidence of the same polar shift inferred for the southern hemisphere, this time causing a change from a humid equatorial zone to an arid trade-wind zone.

One further line of evidence on which Wegener relied should be mentioned. Geodetic observations made early this century seemed to indicate that Greenland was moving westward from Europe at a measurable rate. Such a movement might constitute a direct proof of continental drift. Unfortunately, however, it has not been confirmed by recent measurements employing more sophisticated and accurate techniques.

The geophysical basis of Wegener's theory was closely related to the principle of isostasy. Both assume that the substratum underlying the continents acts as a highly viscous fluid. Wegener argued that if a landmass could move vertically through this fluid, as was widely accepted, it should also be able to move horizontally, as indeed was indicated by evidence of the *folding of strata in mountain belts. Again, the Earth is an oblate sphere, bulging slightly at the equator: this equatorial bulge is of just the size to be expected for a sphere of perfect fluid spinning on its axis at the same rate as the Earth does. Under short-term stresses, such as those that result in *earthquakes, the Earth behaves as an elastic solid, but in the long term it acts as a fluid. Wegener made an analogy with pitch, a material which shatters under a hammer blow but which over a much longer period flows slowly under its own weight. Movement of the continents was thought to be under the control partly of tidal forces, accounting for the westward drift of the Americas, and partly of a so-called *Pohlflucht* (or "flight from the poles") force, causing movement of India and compression of the Alpine and Himalayan mountain belts.

Initially the reaction of geologists and geophysicists to Wegener's theory was mixed, but by the mid 1920s opinion had hardened against him and he had few supporters. A wide variety of criticisms was put forward. The supposed "jigsaw fit" of the Atlantic continents was inaccurate and did not allow for vertical movements of the crust. The similarity of distinctive rock types and geological structures between different continents had been exaggerated; furthermore, mere similarity did not prove former contiguity. The paleontologists upheld land-bridges of reduced size to account for similarities in ancient faunas and floras—despite the geophysical objections. The Carboniferous and Permian tillites of the Gondwana continents were deemed probably not glacial and the northern-hemisphere coals not necessarily tropical.

Part of Wegener's theory involved a contradiction: if the American continents could move laterally by displacing the ocean floor, how could the Cordilleran mountains have been produced by compression, which implies a significant resistance from the supposedly weaker oceanic rock? More-

over, why did the supercontinent Pangaea remain intact for most of the Earth's history and then abruptly break up within a few tens of millions of years?

Perhaps the most serious objection, certainly to geophysicists, was the proposed mechanism of drift. It was not difficult for experts to show that the forces which Wegener proposed were far too weak; polar wandering was likewise regarded as geophysically impossible. Some incensed critics even challenged Wegener's credentials as a scientist: He was a mere advocate, selecting for presentation only those facts that would favor the theory. To quote, he "took liberties with our globe" and "played a game in which there are no restrictive rules and no sharply drawn code of conduct".

Wegener died on the Greenland ice cap in 1930 and it was left to a few enthusiastic disciples to battle on in the attempt to persuade the community of Earth scientists that continental drift was more than mere fantasy. One of these was the South African geologist Alexander *du Toit, who pointed out all kinds of hitherto unemphasized and yet remarkable geological similarities between the Gondwana continents, and also eliminated some of the weaknesses of Wegener's theory, adding modifications of his own. Another was the British geologist Arthur *Holmes, who in 1929 proposed convection currents in the mantle as a driving force for drift, an idea which closely anticipated modern views.

Nevertheless the great majority of geologists remained unconvinced, and opinion in some quarters, especially in the USA, was so hostile that it was advisable for "drifters" to keep their views to themselves if they wished to be considered respectable scientists. It was not until the new discipline of paleomagnetism (see *geophysics) began to produce some astonishing results in the 1950s, followed by equally striking oceanographic discoveries in the 1960s, that the consensus swung dramatically in favor of laterally mobile continents and young ocean basins. With the almost universal acceptance of the theory of *plate tectonics the revolution begun by Wegener over half a century earlier was complete.

The question inevitably arises, "Why was opinion so generally hostile to continental drift for so long?" A number of possible answers can be suggested. Many found Wegener's proposed driving forces totally unconvincing and felt obliged on this ground alone to reject the theory: there are many examples, however, of scientists accepting the existence of a phenomenon without understanding the fundamental cause—the *Pleistocene Ice Age is a good geological example. It has been argued that Wegener's theory was "premature" in that it could not readily be fitted into the framework of existing knowledge. Some might say that for the stability of science any observation or idea that contradicts the established view of the world must be presumed invalid and set aside in the hope that it will eventually turn out to be false or irrelevant. Since geologists like to see things for themselves, more Europeans and Americans might perhaps have been persuaded of the reality of drift if they could have personally investigated the critically important Paleozoic rock successions of South Africa and Brazil. As a final justification, our knowledge of the 70% of the Earth's surface covered by ocean was negligible until the development of new techniques in the last few decades.

But when full allowances are made, it is difficult to avoid the conclusion that much of the hostile reaction to Wegener's ideas was based on prejudice (influenced no doubt by the consideration that he was a geological amateur) rather than on any objective assessment of the facts. The prevailing climate of opinion was such that he was not given a fair hearing. There is perhaps a moral here, that scientists should always strive to avoid dogmatism in their opinions and be receptive to interesting new ideas, however disturbing they may be to the established body of knowledge.   AH

Plate Tectonics

It is rare in the history of science that a single idea or group of closely related ideas can, within a decade of their conception, cause virtually every aspect of their subject to be viewed afresh, and can set into one consistent and intelligible framework the accumulated observations of more than a century. Yet this is no overstatement of the impact of the ideas of plate tectonics upon the Earth sciences as a whole. Geologists, geophysicists and geochemists had for years been studying limited aspects of Earth history, whether it were the distribution of certain kinds of volcano, or the variation in thickness from place to place of the Earth's crust, or the periodic extinctions of groups of ancient organisms. It was as if Earth scientists had been engaged in the completion of an enormous jigsaw puzzle; each making progress with his own group of pieces, but with little consensus on how the groups should be assembled to make the complete picture. And then a crucial piece was found and it at once became obvious how the groups must fit together and what the overall picture must be. Naturally, it has emerged that some earlier pieces which perhaps never fitted too well have had to be lifted out and placed elsewhere; equally there remain parts of the puzzle which have yet to be completed; but at least the broad outlines are now clear.

Early Evidence. In order to discuss and explain plate tectonics it is necessary first to review the background to the discovery. There had long been speculation (see *continental drift) that the continents might have moved with respect to each other during the geological past. Evidence had been of two major kinds—"fit evidence" and "wrong-latitude evidence". The fit evidence was simply that many continental areas today separated by oceans had shapes which suggested that they had at one time fitted together; furthermore, if continents were reassembled according to shape, there were in many cases similarities between the rocks and fossils of the regions thus brought together. This evidence was suggestive—but not conclusive, because both shape and geological similarities could be coincidental.

The "wrong-latitude" evidence was, however, rather different. Various rocks and fossils which were thought to require rather special climatic conditions for their development were found at latitudes at which such conditions seemed impossible; for example, the occurrence of coal near the south pole (*coal forms in warm subtropical conditions similar to those found in the Florida everglades). Conversely, evidence is found in India of the former presence of huge continental ice sheets such as are today restricted to very high latitudes. Clearly it was very difficult to reconcile observations of this kind with the present pattern of climatic belts, and one solution was to allow that the continents had moved so as to occupy different latitudes at different times.

Although probably the balance of geological evidence favored such large-scale movements, much geophysical theory tended against them. The Earth was known from observations of the transmission of seismic waves to be made up of essentially solid, crystalline silicate material with a partially molten metallic core. It was impossible to conceive of any energy source sufficiently great to power these large-scale motions in solid rock.

In any case, the proponents of continental drift left many questions unanswered concerning the details of the process: the crust of the continents was known to be very thick, ranging from 25 to 70km (15–43mi) with an average of about 35km (22mi). Crust under the oceans was much thinner, only 6km (4mi) or so, and, judging from seismic observations, made largely of different material from that in the continents. Was the crust of the continents to be visualized as ploughing through, or overriding, the crust of the ocean basins?

Some geophysical evidence, however, supported the geological observations insofar as it seemed that some rocks had magnetic characteristics (in fact, a "frozen in" magnetic declination and inclination) acquired at the time of their formation and which could not have been acquired at their present latitude. The movements suggested by the magnetic observations were virtually identical to those required to satisfy the "wrong-latitude" evidence of the rocks and fossils.

Thus, about the middle of the 20th century, there was a stalemate with irreconcilable yet apparently unassailable arguments on either side. In passing, however, it is worth noting one other result of the paleomagnetic studies of the 1950s which at the time seemed of only academic interest.

It became established that the polarity of the Earth's magnetic field could periodically reverse: this simply means that, after such a reversal, a compass needle that today points north would point south. A magnetic field such as we have today is conventionally called "normal" and the opposite "reversed". It seems that the Earth's field has changed polarity frequently in the past. A change in polarity takes about 5000 years to accomplish, and the field then remains constant for periods that mostly last for between ten thousand and one and a half million years. It became possible to draw up a "reversal time-scale".

Mid-Ocean Ridges and Sea-Floor Spreading. About the same time that this work was being done, important observations were being made in the ocean basins. The role of submarine warfare in the WWII had led to the development of very effective acoustic sounding devices that made it easy to map the topography of the sea floor—which submarines might use for concealment. By the mid '50s enough information was available to demonstrate that the ocean floors were topographically at least as interesting as the continents and, in some ways, possibly more informative. Because the influences of weathering and erosion, which continuously modify the shape of mountains and valleys on land, do not operate in deep water, the processes that led to the formation of a submarine topography can often clearly be read in its form. In particular, huge ranges of submarine mountains which rise between three and five kilometres above the *ocean floor (the oceans average about five kilometres deep) became known; these ranges were more or less symmetrical and were about 1500km (930mi) wide. Along their crests they often had a curious "median valley"—a rift-like valley with relatively steep sides, two or three kilometres deep and sixty or so kilometres wide. There was a striking similarity to the rift-valley system of East Africa where it had long been thought that the Earth's crust was being stretched apart. In the case of the Atlantic Ocean in particular the overall trend of the mountain range (the mid-Atlantic ridge) was striking. It was central within the Atlantic, and mirrored faithfully the changes in curvature of the margins of the continents on either side. In changing direction to follow the continental shapes the ridge did not simply swing round but was offset repeatedly along great numbers of parallel fractures (now known as transform faults).

It became clear that the mid-Atlantic ridge was part of a more or less continuous world-wide system of ocean ridges more than 40,000km (25,000mi) long; not all divided oceans symmetrically, and not all had rifts along their crests.

Frequently, at the same time as shipboard instruments were being run to survey the ocean floor, another instrument, a seaborne magnetometer, was being towed some distance behind the ship in order to measure the intensity of the Earth's mag-

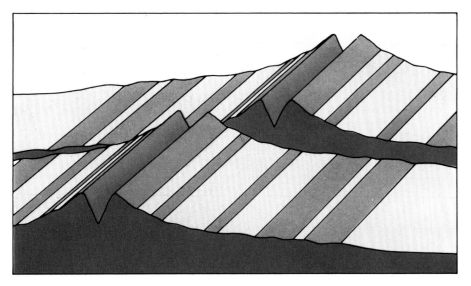

Magnetic stripes across a mid-ocean ridge. The pattern of stripes is symmetrical on either side of the ridge crest. Moreover, the same pattern and symmetry are shown even when two sections of ridge are separated by a transform fault. The distribution of these magnetic stripes is one of the major evidences in favor of plate-tectonic theory.

netic field. It had been expected that in oceanic regions the field would be rather flat and uninteresting; in the event it was found to be extremely variable. It appeared that there were very striking elongate magnetic highs and magnetic lows which ran parallel to the crests of ocean ridges, and that this pattern of parallel highs and lows commonly extended all the way from the ridge to the flanking continents.

It was these observations, and an intriguing suggestion put forward by an American, H. H. *Hess, a couple of years earlier, that sparked the imagination of two Cambridge University geophysicists, Fred Vine and Drummond Matthews. They noticed two features of the oceanic magnetic field which had previously escaped attention. They saw that not only were the magnetic stripes of the ocean floor parallel to ridge crests but that they were *symmetrical* across any particular ridge, that any peculiarities of the pattern on one side of the ridge could be matched on the other. The second feature which they noticed was that, starting at the ridge and working out to one side, the widths of successive magnetic stripes on the ocean floor exactly matched in proportion the known durations of periods of normal and reversed polarity of the Earth's magnetic field going back in time from the present.

To interpret these observations they put forward a bold and completely original hypothesis: they suggested that the floors of the oceans were behaving as the magnetic tape in an enormous tape recorder which was recording the changes in polarity of the Earth's magnetic field. It worked like this: the Atlantic, for example, was imagined to have been initially closed with Africa fitting up against South America – and so on. The continents moved apart and partially molten material from the Earth's interior welled up to occupy the gap in between. Rocks have no magnetic properties while they are

hot, but as they cool below about 450°C (850°F) they can become weak permanent magnets, "freezing" into their mineral structure some "memory" of the ambient magnetic field in which they cooled (see *geophysics). Thus the first material to well up between the separating continents would acquire thermoremnant magnetism which reflected the polarity of the Earth's field at that time. As the separation continued, material would continue to well up and cool along the axis of separation; as soon as a reversal of the magnetic field occurred, however, further cooling would produce rocks which were weak permanent magnets oppositely oriented, and this would continue until the field reverted to its previous polarity. In the case of the Atlantic the axis of separation is believed to be the line of the median valley along the crest of the mid-Atlantic ridge; new material (i.e., lavas) can be observed by submarine photography to be welling up along it today – this will cool with magnetic properties related to the Earth's present magnetic field. Moving out to either flank we find progressively older ocean floor, with the transition from each magnetic stripe to its neighbor recording a change in the polarity of the field at some time in the past, and each stripe recording the polarity of the field at an earlier stage in the opening of the ocean, when that stripe was being formed at the ridge crest. By processes such as these it was possible not only to explain why the crust under oceans was different from that under continents – it owed its formation to quite a different process – but also the way that the magnetic stripes formed and the reason they were symmetrical across each ocean ridge.

When these ideas were first presented in the early '60s they were received with considerable scepticism. The Vine and Matthews interpretation of the magnetic stripes did, however, fit exactly with a

comprehensive sea-floor spreading model for continental drift put forward by Hess a few years earlier. Within two or three years it became clear that virtually all ocean ridges for which there were reliable observations showed the same magnetic characteristics, and that the new interpretation had to be taken very seriously. One important consequence of the model was that, if its magnetic signature allowed an age to be assigned to each piece of ocean floor, it was possible to calculate a velocity for sea-floor spreading. Consistent, and consistently varying, velocities have now been established for ridges on a world-wide basis: oceans are widening at rates ranging from a little over 2 to about 14 centimetres per year (0.8–5.5in/y).

Subduction. Earth scientists now faced something of a predicament. The evidence for sea-floor spreading seemed very strong but the rate of spreading was very high, amounting to a present-day rate of generation of new surface area for the Earth of about 0.5% per million years (the Earth's age is about 4600 million years). Only two possibilities were open: either the Earth was increasing in volume to accommodate this new surface area or in some way surface area was being destroyed as fast as it was being created. Although the expanding Earth hypothesis has had several distinguished advocates, the weight of opinion has been opposed to it for a variety of reasons: if the Earth had been of much smaller diameter in the past, its density and the acceleration due to gravity experienced at the surface would both have been much greater than today; according to one proposal, all surface objects would have weighed four times as much 200 million years ago, with a host of more or less inevitable consequences for the Earth's surface processes (e.g., the weight of certain large land animals would have exceeded the strength of their bones – not *all* of them could have spent their lives up to their necks in water!). Changes in density would also have affected the Earth's rotational speed and hence the length of the day. Studies of growth patterns in fossils of marine organisms which maintain daily and monthly rhythms (e.g., in response to daylight and tides) show that the changes which have occurred in the length of the day are too small to permit any major changes in the Earth's density.

For these reasons attention was paid to the alternative possibility – the destruction of surface area as proposed by Hess. He had pointed to the importance of the so-called deep-ocean trenches which were found in many oceans, but which were particularly well developed around the western and northern sides of the Pacific Ocean. These had been known about in a general way for many years but had become much better understood during the post-WWII era of sea-floor exploration. The trenches are elongate furrows in the ocean floor and in them the water depth ranges between 9 and 11km (5.6–6.8mi), roughly twice the average depth of the oceans: they are normally

250–300km (150–190mi) wide. In the western Pacific they form a semi-continuous chain along the western side and commonly have the form of a series of intersecting arcs. On the concave side (generally the side towards the continent) there is everywhere a parallel ridge, which from place to place breaks the surface to give small islands that are volcanic in origin, although some now carry a surface capping of recent coral growth. Indeed, virtually all of the present-day volcanic activity of the western Pacific lies along such ridges, immediately west of the deep trenches. The strings of volcanic islands are known as island arcs.

Hess drew attention to other anomalous features of the island-arc areas. A series of pioneering experiments carried out by the Dutch geophysicist F. A. Vening Meinesz in the '30s had shown that there were very strong anomalies in the Earth's gravitational field in the neighborhood of island arcs. These suggested that, under the trenches, low-density crustal rocks had been deflected downwards into the denser material of the mantle. Island-arc regions had also long been known to be seismically active; in particular, earthquakes occur on an inclined zone which intersects the surface above the line of the deep ocean trenches, and dips away under the island arcs at somewhat variable angles averaging about 45°. This inclined zone of seismicity extends down into the mantle for about 700km (430mi).

Hess proposed that at these trenches the ocean floor which had been generated some time earlier at a ridge turned downwards and was subducted into the mantle: the inclined seismic zone was thought to track the downward path of the descending material. This proposal could satisfy the gravitational observations and, in a less obvious way, explain the volcanic activity, thought to be triggered by the descending mass.

The next step had to wait until the last years of the '60s. A decade earlier the US government had set up a world-wide network of high-precision seismic observations (checking adherence to any nuclear test-ban agreement requires that the normal pattern of world-wide seismic activity be known). After ten years of observation it became clear that virtually all earthquakes were restricted to a small percentage of the Earth's surface, and that they were distributed along continuous and very well-defined narrow zones. These zones surrounded huge areas which were virtually unaffected by earthquakes. Once more, this had been known in a general way for a long time, but the observations were rather poorly distributed over the Earth's surface and so the patterns had not been very obvious. It became apparent that seismicity was virtually restricted to mid-ocean ridges, island-arc areas and narrow, linear zones which linked arcs to ridges, arcs to arcs or ridges to ridges.

Theory and Proofs. These observations, coming on top of the discovery of sea-floor spreading, led a number of groups of workers separately and simultaneously to the essential ideas of *plate tectonics. The idea was that the surface of the Earth comprised a relatively small number of internally rigid plates and that these were in continuous motion with respect to each other. At mid-ocean ridges they continuously moved apart, new material welling up from the mantle and being welded onto their trailing edges; along the lines of trenches plates converged, one plunging down into the mantle under its neighbor; and along a third sort of boundary (conservative boundaries) plates neither gained nor lost area but simply slipped past each other. This latter kind of boundary frequently had very little topographic expression on the ocean floor except perhaps as an elongate scarp. Virtually all seismicity thus resulted from interactions between plate margins.

The importance of the distinction between continental crust and oceanic crust now becomes clear: plates may be capped by either kind of crust (and in practice most plates are in part oceanic and in part continental) but each plate moves as a coherent entity; continents move apart as if on a conveyor belt, by the generation of new crust at a spreading ridge, and move together by the destruction of oceanic crust at a trench (or *subduction zone*) between them. Oceanic crust has therefore a rather ephemeral existence: the oldest known is about 200 million years old. In contrast, continental crust seems not to be able to be returned to the Earth's interior, at any rate not in any quantity, for it is not only much thicker than oceanic crust but also less

A satellite photograph of the Andes. This range, which contains some of the highest mountains in the world, is a result of the interaction between the westward-moving (oceanic) Nazca plate and the eastward-moving South American plate. As the Nazca plate is subducted, mountain-building, earthquakes and volcanic activity occur.

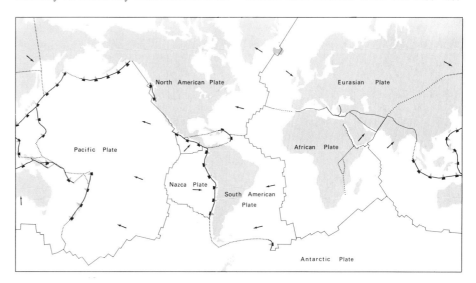

The major plates of the Earth's crust and the directions in which they are moving. Dotted lines show probable plate margins, arrowheads destructive plate margins with underthrusting in the direction of the arrowheads.

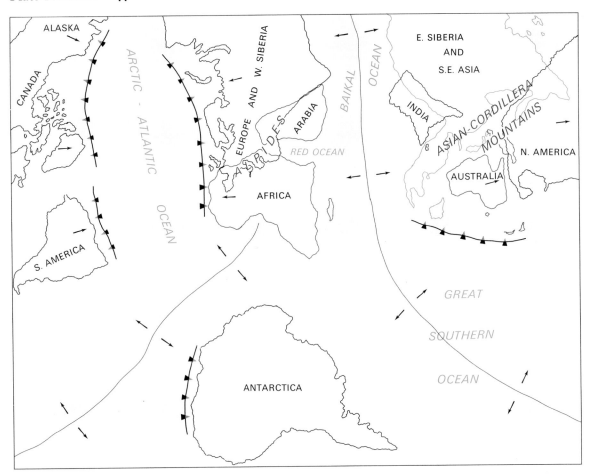

A map of the Earth as it will appear in 150 million years' time.

dense, and it seems to have developed over a long period: the oldest known, in Greenland, is about 3800 million years old.

For any hypothesis to be worthy of serious attention it must be testable, and a great deal of effort has gone into testing the predictions of plate tectonics. One approach was by means of fault-plane solutions. Faults are fractures at the surface of the Earth along which movement occurs between adjacent blocks: sometimes this movement is steady and continuous but occasionally a projection on one block will engage with the opposite wall and the fault will be temporarily locked; locked, that is, until the local stress concentration is sufficient to cause rupture with explosive energy release, giving rise to an ★earthquake. At the instant of rupture, the surface of the Earth may be viewed as four quadrants, two of the quadrants being thrown into instantaneous compression and the other two into extension. Examination of seismic records from stations round the world can reveal the orientation of these quadrants for any particular earthquake.

A photograph from space of the Himalayas. Unlike the Andes, where oceanic plate is being subducted beneath continental plate, the more complex folding and faulting of the Himalayas are a result of the collision of two continental plates. The main folding of the Himalayas took place comparatively recently, in the Miocene.

Insofar as the orientation is quite different for movement on a fault along which blocks are sliding past each other from the situation when blocks are moving apart, and different again from when they are converging, it is possible to assign a type of motion to each earthquake for which there are sufficient data, and to compare that motion with what is expected from plate tectonics. The agreement is striking.

A second and more ambitious test was planned: the *ocean floor is subjected to a steady and continuous, but very slow, rain of fine particulate matter settling through the water and accumulating on the bottom. This fine sediment is partly organic in origin – the skeletons of minute marine organisms such as *foraminifera and *radiolaria – and in part made up of fine clay, either washed out into the ocean basins from the continental shelves or made up of fine volcanic dust. Clearly, other things being equal, the older the ocean floor the thicker should be its veneer of sediment. Furthermore, the age of the lowest layer in the sedimentary pile should be close to that of the formation of the volcanic crust upon which it rests. A program of deep-sea drilling was therefore carried out to core the sediments of the ocean floor and to compare the age of the lowermost sediment (as determined by fossils) with the age of the volcanic rocks of the oceanic crust predicted by plate tectonics and the interpretation of the magnetic stripes.

Sufficient results have been obtained from both the Atlantic and Pacific Oceans to demonstrate that both the thickness of the sedimentary accumulation and the age of the lowest sediment in the pile increase systematically away from ocean ridges.

It is now very difficult to see any way in which the general concept of plate tectonics can be shown to be in error, although undoubtedly ideas will undergo modification in many matters of detail.

The Physics of Plate Tectonics. The silicate materials of which the Earth is largely made have some of the properties of ice. Ice in a glacier may be considered either as a very viscous fluid or as a brittle solid depending on the time-scale of interest: struck with a hammer the ice will shatter; left to flow under its own weight down a valley it changes shape continuously as a fluid. Both ice and rocks "flow" in the solid state most readily when they are close to their melting point, and at temperatures much below their melting point behave in a highly brittle fashion and scarcely flow at all. These observations not only help explain the existence of plates but also resolve the classical continental-drift paradox in which it appeared that on the one hand continents must have moved, yet on the other no forces existed powerful enough to propel them if rocks were as strong as they seemed to be.

The silicate material of the Earth's mantle is close to its melting point a couple of hundred kilometres below the surface: it behaves as a crystalline solid for the trans-

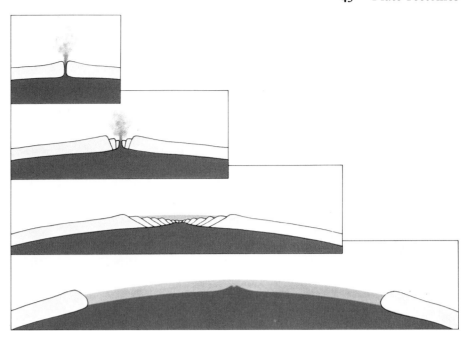

The development of an ocean and a mid-ocean ridge. Initial arching of the continental crust results in fractures and consequent vulcanicity as molten magma from the mantle escapes to the surface. This volcanic activity and slight separation cause downfaulting. As the arch collapses water is able to fill the central area in the form of a shallow sea: by this time major separation is underway and new oceanic crust is in the process of formation. The end result is complete separation of the continental masses, between which stretches an ocean. Centrally placed is a mid-ocean ridge, the site of further sea-floor spreading.

mission of elastic waves, but if subjected to small stress differences will flow readily at the rates required for plate tectonics. As continents move apart, hot ductile mantle material wells up between them to form the new ocean floor. The ocean floor is, however, maintained by the circulation of sea water at a temperature close to $4°C$ (about $40°F$), more than $1000C°$ ($1800F°$) below the temperature required for the start of melting of the mantle. The new ocean floor and the immediately underlying upper mantle are both therefore rapidly chilled, and for that reason become strong, giving rise to oceanic plate. The effects of cooling penetrate the mantle quickly at first and

then more slowly, and thus plates thicken with time, achieving an average thickness of about 80km (50mi). Because oceanic plates are cooler than the underlying mantle they are also more dense and thus tend to sink into it: this is what happens at oceanic trenches.

Plate tectonics is the surface expression of thermal convection within the Earth. Heat is generated in the mantle by the radioactive decay of uranium, thorium and potassium faster than it can be lost by conduction: plate tectonics allows a continuous upwelling of hot material to the surface where it can cool and, once cooled, be "quickly" returned to the mantle.

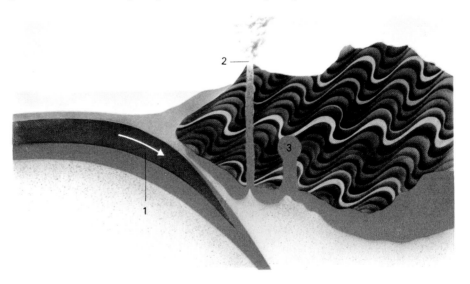

The development of a mountain range at a destructive plate margin, where an oceanic plate (1) is being subducted beneath a continental plate. This results in volcanic activity (2) and igneous intrusion (3) as well as contortion of the continental crust to produce a mountain chain.

Part of the Giant's Causeway near Portrush in County Antrim, Northern Ireland. This columnar structure of basalt consists of tightly packed prisms. Jointing occurred first near the surface of the lava, where there were contractions towards a large number of discrete centers: the result was a polygonal, generally hexagonal, pattern of joints. These developed in depth to produce the columns as cooling penetrated the lava.

justly been described as a "revolution in the Earth sciences". As we have seen, what is essentially a rather simple concept explains satisfactorily many of the problems that have concerned geologists, geophysicists and paleontologists for decades. While it is not in the nature of any scientific discipline to remain static, and so any theory must necessarily be expected to undergo modifications with time, we can say that with the idea of plate tectonics we are close to a full understanding of the physical basis of most, if not all, of the processes that shape the surface of the planet that we live on. ERO

Weathering

Weathering is the breaking down and alteration of rocks at the Earth's surface by the direct effects of local atmospheric conditions – otherwise known as weather. Physical and chemical processes attack the rocks to the extent that they and the products of their weathering may bear little resemblance to the original materials of which they were made up, and may even be completely destroyed. Thus durable granites are reduced to clays, and friable sands and hard limestones simply disappear in solution to leave only minute residues. The biological activities of plants and animals also provide physical and chemical means for rock breakdown.

In the break-up of rocks it is difficult to separate physical disintegration from chemical alteration, and it is true to say that there is little physical weathering that is not also partly chemical. Unlike the situation on the *Moon, where only physical processes operate, atmosphere and water occur everywhere on the Earth and they react chemically with the rock-forming minerals. The new materials which result are relatively more stable and in equilibrium with conditions at the Earth's surface.

Physical Weathering. Rocks are broken down mechanically with little or no chemical change by a variety of causes, some of which can originate within the rock itself while others are of external origin. The deeper a rock is buried beneath overlying material, such as rock or ice, the greater is its strength; but when this material is removed, a rock expands upwards and eventually ruptures, forming a series of cracks or joints. The rock splits along these joints, sometimes parallel to the land surface to form *exfoliation* domes, which are common in granites and sandstones: a good example of their occurrence is in the Yosemite Valley, USA. Joints can also be formed by contraction during cooling of rocks, as in

The development of *mountain chains (*orogenic belts*), within which the rocks have been highly deformed and in many cases recrystallized in the solid state under abnormal temperature conditions, occurs at convergent plate margins (i.e., with one of the plates undergoing subduction) where at least one of the plate margins is continental. At the eastern Pacific margin, the Andes seem to have formed in response to the continued subduction of the Pacific plate in the Peruvian trench; volcanic material has risen from the subduction zone and thickened and elevated the crust. In contrast,

the Alpine–Himalayan mountain belt has developed by the collision of continental masses following the subduction of all intervening ocean floor. Both continental masses are of too low a density to be carried down into the mantle and have come to rest with one having partially overriden the other. The zone of contact is one of enormous geological complexity and the highly deformed rocks within sometimes contain material once deposited on the now vanished intervening ocean floor.

Conclusion. The almost universal acceptance of the theory of plate tectonics has

Large deposits of scree at the foot of Cholatse, Nepal. Scree is composed of rock fragments physically weathered from the mountainside and transported for at most a short distance by gravity. The finest particles have usually been removed by water.

the basaltic lavas of the Giant's Causeway, County Antrim, Northern Ireland. This jointing is important, as it is the means by which all kinds of weathering agents can attack a larger area of the rock.

Rocks can also be exfoliated by granular disintegration, best seen in dry climates with little vegetation, particularly where marked changes in the amount of heat received from the Sun cause thermal expansion and contraction: coarse-grained rocks are affected by this process more rapidly than fine-grained ones. The growth of salt crystals, especially in porous rocks, also causes granular disintegration and such growth can be very disruptive. Most salt weathering takes place in arid areas, where its effect on building materials has been much studied.

Because water expands by about 9% on freezing, frost shattering is probably the single most important physical agent in weathering, particularly (for obvious reasons) in cold climates. If water freezes after entering a joint or crack then its expansion can disrupt the enclosing rock and break it into pieces: angular screes accumulate in this way. In unconsolidated rocks, frost heaving is caused by the movement of water through capillaries from unfrozen ground to ice nuclei.

Chemical Weathering. The main processes of chemical weathering are solution and leaching by water derived largely from rain, and oxidation and carbonation which depend upon the oxygen and carbon dioxide content of the atmosphere. In a general way, chemical weathering is an acid attack on the rocks of the Earth's crust and in particular an attack on the most abundant minerals – *quartz (sand) and aluminosilicates (*clays).

Rainwater, together with soil and groundwater, is a mixed electrolyte containing varying amounts of negatively charged anions (e.g., chloride, bicarbonate and sulfide ions), and positively charged cations (e.g., sodium, potassium, and calcium ions). Ion exchange (a simple reaction between ions in the solution and those held by the mineral grains) is the most important process in chemical weathering. The clay minerals are chiefly responsible for the ion-exchange capacity of rocks, the different clay minerals which are formed by weathering depending upon the parent rock and the degree of leaching. Clay minerals can also change from one to another under suitable conditions; *kaolinite may occur in well drained areas of granitic rocks, *montmorillonite in ill-drained alkaline situations, and *illite in cool temperate climates. The acidity of the weathering environment is highly important, especially to those solution processes that remove minerals like *gypsum, *calcite and

*dolomite. *Feldspars, common minerals in granites, are decomposed by hydrolysis to clays.

Because the Earth's environment is an oxidizing one, many rock minerals combine with oxygen. The best known oxidation process is that of *iron from the ferrous to the ferric state, which, particularly in the presence of water, produces rusting.

There is no absolute scale of degree of weathering, but minerals may be ranked into a series according to general ease of

weathering – quartz, *zircon and *tourmaline being very stable minerals, and *biotite and the feldspars being less stable. Most rocks consist of minerals that are not easily soluble, but over geological time weathering rates can be significant. In general, the rate of chemical weathering is greatest in the moist humid tropics (where it is considered to be 20–40 times that in temperate latitudes) and is least in arid areas (see *arid landscapes). The time taken to weather 1m (3.25ft) of granite in the forest

Rust, an example of the results of chemical weathering. Iron, on exposure to the atmosphere, combines with the oxygen and water present to give hydrated ferric oxide ($Fe_2O_3.nH_2O$). Unlike many tarnishes, rust does not protect the iron beneath from further corrosion.

weathering crust; this process is sometimes called laterization: the laterites of India have been formed in this way (see *bauxites and laterites). Bauxites (aluminum ore) have also been formed under conditions of leaching, usually under alkaline conditions.

Few landforms are produced by the sole agency of weathering. But rocks that have been shattered or rotted by the effects of extreme weathering are easily removed by gravity or worn away by agents of *erosion such as rivers or glaciers. The removal of deeply weathered profiles in granitic areas, for example, is believed to provide the origin of tors, and of dome-like hills (bornhardts) in savannah areas. Many kinds of tunnels and pits are caused by solution of soluble rocks such as limestones, producing distinctive landforms called karren.

The importance of chemical weathering cannot be overstressed. The changes produced in the rocks form the basis, with biological processes, for soil formation and also affect soil fertility and productivity. Furthermore, chemical weathering is the mechanism for the concentration of many valuable substances, including bauxites, opals and iron ores.   MMS

Erosion

The word "erosion" is derived from the Latin, *erodere*, to gnaw or eat away, and refers to the many processes which wear away the Earth's surface. Rock fragments are first loosened by *weathering (the physical and chemical break-up of rocks by the action of weather) and are then eventually removed by gravity, water, ice or wind. It is difficult to say where weathering ends and transport and erosion begin, but erosion usually implies wearing away by moving agents. In general the agents themselves have only a small effect – it is the material that they carry that is responsible for most of the erosion. The combined effects of weathering, transport and erosion are known as denudation (Latin *denudare*, to strip bare) or degradation.

Water. The downslope migration of rock waste and the displacements of bed rock are commonly referred to as *mass movements*. Such movements include falls of dry material, slides, soil creep and various types of flow, depending upon the amount of water or ice present. Sliding requires a slip surface between the moving mass and the underlying stable ground. Where frost occurs frequently, ice needles push up rock fragments (because water expands on freezing) which, when the ice thaws, slide downhill. All these processes reduce the gradient and lower the ground surface; and of course they act with different intensity depending upon the orientation of the slope.

The washing action of rain is an important agent of particle removal, especially on steep slopes and in areas of intensive rainfall and little vegetation: up to one hundred tonnes of soil per acre may be shifted by one storm. Rain may become concentrated into

Chemical weathering at work on a gravestone. Studies of gravestones can be of great value in this context since in most cases the date of erection is accurately known.

zone of the Ivory Coast is considered to be about 22,000 years; in northern England approximately 60cm (1.95ft) of *Carboniferous limestone has disappeared in about 12,000 years, a rate equivalent to 1m/20,000y – an estimate based on the chemical weathering of naturally occurring limestone and of public monuments.

Chemical weathering can produce a rotted rock-form known as a saprolite, which is the product of chemical changes which have

taken place *in situ*. The depth or level to which weathering has taken place, sometimes known as the basal surface of weathering, can be over 300m (1000ft). As a result of chemical changes within this rotted rock, distinct layers or horizons occur, and these make up a weathering profile. Under conditions of heavy leaching, silica is dissolved and insoluble minerals such as alumina and ferric oxide accumulate in the upper part of the profile, forming a

The end-product of fluvial erosion: rounded pebbles in a river pothole in Cape Province, South Africa. The water itself is not the primary agent of erosion, but it contains particles of weathered material and it is these that are responsible for the rounded appearance of the pebbles and the edges of the pothole.

rill wash, which in easily erodible deposits forms intricate gullies and ravines, such as those found in badlands. In mountainous areas where unconsolidated deposits contain large boulders, rain may wash out finer particles, leaving the larger ones behind as protective caps to the less resistant material. This process forms earth pillars which may be several metres high.

Material moved downhill by these means may be channelled into rivers, which are responsible for many types of erosion. One of these is *corrasion*, which refers to the material dissolved in the waters of the river (the dissolved load), representing about 30% of the total removed by river erosion. Rivers also move fragments along their beds which abrade and drill into the bedrock: this process is known as *vertical corrasion* and produces the rounded forms known as potholes. Rock fragments, both suspended and bedload, are jostled by rivers and are broken down by the process of *attrition*. The river's banks are undermined by a scouring process known as *lateral corrasion*. It is estimated that the material brought into the sea by rivers is equivalent to a lowering of the world's land masses by about 1m (3.25ft) in 30,000 years.

A river's volume varies throughout the year, and by far the greater amount of its erosion occurs during its flood stage. The erosion is largely accomplished by turbulent water flow. In steeply graded streams, vertical cutting takes place at rates as great as half a metre (20in) of rock in 3000 years. In mountainous areas or in regions of little weathering, steepsided gorges like the Grand Canyon of Arizona are formed. Most rivers have a winding or meandering habit, eroding the banks on the outsides of the bends and depositing material on the insides. Lateral cutting (i.e., cutting into the banks) is in general much more important than vertical downcutting; it abrades the rock floor to produce rock-cut platforms.

Erosion can also occur when water is not concentrated into river channels. This usually takes place in arid and semi-arid areas where, after storms, flow is in the form of sheet-floods, comparatively shallow floods running over a broad area. Material is transported by such flow over gentle slopes, often at the foot of steep mountains, a delicate balance being maintained between the material delivered to the slope and the rate of its removal. Erosional-transport slopes, pediments, are formed and so the process is called pedimentation.

Ice. Glaciers and ice-sheets are bodies of ice which move slowly over the ground under the action of gravity. Rock fragments loosened by snow and frost become incorporated into the moving ice: if enough of this debris is included in the lower part of

Earth pillars in the Goreme Valley, Turkey. The cappings of rock were originally boulders lying on the surface of the ground: as rain eroded this surface these boulders protected the earth beneath them while the surrounding earth was removed. These particular earth pillars possess an additional point of interest: they have been hollowed out and are used as houses.

Plucking occurs where ice freezes to the bedrock.

the ice, scraping and scratching of the embedded material over the bedrock surface causes *glacial abrasion*, which gives rise to smoothened rock surfaces as well as scratches known as striae (or striations) parallel to the ice movement. The grinding up of rock fragments also forms rock flour, fine sediment associated with rivers draining from glaciers.

At the base of a glacier there are pressure variations which, together with the effects of frost action under the ice, cause joints in the rock to be prized open: blocks become detached and are incorporated into the glacier. This process is called *glacial plucking* or joint-block removal, and is especially important in the coldest glaciers where the ice is frozen to the bedrock, and in rocks where the joints are spaced 1–7m (3.25–23ft) apart. Plucking gives a shattered and broken appearance to the landscape. *Roches moutonnées*, swarms of rocks which have a smooth, gradual slope on one side and a steeper, rougher slope on the other, are caused by abrasion (on the smooth side) and plucking (on the rough).

Glacial erosion modifies river-formed valleys into U-shapes. Relatively small bodies of ice erode corries or cirques, which are amphitheater-like hollows characteristic of glaciated highlands. Glacial erosion is both variable and selective: it is most effective where large volumes of ice are confined in narrow, steeply descending valleys; and in these situations glacial erosion rates can greatly exceed those of rivers. However, in close proximity to areas of intense glacial erosion there may be *plateaux where the effects of ice have been negligible and where the relief is much as it was before the glacier's arrival.

Wind. The wind is an important transporting agent, and this is reflected in the great dust storms which take place in desert areas. The sand grains suspended in the air are the smaller ones, movement of larger particles being along the ground by *saltation* – by a series of jumps. The height reached by such larger particles is rarely more than 2m (6.5ft) above the surface, and so the erosive force of the wind is generally limited to this low level when it meets higher obstacles. The wind picks out structural weaknesses in the rocks, so that their bases are fluted and undercut to form caves, mushroom rocks and irregularly carved ridges known as yardangs.

Sand grains carried by the wind also cause abrasion and polishing of desert surfaces by the natural sandblast – vehicles used in the desert often lose all their paint. Sand grains are blown for vast distances and suffer considerable attrition, becoming almost perfect spheres with frosted surfaces like ground glass, known as millet-seed grains. The wind removes the finer com-

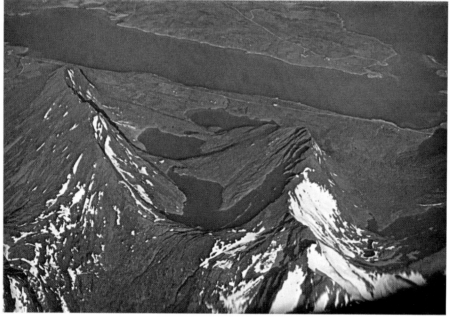

End-products of glacial erosion: a cirque and a characteristic U-shaped glacial valley set high in the mountains of Norway.

ponents of the broken rock, leaving behind a gravelly or bare rock surface, called a desert pavement. Shallow basin-like depressions are formed by this process, known as *deflation*, particularly where soft unconsolidated and friable rocks are exposed: in the Kalahari desert these shallow depressions are known as pans.

In desert areas, much of the wind's energy is used to redistribute sand accumulations. Wind erosion outside these regions becomes important whenever vegetation cover is thin, as in the areas surrounding the polar ice caps: in the Pleistocene period, thick deposits of wind-blown dust and silt (loess) accumulated beyond the glacial limits.

The Sea. Erosion by water, glaciers and wind can act over the whole of the Earth's surface, resulting in an overall reduction of the land (subaerial erosion). But the action of the sea is confined to relatively narrow limits. Wave erosion can occur a metre or so above the level of the highest spring tides; the depth to which wave erosion (as distinct from wave motion) occurs below the lowest tide level is uncertain, though about 7m (23ft) is the usual figure quoted. Clearly, the longer the coastline, the greater the area exposed to the waves.

Though tides and currents may perform erosive and transporting activities, waves are by far the most important marine agents of erosion. This is due both to the hydraulic

action of the water itself and to the action of stones and boulders moved by the waves.

Wave size depends largely on the length of *fetch*, the distance at the disposal of the wind for the generation of the waves, which itself depends largely upon local topography. Two main types of waves are recognized: constructive waves, relatively widely spaced, which tend to build up a beach; and destructive waves, which are closer together and tend to erode and comb down a beach. In a destructive wave, the forward wash is weak but the backwash is powerful and removes material seawards.

When waves break against cliffs during storms, they can exert pressures of over $35MN/m^2$ (5000lbs/in^2), and very high pressures may also result from air compressed between the wave and the cliff. Such wave action concentrates on and opens up cracks and joints in the rocks. Stones and boulders moved by the waves cause attrition; and are hurled at cliffs which become undermined by caves and overhangs. Cliff and cave-roof collapse is aided by surface weathering, and blowholes, arches and eventually stacks (like the famous Old Man of Hoy in the Orkney Islands, off the north coast of Scotland) are produced. In unconsolidated rocks, such erosion may be severe.

Sand and pebbles worked by the waves also abrade the rocks on the foreshore, especially in regions where the difference

Sipapu, the largest of the three natural bridges in the Natural Bridges National Monument, Utah. The bridge, which spans 81m (265ft), was initially formed by one of the two streams that meander through the park, being later enlarged by wind erosion.

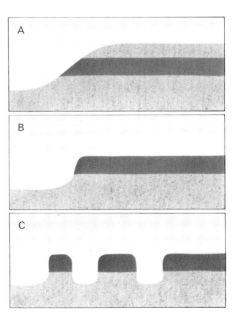

The development of plateau formations through erosion, where a roughly horizontal mass of resistant rock lies in more easily eroded rocks. Initially a table-top is produced, to be fragmented by further erosion so that buttes and mesas are formed.

A dramatic example of marine erosion at work is provided by these cliffs at Bonifacio, Corsica, where the coastline is retreating owing to wave action and subsequent rock falls. Note also the depositional markings displayed in certain of the strata.

between high and low tide is large, and in this way wave-cut platforms are formed. Corrosion (chemical weathering) by sea water and by marine organisms assists in the development of marine platforms.

The Results of Erosion. Erosion is the complement of ★deposition, in which material is laid down to form new land. Given enough time, erosion agents cut across bedrock to form erosion surfaces. The ultimate eroded condition of the land is envisaged as a low-relief plain, the peneplain, across which meander sluggish streams.

The stages in the development of the peneplain have been termed the cycle of erosion. Recent work has been critical of the erosion-cycle concept, but there is no doubt about the existence of eroded surfaces; and, with the help of modern techniques, we can now distinguish between the products of the different agents of erosion and so trace the history of such surfaces.   MMS

Deposition

There are many economic and environmental problems where knowledge of sediment transport and deposition can play a part; for example, the design of harbors to minimize silting or of undersea pipeline routes to minimize the damage done by scouring. Such knowledge is also vitally important in investigating ancient ★sedimentary rocks, where ability to predict the extent and trend of rock bodies can be of value in assessing and exploiting natural resources such as coal, natural gas or groundwater.

The products of the weathering and erosion of older rocks are distributed on the Earth's surface by a variety of processes. In some environments sediment is simply in transit but in many others deposition is significant, leading to accumulations of sediment which, in favorable circumstances, may become part of the sedimentary-rock record. Knowledge of the processes of sediment transport and deposition, as observed in present-day natural environments and in laboratory experiments, enables us to interpret ★sedimentary rocks as the products of such processes.

Agents and Processes. Sediment is moved at the Earth's surface by four main agents, water, wind, ice, and gravity, operating in a variety of processes: these

A blowhole in Corfu. Blowholes are formed when, initially, waves erode deeply into the coastal cliffs until eventually a section of the undermined rocks collapses into the cave beneath it. At high tides waves rush into the blowhole at pressures sufficient for water to be forced out at the landward end, as is happening in this picture. In time the "bridge" of material roofing the seaward part of the cave will also collapse, so that a new inlet is formed.

processes are not exclusive to individual agents, and some agents move sediment in more than one way. It is important to realize that gravity not only plays a part in driving the flow of water and ice but also, in some instances, acts directly on sedimentary particles, causing them to move and be deposited elsewhere. Such direct gravity-produced movement may be aided by the lubricating effect of water or of finer particles between the larger grains.

We will discuss deposition with reference to the four major transport processes, solution, suspension, bedload and mass flow: if we understand these then we are a long way toward understanding deposition, which is merely the cessation of transport.

Solution. Chemical *weathering of unstable minerals causes some of their ions to go into solution. The dissolved ions are transported by rivers into the sea or, more rarely, to an inland lake. In both cases, if addition of ions continues for a long time, there comes a point when the water body is saturated with certain salts, which then precipitate to maintain an equilibrium of input and extraction. The sea achieved an equilibrium in the distant geological past, and this has since been maintained by certain minerals being precipitated from solution in favorable conditions. While sodium chloride (common salt or *halite, Na^+Cl^-) is the most abundant ionic pair in

the sea, it is not the one nearest to saturation: calcium carbonate, the main constituent of limestones and chalk, being far less soluble, is the compound most easily precipitated from seawater, forming the minerals *calcite and *aragonite.

The uppermost layers of the ocean are saturated with calcium carbonate, particularly in tropical latitudes, and so it comes as no surprise to find this material being deposited in great abundance on warm, shallow sea floors such as the Bahama Banks, the Persian Gulf and the Florida Keys. There is also some precipitation of calcium carbonate in colder waters, though here the process is assisted by organic agents. Even in warm seas, such organic agents play a predominant role in precipitation, although straightforward physical processes may also contribute. Many organisms, both animals and plants, secrete calcium carbonate as shells, skeletons or supporting structures, and most naturally occurring calcium carbonate is of this origin. Some aragonite is probably precipitated by inorganic means as ooliths – small rounded particles, a millimetre or so in diameter, which grow by the addition of concentric layers while being rolled about on an agitated sea floor.

Many of the products of organic precipitation are readily recognized as such; for example, the shells of *bivalves, *corals

or *echinoids; but other particles may need to be examined by microscope or electron microscope before their organic origins can be shown. The coccoliths which make up the famous *Cretaceous chalks are a very good example of this.

The only other minerals which precipitate from seawater of normal salinity are certain iron minerals, mainly iron silicates, which accumulate locally off the mouths of some major tropical rivers. Most other minerals that are precipitated from the sea require the salinity to be much greater than normal. In fact a whole suite of minerals is produced by the progressive concentration of seawater. These are collectively referred to as *evaporites. Seawater must have, for example, ten times its normal concentration for sodium chloride to precipitate: clearly, unusual climatic and topographic conditions are needed for this to be achieved naturally. The combination of a hot, arid climate, where the evaporation rate is high, and a basin with a very restricted connection to the open sea allows concentrations to build up as more and more seawater is sucked into the basin to replace

A river delta on the shores of the Arafura Sea. Deltas are a result of the deposition by rivers at their mouths of weathered and eroded material transported by the water. If conditions are favorable (i.e., the water fairly still) the delta will gradually advance seaward.

Weathered and eroded material being transported in suspension colors the water of this river in Sulawesi. Deposition of such sediment at river estuaries may result in formation of a delta. The animals being washed are water buffalo.

that lost by evaporation. The *salinas* of the west coast of Mexico provide examples of these conditions.

Elsewhere, as in the coastal flats (sabkhas) of the Trucial Coast of the Persian Gulf, evaporite minerals precipitate from water within recently deposited carbonate sediments. Evaporation from the exposed sediment surface draws seawater landward through the sediment (rather as coffee seeps into a dry sugarlump), concentrating the water as it moves until *gypsum and *anhydrite (forms of calcium sulfate) precipitate as layers or nodules below the sediment surface.

Suspension. Anyone who has witnessed a dust storm or observed the turbid nature of a river in flood will testify to the ability of both wind and water to carry material in suspension. The sedimentary particles are supported by the turbulence associated with the rapidly moving fluid. Virtually all natural flows of wind and water are fully turbulent – that is, there is a random movement of parcels of fluid superimposed upon the overall fluid movement. The breakdown into discrete clouds of a gently rising column of smoke in still air is a common illustration of the onset of turbulence.

Particles carried in suspension are supported by the upward components of turbulence and, clearly, smaller particles are more likely to be carried than large ones. Normally only the smallest grains are carried in suspension, most typically those in the clay and silt grades (less than 0.0625mm (0.0025in) in diameter), though sand or even larger material may be carried in very strong flows. For particles of silt grade or finer, suspension is the only mode of transport: on *erosion, they go directly into suspension, with no intermediate phase of movement in contact with the bed.

In order for material to be deposited from

suspension, the level of prevailing turbulence must be reduced, so that the flow is no longer able to support such coarse, or so much, sediment. Because of the tendency of larger particles to be deposited more rapidly, the deceleration of a flow carrying a range of grain-sizes in suspension may lead to the deposition of a graded bed; that is, one in which there is a gradual upward diminution in grain-size. In cases where only the finest sediment is transported, fine silt or mud will be deposited with no obvious upward grain-size change, though often with a thin parallel lamination such as that seen in many ancient *shales. All *mudstones are the product of gentle deposition from suspension and originate in environments of low current or wave strength; for example, lakes, lagoons, the sea or ocean floor below the wave base, or the more sheltered parts of tidal flats. Once deposited, fine sediment often develops cohesive strength and may be able to resist erosion by subsequent strong currents.

An unusual but geologically very important type of suspension deposit is that produced by a turbidity current. This is a dense current which flows down a slope beneath a body of clear water, the current's greater density being due to its carrying sediment in suspension: the material is there by virtue of the turbulence which is, in turn, due to the flow's existence in the first place. In other words, the current is a dynamic system with a feedback loop involving sediment, slope and turbulence.

Such currents are a common occurrence in freshwater lakes and reservoirs, where sediment-charged rivers spread material out over the lake floor by under-riding the lake water (see *deltas). Temperature differences may also aid the density current. More important geologically are the turbidity currents which introduce and distribute sediment of quite coarse grain-size

(sands and fine gravels) in deep-ocean settings, where normally only the finest grains are deposited from suspension. These oceanic turbidity currents are generated on the continental slope, often in the heads of submarine canyons, and flow, normally by way of a canyon, to build up a submarine fan at the foot of the slope where they begin to decelerate. Such activity is particularly common off the mouths of major rivers, where high sediment input can cause the accumulation of large volumes of sediment in unstable settings: periodic removal down the slope relieves this instability. The most spectacular present-day example is the Bengal submarine fan which is fed by the Ganges and Brahmaputra Rivers: this fan extends out into the Bay of Bengal for some 3500km (2200mi), testifying to the distances over which turbidity currents can transport sediment.

The layer of sediment laid down by a turbidity current is called a turbidite, and is normally a parallel-sided sheet of sand, which, in the geological record, will be interbedded with finer background sediment, normally mudstone. The sand or sandstone may show evidence of small-scale erosion on its base and internally may show graded bedding, reflecting the gradual waning of the current. There is often some evidence, in the form of lamination and ripples, of the reworking of deposited material by the later stages of the flow or even by more permanent bottom currents. Thick sequences of turbidites are common in the stratigraphic record; for example, the Silurian *graywacke sandstones of the British Southern Uplands.

Bedload. Many of the complex and interesting structures left on a sandy beach or tidal flat by the ebbing tide, or produced on the bed of a sandy river or even in gutters after a storm, are due to transport and deposition of material which has been carried as bedload. Near to the bed, sand and coarser-grained material normally moves by the rolling, bouncing and jumping of the individual particles. In water, grains up to the size of large boulders can be shifted, though in wind only particles less than 2mm (0.08in) across are generally moved. For sand, the processes of movement and the structures produced are similar in both wind and water, and so can be considered together.

Almost as soon as sand begins to move as bedload under a current flowing in a single direction, the bed becomes molded into small-scale ripples which have their crestlines at right angles to the direction of flow. The ripples have an asymmetry in the direction of the flow (i.e., across the crests), with a gently sloping upstream surface and a more steeply inclined downstream surface, the lee face. Ripples move downstream by erosion of material from the upstream surface and deposition on the lee face; and this is reflected in the internal structure of the ripples, which shows inclined laminae representing the successive positions of the lee face. This lamination, known as cross-lamination, is a common feature of ancient sands and sandstones, and its presence makes it possible for us to interpret the deposit in terms of a current of strength appropriate to the formation of ripples, and whose direction of flow can be deduced from the direction of the inclined surfaces.

With stronger flows, a new form of surface structure develops. This has a shape very similar to that of the ripples, but it is considerably larger and is known as a dune. (The word "dune" is often thought to refer only to those forms built by wind, but it is now used equally for water-lain structures.) Dunes range in size from structures with a relief of about 30cm (12in) up to forms several tens of metres high in windblown deserts or on shallow sea floors like that of

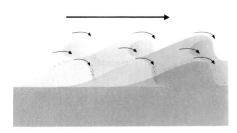

Dune migration. Material is carried up the windward (or upstream) face and drops over the crest onto the lee face. This results in the progressive migration of the dune in the direction of the wind (or water) flow.

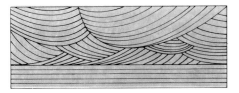

Cross-bedding overlying evenly bedded strata.

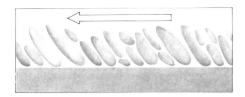

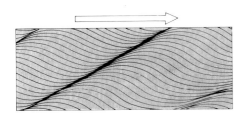

Above, imbricate stacking of pebbles, the planes of the pebbles being inclined in the direction of current flow. *Below*, cross-lamination, resulting from the migration of ripples in the direction of flow by erosion of material from the upstream side and its deposition on the downstream side.

the North Sea. The mechanisms associated with dune migration are very similar to those operating in the case of ripples, and inclined lamination on a larger scale (this time known as cross-bedding) is the characteristic internal structure. This can occur in units up to several tens of metres thick though units of less than 1m (3.25ft) are more common. The largest units are most commonly the products of wind-blown dunes. (See *arid landscapes.)

When the flow strength is greater than that appropriate for the formation of dunes, the bed becomes flattened and a phase of transport is developed where grains close to the bed travel as a rapidly moving sheet (plane-bed transport). Deposition of particles transported in this way produces a parallel lamination in the sand with the laminae, each a few grain diameters thick, showing very slight grain-size differences. The reasons for this sorting are not clear. Parallel laminated sandstone, when split along the lamination, shows a marked linear structure on lamination surfaces, the so-called parting or primary current lineation. This is a common feature of pavement flagstones and is thought to be due to spiral vortices in the flow. It can be used as an indicator of current direction when found in ancient sandstones.

In the case of the most rapid flows, the bed may again be molded into a wave-like form, but the rapid, sheet-like movement of the sand continues and so the waves are much more rounded and symmetrical, and have only a low relief: they often grow and disappear quite rapidly, and can move in an upstream or downstream direction. Such features, called antidunes or standing waves, are often seen in streams running across beaches at low tide or in gutters during heavy rain. They are then obvious because not only is the sediment surface distorted but also the water surface takes on a ·similar form in phase with the sediment waves below it. The internal structures of these forms are poorly understood and probably involve gently inclined internal lamination. They are unlikely to be preserved in deposits and very few examples have been described from the stratigraphic record.

When the sediment in transport has a substantial proportion of pebbles and coarser particles, the transport and depositional mechanisms and structures are somewhat different from those occurring with sand. In pebbly rivers, the gravel is commonly molded into bars which often split the flow to give the stream a braided pattern. The growth of these bars is complex, but involves both vertical accretion of material to the upper flat surface and extension of the downstream end by avalanching of material swept over the top. This avalanching produces cross-bedding, as in dunes, while the vertical accretion produces a crude horizontal stratification in which the more flattened pebbles show an imbricate stacking, like tiles on a roof, with the inclined planes dipping in an upstream direction.

Imbricate stacking of pebbles deposited in a powerful current. These pebbles have been jostled, transported and roughly sorted; and deposited tilting upward in the direction of flow.

Slumps in dune bedding near Route 89, Utah: the bedding is particularly clear in the lower strata. Note the closely horizontal boundaries between the layers of dune-bedded strata, the layers themselves being termed "cosets".

Ripple drift in beach sand. This occurs because deposition is of greater extent on the lee side of ripples, which therefore migrate downstream.

So far we have only discussed the nature of bedload transport by unidirectional flows, but such movement is also important in wave-dominated environments such as shorelines – the oscillatory nature of the water movement in waves is the dominant cause of differences in the structures produced. Ripples are commonly produced by the action of waves on sand, as a walk on any beach will reveal, but the ripples differ in several respects from those produced by currents. They are much more likely to have a symmetrical profile, with rounded or sharp crests. The crest-lines, which are oriented parallel to the wave fronts (i.e., perpendicular to the direction of wave flow), often show great lateral continuity: where two or more wave systems coexist, complex interference patterns may result. Internally, lamination may be wavy with the laminae inclined in opposed directions.

Some beaches are devoid of ripples: here breaking waves on the ebb tide create flow conditions analogous to the plane-bed mode of transport for unidirectional currents. Gravel beaches show good sorting of the pebbles and, if the shapes are suitable, these may be packed with a good imbricated fabric.

Mass Movement. Viscous material on a slope will flow by shearing within the material. Ice is one such material and, flowing in glaciers, it carries and eventually deposits large quantities of sediment (see *glaciation). Similarly, on submarine slopes or on terrestrial hill slopes during heavy rain, masses of sediment may flow as a highly viscous liquid if sufficient muddy material is present to act as a lubricant. In both ice- and mud-flow situations, because of the viscosity, it is possible for large particles, up to the size of very large boulders, to be transported and deposited.

Glaciers transport vast quantities of sediment of a wide variety of grain-sizes, from the finest rock flour to large boulders. All are transported together within the ice with little sorting of grain-sizes and with little abrasion between the grains. Most deposition takes place at the end of the glacier as a dumped mass of unsorted sediment, a moraine of boulder clay, though this can be reworked by meltwater as bedload or in suspension. Where glaciers end in the sea, material is dropped from the melting sole of the ice sheet or from icebergs floating far from the parent glacier. The deposits of such processes are recognized by the presence of "dropstones", incongruently large grains in an otherwise finer background.

High-viscosity flows produced on slopes by an abundance of muddy sediment occur in both submarine and subaerial environments. The best known mudflows occur on alluvial fans in deserts, where deep weathering gives abundant fine sediment which is periodically flushed out in periods of heavy rain. Such mudflows can travel many kilometres and deposit elongate lobes of poorly sorted conglomerate.

Submarine mudflows are less well understood, but are probably generated by slumping on continental slopes, and may evolve downslope until they come under the influence of turbidity currents. The deposits of supposed mudflows are sometimes found interbedded with turbidites in sequences of deepwater sediment in the stratigraphic record.

Conclusion. The transport and deposition of the weathering products of the Earth's surface involve complex and varied phenomena. Environments of deposition are characterized by the processes operating within them, and control of the development of these environments depends on a clear understanding of the processes. Equally, a knowledge of the products of various sedimentary processes is an essential prerequisite to any environmental interpretation of ancient sedimentary rocks, and this can often be an important consideration in planning the exploitation of sedimentary natural resources.   JDC

Glaciation

Today, ten per cent of the Earth's land area is covered by ice with a total volume of twenty-six million cubic kilometres, and 18,000 years ago nearly thirty per cent was covered by ice with a total volume of around seventy-six million cubic kilometres. Some seventy-five per cent of this planet's fresh water is currently in the form of ice.

A fine example of a terminal moraine only a few years old near Athabasca, Canada. Terminal moraines are made up of rock particles eroded from the mountainside by the glacier during its descent and deposited as the ice melts (note the "dirtiness" of the ice at the toe of the glacier) or washed out by subglacial streams. Further in the background a cirque can be seen.

Glacial ice forms in cold areas of the Earth's surface where annual snowfall is greater than the amount of snow which melts during the year. Newly fallen snow has a density of about 0.05g/cm^3 (3lb/ft^3), but as it is progressively buried, individual grains are pressed together and the spaces between them may be filled by frozen meltwater; when the density has risen to 0.83g/cm^3 (51.46lb/ft^3) we have ice. Ice masses build up in this way in cold regions, either near the poles or at high altitudes in more temperate latitudes (there are several glaciers on the equator).

Although glacial ice appears to be a rigid material, it is able to flow to a certain extent and so has a tendency to sag under its own weight. Thus the ice masses which build up in cold regions do not continue to grow upwards, but flow away laterally toward warmer regions. The effect of this slow flow can be seen on the surfaces of many glaciers, where bands of debris may show complicated patterns of folding. Large ice sheets are in fact dome-shaped, having typically a parabolic surface-profile which results from the physical properties of the ice. Where glacier ice moves down very steep slopes, the ice is unable to flow

sufficiently quickly to take up the new shape needed to conform to the changing shape of the glacier bed, and so the ice fractures, producing crevasses.

There are three principal types of glacier: ice sheets, valley glaciers and cirque glaciers.

In polar regions great dome-shaped *ice sheets* such as those of Antarctica and Greenland completely submerge the land. The beds of these ice sheets descend below sea level and include high mountain ranges which do not pierce the ice surface. The discovery of the technique of radio echo-sounding has made it possible to determine the shape of the bedrock surface on which these ice sheets lie.

Valley glaciers occur in mountain areas where ice flows away from the main centers of accumulation. Because of the high rates of accumulation in such areas, valley glaciers need to flow very rapidly in order to discharge the accumulating ice. They are thus often highly crevassed and erode deep valleys.

Very small glaciers often form on mountainsides where the net accumulation is relatively small. These erode small hollows on the mountainside to form *cirque glaciers*.

Glaciation in the Past. We often assume that the modern Earth environment is in some way fixed. In fact there is evidence for all to see that Britain and North America were cold enough to maintain very large glaciers until as recently as 10,000 years ago. In the mountain areas of Britain we find armchair hollows known as *cirques*

(cwms in Wales, corries in Scotland) scooped out by cirque glaciers. We find deep U-shaped trenches typical of those cut by valley glaciers, and over much of Britain we find the typical deposit of glaciers, *till*. This is a mixed deposit ranging from clay to large boulders (often striated by glacial action), and represents material eroded by the glacier from the rocks over which it has passed, transported by the moving ice, and finally dumped when the ice melts. It forms the subsoil of much of Britain. We also find evidence of old *terminal moraines*, which are ridges pushed up by the advancing margin of the ice sheet, and which allow us to reconstruct the position of that margin. (See *deposition; *glacial landscapes.)

If we consider the last 100,000 years in northwest Europe and North America, extensive glaciation has been much more common than the temperate conditions under which we live at the moment. Evidence from North America in particular shows how, time after time, the great ice sheets thrust to the south, then retreated, reaching at their maximum extent some 18,000 years ago the outskirts of what are now New York and Chicago. These great southward advances were associated with southward movement of the climatic zones, with tundra conditions in, for example,

A spectacular example from Strandflat on the Norwegian coast of a U-shaped valley eroded by the glacier that still partly occupies it. On the surface of the glacier can be seen lateral moraine. The fjord in the background is another U-shaped valley that has been invaded by the sea.

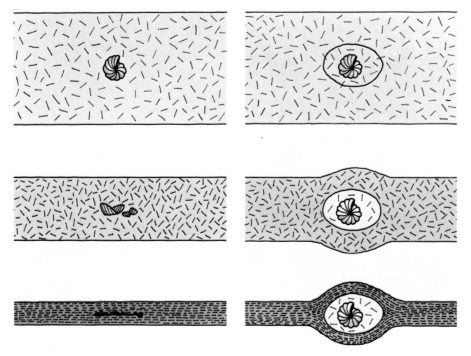

Differential compaction of clay sediments as sedimentary rocks are produced. In the left-hand column, lack of cementation results in the crushing of included organisms during diagenesis of the sediments. In the right-hand column, there is early carbonate concretion around an included organism, protecting it so that even after diagenesis it may be perfectly preserved.

which affect the intensity of solar radiation reaching the planet. However, there are many other possible mechanisms of climatic change, both on this and on smaller scales, which are internal to the Earth and thus more difficult to identify. For instance, it is possible that the Antarctic ice sheet is unstable, and might flow rapidly into the southern oceans, thereby cooling them and raising world-wide sea levels. With more of the Earth's surface covered by water, the planet's reflectivity (albedo) would increase, so that more of the Sun's light and heat would be reflected back into space, causing global cooling and possibly producing glaciation in temperate latitudes.

Many scientists are now searching for clues about the origin of climatic changes, largely in order to be able to predict future changes in the Earth's climate, on which so much of our economic structure depends. The problem is an important one for, if the immediate geological past is any guide to the future, the present interglacial will one day deteriorate into a glacial period. GSB

Diagenesis

When sediments are buried, all kinds of physical and chemical processes take place which may lead to quite radical modifications of the original material. Diagenesis is the term used to cover all those transformations occurring at relatively low temperatures and pressures in environments not too far beneath the Earth's surface: at greater pressures and temperatures there is the rather different process of metamorphism (see *metamorphic rocks).

Perhaps the most obvious effect of diagenesis is the transformation of loose, unconsolidated material into hard, compact *sedimentary rock. This aspect (for it is only one) of diagenesis is termed lithification. Good examples are the conversion of sand into sandstone, mud into shale and peat into coal.

Compaction and Cementation. We can recognize two essentially independent components of diagenesis. The first is physical and reflects the increasing overburden pressure as the sediment is buried. Individual sediment particles pack closer and closer together and, in so doing, forcibly exclude some of the sediment pore-water: this is known as compaction. The second component is chemical rather than physical. New minerals precipitate in pore spaces, displacing yet more pore-water, and bond sediment grains together: this is cementation. The cementing material arises from chemical reactions between unstable sediment particles and pore-waters. It may, as in the case of *limestones, amount to wholesale recrystallization.

Compaction and cementation generally progress side by side throughout diagenesis. Occasionally one is directly related to the other, as in the case of "pressure solution" when physical compaction stress causes recrystallization.

southern Britain and northern and central France, and the prevalence of floating pack ice in the Atlantic down to the latitude of Madrid.

As we have seen, ice sheets have a typical shape, and if we know the areal extent of an old ice sheet from the distribution of till and moraines, it is possible to reconstruct its surface form. Such a reconstruction shows the last ice sheet over Scandinavia to have had a maximum thickness of 2.4km (1.5mi) over its source area. Because of this great weight of ice, the Earth's crust was depressed beneath the ice sheet: when this melted, the crust began to spring upward again, a process which is still continuing. (The process by which the Earth's crust is depressed by superincumbent loads is known as isostasy.)

It has been possible, by radioactive dating of organic material associated with the moraines which mark the successive ice-front positions during retreat, to reconstruct how the great North American ice sheet slowly melted away. We see that by 6000 years ago the ice sheet had collapsed completely in the North American mainland, leaving only remnants in the Canadian Arctic archipelago. The remnants of the European ice sheet are the small glaciers found in the Scandinavian mountains.

Ice Ages. Fluctuations of the Earth's climate and physical environment between glacial episodes (such as that between about 70,000 and 10,000 years ago) and interglacial episodes (such as that between 10,000 years ago and the present day) have been characteristic of the last two million years or so. This is known as the *Quaternary period, during which there have been about seven major glacial epi-

sodes. Before this period, the Earth's climate seems to have been generally warmer, and the great Antarctic ice sheet, which is so important in influencing the Earth's climatic and oceanic circulation, does not seem to have existed prior to about five million years ago.

There are, however, earlier periods of the Earth's history during which we find widespread glacial tills, striated boulders and evidence of glacial *erosion, leading us to think that the Earth has from time to time been subject to ice ages similar to those of the Quaternary. The reason for periodic ice ages, and the glacial and interglacial episodes which occur within them, is still an enigma, but there are two principal hypotheses currently in favor to account for their origins.

Firstly, ice ages could develop when continental plates (see *plate tectonics) move into polar positions so that large ice sheets may build up on them. This hypothesis could explain the initiation of Quaternary ice ages after the Antarctic continent had moved into a polar position in late *Cenozoic times, and might also explain the *Paleozoic ice ages when Gondwanaland lay in a polar position.

However, glaciers seem to have penetrated the tropics during the late Precambrian ice ages, and it may be that for such events we need a second hypothesis, a postulated slow "flickering" of the Sun, whereby its luminosity is (minutely) reduced for a period of time. Astronomical observations of other stars tend, to some degree, to confirm that this is a possibility.

The succession of glacial and interglacial periods within ice ages is most probably related to variations of the Earth's orbit

In other cases the total diagenetic modification of a sediment may be dominated by either compaction or cementation. The two extremes may even occur side by side when very early cementation occurs locally within a sediment: this results in ellipsoidal concretions, or nodules of hard carbonate-cemented sediment, around which are "wrapped" layers of shaley sediment demonstrating differential compaction. This effect is sometimes beautifully illustrated by perfect preservation of delicate fossils within a concretion, whereas the same organisms are crushed or flattened in the surrounding shale.

The extent of differential compaction is large, especially in mudrocks. At *deposition, muds contain 80% or so of water by volume. Early cementation preserves this open fabric and the true three-dimensional nature of any included organisms. Compaction then reduces the overall volume of the mudstone to one-fifth of the original, but does not affect the early-cemented material.

The Starting Point – Sediments at Deposition. Fresh sediments contain a variety of mineral constituents from two principal sources.

*Erosion of the land surface contributes soil matter, coarser mineral particles (especially *quartz) as well as much finer clay particles (*illite, *kaolinite, *chlorite, and *montmorillonite). Also present are amorphous mixtures of iron and aluminum hydroxides together with organic matter. Sedimentary processes tend to sort these components into different classes according to size (or, strictly, hydrodynamic equivalent).

The second main source is the depositional water itself. Calcium carbonate (as *calcite or *aragonite) is precipitated, either directly or as the skeletal remains of marine organisms, from the warmer water regions of the world's oceans. These sediments are the forerunners of limestones.

There is a third component of sediments at deposition – water trapped within the sediment pores. This can be fresh or a strong salt solution, as in marine sediments. The very large volume of water trapped in mud sediments means that dissolved salts may constitute a significant fraction of the initial sediment.

Mineral Stability. Chemically precipitated minerals are stable when surrounded by the waters from which they have been deposited. The fine-grained constituents of soils, however, form under the influence of very dilute aqueous solutions (rainwater) and an oxygen-rich atmosphere. Instability of these components is to be anticipated. The coarser constituents of soils are the more resistant to the

*weathering process. They are less likely to be reactive during diagenesis unless, for some reason, they have had little chance to react in the weathering environment. Glacial or volcanogenic sediments contain coarse particles which can be very unstable.

Diagenetic Environments and Reactions. Recent research programs based on the analysis of sediments and squeezed pore-waters from offshore drilling programs point to the existence of definite diagenetic zones at different depths in buried sediment sequences. Sediments containing organic matter become oxidized just below the sediment/water interface as molecular oxygen is destroyed. From here down to a few tens of metres, sulfate-reducing bacteria are active in marine sediments: sulfate from the overlying depositional waters and organic matter are consumed and hydrogen sulfide and carbon dioxide produced. The former reacts quickly with unstable iron minerals to produce *pyrite (iron sulfide). Reactions between clay minerals, unstable carbonates and bacterial carbon dioxide lead to the precipitation of calcite (or dolomite) cement. Concretionary bodies are often cemented by pyrite and calcite or *dolomite. These "early" precipitated minerals are sometimes termed authigenic.

Sulfate, like oxygen before it, eventually becomes exhausted as one goes deeper in a sedimentary pile. Conditions then favor bacterial fermentation reactions. Organic matter is converted to methane (*natural gas) and more carbon dioxide. Iron carbonate (*siderite) is a common precipitate in this zone. It should be noted that fermentation will occur at much shallower depths in non-marine sediments – as confirmed by the common generation of "marsh gas" in freshwater swamps.

At burial depths of the order of thousands of metres, temperatures are reached which preclude life, and hence, of course, further fermentation. Unstable minerals continue to react with pore-waters, however, and organic matter evolves towards *coal or liquid *petroleum, depending upon its original nature. Some silicates dissolve, others alter: gradual conversion of

montmorillonite to illite is a common phenomenon. Carbonates continue to precipitate or recrystallize to more stable compositions and crystal structures.

Eventually, of course, temperatures and pressures become so high that there is wholesale silicate-mineral recrystallization with total exclusion of pore-waters. The sediment has then become a *metamorphic rock.

The End Product. Sedimentary rocks represent the interaction between some or all of the above processes for variable time intervals, with the vast variety of different sediments produced in different depositional environments. Some general patterns do exist.

Slowly deposited marine clay sediments almost invariably contain pyrite, sometimes in sizable quantities. Ancient limestones are almost always entirely recrystallized to low magnesian calcite. The relative reactivity of carbonate minerals in aqueous systems is amply demonstrated. Sandstones exhibit a bewildering range of diagenetic mineralogy and sequences of alteration. Perhaps this demonstrates most effectively of all the dependence of diagenetic reaction upon the presence and involvement of an aqueous phase. Whereas clay sediments eventually exclude most of their pore-waters and become closed systems, most sandstones remain significantly permeable throughout their geological history. They are therefore migration pathways for aqueous solutions, from any source, and naturally the passage of these solutions leaves its mark.   CDC

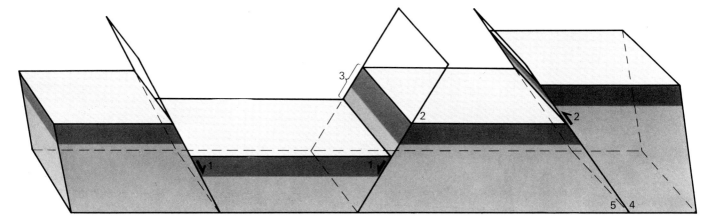

Fault terminology: (1) normal fault; (2) reverse fault; (3) displacement; (4) hanging wall; (5) foot wall.

Faulting

Faults occur on all scales, from the microscopic through to structures that can be traced for a thousand or more kilometres.

The recognition of faults and the terminology used to describe them date from the early days of mining. The faults that concerned the miners were the inclined shear-fractures (dip-slip faults) which disrupted workable seams of coal or metal ores. When these inclined faults were encountered in mine-workings, the rock mass adjacent to the fault plane (the plane along which there had been relative movement) and sloping upward away from the miner was termed the foot wall, the other rock mass adjacent to the fault plane being termed the hanging wall. When the seam in the hanging wall has been displaced downward relative to the same seam in the foot wall, the fault is known as a *normal fault*, because in British mines it was the type normally encountered. When the relative movement has been in the other direction the structure is termed a *reverse fault*.

The orientation of any geological plane is defined by the trend of a horizontal line drawn on that plane, known as the *strike* direction, and the angle of greatest inclination of the plane, the dip angle (or *dip*). Normal and reverse faults both result from slip in the dip direction.

Another common type of shear – which caused little trouble to miners and so was unrecognized for many years – is variously known as a strike-slip, wrench, tear or transcurrent fault. The fault plane is vertical and the sense of movement along it is classified as left- or right-handed, relative to an observer facing the fault. The classification is not as arbitrary as it might appear, since the same sense of movement will be recorded no matter on which side of the fault the observer stands.

Faults which exhibit *oblique slip* – that is, where there has been movement in both the dip and the strike direction – do not fit into the simple system of nomenclature set out here. However, as strike- and dip-slip faults are remarkably common, the terms defined

Small-scale faults. In this rockface can be seen normal faults, reverse faults and strike-slip faults, as well as jointing. Faults may be on far smaller scale even than these – they may be microscopic – or may extend for hundreds of kilometres, as in rift valleys.

The development of an imbricate structure. As the mass of rock flows down the gentle slope its front end encounters a resistance to further motion, while the rearward parts of the block continue to move. This results in compression at the forward edge and the development of a series of faults, often curved.

above are in widespread and frequent use among geologists.

When two faults intersect they may form conjugate fractures. Small-scale normal faults may, in combination, form a graben, and larger fractures and groups of similar fractures having the same general arrangement result in major crustal features known as grabens or *rift valleys: examples are the Rhinegraben and the African Rift System.

During the formation of normal faults and grabens, the main compression which brings about failure in the rock acts vertically (for this reason, normal faults are sometimes termed gravity faults): for both wrench and reverse faults, the main compression acts in the horizontal direction, the type of structure developed being determined by the direction in which the least compression has acted.

Because of limitations set by thickness of the crust and the principle of isostasy, large-scale normal and reverse faults with displacements of more than a kilometre are rare. In contrast, major wrench faults (e.g., the Great Glen Fault of Scotland) exhibit displacements of 100km (60mi) or more. Similar large-scale structures, such as the San Andreas Fault complex, California, and the Anatolian Fault, Turkey, are currently active and have caused catastrophic damage to cities in recent times. Such large structures are thought to be related to plate boundaries (see *plate tectonics).

The other type of fault which exhibits displacements of up to 100km (60mi) is the major overthrust. These comprise thin sheets which have moved over horizontal or gently inclined thrust planes (the whole complex may subsequently be folded). Because these structures are associated with the major fold belts they were originally considered to be the direct result of the compression which gave rise to the fold belt itself – hence the emotive term overthrust. However, it can be demonstrated that many, if not the majority, of such structures were emplaced by a relatively thin (1–3km (0.6–1.9mi)) sheet of rock sliding down a gentle slope with a gradient of one or two degrees. Sliding on such low-angle slopes is possible when the sheet is resting on certain clays (e.g., the Scaley Clays of Italy) and *evaporite rocks such as salt or gypsum. It is now believed that these structures moved downhill because they were supported on weak rocks in which the water pressure balanced the weight of the rock in the sheet such that the frictional resistance to sliding of the sheet was reduced to virtually zero.

When the front end of the gliding block encounters an obstacle of some kind, it slows and stops. However, the rear of the block continues to move, and the compression thus developed at the front end of the block causes the formation of folds (see *folding) or faults. The fault structures which develop at the front end of the block are true thrusts or reverse faults. Often they are curved features which, when they occur close together, are collectively termed imbricate structures.

Joints and Veins. Faults are certainly spectacular structures, but by far the greater majority of fractures one observes in rocks are not faults but joints. These exhibit little or no shear displacement and are mainly the result of minor extensions. Joints usually form well-developed patterns which influence, or even control, topography.

In the undeformed sediments of the

Elaborate patterns of quartz veins decorate these boulders from Cornwall, Great Britain. The veins occupy preexisting faults.

Grand Canyon the causes of the individual sets of joints cannot readily be inferred: indeed, only when joint sets are related to folds can one deduce how the various fractures are related to the direction of compression.

Sediments and *metamorphic rocks at depth in the crust contain water under high pressure. When water at sufficiently high pressure enters fractures in the rock and the water pressure is sustained for long enough, minerals which have been held in solution crystallize on the fracture walls, held apart by the water pressure, so that veins are formed. The vein material is commonly *quartz or *calcite, though sometimes it includes *gold, *silver, *lead, *zinc or other ore minerals (see *ore deposits). Veins of economic importance are less frequently encountered than the non-economic quartz veins widely developed in deformed *sedimentary rocks. However, to compensate for this, economic veins are sometimes of great extent, as for example the Mother Lode of California (the cause of the '49 Gold Rush), which is many metres wide and extends through a significant part of California.

Conclusion. As we have seen, current movement on faults results in destructive *earthquakes; veins are sometimes productive and of economic benefit; while joints are the bane or the blessing of most engineering or mining projects. From this bald statement alone it is clear that faults, joints and veins are of the utmost importance to us all.　　　　　　　　　NJP

Earthquakes

Perhaps 60,000 people died in Lisbon on All Saint's Day, 1755, in what was one of the most violent earthquakes on record. At least two major tremors struck the city, causing enormous structural damage and sending the waters of the River Tagus rushing through the streets. The death toll was especially high because the churches were packed, and because of the fire that ravaged the city after the tremors had passed.

Earthquakes are probably the most dramatic of Man's natural enemies; and for that very reason they are worth our study. By developing an understanding of both the causes and effects of earthquakes we can hope to reduce the horrifying toll of lives in the future, not only by selecting suitable building materials and sites but also by playing an active part in modifying the earthquake itself. Before taking any steps in this direction, we have to be able to predict where earthquakes are likely to occur.

Seismic Waves. When some volume within the Earth fractures or implodes, shockwaves are generated. These seismic waves travel away from the source region (or hypocenter) at speeds that depend on the structures through which they pass.

Although earthquakes occur down to a depth of approximately 700km (400mi), an extension of the initial rupture to the surface would occur only for hypocenters within the upper few tens of kilometres of the Earth. For example, the main shock of the Alaskan earthquake of March 27 1964, one of the largest earthquakes ever recorded, had a hypocenter at a depth of 20–30km (12–30mi).

The shockwaves can be detected by an instrument known as a seismometer, and recorded in the form of a seismogram. The seismogram gives a measure of actual peak-to-peak motions, usually of the order of 2mm (0.1in). The problem of measuring motions while being unavoidably attached to the moving object is overcome by suspending a large weight on a spring. As the ground moves, the mass and the recording

pen remain approximately stationary because of their inertia, while the rest of the equipment moves. Such a device is a seismometer in its simplest form. In most modern instruments the metal weight swings through an electric coil, thus inducing a current that is used to drive an electrical recording system.

In practice, the natural oscillation of the system has to be critically damped so that it responds faithfully to the varying motions of the ground throughout the signal and not just to the first impulse that arrives. Even so, any seismogram will, to a greater or lesser extent, be a reflection of the characteristics of the seismometer used to make it. If high sensitivity is required, the seismometer is tuned to respond vigorously to the natural frequency of a particular phase rather than displaying a high-fidelity broad-band response: in this way amplifications of as much as 200,000 are commonly achieved.

Two types of seismic wave are generated by the rupture, body and surface waves.

Body waves follow ray paths within the Earth that conform to the rules of simple optics: they are reflected and refracted at boundaries of different density; and are

diffracted, or spread out, round the corners of objects that are of a size similar to their wavelength. The motion of the particles in the ground as the wave passes is either in the direction of propagation of the wave front (for the P or compressional waves), or is at right angles to the direction of propagation (for the S or shear waves). In most rocks, P waves travel at about one and a half times the speed of S waves. In the air or a liquid, P waves are identical to sound waves (S waves do not exist in these media since a shearing motion cannot be sustained).

The common feature of surface waves is that all motion is confined to near the surface of the Earth, the amplitude of the motion dying off rather rapidly with depth. They also are of two types, depending on the particle motion: Love waves have a shearing particle motion at right angles to the direction of propagation and in the plane of the surface, and Rayleigh waves have a slightly more complicated, backward elliptical motion with no shearing component. An important feature of their surface dependence is that the long-period (low-frequency) portions penetrate deeper

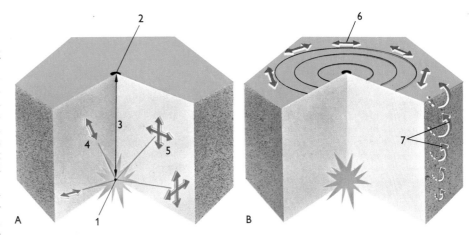

Types of seismic wave generated by an earthquake in relation to the hypocenter (1) and the point on the Earth's surface directly above the hypocenter, the epicenter (2). The distance between these two points is the focal depth (3). Body waves, shown in diagram A, consist of P waves (4) and S waves (5). In B are shown surface waves, comprising Love waves (6) and Rayleigh waves (7).

Major earthquake foci in relation to plate margins.

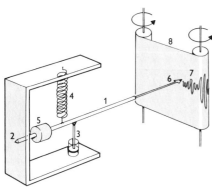

A simplified seismometer. A horizontal bar (1) is pivoted at one end (2) and suspended from a spring (4). A damping mechanism is incorporated (3). As the ground moves, the mass (5) ensures that the bar remains approximately stationary: thus movements are recorded by the pen (6) which draws a seismogram (7) on a chart moving between rotating drums (8).

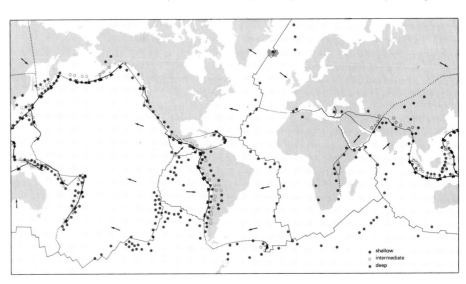

shallow
intermediate
deep

into the Earth than do the short-period (high-frequency) portions. Since seismic velocity generally increases with depth, the long-period surface waves will be traveling through a thicker layer with a higher average velocity than the short-period waves. Within a range of less than a few hundred kilometres from major events, surface waves with periods of 10–20 seconds and amplitudes of a few centimetres may be responsible for such effects as earthslides or the slow swaying of tall buildings.

Earthquake Zones. If the locations of major quakes are plotted on a map of the world, differentiating between shallow, intermediate and deep events, two things become immediately apparent. The first is that the distribution is strictly limited to a number of belts; the second that the deep earthquakes are virtually all associated with ocean trenches and island arc structures bordering the Pacific Ocean. Why, we may ask, should this be so?

The answer lies in the theory of *plate tectonics. This theory had its genesis in the theory of *continental drift as well as in studies of earthquake and volcanic action, and the effects of its general acceptance have in recent years radically altered our ideas about the planet we live on. Briefly, the theory implies that the Earth's crust consists of a number of semi-rigid plates that are in motion relative to each other. At margins where plates are meeting, one edge is forced under the other (subducted); and this is compensated for by the emergence of new material in mid-ocean in the process known as sea-floor spreading.

The belts of deep earthquakes are an indication of where oceanic plates and the lighter continental blocks are meeting: as the oceanic plate is subducted, arcs of volcanic islands, mountains and deep ocean trenches are formed. Coupled with this there is, as one might expect, deep seismic activity. In a similar way, the belts of

The remains of Montevago Cathedral after the earthquake that struck Sicily in 1968. Mild tremors began on January 14, and on the night of the 15th an event of magnitude 5.6 occurred, levelling three towns including Montevago. Hundreds lost their lives and more than 83,000 their homes.

shallow earthquakes indicate areas of active sea-floor spreading. The rather more scattered earthquakes throughout Europe, the Middle East and the Himalayas are a result of the collision of continental blocks.

There is a similar correlation between plate movements and earthquake magnitude. In general, the larger events – those with Richter magnitudes greater than 5 or 6 – do not occur in regions of plate creation but are associated only with the more violent process of plate underthrusting. A clear example of this is the zone of activity along the west coast of North America, which defines a region of lateral slip between the Pacific plate and the American plate. The San Andreas Fault system of

California is a clear expression of this; and the San Francisco earthquake of 1906, in which 500 died and a half billion dollars' worth of damage was done, a direct result.

Magnitude and Intensity. The ascribing of a magnitude to an earthquake is accomplished with reference to data obtained at a number of distant stations. The generally used scale is based on one derived by C. F. Richter to compare the magnitudes of earthquakes in the Californian region. The magnitude M is defined by

$$M = \log_{10} \frac{A}{T} + B,$$

where A and T are respectively the amplitude and the period of either the P pulse or the portion of the surface-wave packet with greatest amplitude; and B is a correction factor that takes account of the distance between the event and the recording station. The use of a logarithmic scale is for convenience, since different events release greatly differing amounts of energy: in the Alaskan earthquake ($M = 8.4$), for example, there was about 10^{13} times more energy (or 100 times more energy than in a 100-kilotonne nuclear explosion) released as in the smallest recordable earthquake.

The original Richter magnitude scale is applicable only to local earthquakes in California, the distance correction factors being based on the local structure. However, the general formula above, as applied to any region of the world, is used to calculate what is frequently referred to as a Richter magnitude. On average, only nineteen events a year have magnitudes greater than 7, and of these only one would have a magnitude greater than 8; so that, in terms of total energy released, one need consider only nineteen or so earthquakes annually.

In the immediate vicinity of the earthquake source it is sometimes found useful to determine isoseismal lines – that is, lines of equal shaking – throughout the region. Use is made of the Mercalli intensity scale which reads not unlike the Beaufort scale for wind forces. There are twelve rather arbitrarily defined levels of intensity, and so we find that Intensity III is "felt by persons at rest on upper floors and favorably placed"; Intensity VII is "Difficult to stand. Noticed by drivers of motor cars. Fall of plaster. Waves on ponds . . ."; and the final level, Intensity XII or "Catastrophic", is described: "Damage nearly total. Large rock masses displaced. Lines of sight and level distorted. Objects thrown into the air."

The intensities experienced in different places are clearly related to their distances from the source; and a pattern of isoseismal lines, which may provide information about distribution of the *faulting, can be deduced. For example, a long horizontal fracture would normally result in a sausage-shaped pattern of isoseismals; and an extremely localized rupture would produce isoseismal lines roughly following concentric circles.

The ground displacement due to the seismic waves near to the source of a destructive earthquake is of the order of a few centimetres or less. Recordings made some 4km (2.5mi) away from the surface faulting of the San Fernando earthquake (February 9 1971, $M = 6.4$) showed displacements of about 25cm (1ft). However, the structural damage to be expected depends upon the acceleration of the ground and the duration of the shaking as well as on the amplitudes of the waves. Gentle heaving of the ground is clearly less destructive than the violent motion of a wave of similar amplitude but very much higher frequency. Moreover, buildings will often withstand high accelerations for short periods but succumb to any prolonged vibration. In addition it should be noted that the horizontal motions of the seismic waves are generally more destructive than the vertical motions.

Fault-Plane Solutions. Detailed information about the present distribution of earthquakes has been a major factor in the study of plate-tectonic processes. In addition, a technique known as the determination of fault-plane solutions has played a large part in confirming the theory.

When a volume of rock fractures along a fault plane, the amount of seismic energy recorded by an observer will depend on the direction, relative to the fault plane, of his observation. Moreover, the sense of the first motion will be either compressional or, in the opposite direction, dilational.

So here we have a way of inferring the orientation of the faulting surface. Observations from stations in different parts of the world can be correlated, and from these it is fairly simple to work backwards to find the direction of the fault plane. There is only one problem: exactly the same results would be obtained for a fault at right angles to this were the sense of the *faulting to be reversed.

This is less of a problem than it might seem, since what we are in general really interested in is the direction of maximum compression, and at right angles to this, the direction of maximum tension.

The fault-plane technique was early on applied to the phenomenon, then only beginning to be accepted, of sea-floor spreading. Spreading occurs along the mid-ocean ridges. At intervals along these there are discontinuities, where sections appear to have been displaced "sideways". It had been suggested that a feature called a transform fault lay between the end of one portion of ridge and the beginning of the next.

If these offsets were due to simple displacement along a *fault the sense of motion would be parallel to the fault in both directions; but if new oceanic crust were being formed at the crests of the ridges and then spreading away, the sense of motion in the region of the fault would be outward in both directions from each of the sections of ridge. Fault-plane solutions were determined for earthquakes located along certain of these faults, and the results conformed perfectly with the hypothesis that new material was being created. Sea-floor spreading is one of the foundations of plate-tectonic theory, and so these studies were a substantial contribution toward our understanding of the nature of the Earth.

Earthquake Prediction and Control. Knowledge of where earthquakes are likely to occur is only part of the battle: we also want to know when; and, if possible, whether or not there is anything that can be done to modify the effects of the fracture.

In 1966 an increase in the number of small local earthquakes was observed in the vicinity of Denver, Colorado and it appeared that this increase was related to the disposal of fluid waste down a deep borehole. At roughly the same time it was shown that underground nuclear explosions at the Nevada test site were responsible for a similar increase there: the explosions were thought to act as a trigger, disturbing an existing distribution of stress that was already close to some critical value.

It wasn't long before somebody suggested that it would be a good idea to bore holes along the San Andreas Fault and pump water down them, on the principle that a number of small earthquakes now is preferable to the catastrophically large one that is bound to happen at some point in the future. The firing of a number of small explosions along the fault seemed an equally good idea, for exactly the same reasons. However, the uncertainties were – and still are – much too great, and the consequences of error too serious, for an exercise of this type to be carried out.

For earthquake prediction to be possible, we need some measurable and unambiguous phenomenon that precedes the fracture. Fortunately there is one. Large earthquakes are often heralded by smaller events, known as foreshocks. These seem to be the first signs of a major stress redistribution, which they in fact trigger. (It is equally logical to expect the period before a major quake to be seismically quiet while energy is steadily accumulated, and this "ominous silence" has also been observed in some cases. It too can serve as an effective warning.)

Foreshocks seem to be due to a phenomenon known as dilatancy hardening. Anyone who has walked across wet sand at the seaside will recall how a dry patch appears round each footprint. The sand grains, which had been closely packed together by the sea, have been disturbed by the pressure of the foot. Space between the grains has increased, and for this reason the intergranular water pressure has dropped, resulting in the apparent dryness.

The same sort of process occurs in rocks, and it is called dilatancy. Other studies have shown that decreasing the pore pressure increases the resistance of rock to fracturing, and this increase in strength is, logically enough, termed dilatancy hardening.

It is believed that, as stress build-up approaches a critical value, cracks begin to open and thus the pore pressure drops. This

Earthquake damage in California, the result of movement along the San Andreas Fault. The most recent large-scale movement, reflected in the devastating San Francisco earthquake, was in 1906 when the land to the west of the fault moved some 6.5m (21ft) to the northwest.

Along shorelines where there are rapidly narrowing U- or V-shaped inlets the amplitudes of the resulting waves can sometimes build up to heights of 20–30m (65–100ft). The east coast of Japan has such a topography, and is surrounded by an earthquake belt which is seismically the most active in the world; and hence this area suffers particularly badly from such waves. Tsunamis can travel a very great distance: for example, the 1960 Chilean earthquake resulted in waves in Sydney Harbor over 1m (3ft) in height.

Consideration of the three effects of visible ground waves, earthquake sounds and earthquake lights is hampered by their being very transitory and poorly documented phenomena. The speed of propagation of seismic waves, measured in kilometres per second, is too fast for the eye. However, it is possible that standing waves could be set up within some structure such as a valley floor, and that this interference pattern then migrates at a visible speed. Similarly, the passage of seismic energy through a corn field would be visible because each blade of corn would sway at its natural frequency.

The conversion of seismic energy in the ground to sound energy in the air is well understood: it would normally give rise to a deep rumbling sound near to the lower limit of our aural range. However, there are still problems in explaining all the reported earthquake sounds; and it must be borne in mind that it is difficult, if not impossible, to separate the sounds produced directly by the earthquake from sounds produced merely by moving structures.

Earthquake lights are even more puzzling. Many reports may arise from hallucination, others through electric arcing when, for instance, a power cable falls down. But there are a sufficient number of sightings for scientists to take seriously the possibility that electrical potentials are set up within the rocks as a result of the stress changes. Some Earth materials are known to possess this type of physical property. The problem is to understand how potentials sufficiently large to cause arcing at the surface can be produced.

Some observers claim to have seen aurorae (polar lights) at the time of earthquakes, and certain of the lighting phenomena may be attributable to this effect. Recently some rather inconclusive evidence has been presented that does suggest a possible causal connection between aurorae and earthquakes. Briefly, the evidence rests on observed correlations between properties of the Earth's upper atmosphere, magnetic field and rotation. By a mechanism that is not yet fully understood, the Earth's magnetic field appears to influence the mean altitude contours of tropospheric pressure in high latitudes. The main source of

results in dilatancy hardening and so the earthquake proper is delayed until water has had time to percolate into the region of low pressure from the surrounding rock. This theory has been tested both by observation of changes in the seismic waves received and by electrical testing of the wetness of local rocks. The results of the tests bear out the theory admirably.

Despite successes of this kind, earthquake prediction is still a long way from becoming a practical reality. Recent researches have shown that precursory phenomena are often very dependent on the orientation of the cracks and the shape of the dilatant region, and for these and other reasons they might go unnoticed. The converse is also true: a region can show all the symptoms of preparation for a quake – and then no quake occurs! As if to add to the present infeasibility of accurate prediction, some workers suggest, and for very good reasons, that the dilatancy-hardening model may be invalid. Nevertheless, we can say with confidence that within a comparatively short space of time geologists will be able to accurately predict and ameliorate the effects of earthquakes.

Associated Phenomena. A variety of phenomena are associated with earthquakes. Effects such as avalanches, earth slumps or the movement of water are relatively easy to explain; but reported occurrences of visible ground waves, sounds or lights in the sky are extremely hard to evaluate. Part of the trouble is that, hardly surprisingly, psychological and physiological effects may well influence observers. However, it is generally agreed that this is at best only a partial explanation for such reports.

Abnormally high-amplitude waves in enclosed stretches of water such as lakes, rivers or narrow channels are known as seiches. When the seismic waves happen to excite the natural frequency of the water in the enclosure, a standing wave results, and this can happen well over a hundred and fifty kilometres away from the epicenter (position on the Earth's surface directly above the hypocenter) of the earthquake due to the long-period surface waves.

Tsunamis or tidal waves, associated with earthquakes, probably come about through a sudden block movement or earthslide on the sea bed displacing a large body of water.

small changes in the length of the day (rate of the Earth's rotation) is believed to be zonal wind circulation patterns, which in turn are influenced by upper atmospheric weather. There is some evidence of a correlation between changes in the rates of rotation and earthquake activity, which is understandable on consideration of the large amounts of elastic energy stored in the Earth as a result of its rotation. A small change in this rate of rotation could trigger an otherwise delicately balanced set of stresses.

Finally, fluctuations in the Earth's magnetic field due to current flows in the ionosphere, at least partly a consequence of sunspot activity, are manifested in the form of aurorae. And it has been claimed that there is a detectable similarity between the periodicity of sunspot activity and earthquake activity. Although this may turn out to be so much geofantasy, the correlations are tantilizing enough to stimulate considerable current research interest.

Moonquakes. Earth is not alone in experiencing quakes. Seismometers placed on the surface of the *Moon during the Apollo missions have detected hundreds of seismic signals believed to be due to "moonquakes". They are all of fairly low magnitude – indeed, earthquakes of the same size would probably go unnoticed, even by people close to the focus.

There is no evidence of any currently active tectonic processes on the Moon similar to those operating on Earth. However, a strong correlation exists between the times of closest Earth/Moon approach and moonquake activity, and it would therefore seem likely that tidal forces act as a trigger to the release of strain within the Moon: the origin of the strain itself is not known.

This correlation suggests an interesting question: does the presence of the Moon influence the pattern of quakes on the Earth? So far there is no evidence of such an effect, but it will probably be some time before a definite answer can be given. MHW

Folding

Though it is easy to recognize a fold in a rock, it is far less easy to define exactly what is meant by the term "fold". A fold is a curved or flexed arrangement of a set of originally parallel surfaces. These surfaces usually define beds (or layers of rock of specific composition) which were originally

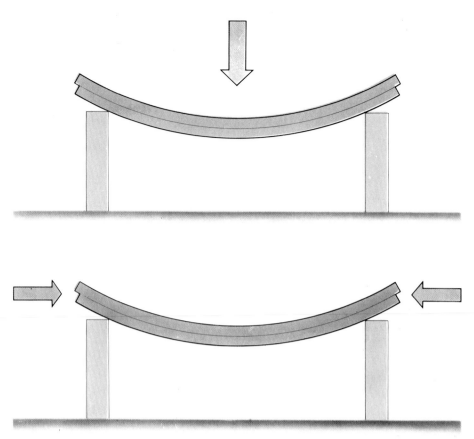

not only parallel but in the form of flat sheets, frequently horizontal. Folds range in size from the microscopic to structures which extend for many tens of kilometres.

From the mechanical viewpoint, folds may result from bending or buckling. Bending is the term used when flexure is induced by pressures applied at high angles to a pile of sheets, whereas buckling occurs when flexure is caused by compression acting inward along the original direction of layering.

Flexures which are the result of bending are known as *drape folds*. Large scale examples of such structures are particularly well developed in the Basin and Range Province of the western USA, where sediments have flexed in response to the vertical component of movements along faults developed in the basement rock beneath. Gentle flexures of considerable lateral extent may develop as the result of differential compaction, as for example where a relatively rigid block of *limestone grades laterally into *mudstone which originally possessed a high porosity (i.e., a high proportion of void spaces). As compaction takes place, the void spaces in

Flexure resulting from bending (*above*) and buckling (*below*).

the mudstone close and the depth of the mudstone layer decreases. Overlying sediments are then flexed as one would expect. In favorable circumstances, drape folds form important traps for hydrocarbons such as *petroleum.

The folds which have received most study are those which have developed in *mountain chains such as the Alps and Rocky Mountains, and the many fossil mountain chains which occurred as the result of earlier orogenies. The dimensions of folds which occur in such chains are extremely varied, their wavelength (i.e., the distance between adjacent peaks or troughs) being largely controlled by the thickness and strength of the individual rock units in

Fold terminology: (1) upright symmetrical; (2) asymmetrical; (3) overturned; and (4) recumbent. A series of folds such as this often culminates with a fold (5) cut by a thrust plane (6) to form a nappe. Also shown is an axial trace, or axial plane (7), the wavelength of one of the folds (8) and the fold axes (9).

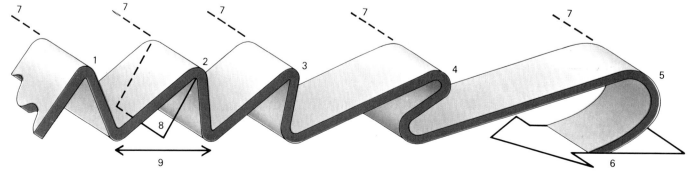

Medium- and high-grade *metamorphic rocks often exhibit complex geometrical forms that are the result of *multiple folding*. Indeed, the unravelling of these complex structures which show the imprint of two, three or even four separate and consecutive phases of folding has absorbed the interest of many geologists for the last two decades. The interference patterns which develop by the folding of folds are many and varied.

One often observes folds with smaller folds "upon their backs". These minor structures were at one time termed drag-folds, but are demonstrably not related to drag and are better termed *parasitic folds*. There are many other forms of minor structures: *pinch and swell* and *boudinage* (so named because it resembles a series of *boudins*, or sausages, resting on a slab) are the result of the extension of beds. These structures occur when relatively strong rocks are distorted, disrupted and spread by the flow of adjacent, weaker rocks. In the environment of high temperature and pressure in which such flow will readily occur, one frequently finds that recrystallization of the weaker rocks results in the development of slaty cleavage, typically aligned parallel to, or gently fanned about, the axial plane of the fold (that is, the plane between the crest and the core of the fold running through the crest of each of the folded beds).

The lateral compression which gave rise to the majority of fold structures in mountain chains developed either as a primary feature of the mechanisms giving rise to *plate tectonics, or as a secondary feature induced as a result of gravity tectonics, the effects of the force of gravity. With regard to the latter, deep-seated disturbances in the lower crust or mantle can result in variations in surface relief. Under suitable conditions, large slabs of rock can glide down gently inclined slopes under the action of gravity – hence the term gravity tectonics (see *faulting). While the sediments slide down the slope they remain undeformed. However, at the front end of the block, where resistance to gliding is first encountered, folds and/or faults develop as a result of induced lateral compression.

Model experiments indicate that, contrary to certain theoretical predictions, such folds form serially, one after the other. These model folds, produced by simple horizontal compression, show a similarity to natural folds which develop in the Jura. Identical model folds can be produced by the gliding mechanism of gravity tectonics. It is exactly this similarity in fold form produced by the two basic types of tectonics which has resulted in the long-lasting controversy regarding the relative importance of primary and secondary lateral compression in producing the fold structures in mountain chains. NJP

Folds may be on the very large or the very small scale – the former case being exemplified by this photograph of the McDonnelly Range of fold mountains in Australia. Note how the folds die away in the direction of their fold axes (that is, towards the front of the picture).

which the structures developed. For example, massive limestones of the type seen in the Jura Mountains develop structures with a wavelength of 1 to 2 kilometres (0.62–1.25mi). *Sandstones and *shales with average bed thickness of less than a metre form the smaller, but still impressive, *box folds*, which have a more or less rectangular cross-section, while thin laminations of sediment give rise to the development of small-scale *kink folds*. This trend is continued into the microscopic level where one can find individual minerals, such as *mica, which contain folds.

Folds are frequently observed and drawn in profile, or section, and it is often tacitly assumed – or even stated – that the fold form can be projected in the direction at right angles to the profile (i.e., along the direction of the fold axis) for long distances. Such folds are termed cylindroidal. In real examples, where folds can be traced in three dimensions, it is found that the structures die out along their fold axis and, where suitably exposed, form *whalebacks*. One of the apparent difficulties in interpreting fold development in terms of cross-sections of a landscape is that the beds in portions of the section may be deformed into folds, yet elsewhere in the section be unfolded or only gently flexured. This relationship is readily understood when one realises that folds not only die out along the direction of the fold axis, but also upward and downward.

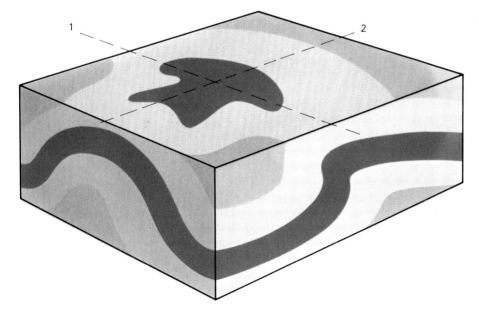

The mushroom interference fold, which develops on a flat outcrop when an initial fold with fold axis (1) is subsequently folded by a second event with axis (2) at right angles to that of the first fold.

A recumbent fold in Carboniferous rocks from Dyfed, UK. In this photograph it can be clearly seen how the degree of flexure dies out both above and below: in the central strata the contortion has been so severe that faulting has resulted. Much of the upper strata has been eroded away.

Dikes are tabular igneous intrusions oriented in such a way as to cut across the bedding planes: they are generally vertical or near-vertical. These dikes, situated near Ullapool, northwestern Scotland, are of quartz pegmatites in gneisses. Similar intrusions, but parallel to the bedding planes, are termed sills.

Igneous Intrusion

Magma which does not erupt as lava or ash from a volcano gradually solidifies below the surface to form an igneous intrusion, or pluton. The rocks so produced are usually revealed only after lengthy *erosion of their cover, but occasionally a new pulse of magma on its way to the surface breaks off fragments that emerge as xenoliths included in lava flows or ash falls.

There are many forms of igneous intrusions, and their classification is generally according to size, shape and their relationship with the country rock around them. A *concordant* intrusion is one whose shape corresponds to the structures of the country rock (for example, following the bedding of strata) whereas a *discordant* intrusion cuts across pre-existing structures. When its thickness is small compared with its other dimensions an intrusion is described as *tabular*; all other types are described as *massive*.

Concordant Intrusions. The commonest concordant tabular intrusion is the sill. Although usually fairly flat-lying, like the filling of a sandwich, sills may be vertical, inclined or undulating depending on the structure, which they follow, of the country rock. Fine examples of sills are the Carboniferous *dolerite sill that forms Salisbury Crags in Edinburgh, Scotland, and the Palisades sill, up to 350m (1000ft) thick, along the west bank of the Hudson River near New York. The Palisades gets its name from the prominent prismatic cooling joints – a common characteristic of large sills – which resemble vertical pillars.

Phacoliths are curved sill-like intrusions emplaced in the crests and troughs of folded rocks. Much larger in size are lopoliths and laccoliths. The lopolith is shaped like a saucer, concave upward: the Duluth Intrusion at the western end of Lake Superior, USA, 250km (150mi) across and 15km (9mi) thick, is an impressive example. A laccolith is a massive blister-shaped concordant intrusion which may arch the overlying rocks upward.

Discordant Intrusions. The best-known type of discordant tabular intrusion is the dike, a wall-like body which has forced its way, perhaps *via* existing fractures, through the country rock. Most dikes are only a few metres wide and a few hundred metres long, but the spectacular Great Dike of Rhodesia measures 530km (330mi) by 5.5km (3.5mi). Dikes often occur in large groups known as *dike swarms*.

Ring dikes are somewhat larger and shaped like hollow cylinders. They are found usually below *volcanoes which have suffered caldera collapse. Often associated with ring dikes is another type of tabular discordant intrusion, the cone-sheet, which is shaped like a hollow cone, apex downward.

Massive discordant intrusions range upward in size from pipes, the filled-in conduits of volcanoes, through plugs, a larger version of the same, bosses and stocks,

Dike swarms intruded during the Tertiary in northwestern Scotland and Northern Ireland. Swarms are composed of dikes radiating from centers of igneous intrusion as shown diagrammatically in B. Centers of such intrusion are Skye (1), Rhum (2), Ardnamurchan (3), Mull (4), Arran (5), Carlingford (6), Mourne (7) and Slieve Gullion (8).

which are intrusions having roughly vertical sides and subcircular outcrops, to batholiths. These are plutons with a surface outcrop greater than about 100km² (40mi²): the largest exceed a quarter of a million square kilometres (100,000mi²) and form the cores of mountain ranges like the Rockies. Batholiths are multiple intrusions, comprising many separately intruded plutons whose emplacement may have occurred over periods of millions of years. They are complex masses of many different rock types and are generally emplaced above subduction zones (see *plate tectonics). The upper surface is generally irregular, with upwardly projecting stocks and dikes that may be the only surface clue to the much larger body at depth.

Emplacement. Intrusions are emplaced in various ways: for example, by making their way along preexisting faults or fracture zones produced by the upward pressure of the magma; or by *stoping*, a process of gradual movement involving the dislodging of the blocks of country rock and their incorporation – perhaps even assimilation – into the magma. Some intrusions, termed diapirs, appear to have burst through overlying rock like rising bubbles, coming to rest at higher levels.

The upward movement of magma occurs for several reasons. Being a liquid it has a tendency to rise through solid (and therefore probably more dense) country rock, the downward pressure of whose weight also forces the magma into any actual or potential zone of weakness. Magma is also hot, and thus has the ability to melt its way through any material that has a lower melting point: this is limited by the fact that melting of solid rock involves progressive loss of heat energy by the magma, and so a lowering of the magma's temperature.

Mineral Formation. Since it is well insulated by overlying solid rock, a magmatic intrusion cools much more slowly than a lava flow. Although there may be some rapid heat loss at the edges, leading to the formation of a fine-grained or even glassy chilled margin, the rest of the magma usually crystallizes only very slowly within this envelope. Settling of crystals of substances denser than the rest of the magma often occurs, so that layers rich in different minerals are formed, the whole mass then being termed a layered intrusion. During such a process the "volatile" components, such as water and certain elements not incorporated in the minerals common earlier, concentrate as fluids which eventually crystallize either within the intrusion or as injections into the adjacent rock, where they form pegmatites and veins. Pegmatites have very large crystals, usually of only *quartz

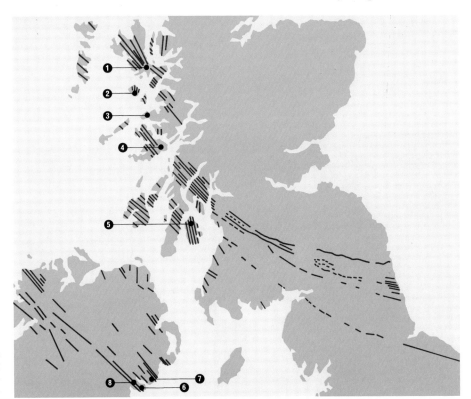

A

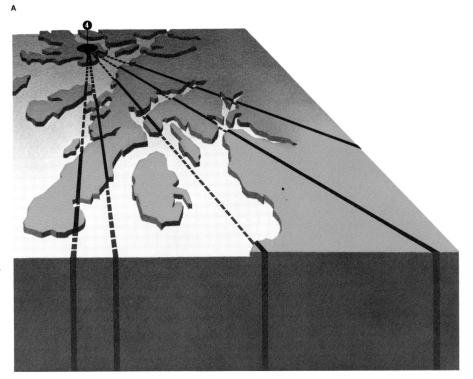

B

and *feldspar, but they sometimes contain rare, economically valuable minerals; while veins may contain also ore minerals, and often show a concentrically zoned structure suggesting several periods of infilling. (See *ore deposits.)

Many geologists, however, regard mineral veins as having been formed by fluids concentrated from the country rock by the heat of an intrusion, rather than as some direct extension of the intrusion itself. Hot, chemically active fluids may diffuse outward from an intrusion to change the

chemical composition of the surrounding rocks, forming new minerals: this process is termed metasomatism. The heat from an intrusion is certainly sufficient to at least harden adjacent rocks, often to recrystallize them and form new minerals in a zone around the intrusion known as a contact metamorphic aureole. Occasionally the heat is great enough to cause partial melting of the country rock – in effect, to produce new magma.

This heat may be of great importance in the future. In many parts of the world solid,

Unusual use made of a volcanic plug in Le Puy, France. Volcanic plugs arise from the erosion of extinct volcanoes, removing the cone but leaving much of the magma that has solidified in the feed channel. They are often termed "puys".

but still extremely hot, intrusions lie sufficiently close to the surface to be reached by drilling. Injecting these with water to produce steam would enable us to harness immensely powerful and long-lived hydrothermal engines, so adding a new energy source for our use at a time when our reserves of gas, oil and coal are dwindling rapidly. JDB

Vulcanicity

A volcanic eruption is perhaps the world's most spectacular way of releasing energy: Krakatoa erupted in 1883 with a force equivalent to the detonation of one thousand million tonnes of TNT; while the relatively unremarkable eruption of Taal in Luzon, Philippine Islands, in 1965, released an amount of energy equivalent to burning nine million tonnes of coal. In both cases violent explosions occurred, but a major fraction of the energy expended was in the form of heat: in the Taal eruption the amount of heat energy was eight times greater than the kinetic energy of the explosions.

The Earth is losing heat continuously over its whole surface, but in certain zones this heat flow is concentrated to critical levels. It is in these zones, principally at the

Forms of igneous intrusion: (1) laccolith; (2) sill; (3) batholith; (4) stock; (5) boss; (6) dike; (7) lopolith; (8) phacolith.

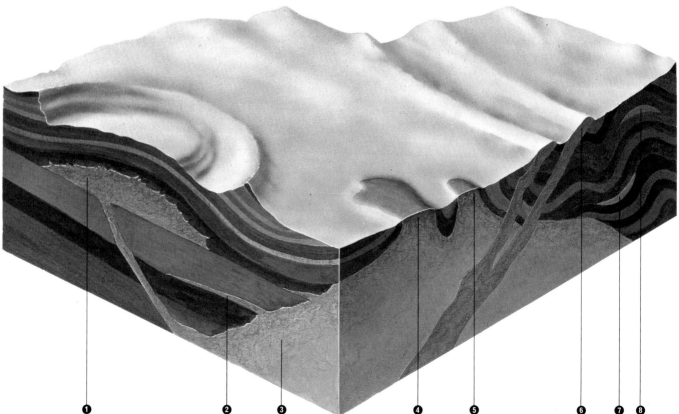

1 2 3 4 5 6 7 8

margins of tectonic plates (see *plate tectonics), but occasionally within them, that volcanic activity occurs – accompanied by, and closely related to, *earthquake activity.

Generally speaking, volcanic activity is less violently expressed when it occurs within oceanic plates – for example, Hawaii – or at constructive oceanic plate margins, when associated earthquakes are restricted to depths of about 65km (40mi) or less. The 1883 eruption of Krakatoa is an extreme example of the explosive nature characterizing volcanoes located along destructive plate margins, or within continental plates, where earthquakes may occur at any depth down to a maximum of about 725km (450mi). The generally violent release of energy in these areas does not necessarily mean that a greater *total* release of energy occurs there. It may be true that most (in fact, 83%) of the world's known active volcanoes are found along destructive plate margins, but a much greater volume of volcanic material is produced, largely un-

seen, at oceanic constructive plate margins, leading to the formation of entire *ocean floors. Oceanic volcanoes may, of course, have spectacular eruptions, producing impressive clouds of ash and steam – as in the case of Surtsey, Iceland, in 1963 – but this is the result of relatively small volumes of lava exploding when erupted into shallow sea water.

The primary energy source of a volcano, then, is heat contained within magma which has concentrated in sufficient volume to be able to move upward through solid rock. Earthquakes detected some 55km (35mi) below Kilauea volcano in Hawaii are thought to result from this movement of magma batches, although the actual formation of the magmatic liquid before concentration could have occurred, undetected, at greater depths. Subsequent earthquakes at successively shallower depths below Kilauea permit "seismic tracking" of magma on its journey through volcanic conduits until its eventual eruption at the summit caldera and/or along the fracture zones traversing the flanks of the volcano. This now well-observed phenom-

An eruption of Vestmannaeyjar in Iceland. Iceland lies on the Mid-Atlantic Ridge and so straddles a center of sea-floor spreading: this eruption is a consequence of that position.

Spatter cones of Mount Etna, Sicily, the highest volcano still active in Europe: its height varies, with an average of about 3280m (10,750ft). Its lower slopes are fertile and therefore much cultivated, and in places densely populated. The name comes from the Greek *aithō*, "I burn".

Principal regions of volcanic activity. Note the close correlation between vulcanicity and plate margins.

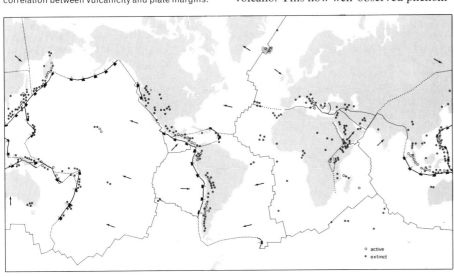

○ active
• extinct

Heimay, Iceland: these houses have collapsed under the weight of recently erupted ash. Volcanic activity takes a high annual toll of property.

Explosive volcanoes disperse their solid material much more widely than do the wholly lava-producing types. Whereas the longest known lava flows, basaltic effusions forming the Columbia River *plateau lavas in the western USA, extend some 200km (125mi), ash from the 1883 eruption of Krakatoa was propelled into the atmosphere round the world, with measurable falls 2500km (1500mi) away in Australia.

Heat may be the primary energy source of volcanism, but gas is the main propellant force behind the high-level eruptive mechanisms of volcanoes. Yet our ideas of the precise compositions and origin of volcanic gas are curiously uncertain. A great deal of water is given off as steam in most eruptions – 15,000 tonnes a day by Paricutin Volcano, Mexico, in 1945 – but how much of this is groundwater sucked into the volcanic system and how much came from the mantle, dissolved in the magma, is not yet known for sure. The same uncertainty applies to other constituents of volcanic gas, such as carbon dioxide, sulfur dioxide and chlorine, since the elements involved in these could have been derived from crustal rocks as well as being present to some extent in the original magma.

We live in a period when volcanic activity is only moderately intense compared with the great outbursts of early *Tertiary or *Ordovician times. Although its intensity has varied through geological time, volcanic activity has been a major force in the evolution of the Earth since its earliest days. Indeed, we owe our very existence to it since largely by its agency the continents on which we live were constructed and the atmosphere we breathe was first generated; and there is evidence that suggests that vulcanicity may have played a major role in the very *origin of life.   JDB

enon renders Kilauea the only volcano whose eruptive activity can be reliably predicted.

Shallow-focus earthquakes below volcanoes at destructive plate margins may well relate to similar magmatic uprise; but deeper-focus earthquakes, while still possibly recording movement of magma, include some shocks caused by the release of tectonic stress as slabs of lithosphere force their way deep into the mantle. Even then, this form of energy release may contribute to magma formation and hence to volcanic activity. Seismic and volcanic activity thus seem to be inextricably linked.

A volcano is rather like a living thing in some ways. The summit regions of Kilauea and Etna, for example, tilt strongly when magma rises high into the cone; and the volcanoes inflate before an eruption and deflate afterwards – almost as if they were breathing. The internal structure of a volcano contains a complex plumbing system of pipes and passageways, through which magma moves, and cisterns known as magma chambers where the magma may be stored temporarily – sometimes permanently if it solidifies as a high-level intrusion. Deeply eroded volcanoes, such as Piton des Neiges on Reunion Island in the Indian Ocean, reveal this internal substructure as a series of cross-cutting dikes and sills.

The products of volcanic activity are lava, tephra (solid particles of any size from fire ash upward), and gas. When the magma involved is very fluid, gas is easily released and eruptions are nonexplosive effusions, sometimes of great volume and extent. The frequent eruptions of *basalt by Hawaiian volcanoes typifies this style of activity. More viscous magmas impede the release of dissolved gas, which may therefore eventually achieve a vapor pressure high enough to blast out magma – and frequently part of the volcano – as clouds of tephra which may be followed by flows of viscous lava. A higher silica content in a magma increases its viscosity, and thus the andesitic (see *andesite) volcanoes characteristic of destructive plate margins are typically explosive in their activity.

A volcanic bomb from Madeira. Bombs are a variety of pyroclastic rocks, rocks that have had their origins in being thrown into the atmosphere by a volcanic eruption. A volcanic bomb is the result of a body of liquid lava having been thrown spinning through the air.

Geosynclines

A geosyncline is an elongate depression in the Earth's crust that fills with great thicknesses of sediment, beneath which the floor of the geosyncline progressively subsides. Subsequent Earth movements may result in the sediments being deformed into a fold-mountain chain (see *mountain chains).

Two types of geosyncline are recognized. A *eugeosyncline* has a great depth of sediment, of the order of 5000m (16,500ft), and contains a large amount of volcanic rocks: it would appear that the sediments were laid down in deep waters. In a *miogeosyncline* there is a far smaller depth of sediment, of the order of 2000m (6500ft), and an almost complete absence of volcanic rocks: the sediments of a miogeosyncline would appear to have been laid down in comparatively shallow waters.

Associated with geosynclines are geanticlines, uplifted areas which supply the sediments for the infilling of the geosynclines: some geanticlines may represent early stages in the folding of the sediments.   AI

Landscapes

Mountain Chains

The Earth has two great systems of mountain chains in which large areas exceed 2000m (6500ft) in altitude – and the two meet in Indochina.

One is a mainly latitudinal system, oriented west-east and comprising the ranges of the Mediterranean lands (such as the Alps and the Taurus range), of Iran, and the Himalayas. In the east this system branches, thereby enclosing a massive area of high *plateaux culminating in Tibet. The highest elevations on Earth are recorded in the Himalayas (Mount Everest, 8848m (29,028ft)), and much of the Tibetan Plateau lies at around 5000–6000m 16,250–19,500ft). One notable characteristic of this system is that volcanoes, though not absent, are rare.

The second great system of mountain chains is far more longitudinal in pattern and surrounds the Pacific Ocean. It comprises the great western cordillera of the Americas, and associated high plateaux, as in Colorado and Central Mexico, which can be regarded as the counterparts of the Tibetan Plateau. In Colombia and Venezuela there is a forking which gives the discontinuous island arc of the West Indies as one branch; and the chains of Indonesia as well as the island arcs of Formosa, Japan and the Soviet Far East, constitute a more or less symmetrical system that is the other branch. Some of these island arcs may be 2000km (1250mi) or more in length, and they are composed of volcanic materials. Indeed this predominantly longitudinal system is characterized by vigorous volcanic and seismic activity.

Most of these great mountain systems are located on the edges of the main continental areas. This is especially clear in the Americas, but the other continents show it to a considerable degree. Thus in Africa, for example, the Drakensberg, the Guinea highlands, and the Atlas Mountains occur on the perimeter.

Both island arcs and mountain chains tend to lie in close proximity to oceanic trenches. The reasons for this will shortly become clear as we describe their origins in terms of the theory of *plate tectonics.

Composition. The sediments of which these mountain chains are composed show very considerable *folding and overthrusting. The sedimentary rocks characteristically show signs of having been formed rapidly, in deep water, and typically consist of "flysch" (sandstone-shale alternations).

Kang Tega in the Himalayas (the name Himalaya is from Sanskrit and means "abode of snow"). From east to west the range extends some 2500km (1550mi), from north to south 80–155km (125–250mi): the total area of the range is about 595,000km² (230,000mi²).

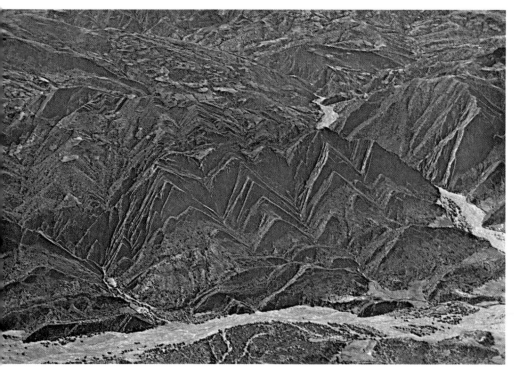

The Andes in Bolivia. The range, which extends from north to south about 8900km (5500mi), contains many of the highest peaks in the world including the highest mountain in the western hemisphere, Aconcagua at 6960m (22,834ft), and a number of active volcanoes, such as Tupungato (6800m; 2075ft).

ance in chains formed in this way.

However, some mountain areas are caused by *faulting, pure and simple. A fine example is the Sierra Nevada of the USA. Most mountains formed in this way are less conspicuous than fold mountains – in fact, they are frequently associated with them in a secondary manner, having formed after a prior episode of strong folding.

Ancient and Modern Mountain Chains. In addition to the present great mountain chains, there is also evidence in some areas of ancient orogenic episodes. The areas in question have now been eroded into hills and plains that are no more than mere stubs of former great ranges. In the eastern USA, for example, the contorted rocks of the Appalachians represent the beveled roots of an ancient range. The same applies to the low Aravalli Range of western India, with its greatly denuded schists, gneisses and granites.

What are now high mountain chains are, by contrast, relatively young. They result from orogenies during the Miocene, Pliocene and Pleistocene epochs – the last 25 million years or so. Indeed, uplift appears to be progressing with some speed at this very moment. In California, movement operates at a rate of up to 13m (42ft) per thousand years, and in New Zealand rates of 11–12m (36–39ft) per thousand years are known. Given rates of this order, a mountain range with a height of 8000m (26,000ft), like the Himalayas, could develop in less than 800,000 years – though rapid rates of *erosion operative in such high-relief situations would extend the time period required. Nevertheless, it is clear, given evidence of rates of uplift and erosion, that substantial mountain chains can, in terms of the great duration of the Earth's history, be very youthful.

Processes of Mountain Sculpture. Although tectonic events are the primary factor affecting mountain development and form, erosional processes are also of great importance. In the global context, denudation generally proceeds at its highest rate in high-relief areas. The Greater Himalayas and Karakoram are associated with some of the highest recorded rates of large-scale water denudation. The values exceed 1000mm(3.25ft)/1000 years for the Upper Indus and Kosi; and information from Alaskan mountains and the European Alps indicates rates in some cases exceeding 600mm(2ft)/1000 years. For comparison, rates for most lowland rivers are in the range 10–100mm(0.4–4in)/1000 years.

These high rates are produced not only by the large amounts of discharge and sediment carried by the mountain rivers. There are also major mass movements created by a combination of high relief, heavy snow- and rainfall, frequent frosts

There may also be intercalations of submarine pillow lavas or welded tuffs indicative of volcanic islands. All fold mountain ranges develop primarily from exceptionally thick piles of sedimentary strata, commonly 15,000m (48,750ft) or more in thickness. The strata must, therefore, have been developed in a subsiding area, what we sometimes call a *geosyncline, on the continental margin.

The sediments have frequently been metamorphosed by heat and pressure to form *schists. In some areas, such as Japan, there are two parallel belts of metamorphism. One belt, always nearer the ocean than is the other, consists of the so-called "blue schists", containing minerals indicative of formation at high pressure but relatively low temperature. Conversely, the other belt possesses granite and metamorphosed sediments containing minerals indicative of low pressure and high temperature. The blue schist probably formed as a result of the high pressures and low temperatures associated with the ocean trenches, while the inner metamorphic and granitic unit represents uplifted island arcs.

The mountain belts also possess masses of basic or ultrabasic rocks occurring as huge thrust slices or slivers. Their composition and structure strongly suggest oceanic crust or upper mantle which has been sheared from downgoing plates and forced upward under compression into the overlying rock. They probably mark the line of joining of continents which have collided as a result of sea-floor spreading.

Origin. The nature of the sediments and the proximity of mountain chains to continental edges and oceanic trenches clearly indicate that plate movements play a fundamental role in mountain-chain development. Active mountain building (orogeny) occurs where a block of continental crust is

carried into collision with another crustal plate.

Most active chains lie on continental plates that are in collision with oceanic crust. For some reason, still imperfectly understood, the lithosphere eventually breaks beneath a geosyncline and starts to plunge into the mantle. The downward-sliding plate of lithosphere butts into the deep water strata, crumples them and rams them against the edge of the continent. The prism of sediment from the geosyncline is compressed into folds and overthrusts. Intensive metamorphism gives schists.

Strata near the bottom of the geosyncline can also be dragged downward by the moving plate. They become heated and the deepest strata may begin to melt. Granitic magma formed by the melting then rises and pushes overlying strata upwards. The melting and stresses associated with the downgoing plate cause the volcanism that produces island arcs in the form of the great Pacific "ring of fire". Their development may be due to the continental plate buckling immediately behind the line of collision where the ocean plate slides beneath it: as the continental plate is lifted and stretched it cracks open and magma bursts through the cracks.

Less common are mountains produced by the collision of two continental plates. When two continents collide they crumple along the colliding edges. However, unlike the situation when a continental plate hits an oceanic plate, a moving continent is too big and its rocks too light for it to be carried down into the mantle with the descending plate of lithosphere. Like a slab of cork the continent bobs up again. This causes marked upward movement, resulting in extremely high mountains. It is this model that explains the development of the Himalayas. Volcanic activity is of little signific-

Part of the Alaska Range, the northeastern end of a chain that includes also the Aleutian Range and the Aleutian Islands. The chain contains nearly eighty volcanoes, many of which are active, and represents the crest of a midocean ridge. The range includes Mount McKinley, the highest peak in North America (6195m; 20,320ft).

massive scale of the whole Earth, mountain chains are mere wrinkles, but their importance lies not only in the contribution they make to our understanding of Earth history and development, but also in the effect they have on the everyday lives of highlanders and lowlanders alike.   AG

River Valleys

Like the veins of a leaf or the branches of a tree, river valleys are means by which water moves through a system. The system drained by a particular valley is called the drainage basin or catchment, and the boundary between it and an adjacent valley is called a watershed or drainage divide. (In the USA the drainage basin is sometimes called the watershed, and so the watershed is occasionally referred to as the water parting.)

Whether they are large or small, drainage basins have similar characteristics and comparable patterns. The individual valleys form a branching system, and in general this will be of roughly the same design as are all other such basins. One example of this can be found by ranking streams according to "order". In this system, fingertip tributaries are described as first-order; when two first-order streams combine the result is a second-order stream. Two second-orders give a third-order, and so on. In any particular basin, the number of streams in each order decreases in a regular way as the order increases, much as one would expect. Similar relationships exist between a stream's order and such characteristics as its length, area and gradient; and these relationships hold true for almost all streams.

Although these topological relationships tend to be constant, other aspects of drainage-basin form vary from basin to basin. Thus dissection or *drainage density* (the number of streams per unit area) depends upon both the climate and the physical characteristics of the area: climate exerts a direct influence since it determines the amount of water that the system has to carry, as well as an indirect one by its effects on local vegetation; and rock and soil types are important since they determine the resistance of the surface to ⋆erosion.

The Development of a Drainage System. It is very unlikely that the surface of an area of land that has just emerged from the sea, and whose structure is complicated by harder and softer beds and by ⋆faults, will be smooth. So one would expect that in the early stages streams would quite naturally follow the initial irregularities. We call such streams *consequents*. Gradually, however, structure and rock character become more important, and rivers are guided by zones of weakness in the rock. Such streams are termed *subsequents*. Another class of streams, those that flow in a direction opposite to that of the geological dip of the beds, are called *obsequents*. Streams which appear to depend neither

and occasional earthquakes. Avalanche activity appears to be able to create and maintain high slope angles. In the Karakoram, for example, perennial slopes and ice slopes above 5100m (16,575ft) have more than 40% of their area lying at gradients between 55° and 60°.

High mountains have been shaped also by glacial erosion, which digs deep U-shaped valleys and creates knife-edged ridges (arêtes), scalloped by corries. Glaciers, unlike rivers, can sometimes "overdeepen" their valleys, creating basins which, after glacial retreat, may give deep lakes like those of the Swiss Alps.

Importance. Mountain chains have a considerable impact on other parts of the environment – and on Man. Their size and location with respect to surface and upper winds affect climate: for instance, the western mountain barrier in North America is responsible for a vast extension of semi-arid climates in the west-central part of the continent, and appears also to cause the development of great waves in the upper westerly circulation of the whole northern

hemisphere, thereby influencing the location of the high and low pressure cells that dominate the weather of western Europe. Likewise, the Tibetan Plateau and Himalayas are instrumental in the establishment of the monsoon, and are sufficiently high to affect the trajectories of the jet streams, the high-speed westerlies at the junction of troposphere and stratosphere.

The peripheral location of the ranges with respect to land masses tends to hinder the drainage of the central parts of continents, and so basins of inland drainage often occur there.

They pose certain problems for Man: steep gradients, rarified air, high precipitation, avalanches, earthquakes and volcanoes. Conversely, they also offer protection, recreation, water power, timber, minerals and climatic variety over short distances. Many of the great life-giving rivers of the world, the Indus, the Ganges, the Tigris and Euphrates, the Orange and the Amazon, owe the bulk of their water to the rain and snow of the high mountains.

In terms of their size in relation to the

upon initial irregularities nor upon weaknesses in the rocks are called *insequents*.

As a river system evolves, one stream, flowing along a more easily erodible course, may capture a stream whose course happens to be more resistant. This process of river capture favors the development of subsequent streams. River capture may also be a result of the relative length of the streams: a stream following a long and circuitous course to the sea will be at a disadvantage compared to one whose course is more direct. Possible signs of river capture that can often be detected in the landscape include windgaps and elbows of capture, incision of the capturing stream below the capture, and the evident misfit nature of the beheaded stream.

Not all valley systems, however, show complete conformity with the local geological structure. Some rivers may, for example, cut across *folds. One possible explanation for this is antecedence, whereby a rapidly eroding stream manages to keep on cutting down through a fold rising only

slowly across its path. This is known as antecedent drainage. Alternatively, in superimposed drainage, the river courses may have initially developed on a cover of rocks whose structure was different to that of the rocks beneath, the upper beds having since disappeared through erosion.

But nevertheless the effects of lithology and rock structure are important. On an absolutely flat surface of relatively homogeneous rock, a drainage net would probably develop a dendritic, branched pattern, with no great linear control apparent. Where the influences of rock resistance and structure are more significant, where streams find it easier to erode channels along lines of weakness, these streams subsequently become dominant, giving rise to a trellised drainage pattern.

Valleys in Plan. When seen from above, valleys, and the channels within them, present a course which is either straight or meandering or braided. Straight valleys and channels are rare, however, both because water flow tends to be turbulent and because there are inevitably obstructions.

Most rivers show a tendency to meander. This seems to be a characteristic of fluid motion – for example, the jet streams of the atmosphere and the Gulf Stream of the North Atlantic show a similar phenomenon. It is not necessary to involve irregularities in valley materials to explain meandering – the most perfect meanders occur in homogeneous alluvium. Another factor which favors meander development is a cohesive bank material. With less cohesive material the river is less constrained by its bank so that it undergoes frequent changes in position, and branching (braiding) may develop. Indeed, braiding streams seem to

A striking example of river meanders near Golden, British Columbia. Meanders are characteristic of a river's mature stage: their net effect is a reduction in the gradient of the river and therefore of the flow velocity. This can result in a near-perfect balance between erosion and deposition.

be characterized by highly variable discharge, abundant bedload and easily erodible banks. Even meandering streams, however, show considerable movement, for they tend to migrate downstream and also to cut off sharp bends to give ox-bow lakes.

If a winding channel should for any reason cut down deeply, it may become what is known as an incised meander. A change in base-level may mean that a pattern of meandering that was developed on floodplain alluvium becomes superimposed on the underlying bedrock.

River Long Profiles. When we talk of a river's "long profile" we mean the direction, in terms of the vertical, of the flow of the water in it, and hence the direction also of its bed.

Most river long profiles are steeper near the source than they are near the sea – exceptions occur, and they include particularly certain streams in arid regions which suffer a decrease in discharge downstream. There are various reasons for this general feature. Perhaps the most important is that grain-size of riverbed material and load tends to be smaller with increasing distances downstream. Thus the gradient of a stream must be steep in its upper reaches, so that the water has a sufficient velocity to move the coarse debris which is supplied to it there. Downstream, because of attrition, material is much finer, which means that the velocity required to carry it is much lower, and so the river can reduce its slope

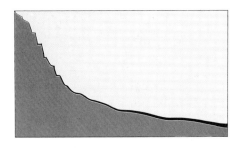

An idealized river long-profile. Close to the source are abrupt changes of gradient: with increasing distance from the source the profile smooths out and the gradient decreases until, in the river's maturity, the direction of flow is horizontal.

and still successfully carry its load.

A second reason is that discharge increases downstream. This makes the stream more efficient, for one large stream will have less bed friction than the smaller tributary streams which join together to make it up. Moreover, it will suffer less turbulence and thus less loss of energy.

However, certain stream long profiles tend to show irregularities: cataracts, waterfalls, and points of locally increased slope – kinchpoints. One cause of this may be the presence of hard bands of rock. Many waterfalls, for example, result from the river flowing on a bed of flat-lying strata which is underlain by weaker beds: the stream gradient is very low on the resistant layer, but when it breaks through the gradient can become very steep. Alternatively, breaks in the regularity of the long profile can be caused by changes in the level to which erosion proceeds – this level is often sea-level. A fall in sea-level can lead to the formation of a steep slope over which the river cascades. Headward erosion would be rapid at this point and the river would attempt to obtain a newly-graded and regular concave profile by cutting back upstream.

Valley Cross Profile. River valleys show considerable differences in their cross-sectional form. Some may be scarcely modified glacial features with a characteristic U-shape; others may be the product of great

Kinchpoints (or knick points) are most generally the result of rejuvenation – that is, the relative rise of the land with respect to sea level. Because of this fall of base level, the river downcuts rapidly towards its previous equilibrium long-profile. The kinchpoint represents the intersection of the new, adjusted long-profile with the old, maladjusted profile; and regresses upstream as adjustment continues.

and rapid vertical incision by powerful streams, possessing a prominent V-shape; while others may owe more to lateral cutting processes and possess a broad, flat bottom.

The form of some cross-profiles may be greatly influenced by differences in rock resistance. Similarly, river valleys cut into unconsolidated sands and gravels will tend to be V-shaped because the loose sandy material tends to fall from the valley sides. In silts and muds, on the other hand, sidewalls are maintained at a steep angle because of the cohesiveness of the materials.

Part of the form of the cross-profile may be depositional in origin: this particularly applies to alluvial floodplains. When a river overtops its channel banks and spreads out, its velocity is reduced and so its ability to transport materials is lessened. Thus deposition takes place, with coarse material being deposited near to the channel as levees, and finer material being carried further away and laid down as backswamp deposits. Also, as rivers meander across the floodplain they alternately scour and deposit.

The flood plain of the Paoshan (or Baoshan) River in Yunnan Province, China. Flood plains, which consist of alluvium deposited by the river, are formed by the downstream migration of meanders. As is evident from this photograph, the alluvium provides fertile soil, and many flood plains are extensively cultivated: particularly famous for its size and richness of soil is that of the Mississippi River.

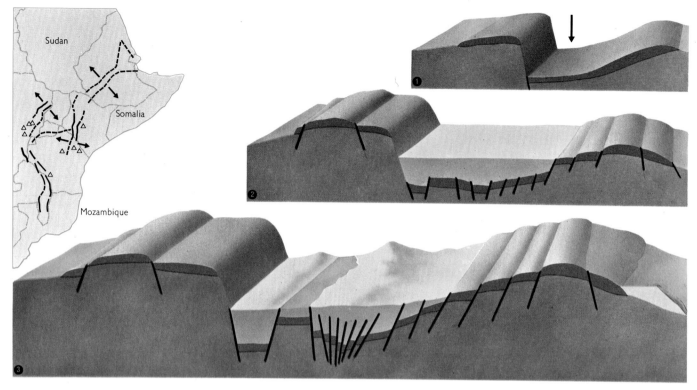

Sometimes a floodplain may be eroded into by a river so that terraces result. Following a fall in the base level of erosion such as that which results in the creation of kinchpoints, the river will tend to cease deposition and will start to cut down into its floodplain alluvium.

Again, some terraces result from a change in climate. Changing climate results in changes of discharge of streams and in the nature of the *weathering and erosion of the valley systems. The latter may well cause the amount and caliber of the load to alter. During the Ice Ages, the advent of a colder phase near an ice sheet often resulted in an increase in frost weathering, glacial deposition and mass movement so that stream load was increased in amount and caliber. Aggradation (see *deposition) took place. Subsequently, however, with a change back to more temperate conditions, incision again predominated, and this resulted in the formation of terraces.

In some rivers the situation may be more complex still. At some point in the past, perhaps because of different climatic or sea-level conditions, rivers may have cut down very low only to be choked by a return to conditions favoring deposition. In such circumstances buried channels may be identified by borehole studies.   AG

Rift Valleys

A rift valley, or graben, is a structural trough formed when an elongated and relatively narrow strip of the Earth's crust sinks between two roughly parallel faults. The valleys so formed are generally fairly straight, and may be hundreds of kilometres in length, although they are characterized by a fairly constant width, of the order of

30–60km (20–40mi), a measurement that probably relates to the local thickness of the crust. The boundary faults generally dip steeply toward the trough, forming steep-sided slopes called scarps; one or other side of the trough may die out in some places to leave a series of fault blocks. An important feature of rift valleys is that they follow the crests of long, low upwarps of the Earth's crust. Volcanic activity is locally associated with them.

The Origin of Rift Valleys. Perhaps the most closely studied rift system is the Rhinegraben, traced by the River Rhine for 300km (190mi) between Basle and Frankfurt. Its average width is 36km (22mi) and it cuts through an elongated dome 190km (120mi) wide and upwarped 2–3km (1–2mi). The floor of the rift began to subside 45 million years ago as the shoulders rose, and the total throw of the faults is some 4.4km (2.7mi).

Many boreholes through the sedimentary fill have enabled a reconstruction to be made showing that the maximum widening during this period has been only 4.5km (2.8mi). A notable feature is that the steep enclosing scarps remain parallel although the valley itself curves. Geophysical observations of the Earth's crust in the Rhinegraben area have been of great importance in attempts to understand the origin of rift valleys, still very much a subject of debate.

The situation of rift valleys at the crests of domes led early to the concept of the dropped keystone of an arch, and experiments showed that downfaulted rift valleys could be produced by upwarping domes in clay models. However, the advent of *plate tectonic theory has led to a considerable clarification.

Geological and geophysical investigation has shown that rift valleys are related to

The East African rift system: triangles on the map indicate positions of recently active volcanoes; arrows the direction of separation. A simplified history of the rift valley shows initial downfaulting in Kenya some 13 million years ago (1), the deposition of sediment in the resulting valley (2), and the complex pattern of faults that has developed over the last million years (3).

anomalous features deep in the lithosphere and upper mantle which resemble those beneath mid-ocean ridges (see *ocean floor); and the projection of continental rift lines into the ancestral oceanic rifts, as reconstructed by fitting the continents together in their past positions, has indicated a close genetic relationship between the continental and oceanic features. It was therefore proposed that the continental rift valleys were zones where active crustal separation was beginning.

However, evidence has since shown that widening, even over long periods of time, has been only minimal – as in the case of the Rhinegraben. Gravimetric, seismal and geothermal investigations of several rift zones indicate the presence about 20km (12mi) beneath the rift valleys of intrusions of what are presumed to be molten or plastic mantle rocks from the seismic low-velocity zone (see *geophysics). These intrusions constitute a heat source which is probably the prime agent of rift-valley formation: the heating of the rocks beneath the rift zone causes expansion, and this accounts both for the updoming and for the limited horizontal expansion as the rift floor subsides.

The characteristic normal rift faults and the local tension indicated by *earthquake records can thus be explained – but other factors may also be involved. Although the *Cenozoic phases of rift faulting are well documented, geological mapping in older rocks has shown that the rift faults fre-

quently coincide with ancient tectonic dislocations, extending beyond the present active rifts, for which dates far back into the *Precambrian have been posited. It has therefore been suggested that a worldwide system of lineaments originated at an early stage of the Earth's history and has periodically been reactivated. Certain segments of these lineaments were suited to become centers of sea-floor spreading, whereas others remained locked in the continents. The East African rifts, for instance, may be unable to open because the continental plate is being compressed by spreading in the Atlantic and Indian oceans.

The East African Rift Valley. The term rift valley was first introduced to designate what is now known as the Great Rift Valley, the East African rift system which forms a complicated pattern extending 4000km (2500mi) south-southwest from the junction of the Red Sea and the Gulf of Aden, in the north, to the Zambesi River, in the south.

It comprises two systems, the eastern and western. The eastern system, from north to south, consists of the Ethiopian Rift Valley and the Gregory Rift Valley, Kenya, with outlying branches. South of latitude 4°S it passes into a series of east-facing fault blocks curving around central Tanzania as far as the Rungwe volcanic massif (9°S), where it is cut across by the western system. A continuation is provided in Zambia by the low scarps of the Luangwa Rift Valley.

The western system sweeps around Uganda and Tanzania in a great curve containing several fjord-like lakes, including Lake Tanganyika. Here its direction becomes southeasterly, cutting across the eastern system before turning southward,

by way of Lake Malawi, to meet the coast near Beira.

The downthrow of the rift faults in the northern sector is 3km (2mi) at Lake Mobutu, and the neighboring horst block of Ruwenzori has been uplifted a similar distance relative to the African *plateau. The floor of Lake Tanganyika is 700m (2300ft) below sea level and a minimum downthrow of 3.5km (2.2mi) is indicated, whereas the marginal horst of Mt Kungwe has been uplifted 1000m (3250ft). It is notable that the greatest uplift and subsidence lie in the same sector. The total length of overlapping rifts in this system is 6800km (4200mi), of which the northern 1500km (950mi), in Ethiopia and Kenya, is accompanied by a profuse belt of volcanoes, mostly extinct. The other, major portion cuts through ancient Precambrian rocks with some local volcanic centers where fault systems intersect each other.   RBM

Plateaux

The term plateau is used more as a description than as a definition since it is applied to any fairly flat high-level region – a tableland or elevated tract of comparatively flat or level land – the surface of which may be uniformly level or have broad summit heights of fairly uniform elevation. There are no scientifically defined lower limits to the height above sea level – although 1000ft (about 300m) has been suggested – and, similarly, one or more of the sides may be steep, but not necessarily so.

Like the other physical attributes the geological structure of plateaux is ill-defined. The strata may be in horizontal or

A view of the East African Rift Valley near Ol Doinyo Lengai. This rift system contains active volcanoes that bear testimony to the fact that much of East Africa (Somali Republic and parts of Kenya, Ethiopia, Tanzania and Mozambique) is splitting away from the rest of the continent.

nearly horizontal beds or be gently or severely displaced, in which case the original summits will have been planed down by *erosion to form high-lying plains.

Classification. Because plateaux are elevated they are usually associated with mountain systems. In this respect three varieties are commonly distinguished:

(i) Marginal plateaux are found where the mountain folds subside into level and almost level areas on their lower flanks. Examples occur in the Appalachians, where the folds die away westward in the Cumberland and Alleghany plateaux, and in the Jura, where the Jura *plié* is distinguished from the Jura *tabulaire*.

(ii) Intermontaine plateaux, which are confined between mountain chains. These are common in mountain cordilleras, famous examples being the Alti Plano of the Andes in Ecuador and Bolivia, the Columbia plateau between the Cascade Range and Rocky Mountains in northwest USA, and the vast plateau of Tibet between the Himalayas and Kumlun mountain chains.

(iii) Steep-sided plateaux formed by the wearing down of ancient mountains into high-level tablelands ringed about, more or less completely, by escarpments. A fine example is the High Karroo in South Africa.

On the deep ocean floor there seems to be a fourth variety of plateaux, upstanding submarine tablelands which are distinct from the continental shelf and the continental masses on it.

Modes of Formation.

Plateaux may also be classified according to their genesis or mode of formation (which includes elevation to a height that precludes the possibility of the plateau being called a plain). The chief genetic categories are as follows:

(*i*) Tectonically uplifted horizontal or gently undulating strata that remain approximately level after uplift. A supreme example is the Colorado plateau in the southwest of the USA.

(*ii*) Plateaux of accumulation of volcanic outflows, especially of *basalt. These are of two main varieties: folds in mountain chains may be overwhelmed locally by outpourings of basalt and under erosion become a series of plateaux; or, secondly, successive outpourings of sheets of lava may form a large plateau, such as occurs on a vast scale in the Columbia plateau, USA, and in the northwest Deccan of India.

(*iii*) Plateaux formed by the wearing down of mountains by erosional agencies. Weathering and other common erosional agents work fastest on steep slopes and tend to lower and eat back into the mountainsides, leaving at their feet a pediment or plateau. Eventually the upstanding areas decrease in size and stand above a plateau sometimes called a peneplain though today more usually regarded as a series of pediments or a pediplain.

These plateaux cut in ancient, contorted strata are often fractured by mountain-building movements and their main fractures or faults are eroded into steep scarps. Sometimes parallel fractures form *rift valleys – as in the plateaux of East Africa and on the middle Rhine between the Vosges and Black Forest. The popular concept of a plateau – as an isolated tableland with steep sides rising above the surrounding countryside – is often due to steep fault scarps, as for example the southern edge of

The edge of Mount Roraima. This plateau, lying on the borders of Venezuela, Brazil and Guyana, is 2810m (9220ft) high and is the source of many rivers, including particularly the Orinoco and Amazon river systems. It was Mount Roraima that inspired Sir Arthur Conan Doyle's famous novel, *The Lost World* (1912).

the Scottish Highlands and the eastern edge of the Massif Central in France. If a plateau has strong fault scarps on most sides it is in Europe called a *horst*.

However, the comparative flatness of all plateaux of this type is due to erosion, mostly subaerial and in some instances marine: a platform caused by marine abrasion or wave action can be uplifted by tectonic forces to form a plateau. More commonly, a mountain system is worn down to its stumps, mainly by subaerial denudation.

Associated Features.

The most ancient plateaux – as in Africa, Brazil, the Deccan, Australia, the Canadian Shield and various

A horst (1), a plateau with steep scarps formed by faulting and later erosion. On the far side of a rift valley (2), step faults (3) lead up to another plateau (4).

shields, massifs, or tablelands in Eurasia – have remnants of former mountains still standing upon their flattened surfaces. Because of the dominance of rainwash over other erosional forces, these features are most common in warmer climates, but they survive in fragmental forms elsewhere, as for example in the granitic tors of Dartmoor.

The term *monadnock*, from the mountain of that name in New England, is used to describe an isolated mountain peak standing above a pediplain or plateau. A monadnock represents the residual mountain peak left upstanding by the reduction of its surroundings by erosion to form a plateau. Such survivals are more common in drier climates and are given special names. *Mesas* are flat-topped, steep-sided tablelands, beyond the edge of which, detached from the main plateau by erosion, may be similar but smaller flat-topped uplands called *buttes*. Combinations and examples of these landforms are common in Spain on the Meseta, itself a large plateau dominated in places by fault blocks.

On tropical tablelands where conditions are humid and arid, residual mountains rising in isolation above surrounding plateaux (pediments or pediplains) are called *inselbergs* or *bornhardts*. Inselbergs (island mountains) vary widely in shape and size and in degree of destruction, their common feature being the abrupt rise above the flat or gently-sloping pediment at their base. Some are large uplands while others are small steep-sided hill-masses: Ayers Rock, Central Australia, and numerous mountains in West Africa are examples. Many inselbergs are dome-shaped and have a granitic structure. They survive for a long time before their eventual diminution, when they may well degenerate into what are called in South Africa *kopjes* and *castle kopjes*. These fantastic collections of massive boulders are reminiscent of the smaller features known as tors in Britain.   R & DB

Karst Landscapes

The word *karst* is the German form of the Slovene word *Krs*, meaning barren stony ground, and refers to the area behind Trieste. Karst landscapes are most noteworthy for their spectacular and distinctive landforms, as is seen by their influence on the art of southern China, the largest area of karst

in the world. They develop on massive soluble rocks, usually limestones or dolomites, in which the dissolving action of water plays a major role in the origin of the landforms.

Limestones are the most commonly occurring soluble rocks on the Earth's surface and comprise about 15% of the *sedimentary rocks. The ideal conditions for karst landscapes occur when the limestones are massive, well jointed and impermeable, to allow the development of secondary permeability; when the relief is high, to permit rapid vertical drainage; and when the rainfall is heavy, to give abundant water to act as a solvent. Not all limestones give rise to karst relief.

Limestones are formed of calcium carbonate, which is only very slightly soluble in pure water; but when the water contains carbon dioxide, as does rainwater, the calcium carbonate is converted into calcium bicarbonate, which *is* soluble – this reaction is reversible, loss of carbon dioxide leading to reprecipitation of calcium carbonate. Carbon dioxide dissolved in water is therefore the most important agent in the solution of limestones: it is derived either from the atmosphere or from biological sources.

Surface solution features are common in all karst landscapes. They include rills, runnels and pits cut into the rock, up to about 2–3m (6.5–10ft) long, called *karren* or *lapies*, terms originated in the Alps where these features were first described and where they are abundant. Karren formed by water containing atmospheric carbon dioxide are sharp-edged and razor-like; those formed by water with biologically-derived carbon dioxide are rounded and smoothed.

Solution penetrates the rock along joints and cracks, and is accelerated under soil and vegetation. At the intersections of fractures, increased solution forms funnel-like hol-

A butte in the Mesa Verde National Park, Colorado, an eroded remnant of a continuation of the mesa visible in the background. Mesas are isolated plateaux formed where a capping of hard strata has resisted erosion and protected the softer strata beneath: on the small scale this same process results in the formation of earth pillars. Continued erosion of the sides of a mesa may reduce it to a butte, or may lead to the appearance of buttes adjacent to the main body of the mesa.

lows which, once formed, are self-perpetuating as they form foci for rain waters, growing to over 100m (325ft) deep and over 300m (1000ft) in diameter. Closed depressions such as these are characteristic of soil-covered karst lands in temperate climates, as in Istria and the Jura, and in Indiana, where they basically replace river valleys. They are known as dolines in Europe and as sinks (or sink-holes) in North America.

The enlargement of bedding-planes and joints by solution enables relatively large streams of water to disappear underground into swallow holes: one of the largest rivers to disappear into the ground is the Trebinjčiča in Yugoslavia, which drains over 900km^2 (350mi^2). Where streams disappear along vertical fractures, deep chasms or shafts can be formed: these are known as potholes, one of the best known being in northern England; Gaping Gill, on the side of Ingleborough, which is over 120m (390ft) deep. Rivers which cut right through a karst area, like the Tarn and the Dordogne, are allogenic – that is, they deposit sediments brought from elsewhere.

Once underground, water dissolves and erodes the limestones to form cave passages. Many passages formed by streams resemble streambeds of the surface, and contain meanders, ox-bows and river deposits: they are known as *vadose* caves, and are normally controlled in detail by the lithology and the structure of the limestones. There is much evidence that solution also takes place at

Karst landforms develop through the solution of limestone by rainwater. Limestone is composed primarily of calcium carbonate ($CaCO_3$), which does not readily dissolve in water. However, rainwater contains dissolved carbon dioxide (CO_2), part of which reacts with the water to give hydrogen and bicarbonate ions (H^+ and HCO_3^-). The presence of these ions permits the $CaCO_3$ to be converted to $CaHCO_3$, calcium bicarbonate, which, being soluble, dissolves in the unreacted water.

in the form of a submarine spring, and this is common in the Adriatic sea and off the coast of Florida.

Karst springs are among the largest springs in the world, because there is a tendency for water to collect into master conduits. There are two main types of karst spring, one where the water issues by means of free flow, the other where the water issues under forced or artesian flow: the latter type is sometimes known as a vauclusian spring after the Fontaine de Vaucluse in southern France, which has at times a discharge as high as that of the Seine at Paris. Since underground water usually has a high calcium bicarbonate content, its emergence to the outside world is frequently accompanied by tufa deposition, mainly the result of loss of carbon dioxide. Such deposition can be assisted by the presence of algae and mosses, which extract lime from the waters and on a large scale may indicate a significant climatic change.

When karst areas occur in climates with marked wet and dry seasons, as in Mediterranean Europe, the limestones may not be able to absorb all the rainfall which floods the surface in the wet season. Such flooding is intensified by the presence of less permeable beds in the limestones or of impermeable unconsolidated rocks like boulder clay. Acidic flood waters can cause planation of the limestones by solution, and this gives

depth, where underground water is under considerable pressure. Caves formed under these conditions are called *phreatic* and are controlled more by the hydrological gradient than by geology. Most large cave systems consist of vadose and phreatic elements, together with large chambers caused by collapse. Cave temperatures normally approximate to the mean annual temperatures of the surrounding area: in caves in mountainous areas where cold air can accumulate, large bodies of ice exist, as for example in the Eisriesenwelt, Austria.

Water not only flows through limestones in defined channels, but also percolates. Percolation water takes up carbon dioxide and dissolves a great deal of calcium carbonate. When percolation water emerges from a crack into a cave passage, carbon

dioxide is lost to the atmosphere and there may also be evaporation, both processes resulting in the precipitation of calcium carbonate. If such deposition is onto the floor of the cave, *stalagmites* are built up; if from the roof, hanging *stalactites* are formed. The shapes and forms of stalagmites and stalactites are controlled by the quantities of carbon dioxide, water and calcium bicarbonate available, and can be indicators of environmental change. Precipitation can also take place from cave streams, forming layers of crystalline travertine or barriers of soft amorphous tufa known as *gours*.

Water which has circulated through the limestones by means of swallow holes and caves is thrown out in springs at the foot of the limestone outcrop. Where the limestone base is below sea-level, the water may issue

Features of karst landscapes: (1) phreatic zone; (2) sinkhole or doline; (3) polje; (4) vadose region; (5) swallow holes; (6) vauclusian spring; (7) limestone pavement; (8) kamenitza.

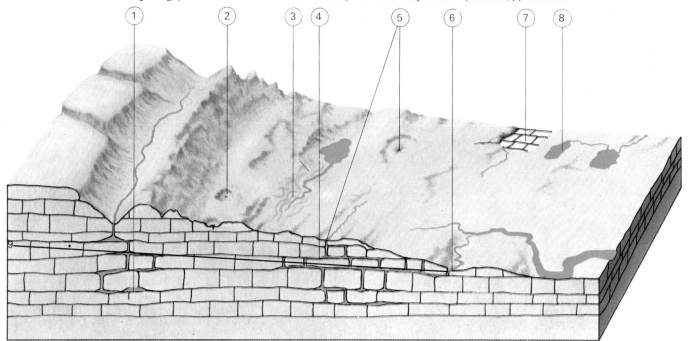

rise to flat-floored, solution-planed areas known as *poljes*: the polje of Popovo in Hercegovina is over 40km (25mi) long.

Though solution is important in all karst landscapes, other *erosion processes occur. Where limestones are only a relatively small part of the total area, fluvial (running-water) erosion is important, forming a fluvial karst with a mixture of karstic and river-eroded features: the limestone areas of the Peak District in England and of Dinant in Belgium are examples. Porous limestones tend also to have fluvial karstic relief, like the Cotswold Hills in England. Glaciated limestones possess abraded surfaces and also shafts formed by melt waters.

Karst landforms develop most slowly in dry areas and in warm or polar deserts, because of lack of water; superficial solution rills and pits are often the only karstic features. Karst develops most rapidly and most fully in the humid tropics, due to the presence of much biogenic carbon dioxide: intensive rainstorms also mean that fluvial activity is added to the solution process in tropical karst. A combination of solution and fluvial erosion produces star-shaped hollows (or *cockpits*) which dissect the limestones into innumerable small drainage basins: such relief, which has often been called cellular or polygonal karst, occurs in Jamaica and New Guinea. Surface repre-

cipitation makes limestone a strong rock in tropical conditions, and so hillslopes in limestone are steeper than in other rock types: hills with slopes of 70° and more occur, the relief being called tower karst.

Much work has been done on the calcium content of karst waters with a view to estimating rates of solutional loss in different areas. The dissolution of the rock depends not only upon the solution rate of calcium carbonate but also upon the volume of water available and its type of flow. Results show that the pattern of limestone removal is similar to the pattern of chemical *weathering in general, and is highest in areas of heaviest rainfall.

Because of relatively rapid solution and collapse of the rock, the underground hydrology of karst areas is continually changing. The behavior of the underground water is different from that exhibited in a porous medium, and the traditional concept of the water table has to be modified. Karst limestones are able to store large quantities of water, but their location and development for human use requires highly specialized study, involving a detailed knowledge of karst terrains. In many areas the actual pattern of underground water flow is unknown; but new methods of water tracing have been introduced, which include dyes and radioisotopes, and these are help-

The interior of Kačna Cave, Yugoslavia, showing spectacular formations of stalactites and stalagmites. These result from the evaporation of water containing dissolved limestone in the form of calcium bicarbonate ($CaHCO_3$), which precipitates as calcium carbonate ($CaCO_3$).

ing the piecing-together of the story.

The most obvious application of the study of karst landforms is to the use of underground water, but caves also act as traps for sediments of economic value and the weathering products of karst areas, such as bauxite, the ore of *aluminum, can also be of great importance.   MMS

Arid Landscapes

Arid areas, or hot deserts, are those where annual rainfall is greatly exceeded by water loss through evaporation from the land and transpiration from plants. Because of the severe water deficit, desert vegetation is limited in its development and desert soils are characterized by soil horizons, clay content and organic matter rather different to those of more humid areas. These two factors, vegetation and soil, combined with the sporadic, limited nature of the rainfall, are reflected in distinctive geomorphic processes – and so, distinctive landscapes.

The effects of *weathering are limited:

The world's hot and cold desert regions.

☐ cold desert

▨ hot desert

there is insufficient water available to lead to thorough leaching of rocks and soils. Thus, while solutional *denudation is of minor importance, there is, especially where there are no rivers flowing away from the desert area, a tendency for soluble salts, such as gypsum and sodium chloride, to accumulate as salt crusts. Similarly, whereas in humid areas limestone tends to be sculptured to give karstic landforms (see *karst landscapes), in arid areas lime tends to be accumulated rather than flushed out, giving rise to lime crusts (or *calcrete*). Where coastal fogs and night dew are of frequent occurrence, enough moisture may be present to produce some chemical weathering: indeed, weathering in arid areas seems to be greatly accelerated when moisture is present, as in shaded areas or near springs. However, much of the weathering in deserts seems to involve the splitting of rock through mechanical or physical processes (see *erosion). Early workers considered that high daily variations in ground surface temperatures (up to 74C°, 135F°) resulted in sufficient expansion and contraction of rock masses to cause disintegration (*insolation weathering*), but laboratory experiments suggest this process is relatively impotent. Recently it has been suggested that rock can be disrupted by salt, for when a solution of salt accumulates in a rock and evaporates, crystals develop which set up considerable forces (*salt crystallization weathering*). Some salt crystals also expand with changes in temperature and humidity (*salt hydration weathering*).

The wind is another important agent of geomorphologic development. The lack of vegetation, together with the rather loose nature of most desert soils, enables wind to be a substantial factor in soil erosion. Satellite- and air-photographs of areas like the Central Sahara show great grooves aligned with the prevailing trade winds; as well as closed depressions with distinctive orientations, on the lee sides of which the wind-eroded materials have accumulated to give a crescentic dune called a *lunette*. Similarly, when wind operates on an unprotected alluvial surface, the fine silt and sand may be blown away, leaving a lag of coarse fragments called a *stone pavement*.

Dunes. Dunes are one of the clearest manifestations of wind action in the desert environment. The wind molds, shifts and deposits sand into a series of distinctive forms, often large enough to show up on satellite photographs. There are great linear sand ridges parallel to the wind (*seifs*), crescentic dunes with horns facing both upwind (*parabolics*) and downwind (*barchans*), star-shaped dunes (*rhourds*), and many others. Some types may reach 300m

Desert landscape in Algeria, with buttes visible in the background. About 18% of the Earth's land surface is covered by deserts of dryness, of which this, the Sahara, is the largest, with an area in the region of 7–8 million km² (2.7–3.1 million mi²), though estimates vary owing to disagreement over its exact boundaries.

1000ft) in height and stretch for tens or hundreds of kilometres. On a continental scale, great sand fields (*ergs*) may be aligned parallel to the wind belts in massive anti-clockwise whorls.

Water. However, despite the low rainfall, the power of water action is important in shaping arid landscapes. When rain does fall it may do so torrentially – in one short storm more rain may fall than, on average, the region receives annually. Because of the lack of vegetation and the poorly structured and ill-developed soils, runoff may take place quickly along ephemeral channels (wadis) as *flash floods* or over shallow-gradient surfaces as *sheet floods*. Such rivers may not, however, always form an integrated system, and there are many examples where they do not flow through to the oceans, but end up in closed depressions (*playas*), where salt crusts develop through the evaporation of the water. Where streams leave their hilly tracts and enter playas, they tend to spread out and give a fan-like plain of ill-sorted debris (an *alluvial fan*).

Desert slope forms owe much to the combined action of weathering and water action. They are little masked by vegetation and soil so that they often appear more angular than those of less arid areas. Frequently there are abrupt breaks of shape, especially between mountain fronts and the plains at their bases: it appears that the basal plains gradually extend as the mountain front, maintaining its steepness, retreats. From these plains (pediplains) may rise isolated residual remnants of the mountain front called *inselbergs* (see *plateaux).

The low-angle (generally less than about 8°) concave surfaces which coalesce to form the pediplains are called *pediments*. They are peculiar because they are essentially bare rock-cut surfaces with only a very thin veneer of detritus, separated from the mountain front by an abrupt change in gradient. The reasons for their development are the cause of much controversy. Some early workers thought they were the result of high-velocity sheetfloods which pared down the surface to give a relatively undissected gentle slope. Other workers have envisaged them as being shaped by streams leaving the mountain front, their courses swinging from side to side and thereby gradually eroding and smoothing the surface (lateral planation). More recently it has been argued that pediments develop through surface and subsurface weathering. At the junction between mountain front and plain there is a natural concentration of water through percolation (see *hydrogeology): this leads to pre-

Sand dunes. In this photograph one can see the difference in gradient between the (steeper) lee side and the (gentler) windward side, as well as the striking differences in their superficial appearance (as shown on either side of the crest in the foreground).

ferential weathering at the break in slope, the weathering products being removed by sheetflow, wind and other processes, so that the mountain front retreats backward and the pediment extends.

However, one cannot interpret all arid landscapes solely in terms of present-day processes. During the Ice Ages of the Pleistocene, when high latitudes were subjected to alternations of glacial expansion and contraction, arid areas also suffered changes in precipitation. At times precipitation was even lower than at present (interpluvials) so that the ergs extended into areas that are today vegetated and too moist

for their development (e.g., northern Nigeria or southern Sudan). At other times (pluvials), average rainfall may have been markedly greater so that wadi systems were more extensive, water levels in playa basins were higher, sand dunes became gullied and weathered, and bedrock became deeply weathered to give striking weathering crusts (*duricrusts*) composed of iron and aluminum oxides (in the case of *ferricretes*) or silica (in the case of *silcretes*). Thus many features of a modern arid landscape may be reflections of climatic conditions of the past.

Man. Notably on desert margins, Man himself may cause geomorphologic

Desert roses (or rock roses) display one of the more unusual modes of occurrence of evaporitic minerals (in the broadest sense of the term). Found, as the name suggests, only in arid areas, these clusters of platy crystals are typically of barite or gypsum, and frequently include also sand grains. Their unusual form caused wonder to early explorers, many of whom suggested that desert roses were genuine flowers that had in some way been petrified.

A characteristic V-shaped river valley in the process of invasion by a glacier. U-shaped valleys are characteristic of landscapes that have been at one time glaciated: they represent river valleys that have been invaded by the ice which, during its passage, erodes the walls and floor to produce a round-bottomed form.

valley glaciers, or short cirque glaciers. A deep crack, called a *bergschrund*, often occurs at the point where the glacier tears itself away from the mountainside, and the glacier may cascade down the flanks of the mountain, exhibiting crevasses where it has flowed over obstacles on its bed, and icefalls – which are equivalent to waterfalls.

In high mountain areas, the mountainsides which flank the glacier are subject to processes of disintegration. In summer, as the sun strikes them in the morning, many such slopes resound to a noise like gunfire, as boulders and rock fragments, firmly frozen during the night, are released from the grip of ice and sweep down the mountainside. Where there is no glacier at the foot of the slope, a scree cone forms at the natural angle of rest of the boulders (about 35°). When this builds up high enough to mask the cliffs from which boulders are falling, there is no further degradation and the slope is stabilized at a much shallower angle than the original mountainside.

Where a valley glacier runs along the foot of the mountainside, the falling boulders accumulate on its surface as a *lateral moraine*, which is transported away. Thus the flanking mountain walls continue to collapse and recede, and may be progressively steepened. Where glaciers occur in adjacent valleys, the intervening ridge may be sharpened to form a knife-edged ridge (arête) due to retreat and coalescence of the mountainsides on both sides. Where valley glaciers meet and coalesce, the two marginal moraines at the point of junction also coalesce to form *medial moraines*.

In the many upland areas of Britain, northern Europe and America which have been glaciated during the last few million years, there are, though these regions are no longer cold enough to support glaciers, abundant signs of glaciers having bitten deeply into the landscape, leaving their own characteristic signature. Not only are there sharp and fractured arêtes and steep glacier headwalls, but many valleys are very deeply incised, with U-shaped cross-profiles and floors composed of smoothed, striated and streamlined rock hummocks (called roches moutonnées – see *erosion).

Some insight into the way in which glaciers produce these features can be found by direct observation of the situation beneath glaciers of today. In recent years, mining and hydroelectricity companies have had cause to construct tunnels which penetrate into the rocks under glaciers. In one such tunnel, beneath the *glacier d'Argentière* in the French Alps, it is possible to make one's way into natural cavities, 185m (600ft) below the surface of the ice, which exist on the down-glacier side of rock hummocks on the glacier bed. In the roofs

changes. Removal of vegetation for firewood or by the grazing of domesticated animals exposes the soil to wind erosion, resulting in dune development and encroachment, and to the erosive action of the infrequent high intensity storms, which produce gully systems (arroyos) with alarming rapidity. These phenomena are known collectively as desertization.

The effect of Man on the landscape of the desert interior is less than that on the margins, for water and vegetation are generally inadequate to support much in the way of agriculture or herding. Only at oases, where groundwater comes sufficiently close to the surface to be exploitable, or on the banks of rivers such as the Nile and the Indus, can farming activities be maintained. Less traditional activities, like mineral exploitation, have slightly modified the situation in some regions, notably in the oil-producing states of the Middle East.

The great variety of desert landscapes must be stressed. Climate is only one control: local tectonic action, for instance, may have had at least as important an impact on landforms. Two contrasting examples are provided by the deserts of the South West United States, where dunes cover less than 1% of the area and alluvial fans and aprons over 31%, and the Sahara, where the situation is almost reversed, well over 25% of the total area being covered by sand dunes and only about 1% by alluvial fans and aprons.   AG

Glacial Landscapes

The deep imprint which glaciers may leave on the landscape is illustrated by the fact that, in southeastern England alone, glaciers eroded and subsequently redeposited a minimum of 300km³ (70mi³) of debris during the last glaciation of the area.

The most obvious signs of glacial activity are produced in highland areas. In these, ice accumulates on the highest peaks and flows away as glaciers which may be long, deep

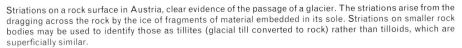

Striations on a rock surface in Austria, clear evidence of the passage of a glacier. The striations arise from the dragging across the rock by the ice of fragments of material embedded in its sole. Striations on smaller rock bodies may be used to identify those as tillites (glacial till converted to rock) rather than tilloids, which are superficially similar.

of these cavities we see the rock-studded glacier sole (the bottom surface of the ice): it is these rock fragments, eroded from its bed by the glacier and then carried along, which are the cutting tools by which rough hillocks are ground down and fine striations are cut into their surfaces.

The debris fragments move less quickly than the ice in which they are contained because of the frictional drag between them and the glacier bed. If the pressure between the glacier and its bed increases, the frictional drag increases and so the debris is retarded even more: however, because of the increased pressure the rate of erosion of the bed increases. If, however, the pressure increases beyond a critical value the increased frictional drag slows down the rate of movement of debris particles sufficiently to cause the erosion rate to begin to slow down. With even higher pressures, the particles no longer move over the bed, erosion has ceased and deposition of the typical glacial material, *glacial till*, has commenced.

Thus, with increasing pressure, erosion

rate increases to a maximum at a critical pressure, and then declines to zero at the second critical pressure, where deposition commences. If a glacier flows for the first time through a typical V-shaped valley in a mountain region, the gradual increase in pressure, from zero at the point where the glacier surface touches the valley wall to a maximum in the valley floor, will tend to widen the valley near to the apex of the V, so that it becomes the typically U-shaped glacial valley.

As we have seen, material eroded from

Five chapters in the story of glacial erratics. *Top*, glacial meltwater splashes round a boulder embedded in the upper surface of the ice, causing local melting. *Upper center*, further melting near the edge of a retreating glacier leaves a boulder straddling an empty channel. *Center*, the ice retreats, leaving this boulder perched on a layer of ice but separated from the main body of the glacier. *Lower center*, the retreat has progressed still further and the ice beneath this boulder has melted away completely, so that it is now stranded on a mass of smaller drift particles. *Bottom*, the end result, boulders lying on terrain of Carboniferous limestones. Study of erratics is important since tracing of them back to source may provide information on past glaciations.

the bed of the glacier is transported in the base of the ice, eroding as it moves, and finally deposited as till. Some boulders eroded by glaciers may be carried many hundreds of kilometres before being deposited. For instance, rocks from the region of Oslo have been found on the east coast of England, transported there by the great ice sheets which have covered northwest Europe several times in the last half million years. Such rocks are known as glacial erratics.

Although it is the highland landscapes that are dominantly scarred by the evidence of glacial erosion, most of the world's glaciated plains are deeply mantled by glacial deposits which mask pre-existing landscapes. Much of this material is till deposited directly by the glacier. These till surfaces may also be built up into streamlined forms, such as *drumlins*, which occur when the moving debris jams against obstructions on the glacier bed; *crag and tail* features, where till squeezes into the cavities left on the down-glacier flanks of rock obstructions; or *flutes*, where till is squeezed into ridges on the down-glacier sides of boulders on the till surface.

Other characteristic depositional features are *terminal moraines*. Many of these are formed when the ice advances and pushes into the till or other soft sediments lying in front of the glacier, thereby pushing up a ridge which lies along the glacier margin – rather as a bulldozer can push up a ridge of earth. Many of these ridges are to be found on the plains of Europe and North America and represent the positions of the front edges of the large ice sheets which have advanced into these areas during the geologically recent past.

The other potent agent in molding glacial landscapes is the water released when the glacier melts. Some of the most powerful and active streams on the Earth's surface are those fed by glaciers. The meltwater from glaciers is generated in two ways: some is produced beneath the ice from the heat generated by friction between the glacier and its bed, and some is produced largely during summer by melting of the glacier's surface, much of it finding its way down to the bed of the glacier *via* crevasses in the surface. Thus, beneath the glacier, we find great torrents of water which erode the dramatically deep gorges often to be seen in mountainous valleys once ice-covered. These streams transport a great deal of debris which they tend to deposit beyond the glacier as *outwash plains*, often dimpled by depressions known as *kettle holes*, the hollows left when ice blocks carried by the streams and deposited with the other sediments finally melt out.

A dramatic example of the enormous power locked up in glaciers is provided by certain glaciers in Iceland which are subject to glacier bursts: large quantities of water, built up beneath the glacier, are suddenly and catastrophically released, giving measured discharges of water up to $2km^3$ ($0.5mi^3$) per day. GSB

Tundra Landscapes

Much of the circumpolar zone of the northern hemisphere is characterized by vast, nearly flat regions with arctic climates and vegetation, in which most physical and biological processes are dominated by the effects of low temperature. This is *arctic tundra*; and it has an equivalent lying above the timberline in mountain areas, *alpine tundra*.

The single most important reason for the particular character of arctic tundra is that the subsoil is permanently frozen, often to some considerable depth. For instance, in many parts of Siberia this frozen earth extends to a depth of up to 100m (325ft). This is not a rare phenomenon: up to one fifth of the Earth's land area is underlain by such permanently frozen ground.

If the climate changes so as to cause the mean annual air temperature to fall below 0°C, the depth of frost penetration in the ground may well exceed the depth of summer thawing, so that there forms a thin layer of frozen ground which does not thaw during summer. This is called *permafrost*. It thickens progressively year by year until a balance is struck between heat flow from the Earth's interior and heat loss from the ground into the atmosphere.

Although air temperature is the principal control on permafrost development, other factors are also important. The high thermal capacity of lakes and rivers tends to inhibit permafrost development, which is also affected by the vegetation cover and the nature of subsurface soil or rock because of their different heat conductivities. In practice we find that the zone of continuous permafrost is limited to those areas where the mean annual temperature is −15°C (5°F), and this also roughly coincides with the timberline, the northernmost point where trees exist, which marks the southern boundary of the arctic tundra. To the south of this line we find zones of discontinuous and sporadic permafrost, and the boreal forests or *taiga* with their spruce, larch, pine, fir and birch trees as well as abundant tree lichens.

During winter on the tundra the ground is completely frozen to some depth; but in summer the surface layer thaws, to a depth of between 20cm (8in) and 1m (3.25ft). This surface layer is called the *active layer*.

Tundra, landscape characteristic of nearly one tenth of the Earth's land surface, is found at high latitudes and on high mountains. Vegetation is generally of low shrubs, mosses, lichens, grasses and similar herbs, and cushion plants. The faunas are similarly restricted in diversity, though seas in tundra regions are rich in aquatic mammals. The term is derived from the Finnish and means "hostile country".

many places in what are now temperate latitudes have experienced an arctic tundra environment for most of the last 70,000 years. Indeed, one might regard arctic rather than temperate climates to be typical of these latitudes: although we have been free from these conditions for about 10,000 years, there is every reason to believe that they will return. GSB

Coastal Plains

The olde sea wall (he cried) is down
The rising tide comes on apace,
The boats adrift in yonder towne
Go sailing uppe the market place.
Jean Ingelow's verse draws attention to several features characteristic of coastal plains: they are of shallow gradient and vulnerable to damage by flooding – on the other hand, they are often densely populated and of great economic importance.

We can define a coastal plain as an area of gentle slope, lying inland from the sea, and adjacent to the coastline. The plain is of either depositional or erosional origin, although the former is more common. An important control on the formation of coastal plains is the movement of sea-level relative to the land, because it is only during a period when sea-level remains more or less stable that the land adjacent to the coast can be reduced to a low slope.

As the shallow gradient can result from a variety of processes, coastal plains can be classified in terms of the differing processes that created them.

Erosional coastal plains. These can be formed by the action of rivers, glacial processes or the sea itself.

River erosion of solid rock is normally confined to the steeper reaches of the river channel toward the source, and is primarily downward; it is, therefore, unusual for a wide coastal plain to be cut by river *erosion. By the time a river has started lateral (sideways) erosion it is normally flowing over deposits on an alluvial plain.

Glacial erosion can produce extensive surfaces of relatively low relief. The best known example of a coastal plain formed in this way is the *strandflat* of western Norway. This wide coastal strip is of generally low relief, rock outcrops occurring widely on the low glacially-smoothed rocky islands and adjacent low-lying coastal zone. Considerable argument has been aroused by this geomorphological feature, but it is generally believed to be of glacial origin, formed by the erosive activity of ice flowing vigorously from the Norwegian highlands to the Atlantic Ocean.

Marine processes can also erode low-gradient surfaces under conditions of slowly rising sea-level. Waves can smooth even the hardest rocks to a slope of about 1 in 100, but they cannot erode effectively in depths greater than about 10–20m

Because of the lack of downward drainage through the underlying permafrost, and because of the water produced by melting of snow and interstitial ice, the active layer is often saturated with water, and therefore small surface pools of water unable to penetrate into the ground provide a common feature of the tundra.

The permafrost surface is in a state of delicate thermal balance, a balance which is easily upset. The growth of trees above the permafrost locally upsets this balance because of their insulating effect. As a result the depth of the active layer varies locally, and there are patches which are much wetter and more unstable than others. The instability of the ground often results in the tree falling over, and trees rarely survive to any great age. Similarly, man-made structures such as houses or pipelines tend to warm the permafrost surface and accelerate melting, so that the structure is liable to sink into the water-saturated subsoil. Large structures in tundra regions are therefore often built on piles, which are held firmly at depth within the permafrost.

Annual freezing and thawing in the active layer causes expansion of the soil, followed by contraction. The forces generated by freezing and thawing tend to segregate particles of different grain size in the soil to produce *sorted polygons*, whose margins are outlined by the coarser soil fragments. Polygons tend to form in the water-saturated active layer, which in summer is extremely unstable and tends to flow down even the gentlest slopes over the underlying frozen surface. Because of this flow, many sorted polygons are streaked out to produce *sorted-stripes* which run down the line of steepest slope.

The churning of the active layer by freezing and thawing prevents the development of a stable soil structure, and its water-saturated state prevents circulation of air so that the soil becomes acid and sterile. Because of these factors, wet tundra does not allow a rich vegetation to become established. Polar desert soils with a low organic content and low productivity develop, and, when their nature is considered together with the short growing season (during which perhaps just a couple of months may have mean temperatures above freezing), it is easy to see why only a sparse flora is maintained. In soils which are particularly susceptible to churning by frost, organic growth is almost completely inhibited and primitive *ahumic soils* (i.e., without humus, organic material) occur. But on well-drained areas which thaw early in the spring, lush tundra grasslands may develop, becoming a favorite nesting ground for birds and grazing area for musk-ox and reindeer.

Flying over the tundra, one frequently sees a large-scale polygonal pattern outlined on the ground beneath. Closer investigation shows this to be composed of large cracks in the ground surface. These are produced by cooling of the ground to temperatures below −6°C (21°F), at which it begins to contract. Cracking due to contraction, as in the dried-up bed of a lake, typically produces a polygonal (generally hexagonal) structure, which may be several hundred metres across.

It is a mistake to assume that tundra conditions have always been restricted to a narrow circumpolar zone. During the coldest parts of the last ice age (see *glaciation), about 20,000 years ago, much of the landscape of the unglaciated parts of Britain, northern Europe and north America was probably very similar to that of modern tundra areas. Even today, low-level flight by airplane over parts of southeast England can reveal the remains of large-scale tundra polygons and striped patterns on slopes. These serve to remind us that

Coastal plain bisected by the meandering Tailings River, Bougainville, Solomon Islands: a fine oxbow lake is clearly visible. The sediment being deposited under the sea surface at the estuary is in fact effluent carried down by the river from copper mines further up stream.

Lagoons, in which salt marsh development is often active, occur on the landward side of the barrier islands, while sand dunes grow on the barrier crest. When lagoons are absent, dunes may stretch far inland: they often form linear ridges, based on beach-ridge foundations, especially in areas of slowly falling sea-level, as for example along parts of the Tasmanian coast.

Raised and submerged coastal plains. Sea-level has fluctuated widely and rapidly in the 25,000 years since ice sheets were last extensive. A rapid rise since 15,000 years ago has submerged some coastal plains, which can be identified by drowned barriers on the continental shelf: examples occur off northern Australia. Raised coastal plains are widespread in those flat, high-latitude areas that have recently risen rapidly as the weight of the ice sheets has been removed by melting: southwest Baffin Island illustrates this type well.

Conclusions. Their low gradient and proximity to the sea make coastal plains desirable areas for development. Many are densely populated, but, because they are so low-lying, they are liable to damage by the sea under surge or hurricane conditions. Sand dunes form the natural defence of many coastal barrier islands, but these are liable to breaching with consequent flooding. For the protection of these densely peopled coasts, it is of paramount importance that we have an adequate knowledge of their character and the processes that operate upon them.   CAMK

(30–65ft). For this reason, when the sea-level is stationary, only a narrow platform can be created by the waves. But when sea-level is rising slowly, a much wider marine-cut surface can be formed. The best examples of such surfaces are found at the base of extensive marine transgressions, such as that at the base of the *Cretaceous rocks in western Europe.

Depositional coastal plains. The commonest type of coastal plain is depositional: these can owe their origins to rivers, glaciers or the sea.

Fluvial coastal plains are those areas where the coastline bounds an area of recent river deposition. The coast of the southeast USA provides a good example of such a plain. The area has undergone a series of depositional phases in which wedges of river-derived sediment were laid down and later tilted down toward the sea, producing an area of low relief for some distance inland. Many fluvial coastal plains are deltaic in character, such as the Mississippi delta area.

Glacial deposits can form extensive low areas that, where they meet the sea, constitute coastal plains. The south coast of Iceland, where glacial outwash from the glaciers draining Vatnajökull has formed a

wide and very flat sandy plain, locally called a *sandur*, provides a good example. The sandy plain is crossed by rapidly-flowing glacial meltwater streams that deposit their load to build up the coastal plain. It is bordered along much of its length by a coastal barrier, often separated from the dry part of the plain by a shallow lagoon.

A slight fall of sea-level can expose a marine-formed coastal plain, while in some areas marine deposits overlie glacial or river sediments. The coast of Lincolnshire (UK) provides an example of a coastal plain of marine silts overlying glacial *tills and bordered by sand dunes. Part of this coast is building actively seawards, thus providing a truly marine coastal plain of low relief. Such accretional coasts are usually characterized by coastal dunes and often by salt marshes and barrier beaches.

Barrier islands are characteristic of coastal plain seaward margins, and they occur around an estimated 13% of the world's coastlines, being especially well developed along the eastern coast of the USA, parts of southeast South America, Africa and Australia, as well as India. The barriers are mostly sandy, and are best developed in areas of low to moderate tidal range where constructive swell waves predominate.

Deltas

The deltas of the world's great rivers have historically been sites of important civilizations and are still centers of population. The early civilizations of the Tigris–Euphrates, Indus and Nile deltas depended upon the irrigation and soil fertility provided by the rivers. Today, the population of Bangladesh relies on the rich soil deposited by the Ganges and Brahmaputra.

It is ironic that the Nile delta, the site of such magnificent early civilization, should presently be retreating as a result of modern civilization. Since the building of the Aswan Dam, much of the sediment which previously fed the delta is now trapped in the man-made lake behind the dam, and marine erosion is, in consequence, winning the battle at the coast. This problem illustrates the delicate balance between the constructive forces of the river and the destructive forces of the sea, as well as the economic importance of understanding deltaic processes. The balance between construction and destruction is not everywhere the same and the variety of shapes of present day deltas, diverging as they often do from the triangular shape implicit in the name, delta (Δ), reflect this varying balance.

Delta Processes. The processes which influence delta morphology can be divided into two classes, those associated with the

An unusual photograph of a arcuate (or fan-shaped) delta in East Greenland, with icebergs out to sea. The structure in the left foreground is an earth pillar. The term "delta" derives from the resemblance of arcuate deltas to the fourth letter of the Greek alphabet, delta: Δ.

river itself and those associated with the body of water, the basin, into which the delta is being built and which may be the sea or a lake.

River Input Factors. Three properties of river input seem important: the density of the inflowing water and the amount and type of sediment carried. When a river flows into fresh water, temperature determines how the river- and basin-water mix. Cold dense river water may flow under the basin water while warm river water will float. In the sea, river water is always lighter (because of the sea's salt content) and floats out at the surface.

The amount of sediment discharged determines the delta's ability to build forward: the Nile shows how critical this factor is. Rivers move sediment in suspension and as bedload and the ratio between the amounts carried in these two ways influences the sediment distribution over the delta. Large rivers, particularly in warm climates, carry virtually all their loads in suspension, while smaller streams have more important bedload.

Basinal Factors. We have seen how salinity and temperature of basinal water may influence the flow pattern over a delta, but other basinal factors are also important. First, the depth of the basin determines how thick a pile of sediment must be deposited for the delta to advance. Second, the "energy" of the basin is important: by "energy" we mean the levels of activity of waves and tidal currents. If these are high, they will be more likely to redistribute sediment brought to the basin by the river.

Major Types of Delta. Many present-day deltas depart from the classical triangular shape, where there is an upstream apex and a system of diverging distributary channels fanning out to the shore. While this delta type exists, it is only a single example of a number of patterns which can only be arbitrarily classified. The classification given here involves morphological and process criteria.

Bedload-dominated, Low-energy Deltas. An

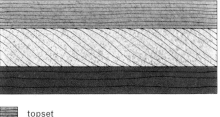

topset
foreset
bottomset

The grading of beds in a deltaic deposit, according to Gilbert. The topset consists of sand and gravel in horizontal layers, the foreset of cross-bedded coarser material; and the bottomset of silt in horizontal layers.

American geologist, G. K. *Gilbert, at the end of the last century recognized ancient deltas around the shores of now-drained lakes in Utah. As well as describing the morphology of the deltas, Gilbert recorded their internal structure as exposed in gullies. These small lake deltas, similar to many seen in reservoirs during drought, were flat-topped lobes of sand and gravel with steeply inclined surfaces in their downstream ends. The top surfaces corresponded to assumed lake-water level. Gilbert recorded three distinct internal un-

its, each having characteristic grain-size and bedding.

The lowest was horizontally bedded fine sediment. This was overlain by inclined layers of coarse sediment; and this in turn was overlain by horizontally bedded sands and gravels. These units, the so-called "bottomset", "foreset" and "topset" beds, represented respectively the sediment carried out in suspension, the delta slope deposits produced by avalanching during delta advance, and the bedload deposits of the stream which fed the slope.

This simple "Gilbert Delta" dominated thinking for some time and only in the last thirty years have geologists appreciated how restrictive this sequence is. Inclined bedding (cross-bedding) is no longer thought diagnostic of deltaic sediments as it is known to form in other ways also. Our broader understanding of large marine deltas shows that the processes, morphologies and sediment sequences are more varied and complex.

Highly Constructive Marine Deltas. Major rivers tend to have suspended loads and, because of the greater density of seawater, river water almost always floats out over the sea. The plume of turbid river water may extend many miles off-shore and as it widens and decelerates it drops its load, which accumulates on the sea floor: the finest material travels furthest while the coarsest sediment is dropped rapidly near the river mouth. In consequence, the sea floor away from a river mouth shows a progressive diminution in grain-size of sediment. The slope of this surface of accumulation, the delta slope, is usually very low – of the order of a degree or two – in contrast to the steep avalanche slope of the Gilbert Delta. These large deltas build forward, therefore, depositing a sheet of sediment which coarsens vertically upwards. Most rivers discharge their sediment from several distributary channels and their patterns on the tops of constructive deltas are variable. Two main types are recognized, elongate and lobate.

The present-day Mississippi Delta is the classic example of the elongate or "bird-

The growth of the Mississippi Delta: (1) about 1890; (2) about 1940.

1

2

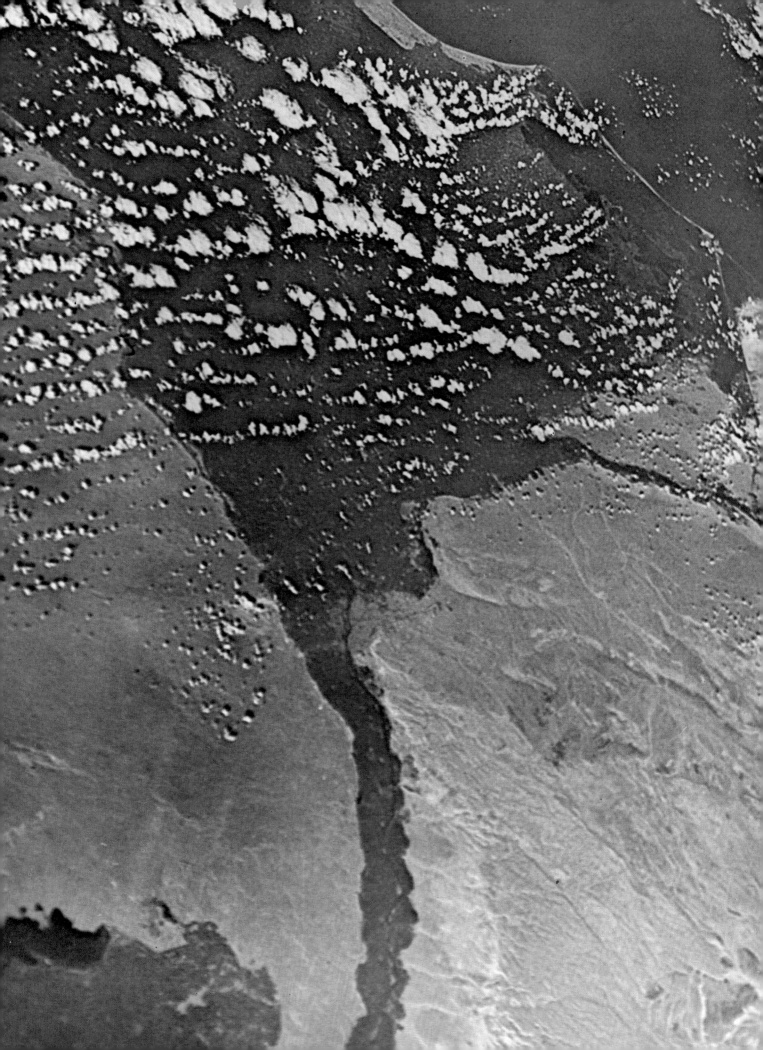

foot" delta. The distributary channels are straight and build out well in advance of the rest of the shore, so that sheltered bays are formed between them. The channels are flanked by levees which accumulate during flood overspill and which extend offshore. At the mouth of each channel is a pronounced shallowing caused by the rapid dumping of sediment at a "mouth bar". As each distributary advances it produces an elongate sand unit with a lens-shaped cross-section, a so-called "bar-finger sand".

Other constructive deltas, lobate deltas, have a pattern of distributaries which fan out from an apex and migrate with time. Each distributary has its own mouth bar, but, because of migration and switching of the distributaries, sheet sands rather than elongate bodies are deposited above the coarsening-upwards unit. The present birdfoot delta of the Mississippi is the seventh to grow there since the post-glacial rise of sea-level, but its six predecessors were all lobate with much less indented shorelines.

Destructive Marine Deltas. While highly constructive deltas are influenced by basinal processes, river processes dominate and determine the delta-top morphology. But this is often not the case; as for example, where input is reduced or where basinal energy is high. Both waves and tidal currents rework river-borne sediment into characteristic morphologies.

Maps or aerial photographs of the Rhône or Ebro deltas in the Mediterranean, or of the western part of the Niger Delta, show how the shore has built forward as a series of ridges accreted on to a beach. Waves have swept material sideways from the river mouths to accumulate as beaches and spits. In the sheltered areas behind these have developed swamps and marshes through which cut the distributary channels. These are typical wave-dominated deltas.

The Brahmaputra/Ganges Delta is a typical example of a tide-dominated delta. The shoreline is primarily influenced by wide-mouthed channels kept open by tidal ebb and flow. Some channels are linked to river distributaries while others are purely tidal. Similarly, the southern part of the Niger Delta has tidal channels as the dominant feature, resulting in a complex pattern of sands at the top of the deltaic sequence.

Rivers Without Deltas. Many of the world's major rivers have deltas which seem small in relation to the size of the river. Sediment – which is undoubtedly delivered by the rivers – seems to disappear. It is often found that submarine canyons occur off these river mouths and traverse the continental shelf and slope. At their lower ends, fans of sediment build out over the *ocean floor. A large proportion of the sediment, therefore, is being transferred direct to the ocean floor, probably by turbidity currents.

The most spectacular example is the River Congo, which has no delta to speak

A photograph from space of the arcuate delta of the Nile, whose fertile soils were cultivated by the fore-runners of the Ancient Egyptian civilization.

The coast on Saint Abb's Head, Berwickshire, Scotland, displaying many of the typical features of a coastline retreating under the effects of wave attack, most obviously the steep cliffs. Retreat of a coastline may be due to erosion (as in this case) or submergence of the land, retreat being fastest when both are operative. Conversely, advance may be due to deposition at a rivermouth or emergence of the land. When opposing processes operate (e.g., deposition and submergence), the advance or retreat is considerably slower.

of. Here a canyon extends right into the river mouth and accumulated sediment is flushed out during floods. The situation is quite common with marine deltas, but is even more common in lakes, where density differences favor the development of underflows, so allowing the river to effectively flow out over the lake floor. The Rhône, entering Lake Geneva, is a well described example of this.

Conclusion. Deltas are shoreline areas with complex and variable sediment dynamics, the understanding of which is important to many human activities such as navigation and agriculture. Such understanding is also important for the oil geologist, as many of the world's major oil fields are located in sands laid down by ancient deltas. The ability to predict the extents and shapes of these can be vitally important, and a detailed understanding of delta processes may often increase that ability.   JDC

Coastlines

The sea has a great fascination for most people, particularly along the coast where its action on the land has produced such a wealth of varied forms. The coast can be an area of great natural beauty, but too often it

has been spoilt by ruthless economic exploitation: in order for us to make the best use of the coast and to preserve it properly for the future, the processes at work on it must be fully understood.

One of the major influences on the coastlines of the world has been the great rise in the world's sea-level, starting about 15,000 years ago and lasting until only a few thousand years ago, since when it has remained fairly static in most areas. Thus the world's coastlines are almost universally young, and the many features of coastal erosion and deposition are still developing rapidly.

Classification. Coasts can be classified using a number of different criteria. One classification is concerned with the broad structure related to *plate tectonic theory; another is based on the dynamic processes operating at the coast, the waves and tides; a third is based on the degree of modification the coast has undergone by reason of the influences of the sea or other agents; and the fourth is related to the advance or retreat of the sea.

Structural classification. There are three main types of coast arising from plate tectonic action. One is found at the trailing edge of separating plates; coasts along the Atlantic Ocean, which is a widening rift, are examples of this. Collision coasts comprise another category: meeting plates create stresses that result in the island arcs and ★ocean trenches that border the northwest and southeast Pacific. A third type is the conservative plate boundary, where plates slide laterally past each other, as exemplified by the coast of California alongside the San Andreas ★Fault.

Dynamic classification. Different regions are characterized by different wave types. In high latitudes storm waves are common and exert a generally destructive effect on the coast, while in lower latitudes long, low constructive swells are usual. A third major category is the low-energy coastal type of wave, which occurs where protection from large waves is offered by the narrow coastal stretch ("fetch") of open water across which wind blows, so reducing the height of the waves. Neighboring land, offshore ice (in polar regions) or coral reefs (in tropical regions) can limit the extent of the fetch. Such coasts are dominated by short, variable waves, generated by the local winds.

Coasts can also be described as macrotidal, where the difference between high and low tides is large; mesotidal, where the range is moderate; and microtidal, where the range is small.

Process classification. Coasts can be subdivided according to the process that plays most part in determining their form. Marine processes can be dominantly erosional or depositional, the latter producing extensive beaches and other accretional forms. Subaerially dominated coasts show the influence of fluvial, glacial, volcanic or other processes in their form. For example, drowned river valleys of the ria coasts, such as those of southwest Ireland, can be differentiated from glacially eroded drown-

ed coasts, exemplified by the fjords of Norway and southwest New Zealand. (See also ★coastal plains.)

Spatial classification. For many practical purposes it is useful to know whether the coast is advancing or retreating. A coast may advance seaward either through ★deposition or because sea-level is falling; the latter situation occurs when recovery from glacial unloading is still continuing (see ★glaciation): such coasts (e.g., the shores of the Hudson Bay), are characterized by raised shorelines. Retreating coasts result from either ★erosion or rising sealevel: the coast of Holderness is a good example of the former and coasts along parts of the southern North Sea, where rising sea-level more than compensates for coastal deposition, are typical of the latter.

High Coasts. Two basic coastal types can be differentiated in terms of coastal form. These are high, usually intricate coasts, and low, often smooth coasts. The action of marine processes on a high, complicated coast usually produces a straightening of the coastline as a result of both erosion and deposition.

Erosional features include various cliff forms, which are determined both by the rock type or drift and by its degree of exposure to wave attack. *Vertical cliffs* occur in resistant rocks along exposed coasts, where the marine attack is vigorous. The 200m (650ft) high cliffs of Moher in western Ireland and the chalk cliffs of Beachy Head are examples. *Drift cliffs* are usually less steep and liable to slumping and landsliding, such as those at the Warren at Folkestone, southern England. *Wave-cut platforms* may extend seaward from the foot of the cliffs, with a typical slope around 1 in 100. Their extent depends on a number of factors: the time that sea-level has remained stable, the resistance of the rock, the vigor of wave attack, and the amount of beach material.

Details of the coastal scene include *blowholes*, *geos* (narrow clefts along joints), *arches* and *stacks*, all of which develop, mainly by hydraulic action, in suitably jointed rocks as the sea attacks the cliffs. In fact, these features often form progressively in the order given, until the stack is finally eroded by the waves to leave a platform below the cliffs.

The straightening of the coastline is aided by wave refraction, which concentrates the wave energy on the headlands, material eroded from them drifting into the bays. The coastal system consists of a number of cells within which movement of material is restricted; these are the bays between adjacent headlands.

Depositional forms are built up wherever more material reaches the coast from offshore or alongshore than leaves it: usually the alongshore movement of material is more important. On an intricate coastline its direction is dependent on both the coastal outline and on the nature of wave attack. Beaches accumulate in the sheltered bays, while spits prolong the coastal direc-

tion where this changes abruptly, or where rivers interrupt the alongshore movement of material. Barriers may be formed across bays, and islands may be tied to the mainland by sand or shingle bars called tombolos. In high-latitude, previously glaciated areas, tombolos are often formed of shingle, the commonest beach material in such regions. Sand is the dominant beach material in low and middle latitudes, where swell waves predominate. Where material can move into an area from two directions, or where shelter is provided offshore, a cuspate foreland (such as Dungeness, the point of which is protected from wave action from the southeast by the proximity of France), may be formed.

Depositional forms tend to be more complex where the fetch is small and the direction of longshore drift is variable owing to changing wave conditions and coastal orientation.

Low coasts. Flat, gently sloping coasts usually consist of relatively unresistant material, generally depositional in character. For this reason, the waves can relatively easily modify the coastal outline to form a fairly straight coast: alongshore movement of material can operate over long stretches of coastline, producing extensive cells within which material can move.

Low coasts are usually sandy in nature, and their commonest features are wavebuilt sandy barriers with sand dunes, lagoons and salt marshes. The sand forming the extensive barriers is derived partly from inshore (*via* the rivers), sometimes from the erosion of sandy cliffs, but often predominantly from the continental shelf, across which the rising sea has swept sandy deposits which are built into barriers by the long, constructive swells. The southeast coast of the United States exemplifies a typical low-barrier coast, with dunes on the barrier islands and lagoons behind, separating an intricate, low mainland coast from the smooth, open-ocean beaches on the seaward side of the barrier.

Organic coastlines. In areas of very low wave energy, as in deep embayments such as the Wash (England), salt-marsh coasts occur in middle and high latitudes; while in similar circumstances in low latitudes, where fine material is available, mangrove coasts are more characteristic. The nature of the vegetation, and in particular its ability to trap silt and fine sediment brought in by the tide, plays an important part in the formation of such coasts.

Another form of organic coast is provided by the reef-building ★corals of the tropical seas. Fringing reefs built up by these animals become barrier reefs as their volcanic foundation subsides. When it has disappeared, a more or less circular atoll remains.

Conclusion. We are living in a time of great geological activity, and so our coastlines are in the process of rapid change. For this reason also we are lucky enough to be able to see a huge diversity of coastal forms, only a few of which, for reasons of space, have been described here.   CAMK

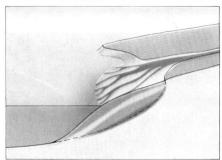

Advancing and retreating coastlines. *Above*, a retreating coastline with steep cliffs carved out by the erosive action of waves; *below*, a coastline advancing as sediment is washed into the sea by a river.

A row of craters testifies to the dominance of vulcanicity as a shaper of Icelandic landscapes. Note the lava plateaux on either side of the craters: such plateaux are characteristic of fissure eruptions. Lavas from such eruptions are generally of basalt and highly fluid.

Volcanic Landscapes

An erupting volcano is a flashback to Earth's primeval dawn, for volcanic landscapes, the most primitive landscapes of all, display the Earth's crust in the very process of formation. The type of landform produced by volcanic activity depends largely upon the style of eruption, which in turn is controlled to a large extent by the chemical and physical properties of the magma involved as well as by the structural nature of the region.

The mountains, hills, plateaux, plains and depressions resulting from volcanic activity are constructional landforms; that is, they are actively built rather than carved out by the forces of *erosion. Of course, erosion is at work – even as the volcanic products pile up – and nowhere was the contest between these opposing forces better displayed than in the eruption of Surtsey volcano off the south coast of Iceland during 1963–66. In its early stages, the volcano kept piling up ash and cinders but at the same time its crater was continually being flooded by the sea and its material washed away. Eventually a barrier of solid lava excluded the sea, and the volcano became a permanent island.

The term "volcano" embraces both the aperture in the Earth's surface from which volcanic materials – lava, tephra, and gas – emerge and the hill which forms from their accumulation. The hole in the ground may be a lengthy fissure if the crust is under tension, as, for example, in central Iceland, and in this case a *fissure volcano* is formed. If the volcanic activity is centered on a more or less circular vent or cluster of vents, a *central volcano* is the result.

Fissure volcanoes emit large volumes of very fluid material, either lava flows which are usually basaltic, or ash flows which have been described as aerosols of incandescent solid fragments suspended in hot gas. On coming to rest these solidify into ignimbrites which commonly have the composition of *rhyolite. In both cases, large areas are rapidly covered with flat-lying sheets of volcanic material, and repeated emissions build up great plains or *plateaux. The *basalt plateau of the Parana basin in southern Brazil and Uruguay is some 1,000,000km^2 (380,000mi^2) in area, the lava having a probable volume of 850,000km^3 (200,000mi^3). Outcrops of ignimbrite now occurring in Nevada and Utah indicate that the original plain constructed there some 30 million years ago must have had an area of about 130,000km^2 (50,000mi^2).

Lava plateaux are composed of overlapping broad flat cones of lava which have slopes of not much more than 1°. This *shield* structure is more obvious in the case of some Icelandic volcanoes, which have diameters up to about fifteen kilometres (about 10mi) and slopes up to about 8°, and which have usually been built up by one lengthy eruption; but the largest shield volcanoes, such as those of Hawaii and the Galapagos Islands, are formed by innumerable fluid outpourings of basalt. Mauna Loa, a Hawaiian shield volcano, is effectively the Earth's largest mountain, rising nearly 10,000m (more than 30,000ft) from the sea floor, but with slopes nowhere steeper than 12°.

The Hawaiian shield volcanoes erupt both from a large summit caldera which may be occupied by a lava lake and from fissure or rift zones along the flanks: in this respect they are transitional between central and fissure volcanoes. Their basalt eruptions are voluminous but rarely explosive – although the projection of red-hot lava to heights up to 450m (1500ft) forming a "curtain of fire" along the rifts is spectacular enough. Older Hawaiian volcanoes display numerous parasitic cones of tephra and lava along the flanks, marring the original symmetry of the shield.

Volcanoes which erupt both lava and tephra build up a layered cone shape and are termed *composite volcanoes* or *stratovolcanoes*. This graceful shape, conforming to the popular image of a volcano, is produced by concentration of activity at the central summit crater, from which short flows may emerge and away from which tephra layers thin out in all directions. Fuji-san in Japan is a well known example, having rather a large proportion of lava, but the most symmetrical of all is Mayon in the Philippines.

Eruptions from central volcanoes are generally classified according to their scale

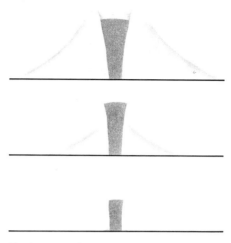

The formation of a volcanic plug. The cone of the extinct volcano is less resistant to erosion than the solidified magma within.

Hot springs at Kerlingarfjoll, Iceland, contrast oddly with the glaciated hillside behind. Such springs are typical of volcanic landscapes: they arise from contact between groundwater and volcanic gases, etc., underground. In several countries, underground hot water is being exploited as a source of heating and electricity.

of violence, the name of each category being for a typical example – e.g., Peléean for Mont Pelée in Martinique. Any volcano, however, is capable of erupting in several different ways, even during a single episode of activity.

The eruptive activity may change a volcano's morphology continuously over a long period or very rapidly in the space of a few weeks or less. In the first case, flank activity involving explosions of viscous magma may erect numerous ash or cinder cones, which have steeper slopes than lava cones (because the natural angle of rest of particulate material is greater). Spatter cones and spatter ramparts form around parasitic vents and fissures when eruptions are less violent and the magma more fluid. Very viscous magma such as *dacite or rhyolite may be extruded as *domes*. Rapid changes in morphology may ensue from landslide collapse of a whole sector of the volcano, initiating rubble and mud flows and leaving gaping depressions. More frequently documented than this is caldera collapse.

A caldera is a large, open depression at or near the summit of a volcano : older calderas may be almost totally obscured by erosion or later infillings. Steep and sometimes terraced walls suggest periodic downfaulting, which may be caused by withdrawal of magma beneath. Catastrophic collapse has occurred in many cases after rapid emission of great volumes of ash flows – the 1883 Krakatoa eruption was of this type. Crater Lake in Oregon, USA, is a waterfilled caldera some 8km (5mi) across and about 600m (2000ft) deep which formed after the collapse of a 3700m (12,000ft) andesitic stratovolcano, known as Mount Mazama, about 6000 years ago. The violence of the event may be judged from the fact that ash flows spread over fifty kilometres (about 30mi) from the volcano and ash falls occurred over 800km (500mi) away in Canada.

A great deal of volcanic activity occurs under water. If the load pressure of the water exceeds the gas pressure of the magma, there is no formation of tephra and piles of pillow lavas are produced. Eruptions in shallower water lead to the formation of ash or tuff cones in which glass fragments are very common. By the nature of things these only rarely become land forms, but similar volcanic forms are produced when eruptions occur under ice sheets, as in Iceland. Ridges or cones of granular glassy rocks and pillow lavas result from subglacial eruption through fissures or vents respectively. Occasionally these are topped by massive lava bodies, erupted subaerially when the overlying ice was melted through, producing volcanic table mountains.

Hot springs, geysers and fumaroles are generally regarded as typifying the waning stages of volcanoes. The heated water and gas cause strong chemical alteration of surrounding rocks and minerals, and the relatively soluble products weather easily to form rounded slopes with characteristically bleached soil and rock colors. Bitter experience, provided, for example, by Vesuvius in AD 79, has shown that it is dangerous to regard a volcano as extinct, and it is by no means certain that this "solfatara" stage of a volcano need be its last, as was shown in the case of the original Solfatara in the Phlegraean Fields, near Pozzuoli, Italy : nearby, in 1538, there was a massive cinder-cone eruption resulting in the formation of Monte Nuovo – the "new mountain".   JDB

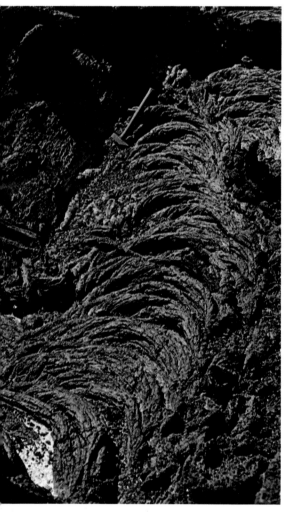

A beautiful example of a ropy (pahoehoe) lava formation is provided by part of a recent basalt flow on the island of Jebel At Tair in the Red Sea. Lava that is instead blocky in form is called aa.

Types of Volcanic Eruption

		Eruption Type	Characteristic Activity	Other Features	Example
No magma involved	Eruption increases in violence	Fumarolic	generally long-lived weak to moderate escape of gas producing mineral encrustations	minor amounts of ash and boiling mud pools	Solfatara, Italy
		Gas eruption	continuous or rhythmic discharge of gas	may precede more violent eruption involving magma discharge	Hekla, Iceland, 1947
		Ultravulcanian	weak/violent ejection of old solid lava blocks		Kilauea, Hawaii, 1924
Magma increases in viscosity	Eruption increases in violence	Basaltic flood	lava fountains; voluminous, widespread flows of very fluid lava	spatter cones and ramparts; flat shields forming lava	Lakagigar, Iceland, 1783
		Hawaiian	lava fountains; thin, widespread flows of fluid lava from craters or fissures	spatter cones and ramparts; broad shield volcanoes	Mauna Loa, Hawaiian Islands
		Strombolian	moderate explosions of pasty lava as bombs and cinders; short flows	cinder cones	Stromboli, Italy. Paricutin, Mexico, 1943–52
		Vulcanian	moderate/violent explosions of lava blocks and ash; rare thick short flows	ash and block cones	Vulcano, Italy, 19th Century
		Peléean	moderate/violent explosions of lava blocks and ash and glowing avalanches (*nuées ardentes*)	ash and pumice deposits; viscous domes extruded	Mt. Pelée, Martinique, 1902
		Plinian	extremely violent ejection of ash to great heights; ash flows of varying size. May be associated with caldera collapse	beds of ash and pumice	Vesuvius, AD 79; Krakatoa, 1883
		Rhyolitic flood	rapid voluminous effusion of hot ash flows from fissures or calderas	ash flows variously welded into ignimbrite plains	Katmai, Alaska, 1912
		Subaquatic	steam and ash explosions in shallow water	ash and cinder cones, pillow lavas below	Capelhinos, Azores, 1957
		Subglacial	lava erupts below or into ice and snow causing floods	mudflows, pillow lavas, glassy fragments	Katla, Iceland

The Ocean Floor

The Ocean Floor and Sea-Floor Spreading

It is well known that the oceans cover some two-thirds of our planet's surface; what is perhaps less well known is that the floors of the ocean basins provide as great a variety of relief as any of the continents. Mount Everest could be completely submerged in many parts of the Pacific Ocean; the grandeur of the Alps is repeated on a global scale by the world-encircling mid-ocean ridge system.

In geological terms, the sea floor is ephemeral. It is being continuously created and destroyed by the sea-floor spreading process (see *plate tectonics), and from initial creation to eventual destruction a given piece of it may last no more than 200 million years – a small fraction of the age of the Earth. However, as it moves like a conveyor-belt across the globe, the sea floor develops a fascinating variety of topographic forms.

The Shaping of the Sea Floor. The sea floor is formed at the mid-ocean ridges by the cooling and solidification of magma and lava welling up from deep within the Earth. Because the newly-formed sea-floor mat-erial is hot, it has a low density and rises high above the average level of the sea floor. It is then carried away from the ridge-crest at a rate of a few centimetres a year, gradually sinking deeper as the newly-formed material cools and shrinks. From an initial average depth of some 2700m (8775ft), it may sink to about 4000m (13,000ft) after 20 million years, and by a further 1500m (4875ft) in the next 50 million years.

During the first few million years the rocks of the sea floor are subjected to strong forces which break them and uplift some relative to others to produce a rugged morphology. As time goes on, the ocean bottom gradually accumulates a cover of sediment derived from the remains of dead marine organisms or washed off the land by rain. This mutes the topographic forms as snow blankets a landscape, until eventually all the undulations, peaks and crevices are entirely covered by a smooth plain of sediment.

Mid-Ocean Ridges. Each of the world's oceans contains a mid-ocean ridge, the site of the spreading center at which all the sea floor in that ocean is produced. There is a virtually continuous system of such ridges around the world, almost completely circ-ling the globe. Starting near the northern coast of Siberia, the mid-ocean ridge system crosses the Arctic Ocean, then descends through Iceland and down the North and South Atlantic as the Mid-Atlantic Ridge, curving south of Africa into the Indian Ocean. There it branches, the Carlsberg Ridge running north between India and Africa before curving westward into the Gulf of Aden; and the other branch running south of Australia as the Indian–Antarctic Ridge and the Pacific–Antarctic Ridge. Thence the East Pacific Rise goes north toward Southern California, and finally the Gorda and Juan de Fuca Ridges run northward off the west coast of Canada.

A typical mid-ocean ridge may be a thousand or more kilometres across, and rise to at least 3km (1.9mi) above its base. The mean depth below the water surface of the world's ridge crests is 2700m (8775ft), but in Iceland and at the western end of the Gulf of Aden ridges emerge above sea-level.

The ocean floor near the northern edge of the Madeira abyssal plain at a depth of 4670m (15,320ft). The level bottom is of globigerina ooze, the small mounds being produced by bottom-dwelling animals. Note the starfish imprints on the right.

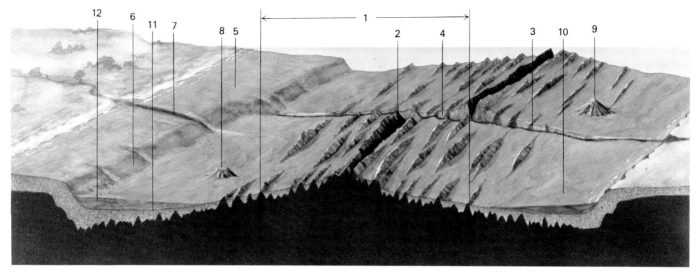

The ocean basin between passive continental plates (not to scale). Occupying a central position is a mid-ocean ridge (1) on whose slopes are abyssal hills and along whose crest runs a median valley (2). At right angles to the line of the ridge is a fracture zone, or transform fault (3), motion along which has resulted in an offset in the ridge (4). Also shown are the continental shelf (5), the continental rise (6), a submarine canyon (7), a guyot (8), a seamount (9) and the abyssal plain (10). Near to the ridge, which is a center of sea-floor spreading, basaltic rocks are found: closer to the continents, pelagic sediments (11) are overlain by turbidites (12).

The detailed form of a mid-ocean ridge depends on the rate at which it is spreading. "Slow" ridges such as the Mid-Atlantic Ridge (separating at 3cm (1.2in) per year) exhibit extremely rough topography, comprising row upon row of individual ridges and valleys, all parallel to the main ridge crest. These minor ridges are about 20km (12.5mi) apart, may be up to 50km (30mi) long, and are 5 to 10km (3–6mi) across. They are usually several hundreds of metres high, and their sides, which are formed by *faulting, are very steep, perhaps almost vertical in places. Near the crest of a slow-spreading ridge these long blocks form what are called the *crestal mountains.*

At the precise center of most slow-spreading ridges is an especially deep valley, known as the *median valley.* This may be over 1km (0.6mi) deep from its floor to the summits of the crestal mountains. The main part of the valley is about 10km (6mi) across, but the tops of the crestal mountains on either side may be 50km (30mi) apart. The median valley is roughly V-shaped, with a floor only a few kilometres across. Within this floor is the actual site of the generation of new sea floor, and recent studies using manned submersibles have shown that the lavas forming the new material erupt into long, low volcanic ridges. As the newly-created sea floor moves away, it is fractured into enormous blocks which are subsequently uplifted to form the crestal mountains.

Fast-spreading ridges such as the East Pacific Rise (up to about 20cms (8in) per year separation rate) are generally much less rugged than slow-spreading ridges, although they still exhibit a subdued ridge-and-valley topography which is the result of faulting near the site of sea-floor generation. They usually have no median valley, and new lavas pile up directly on the crest of the ridge before being spread apart.

Because they consist of newly-formed volcanic rock, the central portions of mid-ocean ridges are completely devoid of sediment, though this begins to accumulate immediately at a rate of a few tens of metres every million years. Because the lavas are extruded under water, they cool rapidly and soon become very viscous. This gives rise to weird-shaped formations, called *pillow lavas* but often resembling, not pillows, but short sections of squeezed toothpaste. These rocks are all *basalts (dark-colored lavas with relatively low silica content). The feet of the scarps formed by faulting are usually littered with scree made up of broken pillow lavas, completing the chaotic aspect of these regions.

Fracture Zones. The crests of the mid-ocean ridges are not continuous over distances of more than a few hundreds of kilometres, but are offset by enormous fractures called transform faults (see *plate tectonics). On either side of a transform fault the sea floor is moving in opposite directions. As the sea floor moves past the end of a transform fault the relative motion ceases, but any topographic forms created within the zone of the transform will be preserved. In this way long features known as *fracture zones* are created which extend, at right angles to the mid-ocean ridge system, across entire ocean basins.

Within the active (i.e., transform fault) part of a fracture zone, the opposite motion on either side of the fault breaks the sea floor into small fragments. Often there is additional volcanic activity within this area, and recent studies have shown that transform faults are the sites of hydrothermal activity, where seawater, circulating within the fractures in the crust, carries minerals such as chromium from the interior and deposits them on the sea floor.

Pillow lavas of the deep ocean floor on the slopes of Mount Pluto in the mid-Atlantic, as photographed from the submersible Alvin. These globular forms are typical of lavas extruded under water.

At the junction between a transform fault and a median valley the sea floor is usually particularly deep. This phenomenon is not well understood, but it does contribute to the shaping of fracture zones, which usually have deep valleys associated with them. There must also be strong vertical forces (again, not well understood) operating within the transform fault region, which result in the sides of fracture zones having been uplifted by hundreds of metres.

Fracture zones typically consist of a deep valley perhaps 10km (6mi) across, flanked on one side by a steep scarp and a high ridge. However, the larger fracture zones often have rather complex structures, and there are a great variety of different forms incorporating various combinations of linear ridges, valleys and scarps. Because the sides of the fracture zones have been faulted, they often expose bare rock, some of which has been uplifted from depths of hundreds of metres beneath the sea floor. The bottoms of the valleys, however, are usually filled with sediment; indeed, since the fracture zones cut through the mid-ocean ridges, they often provide channels by which sediments may be carried from one ocean basin to another.

The width and vertical relief of a fracture zone is related to the distance of offset at the ridge crest, which may range from tens to hundreds of kilometres. The most spectacular fracture zones are found in the northeastern Pacific where they extend for thousands of kilometres westward from the west coast of North America.

Abyssal Hills. As the ocean floor ages, it spreads away from the ridge crest where it was formed and gradually sinks. At the same time, as we have seen, sediment begins to cover it, thickening at a rate of a few tens of metres every million years. By the time the sea floor is several hundred kilometres from the ridge crest and approaching the base of the ridge, the sedimentary cover is considerable. Many of the lower sea-floor ridges have already been completely buried by sediment, and others have sediment draped over them, producing more rounded forms than occur near the ridge-crests. Only the steepest and highest parts of what once were crestal mountains are now free of sedimentary cover. These topographical features are referred to as abyssal hills, but of course there is no clearcut dividing line between these and the crestal mountains. The spacing and horizontal dimensions of the abyssal hills are similar to those of the crestal mountains, but naturally their vertical relief is less, usually only a few hundred metres. In general, the spacing increases with age, as more and more hills become buried and the sediment basins between them expand.

The sediments which accumulate on the flanks of mid-ocean ridges and in the abyssal hill regions are called pelagic or deep-sea sediments. They are generally oozes (consisting of the remains of mainly microscopic marine organisms) and clays. The oozes are classified according to the type of organism forming them; e.g., globigerina ooze (the remains of *globigerina), which is made principally of calcium carbonate, and diatomaceous and radiolarian oozes (the remains of *diatoms and *radiolaria, respectively), which are siliceous. The solubility of calcium carbonate in seawater increases with depth, and below about 5000m (3mi) it is so high that all calcareous sediments are dissolved, leaving

A small-scale turbidity current. Such currents, suspensions of silt and mud in water rapidly moving downslope on the bottom of the sea, are responsible for carving out submarine canyons.

the red clays which are generally found in the deepest water. These comprise very fine particles which have been eroded off the continents and carried either in the sea or by wind out into the deep oceans, together with volcanic dust.

The distribution of sediments depends on the action of currents as well as the age and depth of the sea. Faster-flowing currents can pick up and carry along particles of sediment, but if they change direction or slow down they may redeposit them. Also, in the presence of currents, sediments will be deposited more easily on gentle than on steep gradients, so current action will cause preferential accumulation of sediments on the floors of basins and keep steep slopes free.

Topographic forms can greatly influence currents, which may for example be channelled along deep valleys, accelerated by passing over shallow saddles (so that such features may be scoured free of sediment), or diverted round large hills and ridges. Because of the Earth's rotation, currents meeting a sea-mount will not divide but will be preferentially diverted on one side; this may result in a trench being scoured out on one side, whereas sediment will accumulate on the other side where there is little current flowing.

Sometimes sediments will temporarily accumulate on steep slopes, perhaps supported by small ledges, but will eventually become unstable and slump to the bottom of the slope. Often the sediments become fluidized during their fall, and flow freely

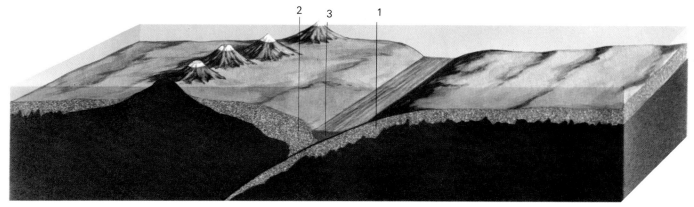

down into basins where they are deposited in almost flat-lying layers.

Abyssal Plains. Gradually the sediment thickness increases until most of the abyssal hills are covered. However, the deepest parts of the oceans are often the sites of large, extremely flat areas whose lack of relief cannot be explained entirely as a result of pelagic sedimentation. These are the abyssal plains, and typically they have slopes of one in a thousand (or less) over hundreds of kilometres. They are usually located just seaward of the continental margins, for the adjacent continents provide rich sources of sediments for deposition in them.

The extreme flatness of the abyssal plains is explained by the fact that the uppermost sediments on them have been deposited by turbidity currents. These currents are caused by the slumping of great masses of sediments off the continental slopes. As they fall, the sediments carry the surrounding water along with them, and become fluidized. In this condition they flow very easily, and turbidity currents may attain speeds of around 100km/h (60mph), persisting for many hours and flowing for thousands of kilometres. Because they are so fluid, they generally reach the deepest parts of the basins and deposit their sediments in level beds called turbidites. During the flow, the largest-grained material is deposited first, the finer grains later, so that an individual turbidite bed, a few tens of centimetres thick, will vary from silt or clay at the top to sand or gravel at the bottom. This makes these deposits easy to recognize in, for example, a cored section.

Turbidity currents are effective agents of *erosion, and may scour out channels where they flow. They may also "overflow" the edges of these channels, depositing sediment and building up levees like those produced in the flood-plains of terrestrial rivers. These channels may extend for thousands of kilometres through the ocean basins, and are known as *mid-ocean canyons*.

Fast-flowing currents also leave their mark on the bottom in the form of ripples and dunes in the sediments. Such structures may be associated with the currents of the general oceanic circulation as well as with turbidity currents, and are not, of course, confined to the abyssal plains.

Continental margins. The deep ocean basins are typically four to five kilometres (2.5–3mi) deep, whereas the average elev-ation of continents is several hundred metres above sea-level. The transition between these two levels usually occurs over a fairly narrow zone, known as the continental margin. Geologists distinguish two types of continental margin: active and passive.

Passive continental margins are found when new ocean basins open up, which takes place owing to the rifting apart of continents. During this process (see *rift valleys) the Earth's crust is subjected to strong tensional forces which fracture it. At the same time dense magma from the Earth's upper mantle is intruded into the crust, increasing its average density. Great slices of this overweighted crust sink along the tensional fractures, and simultaneously the two sides gradually spread apart. This process continues, with the rift floor gradually getting denser and sinking deeper, until eventually all of the old continental crust has been moved away to either side. At this point the rift floor will have sunk below sea level and attained the composition and structure of oceanic crust. Water can now flood in, and a new ocean has been formed. Subsequently the ocean grows by the normal sea-floor spreading process.

The passive continental margin formed by this process clearly will consist of a series of downthrown steps in the continental crust, and is usually also characterized by a seaward thinning of the crust. The changes which have taken place often cause the edge of the continent to sink slightly so that it becomes flooded: this flooded region then forms the *continental shelf*, which is rarely deeper than about 200m (650ft).

The continental shelf has exactly the same underlying structure as those parts of the continent which are above sea-level, though its surface features may be altered by erosion and deposition of sediment under the influence of strong, mainly tidal, currents. These may cause the formation, and perhaps migration, of great bars and banks of sand and gravel. Deposits of this kind can sometimes be usefully exploited, but may also cause inconvenience and perhaps hazard to shipping.

As time goes on, sediments which have been washed off the land gradually accumulate on the continental shelf, eventually forming deposits many kilometres thick. The shelf tends to sink under the weight of these sediments, so that its upper surface still remains below sea-level. These thick sedimentary sequences are excellent sites for the formation of *petroleum, and a great deal of effort is being expended by governments, industry and research institutions around the world to explore the continental shelves for this valuable resource.

Most continental margins have an abrupt edge, usually at about 200m (650ft) depth, where the sea floor falls away quite steeply toward the typical ocean depths of 4 to 5km (2.5–3mi). This steep region is called the *continental slope*. Usually it marks the position of the old rifted continental edge, but in some places, especially opposite the mouths of large rivers, great quantities of sediment are laid down over the continental slope, which consequently grows outward, forming a *sedimentary cone*. An example is the Ganges Cone, which extends from the mouth of the Ganges in the Bay of Bengal far southward into the Indian Ocean.

At the foot of the continental slope is the *continental rise* – a region of gentler slope, marking the final transition to the deep ocean floor.

Continental slopes and rises throughout the world are cut by deep valleys called *submarine canyons*, the largest of which may extend far out to sea to become mid-ocean canyons. They are generally a few hundred metres deep and several kilometres wide, usually with a V-shaped cross-section. Submarine canyons are created by the erosional effect of turbidity currents as they sweep off the top of the continental slope down into the basins. These currents are extremely powerful, and are a potential hazard to any man-made structures on the sea floor in these regions. They provide a great deal of trouble for the engineers who lay and maintain submarine telegraph cables.

Arcs, Trenches and Active Margins. Although sea floor is being continuously created at the mid-ocean ridge axes, the surface of the Earth is not, so far as we know, expanding. Therefore there must also be places where sea floor is being destroyed or consumed. This happens at the great trenches, where converging slabs

A deep-sea trench and its accompanying arc of volcanic islands. The dense oceanic plate (1) is being subducted beneath the lighter continental plate (2), resulting in deep earthquake activity and the formation of the island arc. The basaltic rocks of the oceanic plate are overlain by pelagic sediments, and close to the trench these are in turn overlain by turbidites (3).

of sea floor, or "plates", meet: one slab dips down under the opposite one, plunging at an angle of about 45° into the interior of the Earth. The deepest and perhaps the most famous of the trenches is the Marianas Trench, south of Japan: it was here that Piccard and Walsh dived in a bathyscaphe in 1960. The bottom of the Marianas Trench, 11,022m (35,820ft) below sea-level, is the deepest point on the surface of the solid Earth.

Most of the deep-sea trenches are found around the borders of the Pacific, stretching from the Kermadec Trench just north of New Zealand in a great arc west, north and finally eastward across the Aleutian Trench between Siberia and Alaska, and also off the west coast of Central and South America. Other trenches occur in the Caribbean, the southernmost Atlantic (the South Sandwich Trench) and south of the East Indies in the Indian Ocean.

A trench always runs parallel to the edge of a continent (e.g., South America) or a line of islands. In the latter case, the islands always lie along a curve which is convex toward the open ocean (e.g., the Aleutian Islands). For this reason they are known as island arcs.

Trenches have an asymmetric V-shaped cross-section. The steep side of the V always borders the continent or island arc, and the gentler slope (perhaps 1 in 5 to 1 in 20) leads out to the open sea. Trenches are typically several kilometres deep relative to the surrounding sea floor, which itself is about 5km (3mi) below sea-level. A trench may be several thousand kilometres long and about 50 to 100km (30–60mi) wide at the top. Although sediments mantle the sloping sides, there is usually a narrow plain of flat sediments (probably derived from local slumping and turbidity currents), a few kilometres wide, at the bottom.

The trenches are formed by the downturning of the sea floor prior to its 45° plunge into the Earth's interior. As one plate is pulled down under the other, some of the sediments from the down-going plate are scraped off and piled up onto the other. Other sediments, however, are carried down into the Earth with the descending slab, and this property of the trenches has led to the suggestion that they be used as sites for dumping various waste materials. Research to investigate the feasibility of such proposals is actively under way at present.

As the descending slab moves down into the Earth it heats up, and eventually part of it melts. The molten material may then rise to the surface behind the trench and erupt through volcanoes. It is in this way that the island arcs are formed, all these islands being volcanic. The curved distribution of the islands is a function of the geometrical relation between the dipping, down-going slab and the curved surface of the Earth.

When the trench borders on a continent, it produces an active continental margin. Such a margin (in common with the island arcs and mid-ocean ridges) is characterized by *earthquakes and volcanic activity, in contrast to the quiet sedimentation and erosion which are the only activity on the passive margins.

Seamounts, Guyots and Atolls. Strictly speaking, any large, isolated peak on the sea floor – for example, an especially high peak in the crestal mountains – may be called a seamount. More often, however, seamounts are formed by volcanic activity on sea floor which is shallow enough for volcanoes to rise above sea-level. This may occur where a mid-ocean ridge is particularly shallow (the volcanoes of Surtsey, which first appeared off Iceland in 1963, and Tristan da Cunha in the South Atlantic, are examples), in island arcs, or at "hot-spots". The latter are areas of uplift and local volcanic activity in the Earth's crust situated away from the main active plate boundaries. They apparently lie above particularly hot zones in the Earth's mantle, which some scientists believe to be the sites of rising convection currents of plastic rock. During sea-floor spreading the ocean floor may be carried over a hot spot, which will cause uplift and

An aerial view of Saba Island, Red Sea, part of a chain of recently active volcanoes marking the plate margin where Africa and Arabia are being rifted apart.

formation of volcanoes above sea-level. As the sea floor moves on, the area will sink again, and the volcanoes will be submerged, becoming seamounts. A long line of seamounts may be produced in this way as the sea floor moves over a hot spot.

A good example of the result of this process is the Hawaiian-Emperor seamount chain in the northwest Pacific. At present, volcanic activity occurs only in the Hawaiian Islands: however, stretching away to the northwest is a long line of submerged volcanoes making up the seamount chain. As would be expected, they become progressively older as one goes farther away from the present hot-spot.

Most oceanic volcanoes tend to sink with time, as the Earth's crust and upper mantle on which they were formed cool and sink. However, some seamounts are so heavy that the Earth bends beneath them, and so they tend to sink under their own weight. In this case a shallow "moat", perhaps many tens of kilometres wide, forms on the sea floor around them. Study of this phenomenon can be of interest to marine geologists since it enables the strength of the Earth's lithosphere (the rigid, 80km (50mi) thick outer layer) to be determined.

If a volcano is formed in waters inhabited by reef-building *corals, and subsequently sinks, an atoll may be formed. At first a "fringing reef" grows up around the shore of the volcanic island. As the island sinks, the coral grows upward so that the top of the reef remains at sea-level. Since volcanic islands are generally cone-shaped, the area above sea-level diminishes as the island sinks. Coral thrives best on the outer side of

a reef, where it is adjacent to open water containing plenty of nutrients, and so an outer coral ring forms, containing a shallow lagoon. Eventually the original island may sink entirely, leaving only a ring-shaped atoll surrounding an empty lagoon. This theory of atoll formation was first propounded by Charles *Darwin in the nineteenth century, and after considerable controversy it was finally proved correct when American geologists drilled deep boreholes into some Pacific atolls and found volcanic rocks overlain by great thicknesses of coral.

Some seamounts are shaped like truncated cones, having quite flat tops. These are called guyots. Many guyots are thought to be submerged atolls where coral growth has not kept up with subsidence, either because the island sank too rapidly or because the coral for some reason died. However, some guyots were probably formed by erosion of volcanic islands to a flat *plateau prior to sinking. Around the Azores, for example, in the north Atlantic, are a number of flat-topped seamounts, some of which contain central pinnacles which may be ancient, uneroded volcanic plugs.

Conclusions. In this article we have examined some of the more important topographic forms of the sea floor and seen something of their variety, complexity and origins. The slope of the sea floor varies from the near-vertical cliffs of the mid-ocean ridges and fracture zones to the incredibly flat surfaces of the abyssal plains. Its composition ranges from rough, blocky basalt through shelly sediments to the finest

The formation of an atoll as the relative depth of sea around a volcanic island increases (either by a eustatic rise, as here, or as the island sinks beneath the surface while migrating away from a center of sea-floor spreading). Initially coral is growing around the edges of the island: as the sea level rises, so the coral grows upward to stay above the surface. The end result is an atoll surrounding an empty lagoon.

of clays. The oceans contain mountains and canyons, hills and valleys. We have looked briefly at the important and fascinating mechanisms of sediment deposition and transport, and have seen how currents can modify or sometimes completely change this process.

In the last 20 years, the deep ocean floor has been mapped on a reconnaissance basis, the major features have been identified and many of the processes occurring there are becoming understood. However, much remains to be done. The distribution of sediment types is incompletely mapped. The polar regions are poorly surveyed, and more detailed surveys are needed in many parts of the oceans. We have barely begun to search for minerals in the deep oceans, or even to understand where and how they occur. We are not certain what causes the uplift of crestal mountains or the depression of fracture zone valleys, nor are we sure how stable is the sea floor far from the ridges and trenches. As mankind extends his activities out onto the continental shelves and down into the deep ocean basins, it is becoming increasingly important for us to answer these questions and achieve a detailed knowledge and understanding of all parts of the ocean floor. Perhaps our very survival depends upon it. RCS

Economic Geology

Engineering Geology

On the evening of December 2 1959 the newspaper headlines throughout Europe carried the shattering news of the failure of the Malpasset Dam in southern France and the subsequent loss of over 400 lives in the catastrophic flood wave resulting from the emptying of the reservoir. Less than four years later, on October 9 1963, an even more massive flood wave was formed when part of Mount Toc, in northern Italy, fell into the Vaoint Reservoir causing considerable destruction downstream and the deaths of over 1500 people.

Such events bring sharply to our awareness the hazards inherent in civil-engineering construction, when Man interferes, often unwittingly, with the forces of the natural environment. In both cases the reaction of a rock mass to changed conditions – resulting from the construction of a dam – was not fully appreciated. At Malpasset, a highly stressed arch dam was built on a relatively weak foundation composed of faulted schists; failure occurred in the left foundation as a result of rupture along pre-existing defects within the rock structure. However, at Vaoint, the conditions were different in that flooding by the reservoir of the toe of a very old landslide, forming one side of Mount Toc, resulted in a weakening of the rock mass, followed by movement of about 300 million cubic metres ($1060 \times 10^7 \text{ft}^3$) of rock.

Such dramatic examples illustrate the importance of adequate prediction of geological conditions prior to the construction of engineering works. However, engineering geology, which is concerned with the application of geology in engineering practice, is not directed solely to ensuring that sites are free of natural hazards: it is equally important that the cost estimates for a project are as accurate as possible and that unreasonable delays or cost increase will not occur during construction as a consequence of the late recognition of geological conditions which had not been fully appreciated at the outset.

Rock Properties. Most branches of economic geology are concerned with the extraction of mineral resources and the use of these resources in our industrial society. In distinction, engineering geology is concerned with the engineering properties, behavior and interaction of rocks, soils and water in their natural environment. The main properties which are of relevance to the behavior of rock materials are their strength, deformability and permeability (see *hydrogeology), together with their long-term chemical stability. The strength of a natural rock mass is a function of its ability to withstand imposed loads: the failure of the Malpasset dam foundation and the slope of Mount Toc were both a consequence of rock masses being unable to withstand such loads. Rock failures of this type result from movements along planes of weakness within the rock mass, such as bedding planes, joints and faults, which reduce the bulk strength of the rock.

The deformability of a rock is of direct

Part of the Tarbela Dam on the Indus River in Pakistan under construction. The dam is destined to be the biggest in the world, but flaws due possibly to poor design have delayed completion.

A simple dam being constructed in the hills of China to provide a drinking-water reservoir. Use has been made of the form of the sides of the valley, and most of the stone used for building is local.

relevance to the compression which occurs as a consequence of loading. When a heavy structure such as a power station or multi-storey block is constructed, settlement of the underlying ground takes place. If this settlement is excessive then the building may finish up at a lower level than is required, or it may tilt or even crack.

Many rock types are subjected to chemical and physical changes when exposed to a new environment (see *weathering), and, when those changes are relatively rapid, the life of an engineering structure can be affected. For example, solution of limestone by groundwater can give rise to the opening of joints, and consequent increase in permeability and progressively greater groundwater flow through the rock mass. Such changes have serious implications on the watertightness of a reservoir when the underlying rocks are composed of limestone.

Apart from these properties, there are two essentially environmental factors which can influence the engineering behaviour of rocks and soils; these are the natural stress state and the groundwater conditions. All rocks in the Earth's crust are confined by the surrounding rock materials and are in consequence subject to stress, which may result from the weight of overlying material and/or the aftereffects of mountain-building processes (see *plate tectonics). Artificial excavations, such as tunnels or mines, result in the release of such stresses and this can have adverse effects on the stability of the excavation. Similarly, the deeper an excavation is below the water table the greater is the groundwater pressure, and such a situation can give rise to rupture of the rock by an inrush.

Early Stages of a Project. Any engineering project, whether for construction or extraction purposes, passes through a number of well-defined stages involving investigation, design, construction and operation. Although engineering geology has a role to play throughout this sequence, the subject is possibly of greatest importance during the early stages, as a project is being conceived and passing through phases of investigation and design. Later on, when firm decisions have been made, the role of engineering geology is to confirm that the predictions are correct – or to identify changes in the actual conditions from those predicted, so that appropriate steps can be taken.

The engineering geologist works with experts in the related fields of soil mechanics and rock mechanics. The study of the engineering properties of soft and unconsolidated rocks, such as clays, silts, sands and gravels, is known as soil mechanics (the term soil has in engineering, therefore, a very different meaning from the one it has in geology – see *soils). Rock mechanics deals with the study of the properties of rock masses as engineering materials.

The investigation of an engineering project is the vital first stage to the formulation of any scheme, whether it is a major reservoir and hydroelectric complex or a small housing development. Before construction can commence it is necessary to plan the basic layout of the project and then to design each unit in detail. In consequence an investigation is normally phased, customarily moving from the general to the particular as information is accumulated.

The first stage is to review such data as may be available, and this normally takes the form of a desk study supplemented by a walk-over survey of the project area. A clearer idea will be obtained at that stage as to the topography and geology of the region, the technical problems which may be present and, possibly most important, the type of investigation methods which would be most appropriate. The next stage is to prepare geological maps of the complete area, together with more detailed maps of locations where structures are to be built. It is normal practice for aerial photographs to be used during this mapping procedure; and, under appropriate circumstances, remote sensing systems involving special forms of photography (infrared, false-color) or oblique radar housed in aircraft may also be used. One of the purposes of such mapping is to identify and delimit geological conditions which may give rise to engineering problems such as fault zones, unstable ground or shallow groundwater.

The physical exploration of the subsurface is carried out by excavations, boreholes or geophysical surveying.

Excavations permit access into the rocks and soils in which construction is to take place, so that direct observations can be made, samples taken for laboratory testing and *in situ* tests of the engineering properties carried out. Such excavations include trial pits, trenches, shafts, adits and tunnels. Possibly the commonest form of exploration is based upon the drilling of boreholes, in which cylindrical cores of rock or soil are recovered. Rock cores are obtained by drilling a diamond-tipped core barrel into the ground, progressively recovering lengths of rock – if unstable conditions are encountered, steel casing is installed in order to support the borehole walls. Water is added to keep the bit cool and to remove cuttings as drilling proceeds. Boreholes drilled in softer rocks or soils do not require such powerful equipment and tube samples can be obtained by pressing or hammering samplers into the ground. Once the samples of rock or soil have been recovered the materials are described in detail for record purposes and selected samples sent for laboratory testing.

The techniques of geophysical surveying most commonly used in engineering geology are the seismic and resistivity methods (see *geophysics). The seismic technique is most successfully used for determining the contact between rock and overlying superficial materials, and for providing an indication of the relative quality of the rock in engineering terms, higher seismic-wave velocity being generally associated with denser, more massive rocks. The resistivity method relies upon variations in the apparent resistivity (resistance to the flow of electricity) of the rocks, primarily determined by the degree of saturation of the rocks or soil, and the salinity of the groundwater. Thus dry sands have a high resistivity, whereas sands saturated with seawater, or shales, tend to have lower resistivities. By this means it is possible to delimit, in plan or profile, variations in apparent resistivity and hopefully relate such variations to changes in geology or groundwater conditions.

At the close of an investigation the geological information is prepared in the form of plans and sections, and then related to the results of field and laboratory tests and other observations. Only a tiny fraction of the ground has been sampled or tested, so that it is essential to rely to a considerable degree on the extrapolation of known geological conditions in the interpretation of the results of the engineering tests. Very considerable reliance must be placed upon geological judgment at this stage but, where possible, this can be minimized by appropriate design of the investigations: in the case of a major project it is clearly advantageous to phase the investigations so that the explorations can be modified and developed progressively as new information becomes available.

Building a Dam. The application of engineering geology in practice can probably be best illustrated by considering the range of problems which could be associated with the construction of a major dam in connection with a hydroelectric scheme.

The location of a dam is generally governed by a constriction in a valley, which permits use of the smallest possible structure. Such a constriction is commonly geologically controlled, in that the steeper valley-sides are associated with more massive and resistant rocks – a feature which again favors construction of the dam. The reservoir basin upstream of the dam should preferably open out into a broad flat-bottomed area with gently sloping valley sides.

Once one or more alternative dam sites have been located, consideration needs to be given to the type of dam which can be constructed at the site. Concrete dams require rigid rock foundations, the depth of overburden and weathered cover being at a minimum so that limited excavation will be necessary. If the valley sides are moderately steep it is practicable to construct a curved arch dam, which transmits the loads from the weight of concrete in the dam and of water in the reservoir into the floor and flanks of the valley. Gravity and buttress dams, which are constructed in more open valleys, transmit stresses directly into the valley floor and generally require less rigid foundations than arch dams.

The Volta Dam in Ghana. Lake Volta, formed by the dam, is the largest manmade lake in the world, with an area of 8482km^2 (3275mi^2). The Volta Dam project supplies hydroelectricity and water for irrigation.

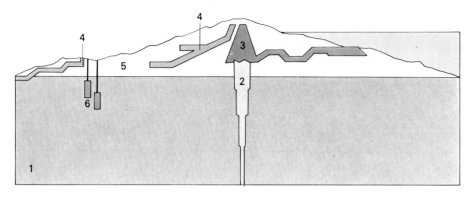

The Aswan Dam in cross-section: (1), alluvium bed; (2), clay and sand grouted with cement; (3), clay; (4), crushed stone and sand; (5), rock and sand; (6), drainage wells.

Various methods can be used to improve, or reinforce, the rock mass or to modify the manner in which the stresses from the dam are transmitted into the rock foundations so that less favorable sites can be adopted. For example, the Kariba Dam has been constructed on massive biotite *gneisses. The upper part of the south bank is composed of fractured, weathered quartzites which include clay seams: it was recognized that the rock was too weak and deformable to withstand the loads from the dam, and so it was decided to construct a series of buttresses underground, carrying the thrust of the uppermost part of the dam into sounder rock at depth.

However, if the foundations are unfavorable for a concrete dam it is more common practice to construct an embankment dam, composed of placed and compacted soil and rock materials. Embankment dams can be built in a wide variety of geological and topographical situations, including relatively weak and deformable conditions, which occur when the valley floor is underlain by thick alluvium composed of clays, silts and sands. An essential feature of any embankment dam is its impermeable membrane, commonly composed of low-permeability soils which contain a significant proportion of clay and silt; artificial membranes can be provided by concrete or bitumen. The membrane is supported by shoulders composed of more free-draining material such as sand or gravel. In order to ensure that groundwater pressures within the dam are controlled, filter and drainage layers composed of aggregates are provided next to the membrane and the foundation. If the membrane is inside the dam (the usual situation) then a protective layer of coarse rock is dumped on the upstream face to minimize *erosion by wave attack.

An additional factor which can influence the type of dam chosen is the availability of construction materials. In the case of concrete dams, coarse and fine aggregates are combined with cement, which must be imported to the site, for the manufacture of concrete. Aggregates can generally be obtained either from existing quarries or by the development of a new source of gravel and sand. If cement is not readily available at the site, or no potential sources of concrete aggregate are present, it may be necessary to construct an embankment dam.

Although more types of construction material are required in the case of an embankment dam it is not uncommon for such a dam to be built upon a site that could equally be used for a concrete dam – the reason is that an embankment dam can be constructed more economically. Although the foundation requirements for embankments are less rigorous than for concrete dams, significant problems can arise if the strength or deformability are suspect. If the foundations are composed of clay, high pore-water pressures will be generated in the foundation as the embankment is heightened. Boreholes drilled in the foundation and back-filled with sand are used to drain the pore-water, thus reducing the pressures and increasing the *in situ* strength of the clay. Soft clays can deform considerably under imposed load, leading to foundation settlement. This can be controlled by appropriate drainage and compensated for by addition of sufficient fill.

Apart from the questions which arise from the location and foundations of the dam, together with the materials which are to be used in construction, one of the major issues which needs to be assessed is the watertightness of the reservoir and the dam-site. Leakage from a reservoir can result from flow through specific geological defects, such as fractured rocks, or by general seepage through a zone of moderate permeability where the preexisting groundwater pressures are less than those which will be imposed by the reservoir. Special measures need to be adopted at the dam to minimize seepage and these generally involve the construction of a treated zone of rock – usually by injecting a cement-water mix (grouting) from boreholes – across the valley below the dam to an elevation above the top water level.

The water contained in the reservoir must be controlled as it passes through, over or past the dam. Water needs to be discharged from the reservoir in a controlled manner if it is to be used for water supply, irrigation or power generation. Similarly water may overtop the dam if the reservoir is filled and, for this reason, a spillway needs to be provided. These hydraulic works can be incorporated within the dam if it is of concrete construction and adequate size, but, in the case of embankment dams, they are commonly underground or excavated into the adjacent hillside.

Tunnelling is commonly associated with the construction of dams, the method of excavation and support being determined by the rock conditions. In fractured or faulted rocks, or where clay seams are present, instability can develop soon after excavation, so that installation of support is urgently required. In soft or water-bearing ground it is necessary to use carefully controlled excavation methods and rapid support. The prediction of geological and groundwater conditions in advance of tunnelling is possibly one of the more difficult problems in engineering geology.

Apart from the structures associated with the construction of a dam, there can be a number of related matters of engineering geological importance. For example: the construction of large reservoirs can give rise to *earthquakes, probably as a consequence of the imposed load of the reservoir associated with changes in groundwater pressures at depth; the instability of valley sides within the reservoir can give rise to damage to engineering structures, dislocation of communications or, in exceptional cases such as Vaoint, catastrophic flood waves which overtop the dam crest; changes in water-level downstream of the dam can give rise to land sliding; and the release of water from reservoirs can cause major erosion in the valley floor downstream of the dam.

This example, the construction of a dam, demonstrates the diversity of activity within engineering geology. Similar scope occurs in other areas – such as highway engineering, where it is necessary to apply engineering geology in initial route selection, to ensuring the stability of embankments and cuttings, exploring sources of construction materials and controlling groundwater flow into excavations.

Conclusions. The most spectacular consequences of the interaction between engineering and geology are, regrettably, failures which cause damage and loss of life. The more routine interactions are hardly noticed – unless they cause some form of inconvenience, however trivial. The primary purpose of engineering geology is to forestall such inconveniences by ensuring that the geological hazards likely to occur during construction and operation are recognized and minimized, that the cost of a scheme is correctly estimated, and that unforeseen delays do not take place. Such aims must inevitably appear optimistic, particularly in light of the difficulties of predicting geological conditions at depth

The Kariba Dam on the Zambesi River provides hydroelectricity for much of Zambia and Rhodesia. The artificial lake that has been created behind it has a total area of more than 5000km² (2000mi²).

and the response of rock and soil materials to engineering operations. Nevertheless, without the essential input of engineering geology, there is little doubt that many engineering projects would be rendered impossible by failures, delays and cost.   JK

Hydrogeology

The importance of water to mankind is generally underrated except at times of flood and drought. Yet these are merely localized, extreme events in the *hydrologic cycle*, which embraces the occurrence and mode of circulation of water from the atmosphere to the Earth and back again in perpetuity. Powered by the energy of the Sun, the cycle cannot be halted, though most of its individual components can be modified by Man on a local scale.

The discipline dealing with the distribution of water beneath the Earth's surface from a strictly resources viewpoint is described as groundwater hydrology or geohydrology, while the term hydrogeology is used in a wider, geological context that recognizes the significance of water as an agent of geological processes as well as the most beneficial of all Earth's resources.

Of the total amount of precipitation that falls as rain, snow, hail or dew, some two-thirds is returned to the atmosphere by evaporation and the remainder is disposed of either as direct surface run-off to streams or through infiltration into the soil, the proportion depending on prevailing geological conditions. The water in active circulation in the hydrological cycle is named meteoric water, and a part of this is called connate water when temporarily removed from the cycle by geological circumstances. Those new additions to the circulation system from volcanic or magmatic sources are known as juvenile water and are infinitesimally small in comparison with meteoric water. In order that one may appreciate the full role of underground water in the subsurface regime, it is necessary to realize firstly that the source of such water is precipitation and, secondly, that the water is never "pure" but rather behaves as a solvent.

Distribution and Occurrence. Following the natural sequence of the hydrologic cycle, the water that penetrates into the ground enters the province of subsurface water. The *zone of aeration* is that part of the ground in which spaces are not permanently filled with water: adjacent to streams or lakes it is less than 1m (3.25ft) thick, but elsewhere it can be 100m (330ft) or more in thickness. Any water infiltrating downward through this zone is subject to the forces of molecular attraction that tend to suspend the water against the pull of gravity: this water is termed vadose, from the Latin *vadosus*, shallow. Below the lower limit of the zone of aeration all inter-

The hydrologic cycle. Water from the atmosphere falls as rain, snow, etc., on both land and sea. Some of the water that falls on the land is carried by surface run-off (i.e., streams and rivers) to the ocean; some travels to the ocean by groundwater seepage; while some is returned to the atmosphere either by straightforward evaporation from bodies of water and from the land surface or by the transpiration of plants. The hydrologic cycle, shown schematically on the right, is completed by evaporation of water from the ocean to the atmosphere.

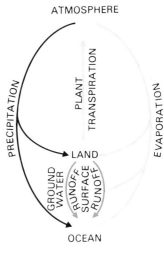

ATMOSPHERE

PRECIPITATION

PLANT TRANSPIRATION

EVAPORATION

GROUND WATER RUNOFF

SURFACE RUNOFF

LAND

OCEAN

Vadose cave formed by the waters of the Quashies River, Jamaica. Vadose water is underground water that is found in the vadose region; that is, below the surface of the Earth but above the water table. Especially in karst it can be responsible for the formation of spectacular caves such as this one.

connected void space is fully occupied by water. In this *zone of saturation* occurs *groundwater*. Whereas infiltration is directed essentially vertically downward, groundwater percolates in any direction in response to hydrostatic forces, though it commonly has a dominant sideways component.

Water occurs underground in a variety of void spaces in a wide range of geological strata, and can move *via* pores and cracks from the surface of the soil down through the unweathered rock to depths of 10km (6mi) and more. Nevertheless, it is the uppermost 500m (1600ft) of the Earth's crust that is of most significance for water-supply purposes. Which materials form natural underground reservoirs (aquifers) depends on the size of their voids and the degree to which these are interconnected.

Two important factors govern the amount and availability of water in an aquifer. The first is the *porosity*, which is the proportion of the total volume of the rock or deposit that consists of voids which

can be occupied by water. This gives some indication of the holding/storage capacity of a geological formation. However, so far as water supplies are concerned, the capacity of a material to yield water is of much more importance than its ability to hold water. For example, it is not possible to completely dewater a saturated deposit by drainage because some water is always retained by molecular attraction. The void space actually emptied is described as the effective porosity of the deposit.

The second factor is the *permeability*, and this determines how easily water can move through geological formations. Although sometimes erroneously regarded as synonymous with porosity, it is time-related and has dimensions of velocity. Permeability depends on the lithology and geological history of the formation. In an unconsolidated deposit, it is a function of the size-distribution, shape and packing of the grains of the aquifer material and depends more particularly on the geometry of primary void spaces. In a consolidated rock,

on the other hand, where cementation has significantly reduced the number and size of original pore spaces, the permeability depends largely upon the presence and scale of secondary features such as fractures or solution channels.

To illustrate the point, one may consider clay, chalk, gravel and limestone, which have porosities ranging from more than 50% for clay to less than 10% for limestone. Their permeabilities are, however, in reverse order: clay's is low because its small pores offer great resistance to flow of water in comparison with the large solution channels that produce the very high permeability of some limestone formations. Since permeability is a relative term it follows that a classification of deposits as permeable or impermeable is only meaningful in a particular context since truly impermeable strata are rarely encountered.

Any permeable formation or group of adjacent formations that yields water readily enough to be significant as a source of supply may be designated an *aquifer*. This includes both the zone of aeration, which functions as the intake area, and the zone of saturation from which the groundwater is obtained. The complementary term aquiclude is used to describe strata such as clays and shales which are impermeable in comparison with adjacent aquifers.

Where the top of the zone of saturation of an aquifer is a free-water surface it is known as the *water table* (or phreatic surface) and the condition is described as unconfined. If a well is drilled from the zone of aeration into the zone of saturation, then water will be struck at the level of the water table and will remain standing at that level even if the well is deepened. An aquifer that is saturated up to the base of an overlying aquiclude is said to be confined (or artesian). When a confined aquifer is pierced by a well the groundwater is under sufficient pressure to rise above the top of the aquifer. The pressure (or piezometric) surface is an imaginary plane extrapolated from the static level of the water in such wells: where the level of this surface is above the ground the wells are described as having artesian flow and need no pumping. Should the pressure surface be lower than the ground surface then sub-artesian flow will take place.

Further information can be obtained by geophysical techniques (see *geophysics): these include studies of density, magnetic susceptibility, and electrical potential and conductivity. Gravity methods, like magnetic methods, are not commonly useful in groundwater studies. Electrical methods, on the other hand, have long been used. Measurements of the ability of Earth materials to conduct electrical energy allow the sequence and nature of strata to be identified. Seismic methods also have great value in hydrogeological investigations. In both electrical resistivity and seismic refraction surveying it is important to recognize that in most cases corroboratory evidence of a more direct nature is required.

Confirmation of the geophysical interpretation is best obtained by drilling wells and collecting representative samples of the strata and water penetrated. The nature of the samples depends on the type of tool used to drill the hole. Percussion and rotary methods are both in common use and each has advantages and disadvantages in specific cases. The only wholly satisfactory sampling method is continuous coring, but this is time-consuming and expensive.

Unless the hydrogeological conditions are particularly simple no single exploration technique will be sufficient, but a combination of several techniques should at least ensure that the conditions are sufficiently well understood to allow predictions to be made of groundwater occurrence, well locations, design, yield and performance, water quality, and especially the effects on the immediate hydrological environment that development of the groundwater resources must inevitably introduce.

Hydrogeology Today. The steadily increasing demand for water in all countries of the world is the result of expanded industrial production, improved domestic conditions and intensive agricultural programs sustained by irrigation practice.

These demands are still being met in conventional fashion from natural reservoirs or wells, and the same sources are likely to be sufficient for many years to come. Nonetheless, the groundwater reservoir is finite; and the serious, sometimes catastrophic, effects of local overdevelopment give plain warning of the larger scale consequences of ill-conceived development.

Understandably, Man has in the past been primarily interested in drinking-water but, by definition, groundwaters include both drinkable and non-drinkable waters. It follows that, in the future, use must be made of brackish and saline groundwaters, either by mixing with drinkable water or by desalination.

One of the newer applications of hydrology is to the study of geothermal energy potential and its beneficial use. Heat is conducted from the interior of the Earth through the agency of *vulcanicity. Groundwater can be trapped in particular geological conditions and become superheated to form steam: wells drilled into such reservoirs tap the steam and conduct it to the surface where it may be used as a source of energy. Despite its apparent wide potential, relatively little power is produced from geothermal sources outside the USA and Italy.

Of more widespread and immediate benefit is geothermal water too low in temperature to be used for the steam-generation of electricity, but which can readily be used for space heating for domestic and agricul-

tural purposes. Rapid progress is being made in this respect in Iceland, Japan, the USSR, Europe and the USA. In many other parts of the world geothermal sources of energy are never likely to reach major proportions.

Because groundwater is an integral part of the hydrologic cycle, any form of development must affect the related components to some degree. The modern approach to hydrogeology emphasizes study of those quantitative aspects that allow predictions to be made of the consequences of Man's utilization of groundwater and subsequent modification of the cycle.   GPJ

Mining Geology

Every year the mines of the free world produce around three thousand million tonnes of ore ready to be processed for metals and minerals – quite apart from coal production and ignoring the necessary waste rock and overburden which accompanies mining. This amount of rock would be yielded by a tunnel 3m (10ft) in diameter driven $3\frac{1}{2}$ times round the world at the equator, and reflects an average demand of about one tonne of ore per annum for every man, woman and child in the free world.

The primary industry of mining must meet the escalating world demand for min-

Chuquicamata copper mine, Chile. Copper dominates the Chilean economy; its control has played a major part in the political history of the country.

erals in ever-wider variety and in ever-greater quantities. In two major facets of the mining industry, mining geology plays an essential role. First, there is the engineering task of extracting the valuable content of a mineral deposit (the ore) economically, efficiently and safely, and with minimum disturbance to the environment. Second, since every ore deposit has a limited life, the future of the industry depends on prospecting and exploration to discover and evaluate new deposits to replace dying mines and to meet increasing demands.

Mining Operations and Methods. *Ore deposits exist in a huge variety of physical sizes, shapes, attitudes and geological environments, and for each ore body these factors must be accurately defined so that the optimum mining method can be designed. Alluvial mining of unconsolidated sediments is by dredging or hydraulic jets. For hard rock the choice lies between open-cut and underground methods. Steeply dipping and narrow tabular ore bodies are

best exploited by underground "stoping" with access to the working faces from horizontal tunnels driven at intervals from a vertical or inclined shaft. Horizontally orientated tabular ore bodies close to the surface are most efficiently exploited by open-cast mining, where the waste stripped from above the ore is cast back into the excavation behind the working face.

Moderately dipping tabular ore bodies and those of irregular, massive or pipelike form are most economically worked by open pits from the surface, the successive "benches" forming an inverted cone. Waste rock must be mined from outside the ore limits to form a stable pit slope; when the ratio of waste to ore reaches a critical figure (which varies for each individual ore body) the economic pit limit is reached, and mining to further depth may be found to be more economic by underground methods. Underground mining methods are relatively labor-intensive and more complex and expensive than the machine-intensive and generally larger-scale open-pit meth-

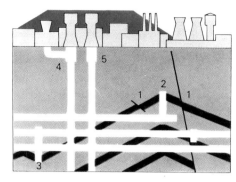

Simplified diagram of the workings of an underground coal mine. Folded coal seams are displaced by faults (1). Shafts are built upward (2) or downward (3) from the horizontal corridors. Access is gained via a vertical pit shaft (4), and coal brought to the surface via another (5).

ods which are so well suited to the modern technology of bulk mining and treatment.

For relatively large and massive ore bodies with suitable rock-strength characteristics, underground mass-mining methods, such as block caving, where the ore is induced to collapse in controlled fashion and withdrawn through workings below, are used. For relatively small ore bodies, the variants of numerous stoping methods are classified as either open stopes, where the strength of the rocks of the excavation walls permits safe working in open excavations with little artificial support, or as filled stopes, where the excavations must be back-filled with rock or sand.

Whatever the working method, the overall economics and engineering efficiency of any mine depend upon the accuracy and timeliness of the observations and predictions of the mining geologist. The following are typical examples:

(i) Estimation of ore reserves, in categories such as "proved", "probable" and "possible", according to the relative certainty of the figures, forms the basis of engineering and financial control and planning throughout the life of the mine. A typical simplified ore reserve statement such as "10.50 million tonnes of proved ore at an average in-situ grade of 2.65% copper" summarizes two related quantitative investigations. The tonnage reflects the geologist's prediction of the detailed three-dimensional shape of the minable ore body. The average grade or content of the valuable metal or mineral in the ore can be calculated from mathematical weighting procedures or statistical treatment of sample data taken.

(ii) The profitability of mining operations depends on clean mining practice, which is enhanced by accurate and timely forecasts of potentially hazardous geological conditions. For example, faults and shear zones cause displacement of the ore body, and underground waters under high pressure can cause disasters.

Smashing pieces of coal on a muck tipple in the Mount Isa coalmines, Queensland, Australia. Much Australian coal, the bulk of which comes from New South Wales and Queensland, is exported to Europe.

Loading ore at Chuquicamata copper mine, Chile. Much of what we can see being loaded is in fact gangue–that is, unwanted parts of the ore–that will be removed during processing.

(*iii*) Most mines require closely coordinated geological and engineering advance planning and operational practice in grade control (or quality control). The blending of heterogeneous ores from various parts of the mine achieves a uniform feed which ensures optimum efficiency in the treatment plant and a uniform quality of product.

(*iv*) It is the responsibility of the mining geologist to discover and explore any extensions of the known ore body which affect the continuity and life of operations. This task draws upon applied geological science and technology, including a working theory of ore deposition and an understanding of the physical limitations to ore.

Prospecting and Exploration. Prospecting is the science of discovering new mineral fields and mining prospects. Exploration is the work of investigating and evaluating the physical characteristics and economic viability of prospects, very few of which, in the end, become mines.

Mineral deposits are geological anomalies, and those few with economic ore grade and deposit size are not only rare but also, usually, well concealed by nature. Modern prospecting is essentially the application of geology to finding the most favorable target areas, assisted by a variety of techniques designed to detect the geophysical, geochemical and/or mineralogical anomalies within the geological environment that betray the presence of an ore body.

The primary guides to potential mineral fields are global or large-scale geological features or rock associations. For instance, for porphyry copper deposits one would search the calc-alkaline intrusives of the volcanic belts above subduction zones such as the Andes.

At successively finer scales, geological indications of many kinds are sought and integrated with the results of the following main prospecting techniques:

(*i*) Photogeological interpretation assists rapid production of geological maps from aerial photographs, and similar interpretations are based on images from "remote sensing" surveys from artificial satellites.

(*ii*) Geophysical methods are designed to detect anomalies in the Earth's physical properties associated with certain rock types, structures and mineralization patterns, and these must then be geologically interpreted. The most useful geophysical surveys for mineral deposits are magnetic (relatively cheap, fast and very helpful in geological mapping as well as for detecting magnetic iron ores, etc.), radiometric (the main prospecting tool for uranium deposits, using Geiger counters, scintillometers and gamma-ray spectrometers), electromagnetic, electrical, gravitometric, and

seismic (of limited use in mineral exploration as contrasted with *petroleum exploration surveys). Magnetic, electromagnetic and radiometric techniques are commonly adapted to airborne surveys in helicopters or fixed-wing aircraft and so afford rapid systematic coverage of large areas. (See *geophysics.)

(*iii*) Geochemical methods are based upon chemical analysis – in the parts-per-million (ppm) range – to detect the anomalous primary or secondary dispersion patterns of the constituent elements around ore deposits. A copper sulfide ore body being weathered in mountainous tropical rain forest can, though invisible to the eye, be detected by the few ppm of copper in the stream sediment and by sampling the residual soil, which may be similarly enriched to a few hundred ppm of copper, above the ore body. Besides stream sediment and soil surveys, geochemical methods are being developed to detect trace amounts of gases such as sulfur dioxide or radon released into the soil or atmosphere from ore bodies that are being weathered. Geobotanical meth-

ods make use of vegetational patterns, which may reflect anomalous soil chemistry over mineralized ground.

(*iv*) Mineralogical methods include the still-important prospector's pan for gold, diamonds and other heavy minerals dispersed downstream from the source ore body. Trains of mineralized boulders give similar evidence in glaciated countries. More sophisticated is the use of patterns of mineralogical alterations such as the potassic alteration of volcanic rocks associated with porphyry *copper deposits.

Once a mineralized area or prospect has been discovered, the more detailed and expensive work of exploration begins. It continues until the prospect has been either abandoned as uneconomic or promoted through a feasibility study for development and capitalization as a new mine. Drilling methods of various kinds, but especially percussion drilling and diamond-core drilling, play an essential part in providing not only geological information concerning the three-dimensional shape and size of the ore body but also samples of its valuable content, on which an estimate of ore reserves can be based. Exploratory underground workings may also be used to provide bulk samples for metallurgical testing, and to determine geological and geotechnical engineering data necessary to design the future mine and forecast its operating costs and mineral output.

Conclusion. Food and clothing excepted, most of Man's essential raw materials are dug from the crust of the Earth. Fossil fuels, construction materials, industrial minerals, metals – all are produced in vast quantities. The continuously improving knowledge and application of geology to the mining industry adds confidence to the view that Man will not waste his planet's precious mineral resources, and that through his understanding he will be able to discover enough to meet his future needs – at an acceptable cost.   GRD

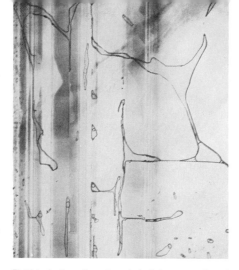

Fluid inclusions (here in sphalerite) are samples of the fluid from which hydrothermal ore deposits were precipitated: each comprises saline water and a gas bubble. The bulk composition of the fluid can be deduced from its freezing point, the temperature of formation of the deposit from the temperature at which the gas bubble disappears.

Banded iron formations are generally of early Precambrian age. They are made up of alternating laminae of chert on the one hand and a variety of iron minerals including siderite, hematite, magnetite, pyrite and greenalite on the other. The texture is usually fine-grained but sometimes hematite occurs as ooliths.

Ore Deposits

In 1964 Ken Philips, a geologist working for Conzinc Riotinto of Australia Exploration, Ltd., led an exploration team over to the southwest of the Crown Prince Range, which runs down the spine of Bougainville Island, Papua/New Guinea. In the unpopulated and mountainous Panguna region, an area where gold had been worked between 1933 and 1951, the team discovered anomalously high copper concentrations in stream sediments. Subsequent drilling of over 200 holes to obtain 80km (50mi) of core proved the existence of one of the biggest porphyry copper ore deposits in the world.

The story is not altogether an unusual one: ore deposits can be found through scientific exploration or just plain luck – and often a combination of the two. The term "ore deposit" is as much a commercial as a geological one, referring as it does to economically workable natural concentrations of elements and minerals.

The total cost per tonne of ore, mined and processed, must be less than the income derived from the element or mineral contained in each of those tonnes. This means that the principal factors which determine the economic feasibility of working a particular deposit are the concentrations of the element of interest (grades), the size of the deposit, its location and the market values of the products.

The concentrations of elements in ore deposits are, typically, a thousand times greater than the average throughout the crust as a whole. But the concentration factors are quite variable and range from as little as 10 (iron deposits) to as much as 30,000 (chromium deposits).

Ore deposits must be large in order to provide a reasonable return on capital investment. Reserve tonnages are variable, but few deposits contain less than one million tonnes of ore, and the largest contain around 15,000 million tonnes.

The depth of the deposit beneath the land surface is critically important because it determines whether a mine is worked by open-cast or underground techniques (see *mining geology).

Geographical location determines transport costs as well as the type of government which controls mining activity. Metal prices are clearly critical when considering the commercial viability of the deposit. Fluctuations in the price, whether caused by changing economic or political conditions, can make and break mines.

When the natural processes which can lead to the formation of ore deposits are considered, diversity is the theme: practically all natural processes can generate elemental concentrations. There are three major generic groupings: magmatic, hydrothermal and sedimentary.

Magmatic Deposits. In deposits of this kind, the minerals have typically crystallized from a silicate melt (magma) and accumulated selectively in layered basic intrusions as magmatic sediments. The chromite seams of the Bushveld igneous complex of southern Africa are the prime examples of magmatic deposits.

Hydrothermal Deposits. As the name suggests, hydrothermal ore deposits are formed by minerals dissolved in hot water (at temperatures of the order of 100–700°C, 200–1300°F) crystallizing out. The minerals contain small inclusions of the solution from which they crystallized.

Modern geochemical work, in particular that based on the use of radioisotopes as tracers, has shown that the natural fluids in hydrothermal ore deposits had diverse origins: magmatically derived water, meteoric water (rainwater), sea water and water contained in deep sedimentary formations have all been found in the cores of porphyry copper deposits, whereas in the surrounding layers evidence of the presence of rainwater has been obtained.

Sedimentary Deposits. Chemical precipitates can also form in the sedimentary environment, being in some, but not all, cases a result of evaporation increasing the concentration of a solution. Crystal accumulations (*evaporites) which formed in response to evaporative concentration of enclosed water bodies in hot climates are particularly important sources of *salt, potassium minerals for the fertilizer industry, and such uncommon elements as boron and strontium.

A good example of straightforward accretion at work is provided by the manganese nodules which cover large areas of the deep *ocean floor. The *Precambrian banded iron formations are also thought to be straightforward chemical precipitates.

There are two main exceptions to this theme of chemical precipitation – residual deposits and placers. The former are the surface residues, similar to soils, which remain after being deeply leached by percolating groundwater in hot, wet climates. Examples include *bauxites, as a source of aluminum, and nickeliferous laterites. Where leaching of sulfide ore deposits occurs, residual red and brown iron hydroxyoxide cappings (gossans) are left, and other elements can be carried down and precipitated in a zone of "supergene" enrichment near the water table. Placer deposits were formed where moving water mechanically sorted heavy and physically resistant minerals in river gravels and beach sands.

Formation Today. As can be seen, most geological processes play a part in the formation of ore deposits. Most of the processes involved in the formation of sedimentary ores can be observed taking place at the moment: for example, on the continental shelf off southwest Africa phosphatization of sediment is actually occurring today. Magmatic and hydrothermal deposits, however, are less commonly observed in the process of active formation for the obvious reason that they generally occur within the Earth.

A striking exception is provided by a hydrothermal exhalite type of deposit ac-

Manganese nodules form on the sea floor by straight forward accretion: they comprise an important mineral resource of the future since manganese ore deposits are relatively scarce on dry land. Ore deposits from the ocean floor (copper, cobalt and nickel nodules are also forming) are today the focus of much research.

A core of sediments from the Atlantis II Deep, Red Sea: the most recent sediments are at the top left, the oldest (i.e., deepest) at the bottom right. Metalliferous sediments have been accumulating on the floor of the Deep for the last 12,900 years, and this sedimentary ore deposit is still today in the process of formation.

tively forming today in the Atlantis II deep at the bottom of the Red Sea. At a depth of about 2km (1.25mi) there is a depression in which are found layered hot metalliferous brines. The lower layer has increased in temperature from 56°C (133°F) in 1966 to 59°C (138°F) in 1971, and the upper layer is at 49°C (120°F). For the past 12,900 years, metalliferous sediments have been accumulating on the floor of the depression, and there now exists an ore deposit of about 200 million tonnes (on a dry basis) which contains 0.34% copper, 1.58% zinc and smaller quantities of silver, gold and lead. ETCS

Crystals

The outer surfaces of nearly all natural objects in the world about us are more or less curved. The bodies of plants and animals, the slopes of hillsides, the horizon itself and the heavenly bodies, all are characterized by regular or irregular curvature. Only in the Man-made environment are the flat plane and the straight line familiar sights – except in the case of crystals.

Man's interest in crystals began with the observation that there exists one unique class of natural objects which, in many instances, do possess plane surfaces and straight, even parallel edges. Well-formed crystals have a geometric regularity which cries out for explanation and understanding. The study of crystals, crystallography, in its classical pre-20th-century sense, was concerned with the description and classification of crystals and attempted to explain the origin of crystal shapes in terms of regular geometric patterns in the fine-scale structure of crystal-building substances. Powerful modern techniques involving the use of X-rays, electron microscopes and nuclear reactors, have enabled the 20th-century crystallographer to explore this fine-scale structure down to the atomic level. Crystallography today is a highly technical and complex subject with appli-

cations in chemistry, metallurgy and biology: it was X-ray crystallography which eventually revealed the double-helical nature of the deoxyribonucleic acid (DNA) molecule, and showed how the mechanism whereby the living cell replicates itself is based essentially on the atomic structure of that molecule. To the geologist who seeks to understand the rocks of the Earth, some knowledge of crystallography is essential, since rocks consist of minerals and *minerals are crystalline substances.

For a substance to be crystalline, and hence potentially capable of forming regular crystals with plane surfaces (faces), it must of course be solid. It must, in addition, be more or less homogeneous and have a definite chemical composition expressible as a formula. Most importantly, it must possess an orderly arrangement of its constituent particles (ions). The best-known example of a non-crystalline solid is glass, which can vary widely in composition and whose ions are more or less randomly arranged.

It is important to realize that crystalline substances, including minerals, do not always form well-shaped crystals with regular faces, although they continually aspire to this condition during growth. For instance, competition for space between minerals crystallizing from a magma usually produces a final mosaic in which the majority of crystal boundaries are uneven, irregular and curved. Nevertheless the rarer well-shaped crystals, formed without constraint (often in cavities), provided the first clues concerning the nature of the crystalline state.

One of the most important concepts in crystallography, although hinted at by earlier workers, was fully formulated by the Frenchman René-Just *Haüy in 1784. The mineral calcite (calcium carbonate) often forms well-shaped crystals, and these are produced in a bewildering, apparently haphazard variety of shapes (called crystal habits). After accidentally breaking a speci-

men of calcite and noting that the plane of fracture was flat and smooth, Haüy succeeded, after several trials, in cleaving the damaged piece along two further sets of planes. Extending the experiment, he found that calcite crystals of all habits produced six-sided rhombohedral cleavage fragments with identical angles between corresponding sides. Each cleavage fragment could be further subdivided into smaller identically shaped fragments down to the limit at which the fragments could be observed. Haüy proposed therefore that all crystals of calcite are built of large numbers of minutely small, identically shaped and identically oriented "molécules constituantes", which he conceived as solid objects.

Haüy went on to demonstrate how different crystal faces and crystal habits of a mineral could all be generated from the same "molécule" by appropriate stacking procedures. In one of his original diagrams he showed how the pointed "dog-tooth" habit of calcite is constructed. This early work was extended to many other minerals: he showed also how a fluorite crystal, a cube with bevelled edges, can be built from "molécules" which in this case are perfect cubes. The "molécule" (now called the unit cell) is in reality so small compared with the complete crystal that surfaces resembling flights of steps when represented diagrammatically are, in the developed crystal, smooth and planar – i.e., they are genuine faces.

Unlike Haüy, we now know the absolute sizes, measurable in Angstroms, of the unit cells of most crystalline substances. In the case of fluorite, for example, a crystal having a volume of one cubic millimetre is built of approximately 6×10^{18} (six million million million) unit cells.

It is now realized that, although Haüy's concept explains the development of crystal

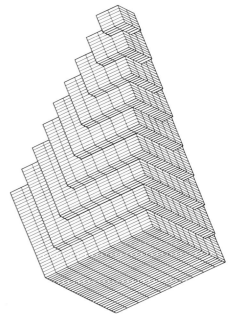

Haüy's diagram to show how a crystal of "dog-tooth" calcite can be considered as constructed of a large number of very small rhombohedral units.

A normal and a "swallowtail" crystal of gypsum. The latter is an example of twinning, which can be of great value in mineral identification. A crystal is said to be twinned when it comprises two or more parts in which the crystal lattices are differently oriented: such disorientation, which leads to the development of complex crystals, is never random; the parts of a twinned crystal of a given substance are related to one another according to certain crystallographic rules. In simplest terms, twinning is in some cases the result of a type of "defect" or "error" that occurs during nucleation (the very earliest stages in the formation of a crystal): this is somewhat analogous to biological twinning. In other cases twinning results from readjustment of the lattice of a completely formed crystal in response to exterior physical stresses.

Below: Crystals in the cubic system are referred to three crystallographic axes which are at right angles to each other (A). An axis of symmetry is an axis about which a crystal can be rotated so that it comes to occupy two or more indistinguishable positions in space in the course of a complete turn. A cube has three axes of 4-fold symmetry, or tetrads (B), four axes of 3-fold symmetry, or triads (C) and six axes of 2-fold symmetry, or diads (D).

Bottom: A cube has nine planes of symmetry that divide it into sections that are mirror images of each other: three (1) plus two (2) plus two (3) plus two (4). The three planes shown in (1) contain the crystallographic axes and are called axial planes.

faces, it oversimplifies in some respects the true nature of the crystalline state. In particular, we now know that most of the volume occupied by "solid" matter is empty space, and we envisage the fine-scale structure of crystalline materials in terms of an orderly, three-dimensional array or pattern of atoms. The unit cell, which replaces the solid "*molécule constituante*" of Haüy, is the geometrical block-like outline of the unit of pattern which, by repetition, builds up the entire structure.

Rigorous geometrical analysis shows that there are seven and only seven essentially different types of unit cell. All are six-sided block-like objects having opposite pairs of faces parallel to one another (parallelepipeds). All crystalline substances have lattices built of one of these types of unit cell and, according to the type, we recognize seven different crystal systems, namely the triclinic, monoclinic, orthorhombic, tetragonal, trigonal, hexagonal and cubic systems.

Crystal symmetry. Fortunately, it is of-

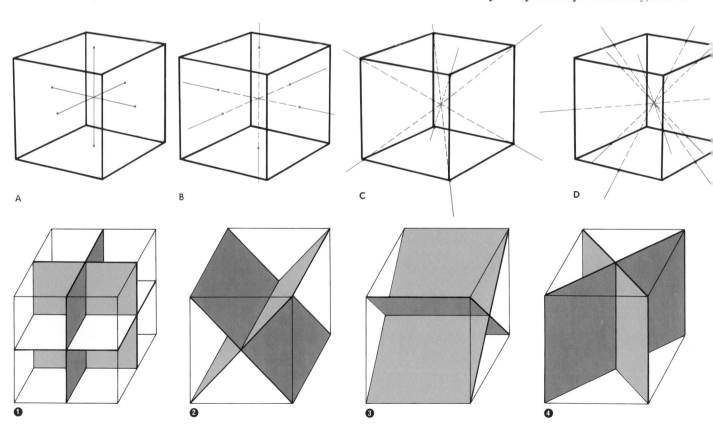

A   B   C   D

❶   ❷   ❸   ❹

Crystal system			
Cubic		Cube	All edges the same length; edges meet at right angles.
Tetragonal		Square prism	Edges meet at right angles; base and top are square, but vertical edges are longer (or shorter) than horizontal edges.
Orthorhombic		Rectangular parallelepiped	Edges meet at right angles; the three edges meeting at any one corner are of unequal lengths.
Monoclinic		Right prism on a parallelogram as base	Base and top are parallelograms, not rectangles, and are horizontal; sides are vertical; the three edges meeting at any one corner are of unequal lengths.
Triclinic		General parallelepiped	The three edges meeting at any one corner are of unequal lengths; none of the angles formed by edges meeting at corners are right angles.
Hexagonal		Prism on $60°/120°$ base	A special case of the monoclinic cell, in which the parallelogram base and top have angles of $60°$ and $120°$, and in which all parallelogram sides are of equal length.
Trigonal		Rhombohedron	A cube deformed along one diagonal; all edges are of equal length.

The seven types of unit cell, giving rise to the seven crystal systems.

ten possible, given well-formed crystals of a substance, to assign it to its correct crystal system without first determining the details of its atomic structure, and hence its unit-cell type. This procedure involves a simple study of the gross morphology of the crystals, and specifically the identification of types of symmetry (called symmetry elements) in the crystals. These symmetry elements are now described.

Plane of symmetry. If a crystal is (theoretically) divided into two equal halves along a plane, such that the two halves are, without displacement, mirror images of one another, the plane is called a plane of symmetry.

Axis of symmetry. If any object is rotated $360°$ about any imaginary line passing through it, it will of course present its original appearance to an observer in any fixed position. It may, however, also present its original appearance from time to time *during* rotation through $360°$. According to whether this occurs every $180°$, $120°$, $90°$ or $60°$, we recognize respectively two-fold, three-fold, four-fold, or six-fold rotation symmetry, and the line about which rotation produces these results is called a two-fold (or diad), three-fold (or triad), four-fold (or tetrad) or six-fold (or hexad) axis of symmetry. No other axes of rotation symmetry are possible in crystals, except for the trivial case of the one-fold (or monad) axis, of which any object possesses an infinite

number.

Inversion-rotation axis of symmetry. Certain groups of crystals possess a more complex form of symmetry element, involving rotation about an axis coupled with inversion of the crystal about its center for the crystal to present its original appearance. For our purposes it is sufficient to note that, for each rotation axis of symmetry, there is an equivalent inversion-rotation axis. These are designated the inverse monad, inverse diad, inverse triad, inverse tetrad, and inverse hexad axes. (An example of an inversion-rotation axis is described under *tetragonal system.)

According to the number and types of symmetry elements present in a crystal, it is assigned to a particular crystal system. The crystal systems are arranged approximately in order of decreasing symmetry, the cubic system being the most symmetrical: a crystal must be assigned to a system as near to the top of the list as is compatible with its symmetry.

Let us take a simple example. Inspection of a box-shaped crystal would quickly reveal the presence of a diad (two-fold) axis passing through the centers of the two largest faces, suggesting that it may belong to the monoclinic system. Further examination, however, would reveal that there are, also, diad axes passing through the centers of the other pairs of faces – that is, three diads in all – and hence the crystal is

correctly assigned to the orthorhombic system. (In addition it should be noted that this crystal also possesses three planes of symmetry. These are insufficient to promote it to a more symmetrical crystal system, say the tetragonal.)

Nevertheless, comparison with a crystal of, for instance, epsomite, which is also orthorhombic (3 diad axes) but lacks mirror planes, suggests that subdivision of each crystal system into smaller categories according to the presence or absence of additional, systematically superfluous symmetry elements may be possible. In fact, as long ago as 1830, J. F. C. Hessel correctly predicted on theoretical grounds that, on the basis of the symmetry elements so far considered, the seven crystal systems could be subdivided into a total of 32 (and *only* 32) crystal classes.

It is worth noting that, when crystalline structures are studied at the atomic or crystal-lattice level of detail, two new kinds of symmetry elements, not observable in complete crystals, may be recognized. A detailed analysis of all possible combinations of symmetry elements and types of lattice produces a total of 230 categories called the 230 space groups. In the same way as an animal belongs to a phylum, an order, a family, a genus, and so on, so a crystalline substance belongs hierarchically to a system, a class and a space group. Unlike the divisions and subdivisions of the animal kingdom, however, which may be added to as new forms are discovered, the divisions and subdivisions of the crystalline state, being prescribed by the laws of mathematics, are precisely known. Although each year sees the discovery of a few new minerals and of many synthetic compounds, each belongs to an already established category of crystal structure.

Examples of real crystals having the geometrical perfection of our idealized examples are very rare. A number of external and often variable factors influence crystal growth, and these normally lead to the unequal development of symmetrically related faces, so that the full symmetry of a natural crystal is partly obscured and appears less than that of its idealized counterpart. In practice, for purposes of classifying a crystal, the crystallographer pays no regard to either the absolute or the relative sizes of faces: instead, he is concerned only with the angular relationships between faces.

A valuable aid in recognizing symmetry elements in even highly distorted crystals is

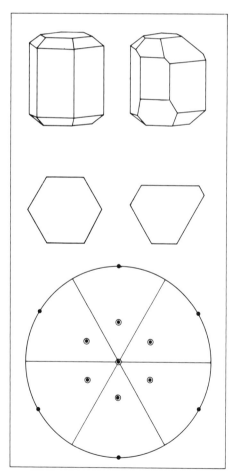

Above left, a "perfectly formed" crystal of apatite, with a cross-section in the shape of a regular hexagon. *Above right*, a "misshapen" apatite crystal, whose cross-section is an irregular hexagon. Angles between corresponding faces are the same in each case. Although the crystals are evidently of different shape, representation of either by a stereographic projection produces the same result (*below*).

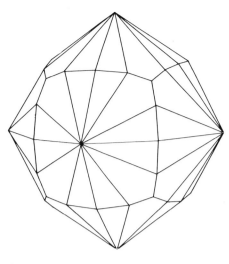

A crystal of silver, a member of the cubic system, having the form of a hexoctahedron.

the device known as the stereographic projection. The angles between perpendiculars to faces are first measured, preferably on an optical instrument called a goniometer, and the perpendiculars are then projected as points or (in the case of faces situated on the underside of the crystal) as small rings on or inside a circular diagram. A full account of the construction of the stereographic projection (stereogram) is beyond the scope of this book, but we can note in passing that, since the projection ignores the shapes and sizes of faces and represents only angular relationships between faces, it simplifies the recognition of crystallographic symmetry elements in the misshapen crystal.

Modern crystallography. The discovery in 1912 by M. von Laue that X-rays are scattered or diffracted in an orderly way when passed through crystalline material proved that such material has a regular internal structure – as had been proposed by Haüy. The physicists W. H. Bragg and his son, W. L. Bragg, quickly realized that the X-ray diffraction patterns produced on photographic plates indirectly represented layers of atoms or ions in the crystal structure, and in 1913, using this new and elegant technique, they were able to determine the detailed atomic structure of the simple substance sodium chloride (*halite).

From these early beginnings, structural crystallography has made great progress, and the precise atomic arrangements of thousands of substances, including most minerals, have now been elucidated. Even newer techniques, involving the diffraction behavior of neutrons and electrons and the use of modern computational methods, have extended the amount of information obtainable from crystalline substances.

A more or less routine geological application of X-ray crystallography is the rapid identification of a mineral from an X-ray diffraction photograph of a small amount of powdered sample. The positions and intensities of the curved diffraction lines seen in such a photograph provide in most cases a unique "fingerprint" for a particular crystalline substance, and its identity is

thus found from files of "fingerprints" using an established search procedure.

Cubic System

The unit cell of this system is a cube, in which all sides are equal and all faces and edges meet at right-angles. Clearly such a unit cell is capable of building a crystal that is itself a cube, and this object contains the maximum number of symmetry elements possible in a plane-sided three-dimensional object, namely 4 triad axes, 3 tetrad axes, 9 mirror planes, 6 diad axes, and an inversion center. Other crystal forms possessing all these symmetry elements, and thus belonging to the highest symmetry class (called the holosymmetric class) of the system, include the octahedron, the dodecahedron and the trapezohedron. Common minerals belonging to this class and crystallizing in these forms or combinations of them are *halite, *galena, *fluorite, *spinels (including *magnetite), *garnet, and the metals *gold, *silver and *copper.

The cubic unit cell may, however, be used to construct other crystal shapes which, while preserving the four triad axes (by definition essential to the cubic system), do not contain all the other symmetry elements of the cube itself. Two examples of crystal forms illustrating this point are the tetrahedron and the pyritohedron. The tetrahedron has four triad axes, but only diads are found in directions corresponding to the three tetrads of the cube, and the number of mirror planes is reduced from nine to six. The pyritohedron, named after the mineral *pyrite, is also less than holosymmetric, possessing four triads, three diads (strictly these are inverse tetrads) and only three mirror planes. Although the symmetry and hence the class of pyrite can be readily determined when it forms pyritohedra, the same mineral frequently crystallizes in the form of cubes, and may thus appear at first sight to belong to the holosymmetric class. However, examination of the faces of such cubes reveals the presence of striations which reduce the symmetry from that of the perfect cube to that of the pyritohedron.

Altogether there are five separate classes in the cubic system. Most common cubic minerals belong to the holosymmetric class: notable examples of minerals belonging to lower symmetry classes are *sphalerite, pyrite and *cobaltite.

Tetragonal System

The unit cell of this system is a square prism which may be envisaged as a cube that has been either stretched or compressed in the vertical direction. The holosymmetric class of the tetragonal system contains a single tetrad axis, 4 diad axes and 5 mirror planes. Although there are no fewer than six less symmetrical classes in this system, only one common mineral, *chalcopyrite, is represented among them. It is worth considering a simplified chalcopyrite crystal since it may be used to illustrate the concept of an inversion-rotation axis of symmetry. Such a crystal, called a sphenoid, possesses no tetrad axis, but passing along its length is an

Both halite (*top*) and spinel (*above*) are members of the highest symmetry class (holosymmetric class) of the cubic system since they possess six diad axes, four triad axes, three tetrad axes, nine mirror planes and an inversion center. Members of this class of the cubic system display the maximum possible number of symmetry elements of a crystal.

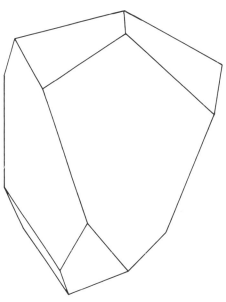

A sphenoidal crystal of chalcopyrite, representative of the tetragonal system.

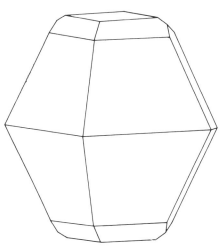

A crystal of sulfur, a member of the holosymmetric class of the orthorhombic system.

axis of symmetry that appears to be a diad (that is, it produces a position of congruence after rotation through 180°). In fact this axis is rather more than a diad: if the crystal is rotated through only 90° and then inverted (i.e., turned upside down), a position of congruence is achieved. This is an example of a four-fold inversion-rotation axis (or inverse tetrad) and its presence is characteristic of two of the lower symmetry classes of the tetragonal system.

Apart from chalcopyrite, the only common tetragonal minerals, *cassiterite, *rutile, *zircon and *idocrase, all belong to the holosymmetric class.

Orthorhombic System

The unit cell of this system is a rectangular parallelepiped, which may be visualized as a brick- or matchbox-shaped block.

The holosymmetric class of this system has three mutually perpendicular diad axes and three mirror planes. The class is represented by a large number of minerals, including *barite, *celestine, *sulfur, *stibnite, *topaz and *olivine.

There are two less symmetrical classes in the orthorhombic system: in one of these the three diads are retained but the mirror planes are lost; and in the other a single diad axis is combined with two mirror planes.

Monoclinic System

The monoclinic unit cell is a right prism having a parallelogram for its base. Such a cell, and crystals constructed therefrom, have rather low symmetry. The holosymmetric class has a single diad axis at right angles to a mirror plane, and this class includes many minerals, especially *gypsum and a number of important rock-forming types such as *orthoclase, *augite, *hornblende, *chlorite and *epidote.

Monoclinic minerals belonging to the two lower symmetry classes, which contain a diad axis or a mirror plane alone, are relatively rare.

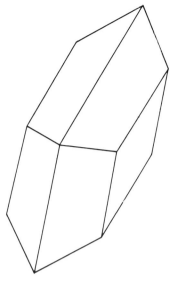

Gypsum, a member of the holosymmetric class of the monoclinic system.

A crystal of corundum, which crystallizes in the holosymmetric class of the trigonal system, having three vertical mirror planes that intersect along an inversion triad axis.

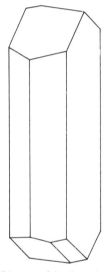

A crystal of albite, one of the plagioclase feldspars, a member of the triclinic system.

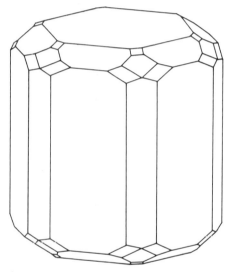

A crystal of beryl, a member of the holosymmetric class of the hexagonal system.

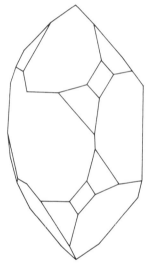

A right-handed crystal of the mineral quartz which, up to around 573 C (1063 F) crystallizes in the trigonal system.

Triclinic System

The triclinic unit cell is the most general type of parallelepiped, in which the three sides meeting at any corner are all of different lengths and in which the angles between edges are not right angles.

Crystals constructed from this type of cell have either no symmetry elements whatsoever or, in the case of the holosymmetric class, possess a single inverse monad. The effect of this axis is that each corner, edge and face of a crystal has an identical counterpart at the opposite extremity of the crystal.

*Kyanite and the important rock-forming silicates, the *plagioclase feldspars, belong to this class.

Hexagonal System

The hexagonal unit cell can be regarded as a monoclinic unit cell in which the parallelogram base has angles of 60° and 120°. The hexagonal holosymmetric class contains a single hexad axis, 6 diad axes and 7 mirror planes, as seen in well-crystallized examples of the mineral *beryl and its varieties *emerald and *aquamarine. Although there are no fewer than six less symmetrical classes in the system, all of which contain at least a hexad or inverse hexad, only two common rock-forming minerals, *apatite and *nepheline, are represented among them.

Trigonal System

The trigonal unit cell is a rhombohedron, which may be visualized as a cube deformed along one diagonal. The holosymmetric class contains an inverse triad axis, 3 diad axes and 3 mirror planes. *Calcite, the three cleavages of which happen to lie parallel to the unit cell sides and which thereby played such an important role in the birth of crystallography as a science, is the best known mineral belonging to this class. Other structurally related carbonate minerals, together with *corundum and *hematite, belong here also.

*Dolomite, *tourmaline and *quartz each represent a separate trigonal class of lower symmetry. It may seem surprising to find quartz in this system. Well-formed crystals having six symmetrically disposed prism faces are common, suggesting that quartz belongs to the hexagonal system – as indeed it does above a temperature of 573°C (1063°F). It can be shown, however, that at normal temperatures quartz is trigonal, and this is also indicated by a consideration of some of the smaller inclined faces on a well-shaped crystal.   FBA

Gemstones

Whether or not a mineral is classed as a gemstone, worthy of mining, cutting, polishing and mounting for the purpose of adornment, is ultimately a matter of supply and demand. Although there is no rigorous definition of a gem, it is possible to list the essential attributes common to the precious stones and shared, at least to some extent, by the semiprecious materials.

The first, self-evident quality is beauty. The beauty of a gemstone may depend on its intrinsic color (e.g., the fiery red of ruby or the verdant green of emerald); or on its dispersion, the property of splitting white light entering the stone into tantalizing gleams of color (e.g., the "fire" of diamond); or on special optical effects such as the iridescence of opal. The second quality is durability. Gem material is ideally extremely hard, and able to resist scratching and abrasion by all common substances, including steel. It should be impervious to chemical attack by all common acids, alkalis and other solvents, and it should be able to withstand accidental percussive blows – that is, it should possess no cleavage or cleave only with great difficulty. Finally the status of a gemstone is greatly enhanced if, as a raw material, it is rare. It seems unlikely that diamonds would be highly prized if they were as common as pebbles of flint.

Transparent and translucent gems are usually fashioned into stones with flat facets. It is important to realize that these facets, although resembling natural crystal faces, are artefacts. The arrangement of facets is chosen to enhance the desirable optical characteristics of the material. More or less opaque gem material such as turquoise or moonstone is usually cut and polished in cabochon form; that is, with a curved upper surface.

A few modern compounds unknown in nature are nowadays made into attractive gemstones, and a number of familiar gem materials such as ruby and emerald are successfully synthesized in the laboratory for use in jewelry.   FBA

Panning in Borneo for the most valuable mineral of all – diamonds. The techniques of panning are simple, though the work is slow and laborious. The pan is filled with water and loose material from the riverbed: larger lumps are broken by hand and the bigger pebbles discarded. The pan is then given a rotary motion, occasionally being tilted from side to side, so that the heavier particles settle towards the center and the unwanted light gangue material is washed over the rim.

Minerals

With the exception only of the relatively rare natural glasses such as obsidian and pitchstone, all the rocks which form the Earth, the Moon and the inner planets are composed of minerals. Minerals may be broadly defined as naturally formed solid substances made up of atoms arranged in an orderly and regular fashion. The chemical composition of a mineral is fixed or varies only within predictable limits, and is thus expressible – unlike that of a rock – as a chemical formula.

Minerals vary greatly in size, shape, color, beauty, abundance and economic importance. Gold, diamonds and rubies are minerals; so are salt, iron ores, asbestos, and the humble constituents of sand and mud. Despite the large number of known minerals (more than two thousand), nearly all types of rock consist almost entirely of a relatively small number, the so-called rock-forming minerals, although a further small number of economically important ore minerals are locally abundant in particular geological or geochemical environments.

Classification. Minerals are normally classified on a chemical basis. Thus we have groups such as the native (i.e., naturally occurring) elements, including copper, sulfur and diamond; the sulfides, which include galena and pyrite; the carbonates (e.g., calcite and dolomite); the oxides; the halides; and so on. The silicate group, which comprises most of the rock-forming minerals, is so large that it is convenient to subdivide it, and this is done on the basis of differences and similarities in the crystalline structures of members of the group.

In some instances, two or more minerals with different names have an identical chemical composition. These *polymorphs* have, however, different crystal structures and, in consequence, different physical properties; they rarely occur together, forming under different temperature and/or pressure conditions.

Another important concept in mineralogy is that of solid solution. The magnesium silicate, forsterite (Mg_2SiO_4), and the iron silicate, fayalite (Fe_2SiO_4), are rare minerals; they are, however, the extreme end-members of a whole series of common minerals, the *olivines, which contain both iron and magnesium in all possible ratios between these extremes. Such a series is called a solid solution series, and all members have the same crystalline structure. This phenomenon is common, especially among the silicates.

The Properties of Minerals. Some of the various physical properties of minerals useful for their correct identification in a hand specimen will now be discussed. Many of these depend directly or indirectly on the crystalline structure.

Crystal Shapes. When a crystalline substance is free to grow without constraint, it usually forms individual "bits" – i.e., crystals – bounded by flat faces which are disposed and related to one another in a way

Iron pyrites crystallizes in the cubic system. It is best known as fool's gold – because of its color, which is slightly similar to that of gold. However, pyrites and gold can be readily distinguished even in the field.

A polished slice of "crazy lace" agate. A form of chalcedony, agate is characterized by layers or bands of different color: these are generally irregular, sometimes concentric, and form by the progressive lining of a cavity.

Biotite displays a single perfect cleavage, as shown by the insertion of a razor blade between two "pages" of this biotite "book". Perfect basal cleavage is typical of the mica family, whose members include biotite, phlogopite, muscovite, lepidolite and glauconite.

determined by the underlying atomic structure. Thus a mineral will usually develop a shape characteristic of the species and called the "crystal habit". In some instances the way in which individual crystals of a mineral are joined or aggregated together is especially characteristic of the species.

Cleavage and Fracture. The atomic structure of many minerals is such that there may be one or more sets of planes within the structure across which the binding forces are weakest. If roughly handled, such minerals will break preferentially along such a plane or planes, giving rise to flat, usually shiny, cleavage surfaces. The number of cleavage directions in a mineral and their angular relationship to one another (if there are more than one) are important diagnostic properties.

Some minerals, having structures of more or less equal strength in all directions, do not exhibit cleavage. When roughly handled these substances fracture rather than cleave. Fracture surfaces are less angular than cleavage surfaces and may occur in any direction since they are not structurally controlled. They may, however, be of value in mineral identification.

Hardness. Different minerals vary greatly in their resistance to scratching and abrasion. This property, hardness, is constant for a particular mineral, and is easily tested by observing whether or not an unknown mineral scratches or is scratched by test minerals. The ten test minerals, originally selected by Friedrich Mohs (1773–1839) in 1812, are arranged in order of increasing hardness and assigned numerical hardness values as follows:

1. talc	6. orthoclase
2. gypsum	7. quartz
3. calcite	8. topaz
4. fluorite	9. corundum
5. apatite	10. diamond

All other minerals have Mohs' hardness

values within this range. If test minerals are not available, unknowns can be compared with the fingernail and a steel pocketknife blade which have hardness values of about $2\frac{1}{2}$ and $5\frac{1}{2}$ respectively. Minerals of hardness 1 feel soapy or greasy.

Density. The density of a substance is usually expressed in grams per cubic centimetre (g/cm^3). Density can, of course, be measured accurately using suitable equipment, but some minerals are so dense that even hefting a sample in the hand helps to identify them.

Optical Properties. The color of a mineral, its opacity or degree of transparency, and its luster (the way in which light is reflected from its surface) are immediately obvious properties. Unfortunately, however, these properties are not always consistent from one specimen to another of the same min-

Double refraction displayed by a crystal of Iceland spar (calcite). Certain crystals can split a ray of unpolarized light into two rays plane-polarized at right-angles to each other. One is refracted in the normal way, the other with a refractive index that depends on the original direction of the ray.

eral. Thus, although sulfur is always a distinctive yellow, quartz can occur in a wide range of colors resulting from minute traces of different impurities. But, used with experience and caution, the optical properties of minerals are of great diagnostic value.

Other Properties. Some minerals have distinctive properties which serve to identify them. These include the property of fluorescence, whereby certain substances emit light in the visible part of the spectrum when bathed in ultraviolet light; the property of effervescence with dilute acids, exhibited by many of the carbonate minerals; and the properties of solubility in water and taste. In addition, magnetite has distinctive magnetic properties, and uranium-bearing minerals are radioactive.

Of the greatest importance in mineral identification is a knowledge of likely mineral and rock associations. For example, zinc minerals often occur together with lead minerals; zeolites usually form in cavities within basalt; and gypsum and anhydrite frequently crystallize in *evaporite sequences. Conversely, a knowledge of incompatibilities is also important: for example, chromite is never found in granites.

Conclusion. The study of minerals can be considered on the one hand as a science in its own right – much modern research is concerned with determining the detailed structures of minerals using sophisticated X-ray and electron-microscopic techniques – and on the other as an essential part of the science of petrology – the origin and history of a rock is often deduced from a detailed study of its component minerals. Above all, the importance of minerals to mankind, past, present and future, cannot be overemphasized. As minerals are the raw materials which have given rise to and support our

Iridescence displayed by plagioclase feldspar. This phenomenon, the production of colors of varied hue, is due to the interference of light reflected from either side of faults and boundaries within crystalline solids (or from front or back of thin films).

complex civilization, we require to understand their nature and origin, and to exploit the Earth's not unlimited resources with wisdom and foresight.

Actinolite

Actinolite is a member of the *amphibole group, and may be regarded as *tremolite in which some of the magnesium has been replaced by iron. It forms green blade-like crystals or fibrous aggregates. If the fibers are long, visible to the unaided eye, and separable, then the mineral is a member of the *asbestos family; if they are minute and interlocking to give a fine-grained rock, the material is called *nephrite or *jade.

Actinolite is a metamorphic mineral found typically in low-grade *schists. With increasing temperatures of metamorphism it reacts with other components in the rock to form *hornblende.
Formula: $Ca_2[Mg,Fe'']_5(Si_8O_{22})[OH,F]_2$.

Amethyst, a colored variety of quartz, crystallizes in the trigonal system.

Agate

Agate is perhaps the best known variety of *chalcedony (compact submicroscopic quartz). It forms at low or moderate temperatures in cavities, and is characterized by possessing alternating layers or bands: these are arranged concentrically, paralleling the cavity walls. The bands vary in their color or degree of transparency, or both.

Agates are slightly porous and may be "improved" for use in jewelry by immersion in pigmenting solutions of the desired color. If the bands are plane and parallel the material is known as onyx. Black and white onyx is used for cameos, the piece being so carved that the figure in white stands in relief on a black background. Sardonyx is red onyx.

Moss agate is misleadingly named, for it shows no layering or banding, though the adjective "moss" is well chosen: set in white chalcedony is darker material (manganese and iron oxides), forming patterns much resembling delicate mosses, lichens or ferns.
Formula: SiO_2.

Amber

Amber is fossil resin or gum exuded by coniferous trees. The most celebrated amber, used for necklaces and bracelets, occurs as yellow-brown translucent nodules in *Oligocene deposits on the southern shore of the Baltic Sea, whence it is redistributed by ocean currents and washed up on the beaches of Norway, Denmark and eastern England. Some of this amber entrapped contemporary insects and spiders, preserving even the soft parts of these delicate organisms. There is one remarkable instance of a spider caught while manufacturing threads for its web; under the microscope it is possible to trace each thread back to the corresponding spinneret of the animal.

Amethyst

Amethyst is well crystallized *quartz having a pale violet to rich purple color due to its containing small amounts (up to 0.1%) of ferric *iron. On close examination it can be seen that the color is concentrated in thin sheets lying parallel to particular crystal faces.

Amethyst was known and valued in an-

Moss agate derives its name from the mossy patterns of iron and manganese oxides in white chalcedony.

tiquity. *Exodus* records that amethyst was one of the twelve stones inscribed with the names of the twelve tribes of Israel on the breastplate of the High Priest. It was believed to possess various magical properties, including the capacity to cure or prevent drunkenness (the name is derived from a Greek word meaning "not drunken").

Since amethyst is both attractive and durable (7 on Mohs' scale), its modern status as merely a lesser gemstone can only be ascribed to its relative abundance.

On heating to 550°C (1022°F), amethyst changes to a yellowish-brown color; and many of the stones marketed as citrine (naturally occurring yellow quartz) are in fact heat-treated amethysts.

Formula: SiO_2.

Amphibole Group

The amphiboles are an important group of rock-forming silicates widely distributed in igneous and metamorphic rocks. They all consist of double chains of silicon and oxygen atoms, the chains being linked by a wide variety of other elements. They form orthorhombic or monoclinic crystals elongated in the direction of the chains and having two cleavages parallel to the chains and at an angle of 120° to each other. This important property distinguishes the amphiboles from the *pyroxenes. Amphiboles further differ from pyroxenes in that they are silicate hydroxides, the hydroxide (OH) group being an essential part of the structure.

All amphiboles have nearly identical amounts of silicon, oxygen and water, but are otherwise very variable chemically because of the extensive atomic substitution that takes place. Altogether some thirty amphibole minerals are known, the more important being *actinolite, *glaucophane, *hornblende, *riebeckite, *tremolite, *asbestos and *nephrite.

Andalusite

The minerals andalusite, *kyanite and *sillimanite are aluminum silicates with identical chemical composition, but each has its own individual structure and related physical properties. This is an example of polymorphism, and the minerals are described as polymorphs. All three form in aluminum-rich rock, typically pelites (sediments in which *clay minerals are abundant), during metamorphism, the pressure and temperature conditions controlling which polymorph is formed.

The formation of andalusite is favored by low pressures, and it is thus found typically in contact metamorphic pelites, although it occurs also in some pelites which have been regionally metamorphosed at depths less than 30km (18mi). The crystals are elongate, with square cross-sections, and are usually reddish or green in color. Cross-sections of the variety of andalusite called chiastolite show a black cross formed by the crystallographic arrangement of *carbon inclusions: such sections are sometimes worn as amulets.

Andalusite has been mined as a raw material for refractories and high-grade porcelain. In particular, for many years it was extracted with difficulty in the Inyo Mountains of California at an elevation of 3000m (10,000ft) for use in sparking plugs – until it was realized that aluminum oxide is a superior raw material for this purpose. Transparent green andalusite is occasionally used as a gemstone.

Formula: Al_2SiO_5.

Anhydrite

Anhydrite (calcium sulfate) is produced when a body of sea water is evaporated to about 15% of its original volume (see *evaporites): it has been estimated that the evaporation of a depth of over 700m (2300ft) of sea water is necessary to yield a 0.5m (1.6ft) bed of anhydrite. Large crystals are rare, the mineral usually forming as colorless or white sugary masses which readily break into rectangular cleavage fragments.

The name alludes to the lack of water in the crystal structure. The mineral hydrates easily, however, to form *gypsum. The reverse of this process may also occur, and the extent to which gypsum and anhydrite in evaporite deposits are primary minerals or have been secondarily derived from one another is not always clear.

The major uses of anhydrite are as a soil conditioner and as an additive to Portland cement to delay its setting time.

Formula: $CaSO_4$.

Apatite

Most magmas contain a small amount of phosphorus which, on crystallization, largely combines with calcium and fluorine to produce apatite, which is thus an important accessory mineral in many igneous rocks. It also occurs in a wide range of metamorphic rocks and, as *collophane, in fossil bones and other organic matter.

Crystals are usually elongate with hexagonal cross-sections: although commonly green or brown they may be translucent in a wide range of colors. Magnificent gems have been cut from the purple crystals of Maine, USA, and the yellow-green crystals of Durango, Mexico ("asparagus stone"), but apatite's low hardness (5 on Mohs' scale) makes it an inferior gem.

The mineral weathers rather easily, contributing to soils the phosphorus essential for plant life and necessary for the building of animal bone and tooth material.

The fluorine present in common apatite (fluorapatite) may be replaced partially or entirely by chlorine (chlorapatite) or by oxygen and hydrogen (hydroxyapatite).

Formula: $Ca_5[F,Cl,OH]_3(PO_4)_3$.

Aquamarine

The lovely transparent sky-blue to blue-green gemstone aquamarine is a variety of *beryl, the color being ascribed to minute amounts of iron or possibly to slight defects in the crystal structure.

Aquamarines occur in cavities in granite. Larger crystals – including the largest gemstone ever discovered, a 104kg (229lb) Brazilian crystal – are obtained from granite pegmatites. They are much less costly than *emerald, the green variety of beryl, doubt-

Aragonite crystallizes in the orthorhombic system: heat and pressure can change it to trigonal calcite.

less because of their greater abundance as large, relatively flawless, crystals.

Formula: $Be_3Al_2Si_6O_{18}$.

Aragonite

Aragonite has the same chemical composition as *calcite, but a different crystal structure. Although some of its properties are the same (for instance, reaction with dilute acid), those which depend upon its orthorhombic structure differ from those of trigonal calcite. Thus aragonite has only a poor cleavage and forms very slender pointed crystals; twisted coral-like aggregates known as "flos ferri"; or, when twinned, distinctive stout, six-sided crystals. The best examples of the latter are found in the Aragon district of Spain and in the sulfur deposits of Girgenti, Sicily.

Aragonite is less common and less stable than calcite. It usually forms near or at the surface in hot spring deposits, in beds of clay or *gypsum and in veins and cavities with other carbonates – calcite, *dolomite and *siderite. It has been noted in the upper oxidized zone of ore deposits.

The shells of certain mollusks are made of aragonite, a constituent of both mother-of-pearl and pearl itself. Many fossil shells were probably originally made of aragonite which has, during prolonged burial, gradually become calcite.

Formula: $CaCO_3$.

Arsenopyrite

The mineral arsenopyrite may be regarded as *pyrite (FeS_2) in which half the sulfur atoms have been replaced by arsenic. It forms monoclinic crystals, less symmetrical than the cubes of pyrite, and these may be distinctive. However, it is the color of the mineral, a metallic silvery-gray, which is its most diagnostic feature. It forms as a high-temperature vein mineral, often with *gold and the ores of *tin and tungsten.

It is an important ore of *arsenic.

Formula: FeAsS.

Asbestos

It is said that Charlemagne enjoyed impressing guests after a banquet by throwing the tablecloth into the fire, where debris was consumed without the cloth itself being affected. His party trick depended on the incombustible nature of the asbestos fibers making up the fabric.

Asbestos is the name given to a number of natural substances of widely differing composition, all sharing the property of crystallizing as long, thin, more or less easily separated fibers. Most important from a commercial point of view is the *serpentine mineral chrysotile, which accounts for over 90% of the world's asbestos production. Other types of asbestos are fibrous varieties of the amphibole minerals *actinolite, *tremolite and *riebeckite. Of these, the best known is the variety of riebeckite known as *crocidolite or "blue asbestos", which forms long, silky fibers. It is this asbestos which has achieved notoriety as a cause of lung cancer.

Sometimes crocidolite is replaced by *quartz, which mimics the silky fibers of the asbestos; iron originally present in the crocidolite structure is oxidized to give a golden brown color. This material is cut and polished and used in inexpensive je-

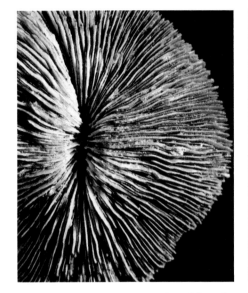

A rose of barite, whose name is derived from the Greek word for "heavy". Barite is found all over the world, fine crystals being known from several countries in Europe and North America.

Tiger-eye is formed when crocidolite, a type of asbestos, is replaced by quartz. The golden brown color is due to the oxidation of iron originally present in the crocidolite.

Top, asbestos mine at Wittenoom, Western Australia. *Above*, chrysotile, a form of asbestos, is fibrous serpentine. Asbestos is most used for fireproofing and for brake linings.

welry under the name "tiger-eye".

Major uses of asbestos are in the building industry, where it is a constituent of wall boards, and in the manufacture of brake linings for automobiles.

Augite
Augite is the most abundant and widely distributed *pyroxene, being an essential component of the basic *igneous rocks basalt, dolerite and gabbro, and a common component of the intermediate rocks andesite and diorite as well as some ultrabasic peridotites. It is also present in some high-grade *metamorphic rocks.

Although so abundant, and familiar to the professional geologist who studies his rocks with the microscope, augite only rarely forms crystals large enough to attract the attention of the amateur. Conspicuous stout black monoclinic crystals with eight-sided cross-sections are found in some porphyritic basalts and andesites, and in

some volcanic tuffs. Usually, however, the rock has to be examined with a hand lens and the right-angled cleavages, characteristic of all pyroxenes, must be sought. Augite in basic igneous rocks of certain types coexists with a second pyroxene, usually orthorhombic *hypersthene.
Formula: $[Ca,Mg,Fe,Al]_2[Si,Al]_2O_6$.

Azurite
The intense azure-blue of azurite makes it one of the more spectacular of minerals, especially when it occurs, as it often does, with bright-green *malachite. Both are copper carbonate hydroxides formed secondarily in the upper zone of copper ore deposits. Azurite may occur as tabular crystals or radiating aggregates, or may be fine-grained and interbanded with malachite. Often malachite completely replaces crystals of azurite, perfectly preserving their original shapes. At one time azurite was ground for use in paint, but the

gradual conversion of the azurite blues to malachite greens proved most unsatisfactory for the artist!

Azurite is a minor ore of copper, and is occasionally used as a decorative material.
Formula: $2CuCO_3 . Cu(OH)_2$.

Barite (Barytes)
Barite (barium sulfate) takes its name from a Greek word meaning "heavy", and its most characteristic property is its density, which is unusually high for a non-metallic mineral. It commonly grows as tabular crystals or as thin, divergent plates forming a "crested" or "cockscomb" aggregate. Barite concretions in some sandstones have a spectacular rosette-like appearance and are called "desert roses". Its crystals are normally colorless to white but may be tinged with a variety of hues.

Barite is a common mineral and occurs in a variety of deposits. It is frequently found with ore minerals in metalliferous veins, and in cavities in limestones and other sedimentary rocks. Less commonly, it is found in hot-spring deposits and as the petrifying mineral in some *fossils. It is the principal source of *barium for the chemical industry, but its chief use is in the petroleum industry, where more than a million tonnes a year are used in the search for oil. Mixed with mud it provides a high density slurry which lubricates the drilling bit in deep oil wells, but which has sufficient weight to prevent oil and gas being blown from the hole.

Formula: $BaSO_4$.

Beryl

Beryl's chief claim to our attention is provided by its famous gem varieties, *emerald and *aquamarine.

Beryls typically grow as hard, more or less elongated crystals with hexagonal cross-sections. The ordinary variety is colorless or pale green, but other colors include yellow (heliodor), pink (morganite), blue-green (aquamarine) and deep green (emerald). All types other than emerald are chiefly formed as accessory minerals in granitic rocks, frequently occurring in cavities and pockets in the rock. Beryl crystals in some granite pegmatites grow to very large sizes, up to 2m (6.5ft) in diameter and 6m (19.5ft) in length.

Formula: $Be_3Al_2Si_6O_{18}$.

Biotite

Biotite is one of the most important members of the *mica group of minerals. Chemically complex, it shares with the other micas a sheet-like crystal structure which confers upon them a perfect cleavage, so that individual crystals can be split and opened up, like the pages of a book, by insertion of a knife blade. The cleavage plates thus produced are flexible, elastic and extremely smooth and shiny. Biotite is readily distinguished from *muscovite mica by its color, which is essentially black with sometimes a hint of green or brown.

Biotite is a common rock-forming min-

Bornite, an important ore of copper, is often called "peacock ore" for its iridescent tarnish.

eral. In igneous rocks it is usually associated with *quartz, *feldspars (especially *orthoclase), muscovite and *amphiboles in granites, syenites, diorites and their fine-grained equivalents: very large crystals are found in granite pegmatites. It is also an important constituent of many metamorphic rocks, especially *schists formed at low to medium grades of metamorphism. In contrast with their random arrangement in an igneous rock, the biotites of a micaschist are all aligned parallel, the result of growth under directed pressure.

A somewhat paler-colored mineral, richer in magnesium and poorer in iron than biotite, but in other respects identical to it, is called phlogopite. This occurs in ultrabasic igneous rocks (notably *kimberlite) and in some pegmatites.

Biotite and phlogopite are susceptible to chemical *weathering and are therefore much less common than muscovite in sedi-

ments, unless these are highly immature.

Formula: $K[Mg,Fe'']_3AlSi_3O_{10}[OH,F]_2$.

Bornite

Bornite is an important ore of *copper. A freshly broken surface has a metallic reddish-bronze coloration, but the mineral is usually identified by the purplish iridescent tarnish it develops on exposed surfaces. It is this property which gives rise to its popular name, "peacock ore".

Formula: Cu_5FeS_4.

Bronzite

Bronzite is an orthorhombic magnesium iron *pyroxene. Named for its bronze-like color and luster, it commonly occurs as stubby crystals or fibrous masses in basic and ultrabasic *igneous rocks. It is found in the early-formed rocks of many layered *igneous intrusions where it has accumulated by sinking through the magma; and it is an important constituent of many charnockites and granulites, and of the stony *meteorites called bronzite-chondrites.

Formula: $[Mg,Fe'']_2Si_2O_6$.

Calcite

Of the few non-silicates sufficiently abundant and concentrated to be regarded as rock-forming minerals, calcite is the most important.

Well-formed crystals occur in a bewildering variety of shapes; but all break in the same way along three mutually oblique cleavage directions to yield rhomb-shaped cleavage fragments. It was this observation, first made in 1782 by *Haüy, that led to the concept that all crystals of the same substance are built of identically-shaped minute building blocks.

Calcite is usually milk-white but it may be tinted in a variety of hues due to impu-

Calcite crystallizes in the trigonal system. Twinning is common (*below*). Though calcite adopts a wide variety of crystal habits – the most common being tabular, scalenohedral (dog-tooth), prismatic and rhombohedral (*below left*) – all of these display a perfect rhombohedral cleavage. Calcite is a major constituent of the limestones.

rities. When, however, it is clear, transparent and colorless ("Iceland Spar"), calcite exhibits better than any other common mineral the property of double refraction; that is, the splitting of light into two rays so that when an object is viewed through the mineral two images are seen.

Calcite is relatively soft (3 on Mohs' scale) and, like most of the carbonates, effervesces readily in dilute acid, giving off bubbles of carbon dioxide.

*Limestone and chalk are sedimentary rocks whose chief or only constituent is calcite. The mineral may be precipitated directly from seawater, but, as it also forms the shells of many living organisms, these on death may accumulate to form bioclastic limestones. Chalk consists of the calcite shells of the microorganisms *coccoolithophores and *foraminifera. Rainwater which has absorbed carbon dioxide during its descent is capable of dissolving calcite, and spectacular cave systems often develop in limestone country (see *karst landscapes).

If limestones are metamorphosed, the calcite recrystallizes in larger grains and the white granular rock produced is called marble. *Carbonatite is an unusual type of igneous rock rich in calcite and other carbonate minerals, and calcite is common in high, medium and low temperature veins where it may accompany metallic ores.
Formula: $CaCO_3$.

Cassiterite
The Phoenicians satisfied the needs of the ancient world for tin – a component of bronze – by purchasing it from islands "in the Atlantic" known as the Cassiterides. It seems almost certain that the islands were the British Isles, specifically Cornwall, where the mineral cassiterite has been mined intermittently for thousands of years.

Cassiterite forms tetragonal, often twinned, brown crystals in high temperature ore deposits. Some granites, in Cornwall and elsewhere, are surrounded by zones of mineralization, cassiterite together with *wolframite, *topaz, *tourmaline and *arsenopyrite forming in veins in the innermost, highest temperature, zones. It seems very probable that Cornish tin was obtained initially not from the veins themselves but from ancient stream channels where the ore was concentrated in an easily extractable form.
Formula: SnO_2.

Celestine (Celestite)
Celestine is usually found disseminated in limestones, sandstones and shales, or forming thin layers between them. The most spectacular crystals come from cavities in sedimentary and volcanic rocks.

The mineral is difficult to distinguish from *barite, having an identical orthorhombic structure and similar physical properties. The rather rare, delicate sky-blue variety to which the mineral owes its name is distinctive. Celestine is mined for its *strontium.
Formula: $SrSO_4$.

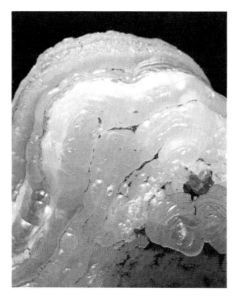
Chalcedony is silica in the form of minute quartz crystals having submicroscopic pores. Here it is seen in its mammillated form.

Chalcopyrite (copper pyrites), a brassy yellow mineral, crystallizes in the tetragonal system, its crystals having a more-or-less tetrahedral appearance.

Cerussite
Cerussite, lead carbonate, often forms well-shaped white or gray orthorhombic crystals in a variety of shapes, the distinctive features being high specific gravity and a brilliant luster. The mineral, which is an ore of lead, is usually of secondary origin, forming in the upper oxidized parts of mineral veins by the action of circulating waters on the primary lead sulfide, *galena.
Formula: $PbCO_3$.

Chalcedony
Chalcedony is the variety of *quartz that lacks all external evidence of crystallinity. Although X-ray and chemical evidences prove that chalcedony consists of quartz grains, these are usually so minute as to be invisible in even the most powerful optical microscope: the word "cryptocrystalline" is applied to this condition.

It is somewhat less dense than ordinary quartz owing to the presence of submicroscopic pore spaces in the material. It is precipitated from silica-rich solutions to form cavity linings, veins and the cementing material for many sediments; and it may replace, partially or entirely, fossils, wood and other minerals. The most familiar examples of chalcedony are chert and flint nodules formed in chalk and limestones by deposition of silica on the sea floor.

A large number of semiprecious and decorative materials are subvarieties of chalcedony. These include sard and cornelian (translucent red), *jasper (opaque red, brown or yellow), chrysoprase (translucent green), prase and plasma (opaque green), bloodstone or heliotrope (green with red spots), *agate and onyx (banded), moss agate (whitish with dark moss-like patterns), and tiger-eye (replacing blue *asbestos).
Formula: SiO_2.

Chalcopyrite
Chalcopyrite is a brittle, brass-yellow metallic-looking mineral somewhat similar in appearance to *pyrite and *gold. It is easily scratched to produce a greenish-black powder. The mineral is tetragonal, although well-shaped tetrahedral crystals are rare: it commonly forms as irregular masses, often with a slight iridescent tarnish.

Chalcopyrite crystallizes in *igneous rocks and in hydrothermal veins with other metallic sulfides. Although at least six other ore minerals contain greater amounts of *copper, chalcopyrite is so abundant that it is one of the major sources of the metal.
Formula: $CuFeS_2$.

Chamosite
Chamosite is closely related in structure and composition to *chlorite. It is the most important primary iron silicate in *ironstones other than those of Precambrian age. It occurs as small green spherical aggregates (ooliths) commonly with *siderite, *calcite, *kaolinite and *goethite, and is also a constituent of some laterite deposits (see *bauxites and laterites). In some cases chamosite is thought to be an original precipitate; in others it appears to have progressively replaced clay or fossil material. It is a major *iron ore.
Formula:
$[Mg,Fe'']_3Fe'''_3(AlSi_3)O_{10}(OH)_8$.

Chlorite
The characteristic green color of many slates and other fine-grained rocks metamorphosed in the greenschist facies is due largely to the presence of the mineral chlorite. The name derives from the Greek word for green, *chloros.*

Chlorite forms platy crystals, scaly aggregates and earthy masses. Crystals have a single perfect cleavage reminiscent of the *micas, but cleavage flakes are brittle and inelastic.

In addition to its principal occurrence in low-grade regionally metamorphosed *schists, chlorite also forms in igneous rocks as an alteration product of *pyroxenes, *amphiboles and micas, and is present in the clay fraction of many sediments.

Formula:
$[Mg,Fe'',Fe''',Mn]_6(AlSi_3)O_{10}OH_8$.

Chromite

*Chromium is present in small amounts (a fraction of a percent) in basic magmas. As the magma cools, the metal combines with iron and oxygen to form chromite, which can often be seen as scattered accessory grains in the resulting rocks. Fortunately, since it is the only ore of the metal, chromite occasionally becomes concentrated by the process of crystal settling: if the magma is solidifying slowly, chromite grains, being denser, are able to sink to the floor of the magma chamber where they form thick layers and lenses.

The mineral is black with a metallic luster and forms equant cubic grains somewhat resembling those of *magnetite. The grains are durable and, when released from their parent rock by erosion, may become concentrated in alluvial sands and gravels.

In addition to its importance as the ore of chromium, chromite is made into refractory bricks used for lining blast furnaces.
Formula: $FeCr_2O_4$.

Cinnabar

Because of its bright vermilion-red color, cinnabar attracted attention in early times, and was powdered for use as a pigment before it was known to contain mercury – a dangerous poison. It is the commonest *mercury mineral and is the only important mercury ore. Cinnabar forms around hot springs and in fractures in areas of recent volcanic activity.
Formula: HgS.

Clay Minerals

To the mineralogist, "clay" is a generic term for a group of related minerals. They are related chemically (being all essentially hydrous aluminum silicates); structurally (being all sheet silicates similar to the *micas); and in their modes of occurrence (being aggregates of extremely small particles usually produced by the breakdown of aluminosilicates such as the *feldspars). The most important are *kaolinite, *halloysite, *illite, *montmorillonite and *vermiculite. They are not readily distinguishable from one another, and sophisticated techniques such as X-ray analysis or electron microscopy are usually necessary to identify individual species.

Collophane

Collophane is the name given to fine-grained phosphatic material occurring as a constituent of fossil bones and other organic matter. Mineralogically, it is a variety of *apatite. Extensive bedded deposits of collophane supply the market for crude phosphate required in the manufacture of agricultural fertilizers (see *phosphorus).
Formula: $Ca_5[F,Cl]_3(PO_4)_3$.

Cordierite

Cordierite usually crystallizes as irregular quartz-like masses or grains which may be colorless, blue or violet-blue. The trans-

Diamond mining employs a sophisticated technology in order to extract the precious stones – but not always, as shown by this diamond mine in Borneo.

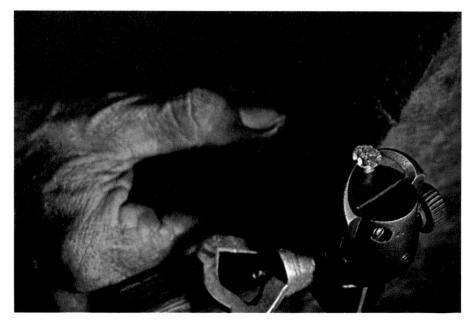

The faces of a gem diamond are not natural crystal faces: skilled craftsmanship must be used so that the "fire" of the stone is best enhanced. Here is a diamond in the final stages of cutting.

parent violet-blue variety, iolite, is used in jewelry.

Cordierite is formed by the medium- or high-grade metamorphism of aluminum-rich rocks, and is found in *schists, *gneisses and *hornfelses, as well as in alluvial gravels derived from these rocks.
Formula: $[Mg,Fe'']_2Al_4Si_5O_{18}$.

Corundum

Corundum is important for two main reasons. Firstly, it is second in hardness only to diamond among natural substances. The mineral and its synthetic equivalent (alundum) are widely used as industrial abrasives: emery, a fine-grained natural mixture of corundum and *magnetite, has been mined on the Greek Island of Naxos for thousands of years. Secondly, the vari-

eties *ruby and *sapphire combine beauty with durability, and so have been used as precious stones since earliest times.

Common corundum is typically gray or weakly tinged in any of several colors, and forms hexagonal crystals tapered at their extremities. It is of widespread occurrence, forming during metamorphism in crystalline limestones and dolomites, gneisses and schists: in some nepheline *syenites it has formed directly from a magmatic melt.
Formula: Al_2O_3.

Cristobalite

Silica occurs naturally in several different structural states (polymorphs). Of these *quartz, with its variations, is by far the most abundant, while others form only under enormously high pressures and are known only from large *meteorite impact craters.

Cristobalite lies between these extremes, being of fairly widespread occurrence but seldom abundant. It forms as small white or gray crystals in some lavas (e.g., *rhyolite, *obsidian) at temperatures higher than those appropriate for the crystallization of quartz. *Opal is essentially a variety of cristobalite, but where temperatures of formation have been relatively low. Formula: SiO_2.

Crocidolite
Crocidolite, also called "blue asbestos", is a variety of the amphibole mineral *riebeckite (see also *asbestos). Formula: $Na_2Fe''_3Fe'''_2Si_8O_{22}[OH,F]_2$.

Cryolite
Although an uncommon mineral, cryolite is important as a flux in the electrolytic process of refining bauxite ore to obtain *aluminum. It occurs in large amounts in only one location: Ivigtut, at the head of Arksukfjord in southwest Greenland. Here the cryolite is found in a unique pegmatite, rich also in *siderite, *quartz, *fluorite, *galena and other minerals. Nowadays much of the cryolite used in the production of aluminum is made artificially from *fluorite. Formula: Na_3AlF_6.

Cuprite
Cuprite is an important secondary ore of *copper, forming in the oxidized zone of copper deposits together with the copper carbonates *malachite and *azurite, and other less common oxides and silicates of this valuable metal.

Cuprite crystals belong to the *cubic system and are usually well-shaped octahedra or cubes, although the variety chalcotrichite (or "plush copper") consists of fine, hair-like fibers. Fine-grained earthy masses of cuprite are called "tile ore". Its color, usually a fine dark red, contrasts splendidly with the vivid greens and blues of associated copper carbonates. Formula: Cu_2O.

Diamond
No mineral excites our imagination more than diamond. Rare and beautiful, it holds pride of place among gemstones. In the modern industrial world, diamond achieves a strategic importance through its unique durability. As the hardest known substance, and quite impervious to attack by acids and alkalis, large quantities are used for grinding, cutting and polishing hard materials. Industrial diamonds are today in increasing amounts prepared synthetically.

In early times diamonds were obtained only from river deposits which, together with certain beach and off-shore gravels, are still important sources. The richest diamond workings, however, and notably those of South Africa, are within pipe-like bodies of *kimberlite. These kimberlite pipes represent magma squeezed up from great depths: from them the secondary deposits are derived by weathering and erosion.

Chemically, diamond is pure *carbon crystallized under great pressure and at high temperatures within kimberlite. Carbon crystallizing under less extreme conditions forms *graphite which has entirely different properties.

Crystals are usually octahedra or cubes, sometimes with curved faces. They may be colorless and transparent, or yellow, brown, reddish or even black. Uncut crystals have a distinctive, rather greasy, appearance.

Gem diamonds are cut so that white light entering the top of the stone is reflected internally and returned, being split into the spectral colors during its passage through the stone. The flashes of delicate colors seen in a well-cut stone are referred to as the "fire" of the diamond. Formula: C.

Diopside
Diopside is a calcium magnesium *pyroxene occurring chiefly in limestones and dolomites which have been thermally metamorphosed. The calcium of the limestone and the calcium and magnesium of the dolomite combine with silica – either supplied by the magma or originally present in the rock as sand grains – to form white or pale-green crystals of diopside that are occasionally large enough to be cut as gemstones. Less commonly, diopside is present in some ultrabasic rocks (where it may be bright green due to a small amount of chromium); and in some *meteorites. Substitution of iron for some of the calcium and magnesium of diopside changes it to the commonest pyroxene, *augite. Formula: $Ca[Mg,Fe'']Si_2O_6$.

Dolomite (Pearl-Spar)
Dolomite occurs widely as a rock-forming mineral. It may be mistaken for *calcite, especially when it forms white rhombohedral crystals with calcite-like cleavages, but it dissolves only slowly in cold dilute acid. It is more readily identified when its crystals develop curved composite faces to produce saddle-shaped aggregates often having a pink tinge and a pearly luster (whence its alternative name "pearl-spar").

Dolomite is formed in a variety of ways. As massive sedimentary deposits (*dolomite is also a rock name), it is probably largely of secondary origin, the result of magnesium-bearing solutions acting on calcite limestone. It occurs as a primary mineral in lead-zinc hydrothermal veins, and is found in cavities and geodes.

The mineral is used as a source of magnesium and in the production of cements and refractory bricks for furnace linings. Formula: $CaMg(CO_3)_2$.

Emerald
Emerald, the rare green variety of *beryl, is weight for weight more costly than *diamond. In contrast with other varieties of beryl, which form in granites and pegmatites, emerald is found in mica *schists and in *calcite veins.

Its green color is due to traces of chromium. Material of the finest deep green color is often badly flawed, and large stones of good quality are rare and extremely costly. Almost indistinguishable synthetic

Emerald's bright green is due to the presence of small quantities of chromium. This rare variety of beryl is weight for weight the most valuable gemstone of all, surpassing even diamond.

Epidote, also known as pistacite, crystallizes in the monoclinic system. Related minerals include clinozoite, piemontite and zoisite (orthorhombic).

emeralds are now made by a secret process: they arouse the suspicions of the expert only because they tend to be too free from flaws to be natural stones. Formula: $Be_3Al_2Si_6O_{18}$.

Enstatite
Enstatite's chief occurrence is in *meteorites (enstatite chondrites): although rare in the Earth's crustal rocks, it may be a major constituent of the upper mantle. It is an end member of the important series of orthorhombic magnesium-iron *pyroxenes. The minerals *bronzite and *hypersthene, which contain increasing amounts of iron substituting for magnesium, are more abundant members of this series. Formula: $MgSiO_3$.

Epidote
Epidote is formed mainly by low- to medium-grade metamorphism of calcium-rich igneous and sedimentary rocks. Individual crystals are green to black striated prisms which may be mistaken for *tourmaline unless the perfect epidote cleavage, parallel to the crystal's length, is noted. When the mineral occurs as disseminated grains in a metamorphic rock its yellowish-green color, similar to that of the pistachio nut, is unique among minerals and quite distinctive. Epidote is occasionally encountered as a gemstone. Formula: $Ca_2Fe'''Al_2O.Si_2O_7.SiO_4(OH)$.

Feldspar Group

The feldspars are the most abundant minerals in the Earth's crust and are widely distributed in igneous, metamorphic and sedimentary rocks: indeed, the precise classification of many rocks depends upon the type and composition present.

They are aluminosilicates containing potassium, sodium and calcium, the group being subdivided into alkali feldspars (*microcline, orthoclase and *sanidine) in which potassium is dominant, sodium subordinate and calcium virtually absent; and the plagioclase feldspars, which vary continuously in composition from pure sodium feldspar to pure calcium feldspar, these having negligible potassium. (See also *feldspathoids.)

All the feldspars form colorless, white or pink (or rarely green) crystals of Mohs' hardness 6, with two cleavages at right angles. Crystals are blocky or, especially in the case of plagioclase, elongated and "lath"-shaped. The two groups of feldspars are distinguished firstly by the "company they keep" and secondly by the type of twinning they exhibit. (These differences are described under *plagioclase and *orthoclase.)

Feldspathoid Group

The feldspathoids are a group of silicate minerals with chemical compositions broadly similar to those of the *feldspars. However, they are relatively deficient in silica and richer in alkalis. The chemistry of the most common feldspathoids, *nepheline, *leucite and *sodalite, thus reflects the composition of the undersaturated magmas from which they have crystallized.

*Lazurite and *scapolite also belong to the feldspathoid group, but form under metamorphic conditions.

Fluorite (Fluorspar)

Fluorite is a common and widely distributed mineral. It crystallizes in veins, either alone or with metallic ore minerals, especially those of lead and zinc; and also forms as an accessory mineral in *igneous rocks. In addition, some of the finest crystals occur in solution cavities in *limestones. It commonly crystallizes as well-formed cubes with four perfect cleavages so disposed that a damaged cube will have one or more of its corners symmetrically broken off along a flat cleavage plane.

The color varies greatly. It may be blue, purple, green or yellow; less commonly colorless, pink, red or black. Although attractive, the mineral is too soft (4 on Mohs' scale) and too readily cleaved to find use as a gemstone. However, an interesting variety called "Blue John", found in Derbyshire, England, consists of fibrous flourite color-banded in shades of blue, purple and white, and this has been carved into ornamental vases and bowls since Roman times.

Fluorite gives its name to fluorescence. A variety of substances, including some varieties of fluorite and about a hundred or so other minerals, respond when bathed in ultraviolet light (which is invisible to the human eye) by glowing in bright colors that lie within the visible spectrum.

The name fluorite derives from the Latin *fluere*, to flow, and refers to its low melting point. It is this property which makes fluorite an important industrial mineral, used as a flux in the open-hearth smelting of iron. Other uses are in the production of hydrofluoric acid, refrigerant liquids and inert fluorocarbon resins such as those used to coat non-stick cooking vessels. Clear, colorless fluorite is made into lenses for use in special types of optical equipment.
Formula: CaF_2.

Galena

Galena, the most important ore of *lead, occurs chiefly in hydrothermal veins, often associated with the zinc sulfide *sphalerite. It forms also as a replacement mineral in *limestones and *dolomites.

Galena crystallizes as cubes or octahedra, or as a combination of these. It is soft ($2\frac{1}{2}$ on

Fluorite crystallizes in the cubic system, twinning (with one crystal interpenetrating the other) being common. The word "fluorescence" comes from fluorite's exhibition of that property, though the phenomenon is not especially pronounced. The most valued variety is Blue John (*above left*).

Mohs' scale), with a high density, a characteristic metallic lead-gray color, and three cleavages at right angles to one another and parallel to the cube faces. It is readily oxidized to a number of secondary lead minerals including *cerussite.

The extraction of lead from galena is one of the simplest metallurgical processes. The temperatures required are merely those of a coal fire, and indeed lead was smelted in this way by, among others, the early American settlers for lead bullets. Much galena contains significant quantities of silver: in extreme cases, galena is sufficiently rich in precious metal (up to 0.25%) to be mined specifically as a silver ore.

The abundance of the various isotopes of lead in galena specimens have been used by geochronologists for dating purposes (see *age of the Earth).
Formula: PbS.

Garnet Group

The garnets are a family of rock-forming silicate minerals which vary greatly in the metal atoms they contain. An individual can be regarded as an intimate mixture of two or more of the following end-members: pyrope (magnesium-aluminum garnet), almandine (iron-aluminum garnet), spessartine (manganese-aluminum garnet), grossularite (calcium-aluminum garnet), andradite (calcium-iron garnet), and uvarovite (calcium-chromium garnet).

All the garnets, whatever their composition, typically form hard, equidimensional 12-, 24- or 36-sided crystals, but in other respects they are extremely diverse. Garnets of all colors except blue are known, red garnets being the most commonly encountered.

The common garnet of *schists and

Galena crystallizes in the cubic system, octahedra being more common than cubes. In this photograph cubes of galena (dark) can be seen with minor calcite (white) and ankerite (buff).

*gneisses approximates to an almandine composition. Grossularite- and andradite-rich types are characteristic of metamorphosed impure limestones, and spessartine-rich garnets are found in metamorphosed manganese deposits. Garnet approximating to pyrope composition occurs in some ultrabasic *igneous rocks, and is probably an important constituent of the upper mantle. Uvarovite, the green garnet of some chromium deposits, is rare.

The hardness of garnet and its tendency when crushed to form sharp cutting-edges make it an important abrasive material. Some varieties, notably ruby-red pyrope, red almandine (carbuncle), orange grossularite (cinnamon stone) and green andradite (demantoid), are used as gemstones.

Glauconite

Glauconite is a member of the *mica group: however, it occurs not in typical mica-like "books" but rather as minute rounded aggregates of green platelets in sedimentary rocks of marine origin. It is the classic example of a mineral forming directly on the sea floor. Sediments rich in glauconite are distinctly green in color and are known as greensands. Since glauconite forms *in situ* in a sediment and contains potassium, it enables a greensand to be dated directly using the potassium-argon method of age determination (see *age of the Earth): as glauconite occurs in rocks of nearly all geological ages, its importance in establishing the time-scale for sedimentary rocks is obvious.

Formula: $K[Fe''', Mg, Al]_2 Si_4 O_{10} (OH)_2$.

Glaucophane

Glaucophane is an *amphibole. It occurs as slender bluish crystals or as massive or fibrous aggregates, typically in sodium-rich *schists which have undergone regional metamorphism at low temperatures and high pressures. These schists, often called blueschists because of their abundance of blue glaucophane, are often derived from geosynclinal sediments, and the necessary metamorphic conditions for their formation are typical of destructive plate margin environments (see *plate tectonics).

Formula:
$Na_2 [Mg, Fe'']_3 Al_2 Si_8 O_{22} (OH)_2$.

Goethite (Limonite)

Goethite, named after the German poet, Goethe, who collected minerals in his spare time, is produced by the oxidation and hydration of iron-bearing minerals such as *pyrite and *magnetite. Like *hematite, it forms black, rounded or stalactitic aggregates with a fibrous radial inner structure, sometimes showing beautiful internal color-banding in shades of brown and yellow.

Limonite is finely divided goethite with variable amounts of water of hydration, and is a distinctive ocherous yellow. Both goethite and limonite are natural pigments imparting yellow and brown hues to many rocks. With *quartz, they constitute the gossans or residual weathered cappings on deposits rich in iron-bearing sulfide minerals.

The greatest accumulations of goethite and limonite, those worked as ores of iron, have formed by direct precipitation from marine or fresh water in shallow seas, lagoons and bogs.

Formula (goethite): $FeO.OH$.

Graphite

Graphite is one of the natural forms of pure *carbon, the other being *diamond. It occurs typically in metamorphosed sedimentary rocks such as recrystallized limestones and coal beds, schists and quartzites, where it probably represents carbon of organic origin.

The contrasts in physical properties between carbon and diamond are extreme. Graphite is very soft, opaque, black and dull with a low density – diamond is more than 1.5 times as dense. The contrasts reflect grossly different atomic (and hence crystalline) structures, diamond requiring very high pressures and temperatures for its formation and graphite being the stable form of carbon under normal conditions.

Graphite is used, in bonded form, in pencils (its soft black flakes cleave away and adhere to paper), as well as in the manufacture of high temperature alloys, lubricants and electrodes.

Formula: C.

Gypsum

Gypsum is the most common sulfate mineral. Its main occurrences are in sedimentary deposits, particularly *evaporite sequences. Having a low solubility, it is one of the first minerals to crystallize from evaporating seawater.

It is unusually soft and can be scratched by the fingernail, a property which makes the compact, fine-grained variety alabaster suitable for ornamental carvings. Large crystals have characteristic parallelogram or lozenge shapes, and are sometimes intergrown to form twin crystals having the shape of arrowheads or fishtails. Transparent varieties are known as selenite, and fibrous types, which may form curved or twisted aggregates, are called satin-spar.

Gypsum transforms to and from *anhydrite by the loss and gain (respectively) of water. Partial loss of water, induced by carefully heating powdered gypsum, produces plaster of Paris: when mixed with water this reverts to gypsum, setting as a rigid mass. Such plasters are used widely in the building industry and in the making of surgical casts and molds for ceramics.

Formula: $CaSO_4 . 2H_2O$.

Halite (Rock Salt)

Halite is common *salt, most familiar as a small but essential part of our diet. The sea owes its taste chiefly to dissolved sodium chloride, and the mineral crystallizes when sea or salt-lake water evaporates to 10% or less of its original volume (see *evaporites).

Most is colorless or white: gray, pink, red or brown material owes its color to impurities, while a striking blue-and-purple type from Stassfurt, Germany, is chemically pure but has structural imperfections. Most halite forms as massive aggregates of interlocking crystals; individual crystals are in the shape of cubes.

Halite, like ice, is able to flow; but unlike ice, which moves downward as glaciers, salt deposits buried in the Earth's crust are lighter than the enclosing rocks and so tend to flow upward to produce structures called salt domes. Valuable concentrations of *petroleum and *natural gas are often trapped by such structures.

The human consumption of salt and its role as a food preservative originally provided its prime uses; and in ancient times it was an important form of currency. Nowadays the chemical industry consumes three-quarters of the world's salt pro-

Kidney iron ore is the name given to hematite in mammillated form. Hematite's distribution is wide, and large concentrations are found in many parts of the world: it is the most important ore of iron.

duction, and it is also used in large quantities to melt winter ice on roads.
Formula: NaCl.

Halloysite
A *clay mineral very closely related structurally and chemically to *kaolinite, halloysite is, like kaolinite, of widespread occurrence; and forms by the weathering or hydrothermal alteration of *feldspar and other aluminum silicate minerals.
Formula: $Al_4Si_4(OH)_8O_{10}.4H_2O$.

Hematite
Hematite is the principal ore of *iron and for this reason can be regarded as perhaps the single mineral of greatest importance to industrial civilization. Its name, derived from the Greek word for blood, refers to the red color of finely divided or earthy hematite, and most red rocks owe their color to this mineral. The largest deposits, those of economic value, are of sedimentary origin, the iron having been originally deposited on the floors of shallow seas. Many of these bedded ironstone formations are of Precambrian age: they are found on all continents.

Well-crystallized hematite, formed at high temperatures in igneous and metamorphic rocks and in hydrothermal veins, occurs as platey hexagonal steel-gray to black crystals with glittering mirror-like surfaces, quite unlike the red material of sediments.

In addition to being the chief source of iron, the metal on which modern civilization is founded, powdered hematite is employed as a fine polishing agent (jeweler's rouge), and has been used continuously as a pigment since earliest times: its stable red color is seen in the Paleolithic paintings in the caves of the Pyrenees, and in the hardy red oxide paints of the present day.
Formula: Fe_2O_3.

Hornblende
Hornblende is the most abundant and widely distributed *amphibole, occurring as a major component of many *igneous rocks, especially those of intermediate composition. It is also the dominant amphibole mineral in many medium-grade regionally metamorphosed rocks, its importance being reflected in the rock names hornblende-schist and amphibolite.

Hornblende forms more or less elongated, even fibrous, monoclinic, dark-green to black crystals, sometimes with a brownish tint due to the presence of titanium. It is most easily mistaken for *augite, but tends to form more elongated, six- rather than eight-sided crystals. Broken grains should be examined for the cleavage angle of 120°, characteristic of the amphiboles.
Formula: $[Ca,Na]_{2-3}[Mg,Fe'',Fe''',Al]_5$ $[Si,Al]_8O_{22}(OH)_2$.

Hypersthene
Hypersthene is the most abundant orthorhombic *pyroxene, containing more iron and less magnesium than does *bronzite. It occurs in somewhat more iron-rich (and less basic) rocks than does bronzite: thus the hypersthene-chondrites are a class of stony *meteorites more iron-rich than the bronzite-chondrites. The presence of hypersthene in a gabbro (or the closely related pyroxene, pigeonite, in a basalt) characterizes the rock as tholeiitic. Hypersthene is also an important constituent of many charnockites and granulites.
Formula: $[Mg,Fe'']SiO_3$.

Ice
Naturally formed ice can be classed as a mineral – and, indeed, is the most abundant mineral on the surface of the Earth! In areas where annual snowfall exceeds annual melting, ice caps and glaciers develop. These are of great importance as agents of erosion, as superimposed loads capable of depressing continental crust, and even as modifiers of worldwide sea levels. Frost too is significant. The splitting and shattering of rock caused by the expansion of water as it freezes in cracks and joints is an important weathering phenomenon. (For further discussions, see *glaciation, *glacial landscapes, *erosion and *weathering.)
Formula: H_2O.

Idocrase (Vesuvianite)
The alternative name for idocrase, vesuvianite, derives from the occurrence of the mineral in dolomitic limestone brought up from depth by the lavas of Mount Vesuvius, Italy. Reactions between igneous magmas and impure limestones produce a variety of new minerals of which idocrase is one. It forms well-shaped tetragonal crystals with square cross-sections in various colors, usually dark green or brown, occasionally blue, yellow or red. The brown transparent crystals from Vesuvius are cut as gemstones. A bright green massive variety, californite, is made into carvings and ornaments, and has been fraudulently marketed as jade.
Formula: $Ca_{10}Al_4[Mg,Fe'']_2(Si_2O_7)_2(SiO_4)_5(OH)_4$.

Illite
Illite, one of the *clay minerals, is structurally very similar to the *micas: it is the dominant mineral in shales and mudstones. Although some sedimentary illite may have been deposited as such after its formation by surface weathering of *feldspars, much of it is formed in place from other clay minerals during the subsequent compaction and lithification of the sediment (*diagenesis). It is extremely fine-grained, dull, and white or pale in color. Less commonly, illite is formed in certain hydrothermal ore deposits, and in alteration zones around hot springs.
Formula (simplified): $KAl_4[Si,Al]_8O_{18}.2H_2O$.

Ilmenite
Ilmenite is a common accessory mineral present as scattered grains in many *basalts and *andesites and their coarser-grained equivalents, *gabbros and *diorites. Occasionally it forms segregations of size sufficient to be ranked as ore bodies, and these are worked for their *titanium. Ilmenite crystals are tabular but rare, the mineral more commonly occurring as irregular grains and masses.

Black and with metallic luster, it resembles *magnetite, with which it often coexists. Unlike magnetite, however, it is non-magnetic and this is used to distinguish the two and to separate the ores. Ilmenite is resistant to weathering and so survives as a major constituent of some black beachsand deposits.
Formula: $FeTiO_3$.

Jadeite
The rarer of the two minerals popularly known as jade (see also *nephrite), jadeite belongs to the *pyroxene group of minerals. White or gray when pure, the green variety regarded by the Chinese as the choicest of all gemstones owes its color to traces of chromium.

Crystals are rare, the mineral usually

Two varieties of chalcedony, a cryptocrystalline form of quartz. On the left is jasper, in this case red though brown and yellow colorations are also known. On the right is chrysoprase.

forming tough, compact aggregates of microscopic grains. Such material, normally found as stream-worn boulders, has been used from early times for the fashioning of implements and ornamental objects of all kinds.

Jadeite is a metamorphic mineral, requiring conditions of high pressure and relatively low temperature. For this reason it forms typically at destructive plate margins, where underthrusting lithospheric plates generate high pressures but where geothermal gradients are low (see *plate tectonics).

Formula: $NaAlSi_2O_6$.

Jasper

Jasper is hard, opaque *chalcedony colored red, brown or yellow by finely divided particles of iron oxides (*hematite, *goethite and *limonite). It forms at moderate to low temperatures, sometimes as thick beds of considerable area.

Jasper is the most common form of silica found in petrified wood. Under certain conditions, buried wood may become saturated with water, rich in silica and iron oxides, and the cellular structure may be replaced and perfectly mimicked by jasper to the extent that even the species of tree can be identified.

Formula: SiO_2.

Kaolinite

Kaolinite has a sheet structure, and is one of the most common of the *clay minerals. It occurs as snowy white earthy masses and loose aggregates of submicroscopic hexagonal plates. It forms secondarily by the breakdown and alteration of aluminous silicates, especially potassic feldspars such as *orthoclase.

The most famous deposits are in Cornwall, England. Here the feldspars of the granites have been severely attacked by hot vapors rising through the rock. The kaolinite (china clay) thus produced is quarried with high-pressure water jets. The pure white clay is used in the manufacture of high-grade porcelain, and as a filler in paper.

Formula: $Al_4Si_4O_{10}(OH)_8$.

Kyanite

The minerals kyanite, *andalusite and *sillimanite have identical chemical composition. Each, however, has its own individual structure and related physical properties. This is an example of polymorphism. All three form in aluminum-rich rocks, typically pelites (sediments in which clay minerals are abundant) during metamorphism, the pressure and temperature conditions of metamorphism controlling which is formed.

The formation of kyanite is favored by high pressures, and it occurs in some *eclogites and in pelitic *schists formed at considerable depth in the Earth's crust. Crystals are long and blade-like and usually of a patchy blue color (the name derives from a Greek word meaning blue). They have the unusual property of being considerably softer parallel to the crystal length than at right-angles to the length: thus a knife will scratch a crystal in the former direction but will itself be blunted by the crystal in the latter. Transparent varieties are occasionally cut as gemstones.

As a commercial refractory mineral, kyanite is the most important of the aluminum silicate polymorphs.

Formula: Al_2SiO_5.

Lapis Lazuli (Lazurite)

The chief interest in the mineral lazurite, a member of the *feldspathoid family, lies in the attraction of its color (a deep azure-blue) when it occurs in the composite substance called lapis lazuli. Lapis, the commonly abbreviated name, is lazurite embedded in a matrix of white *calcite and almost always embellished with small golden specks of *pyrite. Prized since very early times, lapis lazuli occurs in only a few places, and only as a contact metamorphic mineral in limestones close to igneous intrusions. The most famous source is in the Kotcha Valley in Afghanistan. Here the lapis-rich marble mines were already ancient when visited by Marco Polo in 1271, and are still worked today. Lapis of fine quality is also obtained by the Russians from a dolomite limestone near Lake Baikal in Siberia. The artist's pigment ultramarine, responsible for the glorious blue colors in the Old Masters, was made from lapis lazuli until late in the 18th century: in more recent times synthetic lazurite has been used for this purpose.

Formula (lazurite):
$[Na,Ca]_8[Al,Si]_{12}O_{24}[S,SO_4]$.

Leucite

Conspicuous in some of the fine-grained lavas of Vesuvius, Italy, are relatively large, dull, white equidimensional crystals with 24 more-or-less equal faces. These are crystals of leucite, a member of the *feldspathoid group. It crystallizes in place of *feldspar from magmas relatively deficient in silica and rich in potassium; but, being unstable at high pressures, is not found in igneous rocks formed at depth, being virtually restricted to lavas of appropriate composition (e.g., certain trachytes). Moreover, since leucite is easily altered, it is rarely found in lavas older than the Tertiary.

Formula: $KAlSi_2O_6$.

Magnesite

Magnesite is found in compact porcelain-like or coarser marble-like masses or, rarely, as rhombohedral crystals which may be colorless, gray or yellowish. Although similar in appearance to *calcite, it is much less common, and is hardly affected by dilute hydrochloric acid unless this is first heated. It usually forms as a replacement mineral in one of two ways: by the action of carbonate-bearing solutions on magnesium-rich rocks (serpentinite, dunite, peridotite); or by the complementary process involving magnesium-rich solutions interacting with carbonate-rich rocks.

Magnesite mine in Austria. Magnesite is mined both for itself and as an ore of magnesium.

Magnetite (also known as magnetic iron ore), one of the spinel group of minerals, is an important ore of iron found in placer deposits, contact metamorphic aureoles, replacement deposits and intrusive igneous rocks. Its name derives from its property of being attracted by a magnet.

Magnesite, an ore of *magnesium, is used in the manufacture of certain cements and in the production of magnesium oxide.
Formula: $MgCO_3$.

Magnetite

The outstanding property of magnetite, one of the major ores of *iron, is its strongly magnetic character: it is attracted by a magnet, and some rare specimens called lodestones are themselves natural magnets. The fact that grains of magnetite forming in an igneous rock adopt an orientation and polarity determined by the Earth's prevailing magnetic field has been of crucial value in the development of modern geological concepts of sea-floor spreading.

Magnetite crystallizes in the cubic system, usually as opaque black octahedra, and is very widely distributed – as an accessory mineral in almost all igneous and many metamorphic rocks, it is well-nigh ubiquitous. Its high density sometimes results in its sinking in slowly cooling magmas to form accumulations of economic value. New deposits are discovered at depth using magnetometers. To obtain the iron, the ore is reduced at high temperatures in a blast furnace, where oxygen is liberated, impurities combining with a flux such as limestone to be removed as slag.
Formula: Fe_3O_4.

Malachite

The vivid green color of malachite makes it one of the more spectacular minerals, especially when it occurs – as it often does – with bright blue *azurite. Both form secondarily in the upper zone of copper ore deposits. Malachite is the more abundant of the two, sometimes completely replacing crystals of azurite with perfect preservation of their original shapes. Otherwise it usu-

Mammillated malachite lining a cavity and cementing limonitic gossan. When masses of malachite are cut across they show concentric bands of different shades of green: for centuries such material has been polished for decorative use.

ally occurs in aggregates of fibers forming encrusting masses with curved surfaces. When these are cut across, they show concentric color bands of different intensities of green.

Malachite is sufficiently abundant to be mined as an ore of *copper. It is often seen as a green patina or coating on copper roofs which have been attacked by carbon dioxide and moisture in the atmosphere.
Formula: $Cu_2CO_3(OH)_2$.

Marcasite

When a piece of domestic coal is split, a golden-colored film is occasionally seen on the split surface. This is marcasite. More spectacular forms of marcasite – ball-, spear- or arrow-shaped aggregates found in chalk and clay – often attract attention and are brought to museums in the mistaken belief that they are *meteorites.

Marcasite is chemically identical to *pyrite, but crystallizes in the *orthorhombic system under conditions of lower temperature and higher alkalinity. If well crystallized the two minerals can be readily identified from their shapes, but otherwise simple identification techniques are inadequate to distinguish them.
Formula: FeS_2.

Mica Group

The micas are a small but important group of silicates characterized by their atomic structure. Silicon and oxygen form continuous, well-defined layers or sheets bound only weakly to adjacent layers by other elements, so that the minerals have a perfect cleavage yielding thin, generally flexible flakes. Mica "books" are single crystals (often six-sided), the "pages" of which are incipient flakes defined by the cleavage. Cleavage flakes can be easily detached from a mica crystal with the blade of a pocket knife.

The most important micas are *muscovite, *biotite and *phlogopite.

Microcline

Microcline, a member of the *feldspar

family, is chemically identical to *sanidine and *orthoclase but has a more highly ordered structure, the aluminum and silicon atoms being arranged in a regular and predictable fashion. Microcline crystallizes at somewhat lower temperatures than do the other alkali feldspars, and it occurs typically in pegmatites and hydrothermal veins. Like orthoclase it is found as grains in sedimentary rocks.

Microcline forms stubby triclinic crystals very similar in their physical properties to orthoclase – indeed it is difficult to distinguish these two feldspars unless the crystals are green, since the only green feldspar is a variety of microcline called amazonite, or amazon stone, which is sometimes used as a gemstone. Microcline may also be white, pink, or gray.

The mineral may be intimately intergrown with *albite (this mixture is called perthite) or with *quartz, producing a texture reminiscent of ancient cuneiform writing – hence the name "graphic granite".
Formula: $KAlSi_3O_8$.

Monazite

Monazite is a phosphate of *thorium and the *rare earth elements and is exploited as a source of thorium. Although monazite is formed as yellow-brown crystals in granites, pegmatites and gneisses as an accessory mineral only, it is resistant to weathering and becomes concentrated in heavy beach sands and stream placers where it attains the status of an exploitable ore.
Formula: $[Th,Ce,La,Y]PO_4$.

Montmorillonite

Montmorillonite is a *clay mineral occurring only in gray or greenish gray earthy masses of such fine grain-size that crystals cannot normally be distinguished, even under the electron microscope. It owes its origin chiefly to the alteration of fine volcanic ash; and entire beds of volcanic debris may be converted to a montmorillonite rock called bentonite. When placed in water, montmorillonite absorbs water, swelling to

a gel-like mass, and this makes it an important industrial material. It is used as a plasticizer in drilling muds and as a catalyst. Formula:
$[Na,Ca]_{0.33}[Al,Mg]_2Si_4O_{10}(OH)_2 . nH_2O.$

Muscovite

Muscovite is one of the most important of the *mica group. Chemically complex, it shares with the other micas a sheet-like crystal structure which confers upon them a perfect cleavage, so that individual crystals can be split and opened up like the pages of a book. These cleavage flakes are flexible and extremely smooth and shiny. Muscovite is readily distinguished from *biotite mica by its color, which is silvery-gray, sometimes tinged with green, brown or pink.

It is a common rock-forming mineral, widely distributed in igneous, metamorphic and sedimentary rocks. In granites and pegmatites it is associated with *quartz, feldspars (especially *orthoclase and *microcline) and biotite. It is a common constituent of many low- to medium-grade *metamorphic rocks, especially micaschists, in which muscovite grains are aligned parallel due to growth under directed pressure. Muscovite is relatively stable during weathering and transport, and survives as a common constituent of clastic sedimentary rocks such as sandstones and siltstones.

Muscovite is an essential industrial material, due to its unique combination of properties: perfect cleavage, flexibility, low thermal and electrical conductivity, infusibility and transparency. Large muscovite sheets obtained from pegmatites are used in the electrical industry for capacitors and high-temperature insulators. Before the manufacture of glass, it was used for window panes (Muscovy glass), and it remains the best material for furnace windows. Finely ground muscovite is used as a filler in roofing materials, wallpapers and paint. Formula: $KAl_2(Si_3Al)O_{10}(OH)_2.$

Nepheline

Nepheline is the most important member of the *feldspathoid group. It crystallizes as hexagonal crystals or irregular grains with a distinctive greasy luster in silica-poor alkaline-rich *igneous rocks in place of *feldspar: it is found in both volcanic and plutonic rocks of appropriate composition (e.g., nepheline *syenites and *phonolites). Nepheline is mined for use in the manufacture of glass and ceramics. Formula: $NaAlSiO_4.$

Nephrite

Nephrite is the compact, fine-grained fibrous variety of *tremolite and *actinolite. One of the two substances popularly known as jade (see also *jadeite), it is highly prized as a gem and ornamental stone. From early times, Chinese craftsmen have worked the stone, and it has long been used by the Maoris for weapons and decorative objects.

The color of nephrite is variable. White or gray material is relatively free from iron (tremolite); spinach-green material is iron-bearing (actinolite); and other colors due to

Adamite is a zinc olivenite. The olivenites, which crystallize in the orthorhombic system, are hydrous copper arsenates, and all of them are rare. They are most commonly found in copper ore deposits, though even here their occurrence is only very occasional.

various impurities are known. The interlocking microscopic fibers impart to the material great cohesive strength, so that it is extremely difficult to break a specimen, even with a geologists' hammer. Formula: $Ca_2[Mg,Fe]_5(Si_8O_{22})(OH)_2.$

Olivine

Olivine is one of the most important rock-forming minerals. Its composition varies continuously from pure magnesium silicate (forsterite) to pure iron silicate (fayalite), most natural olivines being mixtures of these two end-members. Well-formed crystals are rare, the mineral usually occurring as rounded or irregular grains or as granular masses in igneous rocks. The color is usually a characteristic olive-green (hence the name), but iron-rich types incline toward yellow and brown.

Olivine is a common constituent of many ultrabasic and basic *igneous rocks, such as peridotites, gabbros, dolerites and basalts. Dunite, a rock originally described from the Dun Mountains of New Zealand, consists almost exclusively of olivine. Some dunites have formed by the sinking of olivine crystals and their accumulation on the floor of a magma chamber. Basalt lavas forming oceanic islands often contain rounded fragments of coarse-grained olivine-rich rocks which are believed to have originated in the upper mantle, where olivine is an important – perhaps even the dominant – mineral, as it is in many stony and stony-iron *meteorites. At depths of about 400km (250mi) within the Earth, the orthorhombic crystal structure of olivine is changed by pressure to a denser, cubic arrangement of atoms.

Gem-quality olivine is called *peridot. Formula: $[Mg,Fe'']_2SiO_4.$

Opal

Opal is a variety of *cristobalite which, unlike *quartz and other forms of silica, contains varying amounts of water. It forms at low temperatures from circulating ground waters or hydrothermal hot spring solutions in cavities and cracks, occasionally even impregnating unconsolidated sediments. Pure opal is usually colorless or milky white, but the mineral may be colored by impurities. These types are, however, of only limited interest compared with gem opal, the variety which shows the brilliant flashing play of colors unique among gemstones.

It has long been known that the colors are due to interference effects on light penetrating the surface layers, but the cause of the interference was discovered only in 1964, when the electron microscope revealed that

Massive green orthoclase feldspar from Australia. Orthoclase has two cleavages at right-angles to each other, one of these cleavages being visible in this photograph. Orthoclase is the most important of the alkali feldspars.

opal consists of minute spheres of cristobalite packed together in an orderly fashion. If the spheres are of appropriate size, the play of interference colors results.

Precious opal has been esteemed as one of the noblest of gems since early times, although it suffered a temporary decline during the Victorian era. The discovery in 1872 of Australian opals of surpassing beauty reestablished the former eminence of the stone.

Formula: $SiO_2 . nH_2O$.

Orthoclase

Orthoclase is chemically identical to *microcline and *sanidine but has a crystal structure intermediate between the other potassium feldspars. In microcline the aluminum and silicon atoms are well-ordered and regular; in sanidine they are disposed more-or-less randomly; whereas in orthoclase they are partially ordered. These differences reflect the temperatures at which the three polymorphs have crystallized – the lower this temperature, the more highly ordered the structure.

Orthoclase is the most important and abundant alkali *feldspar. It is a major constituent of granites and syenites and is the most common potassium feldspar in metamorphic and sedimentary rocks. It forms blocky or somewhat elongated monoclinic crystals, which may have square cross-sections, are white or pink in color, are slightly softer than *quartz, and show two cleavages at right-angles. Crystals are often twinned, this phenomenon being best seen on the freshly broken surface of a coarse-grained rock, such as granite, where twinned crystals show a flat cleavage surface extending over only approximately half the width of the grain. This so-called "simple twinning" is very distinctive and quite unlike the "multiple twinning" striations of *plagioclase.

Orthoclase is used as a raw material for the manufacture of porcelain, enamel and glass. A variety called moonstone shows a bluish *opal-like play of colors and is fashioned into gemstones. Perthite is an intimate mixture of orthoclase with *albite.

Formula: $KAlSi_3O_8$.

Peridot

Peridot is simply gem-quality *olivine. It has a bottle-green color and a somewhat oily luster, and makes handsome, if rather soft, cut stones.

Although olivine is a common mineral, large, clear inclusion-free crystals of suitable color are rare. The most famous source is the small island of St John in the Red Sea. St John's Island peridots were known in Biblical times and used by the ancient Egyptians, but no record of their source survived. This led to the speculation that ancient peridots were extraterrestrial, extracted from the *meteorites known as

pallasites. The true source was rediscovered in the late 19th century, and new pits have yielded superb specimens.

Formula: $[Mg,Fe'']_2SiO_4$.

Plagioclase

The plagioclases, abundant and widely distributed rock-forming minerals, belong to the *feldspar group, and vary in chemical composition from pure sodium feldspar (albite) to pure calcium feldspar (anorthite), different names being assigned as the sodium content decreases and the calcium content increases: albite, oligoclase, andesine, labradorite, bytownite, anorthite. Individual members are not normally distinguishable without chemical analysis or use of a special microscope.

The triclinic crystals are often somewhat elongated and "lath"-like, colorless, white or off-white, and show two cleavages at right-angles to one another. Their most distinctive feature is the appearance on cleavage surfaces of numerous parallel striations of different reflectivity due to multiple twinning. This property distinguishes plagioclase from the alkali feldspar *orthoclase.

Plagioclases occur in most igneous rocks: indeed, their precise composition is used as a mineralogical basis for *igneous rock classification. Albite is characteristic of granites, granite pegmatites and spilites; andesine of diorites and andesites; and labradorite of gabbros, dolerites and basalts. Very calcium-rich plagioclase is found in some ultrabasic and lunar rocks. Anorthosites are large bodies of oligoclase-andesine rock found in ancient shield areas of the Earth's crust. Plagioclase is also common in *metamorphic rocks.

Cut and polished labradorite shows an iridescent play of colors and so is used as an ornamental stone.

Formula: $Na(AlSi_3O_8)$ *to* $Ca(Al_2Si_2O_8)$.

Polyhalite

Polyhalite usually forms as pink or red masses (due to iron oxide inclusions) which may be fibrous or foliated. It occurs in bedded *evaporite deposits formed by the evaporation of sea or salt-lake water, and is one of the last minerals to be formed in this way owing to its high solubility. Polyhalite and similar potassium and magnesium salts crystallize only when evaporating seawater has been reduced to 1.5% of its original volume.

Formula: $K_2Ca_2Mg(SO_4)_4 . 2H_2O$.

Pyrite (Fool's Gold)

Pyrite, fool's gold, is a yellow metallic mineral which superficially resembles gold and hence may deceive the foolish or inexperienced prospector. Criteria for distinguishing the two are in fact quite simple: pyrite is brass-yellow, paler than *gold; is brittle and cannot be scratched by a knife, whereas gold can be cut like lead. Pyrite often forms regular crystals, the most common form being a cube with grooved or striated faces: other shapes are the eight-sided octahedron and the twelve-sided pyritohedron.

Pyrite is one of the most widely distributed of the sulfide minerals, forming in a wide range of environments and over a temperature range of $0-1000°C$ $(32-1800°F)$. It occurs in igneous rocks as an accessory mineral; in metamorphic rocks; and as nodules on the sea floor or as grains in black shales formed in stagnant, oxygen-deficient water. It is the most widespread and common metallic mineral in ore veins, being especially abundant with the sulfides of copper, lead and zinc. Pyrite is mined for its *sulfur.

Pyrite in mineral collections requires care since it is attacked by sulfur bacteria: specimens left untreated may be rapidly reduced to piles of white powder.

The mineral *marcasite has the same chemical composition as pyrite but a different crystal structure.

Formula: FeS_2.

Beads of pyrite seen here against a background of white calcite. Pyrite (iron pyrites) is the most widespread sulfide mineral, occurring in igneous, sedimentary and metamorphic rocks. It can be distinguished from chalcopyrite (copper pyrites) by its lesser hardness and paler color.

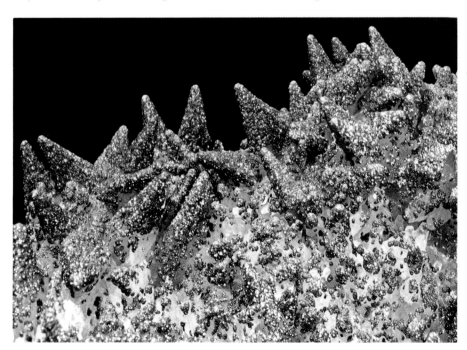

Quartz is one of the most widespread minerals on the surface of the Earth, and is found in a wide variety of forms. It crystallizes in the trigonal system. *Below*, well formed crystals of colorless quartz, rock crystal. As rock crystal is piezoelectric (when pressure is applied to a crystal, positive and negative electric charges appear on opposite crystal faces) it is finding use in oscillators for clocks, radio and radar. *Bottom*, a crystal of smoky quartz.

Pyroxene Group

The pyroxenes are a very important group of rock-forming silicates, found widely distributed in igneous and metamorphic rocks. They consist of single chains of silicon and oxygen atoms, the chains being linked by a variety of other elements. They form stubby or somewhat elongated orthorhombic and monoclinic crystals, having two cleavages at right-angles. This important property distinguishes pyroxenes from *amphiboles. Pyroxenes further differ from amphiboles in that they are anhydrous and thus occur in the water-free rocks of the *Moon and in *meteorites.

The most common rock-forming pyroxenes, *diopside, *augite, *enstatite, *hypersthene, *bronzite and *wollastonite, are silicates with differing amounts of magnesium, iron and calcium. The less com-

A mountain of quartz waste from kaolin workings in Devon, UK. Quartz is the most common gangue material in mineral veins.

mon members of the group, *jadeite, *spodumene and *rhodonite include additional elements.

Pyrrhotite

Pyrrhotite, a distinctive metallic, pinkish-bronze colored mineral, is an iron sulfide approximating to the formula FeS but showing a variable (up to 20%) deficiency in the amount of iron indicated by the formula.

Apart from *magnetite, pyrrhotite is the only appreciably magnetic mineral. It occurs principally as disseminated grains in some basic igneous rocks and in contact metamorphic deposits and pegmatites. Troilite, the sulfide found in *meteorites, is a variety with a negligible iron deficiency.

In recent years, pyrrhotite has been exploited as an ore of iron and sulfur.

Formula: $Fe_{1-x}S$ ($x \leq 0.2$).

Quartz

Quartz, one of the most abundant minerals in the surface layers of the Earth, is enormously diverse in origin, occurrence, varieties and uses. It is chemically extremely simple, consisting only of *silicon and *oxygen in the ratio 1:2. Impurities are never present in more than minute amounts. Well-formed crystals are six-sided prisms terminating in six-sided pyr-

amids. Quartz is hard (value 7 on Mohs' scale) and capable of scratching steel, is usually colorless or white, has a glass-like luster, and lacks a cleavage, forming curved fracture surfaces when broken.

Quartz forms over a wide range of temperatures, being an essential constituent of granites and granite pegmatites. It is the most abundant gangue mineral in ore-bearing veins; is common in a variety of metamorphic rocks, especially *gneisses; and is ubiquitous as a clastic sedimentary mineral. Common sand consists chiefly or entirely of rounded, water-worn grains of quartz. Consolidated *sandstones are extremely rich in quartz and it is virtually the sole constituent of the rock quartzite. In addition, quartz in the form of flint and chert is precipitated in sediments and on the sea floor.

There are many named varieties of quartz, of which *amethyst and *chalcedony, and the subvarieties *agate and *jasper, are the most important. Rock crystal is clear, colorless quartz, and is of great technological importance as the material for oscillator plates; milk quartz, smoky quartz, rose quartz and citrine are respectively white, gray-brown, pink and yellow. Rutilated quartz contains orientated needles of *rutile. Petrified wood and tiger-eye are respectively wood and fibrous *asbestos entirely replaced by submicroscopic quartz.

Formula: SiO_2.

Realgar

Realgar is a monoclinic mineral forming short, striated crystals or granular, compact masses. Its color, dark red to orange-red, attracted attention in early times, and it was used for centuries as a pigment until the poisonous nature of arsenic was discovered. It normally forms in hydrothermal veins with *stibnite and other arsenic minerals, around hot springs and in some limestones: it is a minor ore of arsenic. Fine specimens should be kept enclosed, since prolonged exposure to light reduces the mineral to a yellow powder.

Formula: AsS.

Rhodochrosite

Rhodochrosite has an attractive pink to rose-red color and would almost certainly be used as a gemstone if it were not far too soft. Crystals are rare, the mineral usually forming coarse cleavable masses or fine-grained color-banded aggregates in hydrothermal silver ore veins. In this association, rhodochrosite is believed to be a primary mineral, but it also forms secondarily in some high-temperature *metamorphic rocks and in residual manganese oxide deposits. Massive rhodochrosite is mined as a manganese ore.

Formula: $MnCO_3$.

Rhodonite

Rhodonite is closely similar in structure to the *pyroxene group (for this reason it is sometimes called a pyroxenoid). It occurs usually as pink masses or crystal aggregates in manganese-rich rocks. Its superior hardness, which allows a fine polish, and attractive pink color account for its use as an ornamental stone.

Formula: $MnSiO_3$.

Riebeckite

Riebeckite is a member of the *amphibole group. It occurs as discrete blue or blue-black crystals in some granites, syenites, rhyolites and trachytes, especially if these are rich in alkalis. It also forms veins in metamorphosed ironstones and sandstones, where it consists of fibers and is called *crocidolite.

Formula: $Na_2Fe''_3Fe'''_2Si_8O_{22}(OH)_2$.

Ruby

Ruby is the red, transparent variety of *corundum, highly prized since earliest times as a gemstone. It owes its color – ideally a dramatic deep red tinged with purple and described as "pigeon's blood" – to trace amounts of chromium. Although common corundum is abundant, ruby is rare, especially in the form of large unflawed crystals, and this doubtless influences the prices which quality stones command, weight for weight surpassing *diamond. Corundum and its varieties, ruby and *sapphire, are exceedingly hard, inferior in hardness only to diamond, and for this reason small rubies are used for durable bearings in watches and other instruments.

Large quantities of ruby are now produced synthetically by the Verneuil process (see *sapphire). Except to the expert, these are identical in all respects to natural stones, even surpassing many of them in quality of color and freedom from flaws. Market forces, nevertheless, impose much higher prices on the natural gems.

Formula: Al_2O_3.

Rutile

Rutile occurs as an accessory mineral in a wide variety of igneous and metamorphic rocks. When well-crystallized it forms elongated tetragonal crystals with square or octagonal cross-sections and pyramidal terminations. Needle-like crystals of rutile sometimes occur entirely enclosed within transparent grains of *quartz. Rutile concentrated in beach sands is exploited as an ore of *titanium.

Although the mineral is invariably dark red to black (due to iron impurities), chemically pure, colorless rutile is synthesized in the Verneuil process (see *sapphire) for use as a gemstone. This synthetic substance has refractive indices and "fire" much superior to those of diamond, and makes dazzling, if somewhat soft, gems.

Formula: TiO_2.

Sanidine

Sanidine is a *feldspar, chemically identical to *orthoclase and *microcline but having a less well-ordered structure, the aluminum and silicon atoms being arranged in a more random fashion. Sanidine forms at higher temperatures and is less common than orthoclase, occurring as flat, tabular, glassy crystals in some rhyolite and trachyte lavas.

Formula: $KAlSi_3O_8$.

Sapphire

Sapphire is the name given to all colored varieties of *corundum (except red, which is *ruby), and of these blue is the most abundant and popular. Blue sapphire owes its color to traces of titanium, and yellow sapphire to traces of iron. Sapphires are among the five most prized gemstones, being extremely hard and durable, attractive and rare, though less rare and hence somewhat cheaper than are rubies. A variety known as star sapphire contains microscopic needle-like inclusions arranged in three directions at angles of 60° to one another. When such a stone is fashioned to a smooth rounded shape (cabochon), light is reflected from the inclusions in the form of a six-rayed star.

Large quantities of sapphire and ruby are now produced synthetically by the Verneuil process, invented by a French chemist in 1902. The process involves feeding finely powdered aluminum oxide into an oxyhydrogen flame at 2000°C (3600°F), where it melts. The droplets fall onto a support beneath, crystallizing there as a pear-shaped mass called a boule. Addition to the alumina powder of small amounts of chromium oxide or titanium oxide produces the red and blue of ruby and sapphire.

Formula: Al_2O_3.

Scapolite

Scapolite refers strictly to a series of *feldspathoid minerals, mixtures of two complex aluminosilicates, one containing calcium and carbonate, the other sodium and chlorine. Members of the series crystallize in the *tetragonal system as normally white or bluish-gray prismatic crystals. A transparent yellow variety found in Brazil and Madagascar is cut as a gemstone. Often, however, scapolite is granular and massive with a splintery appearance. It occurs in *metamorphic rocks, particularly in metamorphosed impure limestones and as a replacement for *feldspars in altered *igneous rocks.

Formula:

$Na_4Al_3[Al,Si]_3Si_6O_{24}[Cl,CO_3,SO_4]$ to
$Ca_4Al_3[Al,Si]_3Si_6O_{24}[Cl,CO_3,SO_4]$.

Red crystals of realgar, which crystallizes in the monoclinic system. Realgar (AsS) is often found with another sulfide of arsenic, orpiment (As_2S_3).

Siderite crystallizes in the trigonal system. When its habit is massive (as shown here) it is generally described as clay ironstone.

Serpentines

The serpentines are: chrysotile, a fibrous mineral, the most important form of *asbestos; antigorite, which forms compact masses of small plates; and lizardite, extremely fine-grained matrix material usually associated with veins of chrysotile. Since the serpentines are sheet silicates like the *micas, it was not clear until recently how chrysotile could form fibers: X-ray and electron microscope studies have now shown that the fibers are sheets tightly rolled into concentric hollow cylinders and spirals.

The serpentines form secondarily from magnesium-rich silicates such as the *olivines and *pyroxenes of ultrabasic rocks: the alteration involves addition of water and a considerable volume increase. Serpentinites are rocks originally very rich in olivine that has been entirely converted to serpentine.

Chrysotile is the chief source of the asbestos used in the building industry. Massive varieties, especially those showing variegated colors, are carved as decorative or ornamental material.
Formula: $[Mg,Fe]_3Si_2O_5(OH)_4$.

Siderite (Chalybite)

Siderite is closely related to *calcite, its crystal shapes, three oblique cleavages and relative softness – those of its physical properties that depend on its trigonal crystal structure – being the same. The presence of iron rather than calcium, however, gives siderite a brown or yellow color, a higher density and a greater resistance to acid attack: hot, rather concentrated, hydrochloric acid is needed to make siderite effervesce. In addition, siderite becomes magnetic when strongly heated.

Massive siderite is common in sedimentary deposits of clay and shale, where it may form concretions called clay ironstones – it is extensively worked in these deposits as an ore of *iron. It also forms with sulfide minerals in metalliferous veins of hydrothermal origin.
Formula: $FeCO_3$.

Sillimanite

The minerals sillimanite, *andalusite and *kyanite all have identical chemical composition. Each, however, has its own individual structure and related physical properties. This is an example of polymorphism. All three form in aluminum-rich rocks, typically pelites (sediments in which clay minerals are abundant), during metamorphism, the pressure and temperature conditions determining which of the polymorphs is formed.

Sillimanite is relatively rare, but is of interest to geologists in that it denotes the highest temperature conditions in regionally metamorphosed pelites. It is usually distributed in *schists and *gneisses as long, slender – even fibrous – crystals, usually colorless. Pale blue transparent crystals, occasionally found in Burma and Ceylon, are cut as gemstones.
Formula: Al_2SiO_5.

Sodalite

Sodalite is a *feldspathoid. It occurs usually as blue masses with *feldspars and other feldspathoids in silica-poor igneous rocks, such as nepheline *syenites. It is similar in composition to, and often resembles, *lapis lazuli, but is distinguished by its mode of occurrence and lack of associated specks of *pyrite.
Formula: $Na_4Al_3Si_3O_{12}Cl$.

Sphalerite (Zinc Blende)

Sphalerite is the most important ore of *zinc. It occurs chiefly in hydrothermal veins, almost always with the lead sulfide, *galena. Apart from this, it also forms as a replacement mineral in limestones.

Until the discovery late in the 16th century of the means of extracting zinc from its ore, sphalerite in lead workings was a major inconvenience, since it was often mistaken for galena – but yielded no lead: the name sphalerite derives from a Greek word meaning treacherous.

It is not always easy to recognize. Its color varies, as the iron impurity content increases, from light amber with a resinous appearance to black with an almost metallic luster. Although it crystallizes in the *cubic system, crystals vary greatly in shape and may be complex and distorted with curved faces. An unusually large number of cleavages, six in all, is perhaps its most distinctive feature: cleavage and resinous luster together provide a reliable diagnosis.

While it is the chief ore of zinc, it often contains small amounts of the useful by-products gallium, cadmium, iridium and thallium.
Formula: ZnS.

Sphene (Titanite)

The secondary name for this mineral, titanite, is an allusion to its chemical composition: it is a minor ore of *titanium. The primary name derives from the Greek *sphenos*, an allusion to the mineral's typically flat wedge-shaped monoclinic crystals. Its color is usually yellow-green or brown, occasionally gray or black. Sphene crystallizes at high temperatures and is a common accessory mineral in many igneous and metamorphic rocks.

Although transparent sphene crystals from the St Gotthard district of Switzerland are cut as gems, having a brilliant luster and a "fire" surpassing that of diamond, they are too soft to be widely used.
Formula: $CaTiSiO_5$.

Spinel

The name spinel broadly refers to a group which includes the minerals *magnetite and *chromite. In a narrower sense, spinel is the magnesium aluminum oxide member of the group. It occurs as an accessory mineral in basic igneous rocks, in thermally metamorphosed limestones, and in very aluminum-rich schists.

Spinel crystals are hard and usually octahedral, and form in a variety of colors. Transparent crystals are used as gemstones, many of the historic "rubies" (such as the centerpiece of the British Imperial State Crown) proving to be in fact red spinels. Today synthetic spinels are produced in many colors by the Verneuil process (see *sapphire).

At depths between 40 and 70km (25–43mi), it seems probable that many peridotites of the mantle contain spinel; and at depths of about 400km (250mi), *olivine itself adopts a spinel-like structure.
Formula: $MgAl_2O_4$.

Spodumene

Spodumene is a somewhat rare *pyroxene. It usually occurs in long, flat, striated crystals of various colors in granite pegmatites associated with other lithium-bearing minerals. It is of importance both as an ore of *lithium and as a gem. The crystals are often very large, and the delicate lilac-pink variety kunzite and the rarer green variety hiddenite are especially prized as gems. Both display the interesting phenomenon called pleochroism, in which the color of the stone varies according to the crystallographic direction from which it is viewed.
Formula: $LiAlSi_2O_6$.

Staurolite

Staurolite is exclusively a metamorphic mineral, forming typically as large brown crystals (porphyroblasts) in medium-grade *schists and *gneisses. Crystals are well-formed and, although monoclinic, appear to have higher symmetry. Twins composed of two intergrown crystals are common: they take the form of a cross with the two components crossing either at right-angles or obliquely. These cruciform twins, especially the right-angle type, are known as "fairy stones" or "fairy crosses", and in certain Christian countries are mounted without cutting or polishing to be worn as pendants.
Formula: $Fe_2Al_9Si_4O_{22}(OH)_2$.

Stibnite

Stibnite is the major ore of *antimony. It forms with *quartz in hydrothermal veins, in hot-spring deposits and as a replacement mineral in some limestones. Stibnite can

Stibnite is the commonest antimony mineral. It crystallizes in the orthorhombic system, and has a single perfect cleavage in a direction parallel to the length of the crystals. It can be melted even in a candle-flame.

form spectacular orthorhombic crystals: the finest are elongated, often gracefully bent, striated, and metallic steel-gray.
Formula: Sb_2S_3.

Strontianite

Strontianite usually forms white, gray or pale green needle-like crystals, which may be aggregated together into a fibrous mass. Typically, the mineral occurs in low-temperature veins in limestones together with *barite, *celestine and sulfide ore minerals. It is a minor source of *strontium but is less important commercially than the more abundant and usually purer celestine.
Formula: $SrCO_3$.

Sylvite

Sylvite is closely related to *halite. Not only are the two minerals chemically similar and structurally identical, crystallizing as colorless or white cubic crystals with cubic cleavages, but they also both form when a body of seawater or a salt-lake is evaporated (see *evaporites). Sylvite is less abundant and, being even more soluble, does not crystallize until almost all the water has evaporated. It may be identified by its unpleasantly bitter salt taste.

Millions of tonnes of sylvite are mined annually – for use almost entirely in the manufacture of potash fertilizers.
Formula: KCl.

Talc

Talc forms as a secondary mineral by alteration of magnesium-rich *olivines, *pyroxenes and *amphiboles of ultrabasic rocks: it is often associated with other secondary minerals such as *chlorite and *serpentine. It also occurs in schists produced by metamorphism of magnesium-rich rocks.

Commonly in white or pale-green foliated masses with a pearly luster, soap-like to the touch, talc is easily identified. Its extreme softness accounts for its soapy feel. Massive fine-grained talc is called soapstone or steatite and is carved into ornaments and images.

Powdered talc is used as a lubricant, in plasters and as a filler in paper, and is most familiar in the form of domestic talcum powder and French chalk.
Formula: $Mg_3Si_4O_{10}(OH)_2$.

Topaz

Topaz is found in granites, in cavities in rhyolites, in high-temperature veins and, most abundantly, in granite pegmatites. Well-shaped orthorhombic crystals which are elongated, blunt-ended and striated are particularly well developed in pegmatites, and these may be very large: a single transparent crystal weighing 270kg (596lb) may be seen at the American Museum of Natural History.

The transparent quality of topaz, together with its very wide range of attractive colors and its great hardness (8 on Mohs' scale), make it an important gem material. The most highly valued are sherry-brown: the lovely rose-pink topazes, much in demand, are obtained by heating these brownish stones, the color change resulting from a rearrangement within the topaz structure of trace impurities. The finest blue stones are found in the Ural Mountains.

Topaz is resistant to chemical *weathering and mechanical abrasion and thus also occurs as pebbles and grains in river gravels.
Formula: $Al_2SiO_4[OH,F]_2$.

Tourmaline

When a granite crystallizes, residual liquids may become enriched in a wide variety of elements which are not incorporated in the ordinary silicate minerals of the granite. Tourmaline forms at these later stages, either in the granite itself or in associated pegmatites and veins, and plays a major part in accommodating some of these residual elements, notably boron, lithium and fluorine.

Tourmaline crystals are trigonal, elongated and striated, often with curved triangular cross-sections. The two ends of a crystal are often different, and almost any color is possible. Particularly attractive are crystals that change in color from pink to green along their length.

Tourmaline belongs to a class of crystal structures which lack a center of symmetry. When such crystals are subjected to pressure, they develop an electrical charge; and tourmaline is used in this way in some pressure gauges. It is also used as a gemstone and, in view of its variety of colors, may superficially resemble a number of other gems. A hemispherical tourmaline stone in the Ashmolean Museum at Oxford is carved with the profile of Alexander the Great, and there is some reason to suppose that this may be a rare portrait from life.
Formula:
$Na[Mg,Fe'']_3Al_6(BO_3)_3(Si_6O_{18})[OH,F]_4$.

Tremolite

Tremolite is a member of the *amphibole group. Iron may replace some of its magnesium, the mineral then changing in color from white to green. With increasing amounts of iron the color changes to dark green, and the mineral is then known as *actinolite. Crystals are long and blade-like, but more commonly the mineral occurs as masses of parallel or radiating fibers. In this form it is exploited as an *asbestos. Very fine-grained and compact fibrous tremolite is called *nephrite. Tremolite is formed when impure dolomites or magnesium limestones are invaded, heated and metamorphosed by *igneous intrusions.
Formula: $Ca_2[Mg,Fe'']_5(Si_8O_{22})(OH)_2$.

Turquoise

Turquoise owes its "robin's egg" blue color to the presence, in varying amounts, of *copper. Turquoise usually occurs in veins as massive fine-grained material formed by the action of surface waters on aluminum-rich rocks. It is opaque and is valued for its color alone.

As a gem and ornamental material it has an ancient history. The world's oldest known examples of jewelry may be the carved turquoise bracelets of Egypt's Queen Zer (First Dynasty).
Formula:
$Cu[Al,Fe''']_6(PO_4)_4(OH)_8 . 4H_2O$.

Uraninite (Pitchblende)

Uraninite is uranium oxide, a hard, heavy, black radioactive mineral which often has a pitch-like luster (hence the alternative name, pitchblende). The element *uranium was discovered in the mineral in 1789, but it was not until 1898, when the Curies isolated from it minute amounts of a further new element, *radium, that it attracted interest. Since WWII the search for uranium, the chief fuel of the atomic energy industry, has been thorough and at times frenetic. In the 1950s hundreds of prospectors, working independently, scoured the Colorado Plateau, USA, where brightly-colored secondary uranium minerals and portable Geiger counters aided them to locate deposits.
Formula: UO_2.

Vermiculite

Vermiculite is a *clay mineral. Although it occurs as minute particles in soils, it can

Wavellite, a common member of the phosphate group of minerals, crystallizes in the orthorhombic system: however, crystals are uncommon, the habit of the mineral being more usually as globular (or hemispherical) aggregates displaying a fibrous, radiating structure.

also form large yellow and brown platy crystals as an alteration product of *biotite – such large grain-size is unusual for a clay mineral. A unique property is the rapid and large (up to thirty times original volume) expansion when it is heated quickly to 250–300°C (480–570°F). Large quantities are mined and expanded in this way for use as an insulating material.
Formula: $[Mg,Fe,Al]_3$ $[Al,Si]_4O_{10}(OH)_2.4H_2O$.

Witherite
Witherite is much less abundant and widespread than *barite, but both sometimes occur in association together, especially in hydrothermal lead-bearing veins. Crystals are usually white or yellowish and, although orthorhombic, are often twinned to give a hexagonal (six-sided) appearance.
Formula: $BaCO_3$.

Wolframite
Wolframite, the principal ore of *tungsten, occurs as blackish, tabular crystal groups in *quartz veins and pegmatites close to certain granites, and is often associated with other ore minerals, particularly *cassiterite. It is heavy and has a single, good cleavage.
Formula: $[Fe'',Mn]WO_4$.

Wollastonite
Wollastonite is closely similar in structure to the *pyroxenes. It usually occurs as white, fibrous and splintery masses in siliceous *limestones which have been thermally metamorphosed.
Formula: $CaSiO_3$.

Zeolite Group
When lavas are erupted at the Earth's surface, gases are released from the fluid magma to form holes, called vesicles, in the solidified rock. If such lavas are subsequently buried at shallow depths, heated between 100°C and 500°C (212–932°F), and leached by alkaline waters, the vesicles

may become filled by a variety of minerals, the rounded white fillings being called collectively amygdales. The common minerals of amygdales are *chalcedony (agate), *calcite and a related group of closely similar hydrated aluminosilicates called zeolites.

These are white or colorless (when pure), soft, low-density minerals, often forming as delicate well-shaped crystals or crystal aggregates. Individual species can be identified often on the basis of crystal shape (zeolites are variously blade-like, equant, fibrous, etc.), but X-ray methods may be necessary to distinguish rarer or irregularly formed material. Altogether some thirty different zeolite minerals are known, the most common being stilbite, natrolite, heulandite, analcite and chabazite.

They all have open, sieve-like crystal structures and this accounts for their curious capacity for base exchange: if a sodium zeolite such as natrolite is soaked in water containing calcium ions, the calcium and sodium will change places; it is this property which is employed in water softeners. In addition to their base exchange capacity, zeolites, when dehydrated, are able to absorb a range of substances preferentially, and they are used as "molecular sieves" in many industrial processes.

Zircon
Zircon is one of the most widely distributed accessory minerals, and is found in acid and intermediate *igneous rocks; in metamorphic schists and gneisses; and, owing to its density, hardness ($7\frac{1}{2}$ on Mohs' scale) and resistance to *weathering, as detrital grains concentrated in river and beach sands and gravels.

Zircon forms distinctive brown prisms with square cross-sections and pyramid-shaped ends. Greenish zircons usually owe their color to radiation damage from small amounts of radioactive uranium and thor-

ium present in the structure. In addition to being the major source of *zirconium, zircons are in demand as gems. Cut stones have a brilliance similar to diamond, the most popular color being blue, produced by heating brown stones in the absence of air.
Formula: $ZrSiO_4$.   FBA

Elements
The Periodic Table of the Elements was first published in 1869 by the Russian chemist Dmitri Mendeléev. He found that, if he set down the elements in order of increasing atomic weight, there was a periodic variation in their properties: for example, starting with lithium, the valences of it and the succeeding seven elements read 1,2,3,4,3,2,1,0 (the valence of an element is a measure of the "combining power" of its atoms: e.g., the valence of hydrogen is 1, that of oxygen 2, and so the formula of water is H_2O). When Mendeléev had tabulated the known elements in this way he found gaps in his table: he correctly deduced that these represented elements, whose properties he predicted with remarkable accuracy, that had not yet been discovered.

More recently it was realized that it was an element's atomic number – the number of protons in its nucleus – that determined its position in the table. As the number of protons determines the (equal) number of electrons associated with the atom when it is not ionized, and as it is these electrons which determine the element's chemical properties, the reason for the periodic variation in properties becomes clear.

All of the elements treated below are of economic importance in one way or

A metalware shop in the bazaar at Shīrāz, Iran. Roughly 75% of the chemical elements are metals and, more than any others, they have played a major role in shaping our civilization.

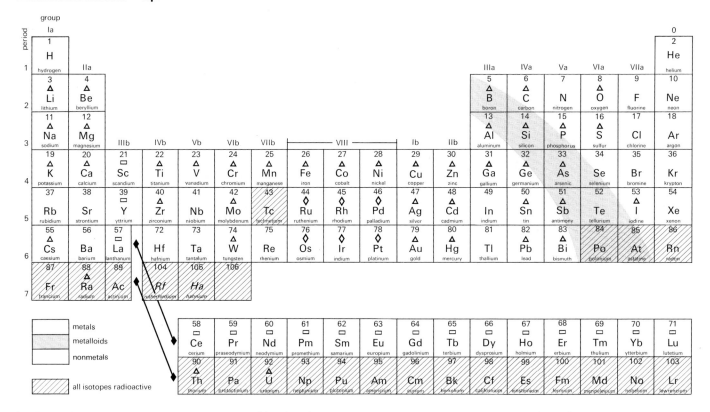

The Periodic Table of the Elements. Elements marked with a triangle (△) are treated in this section, those with a diamond (◇) being grouped together under the heading "Platinum Group" and those with a rectangle (▭) under the heading "Rare Earth Elements". (Element 106 has not yet been officially named.)

another. Most are metals, some are metalloids and some, like oxygen, are neither. They are arranged in order of their appearance in the Periodic Table (i.e., in order of increasing atomic number).   AI

Lithium

The main occurrence of lithium is in the mineral *spodumene, a member of the pyroxene group. It also occurs in the lithium *micas lepidolite and zinnwaldite, and in the mineral petalite ($LiAlSi_4O_{10}$). It is produced either from pegmatites of magmatic origin (see *igneous rocks) which contain spodumene or petalite, or from natural subsurface brines which contain high lithium concentrations (of the order of 300 parts per million).

Nearly all the world's lithium is supplied by the USA and USSR, though that obtained from the USSR is depleted in the ^6Li isotope, which has been extracted for its applications in processes based on nuclear fission and fusion.

Lithium hydroxide is used in the production of greases for motor vehicles and aircraft – in the USA about 50% of all grease currently produced is lithium based. Lithium carbonate is increasingly being used as an additive in the electrical furnaces used for producing aluminum metal. Organo-lithium compounds are used as catalysts in the production of certain synthetic rubbers; and lithium salts are finding a medical application in the treatment of schizophrenia.

Lithium has atomic number 3, atomic weight 6.94, and chemical symbol Li. ETCS

Beryllium

The most common occurrence of beryllium is in combination in the mineral *beryl. Workable concentrations occur only in pegmatites associated with granitic intrusive rocks.

Over half the total world supply of beryllium is consumed in the production of copper-beryllium alloys, used in the manufacture of non-magnetic tools, springs, clips, connectors and wire. Beryllium oxide is used extensively in the electronics industry, and also in nuclear reactors to moderate (i.e., slow down) neutrons.

It has atomic number 4, atomic weight 9.01, and chemical symbol Be.   ETCS

Boron

In nature, boron occurs principally in borates such as borax ($Na_2B_4O_7 . 10H_2O$), colemanite ($Ca_2B_6O_{11} . 5H_2O$) and ulexite ($NaCaB_5O_9 . 8H_2O$). It is also found combined in silicates such as *tourmaline.

Workable borate deposits are of sedimentary origin, formed by the persistent evaporation of lakes whose waters held an unusual combination of substances in solution (see *evaporites).

Boron compounds are used as fluxes to reduce the melting points of glasses and ceramics; and their use in the manufacture of fiberglass for insulation and reinforcement is particularly important. Boron nitride and boron carbide are the hardest substances known, after *diamond.

Boron has atomic number 5, atomic weight 10.82, and chemical symbol B. ETCS

Carbon

Of all the elements, carbon is the most important to us – and in the most basic of ways: were it not for the ability of carbon to combine with other elements in molecules that are either rings or long, complex chains, then life as we know it could not exist.

We find the element in uncombined form as the native element minerals *graphite and *diamond; in combined form in carbonate minerals such as *calcite, *siderite and *aragonite; and as a constituent of *coals and natural hydrocarbons (*natural gases and crude oils, as well as solids such as *amber).

Carbon has atomic number 6, atomic weight 12.01, and chemical symbol C. ETCS

Oxygen

This element constitutes almost 50% by weight of the Earth's crust, oceans and atmosphere. Gaseous elementary oxygen, which now makes up 21% by volume of the atmosphere, is essential to respiration and so to all animal life (see *origin of life). It is used up in combustion and oxidization processes: highly reactive, it forms oxides with almost every other element. Atmospheric oxygen is replenished by plant photosynthesis and by the decomposition of water vapor in the upper atmosphere. In combined form, oxygen occurs as water, in the oceans and in rocks, as carbon dioxide in the atmosphere, and as a major constituent of most minerals. The oxide minerals themselves are also of great importance, especially silica (SiO_2). The metal oxides, generally containing the oxide ion (O^{2-}), include alumina (Al_2O_3) and the various iron oxides. Many mixed oxides containing several metal ions are known.

Elementary oxygen normally occurs as molecules of two oxygen atoms (O_2), but forms also an allotrope with three atoms per molecule, ozone (O_3). Ozone is formed by

the action of ultraviolet radiation on oxygen in the upper atmosphere, where it occurs as a layer that protects the Earth from the Sun's ultraviolet rays.

Oxygen is used chiefly to support respiration or combustion, as in oxyacetylene torches for cutting and welding. In metallurgy vast amounts are used in smelting and refining to burn away impurities, especially in iron and steel production.

Oxygen has atomic number 8, atomic weight 16.00, and chemical symbol O. PCG

Sodium

A major element of the Earth's crust, sodium occurs combined in a variety of minerals, including silicates, such as the *feldspar albite ($NaAlSi_3O_8$); halides, such as *halite (NaCl); and hydrated carbonates, such as trona ($Na_2CO_3 . NaHCO_3 . 2H_2O$).

The principal minerals which are mined for sodium and sodium chemicals are halite and trona, and both occur as evaporitic deposits (see *evaporites). Sodium chemicals are also obtained from natural saline solutions, such as seawater (salt, NaCl), saline lakes (crude salt cake, Na_2SO_4), oil-well brines (salt and crude salt cake) and dry-lake brines (crude salt cake and soda ash, Na_2CO_3).

Sodium is of minor industrial importance compared with its compounds. It is, however, used in the manufacture of tetraethyl *lead, the anti-knock additive for gasoline, and as a core for electric cables. It bursts into flame on contact with water, and so must be stored carefully.

Sodium has atomic number 11, atomic weight 22.99 and chemical symbol Na. ETCS

Magnesium

An abundant element in the Earth's crust, magnesium is found combined in a large variety of silicates, such as *olivine; in oxides, such as periclase (MgO) and *spinel; in carbonates, such as *magnesite and *dolomite; and as the hydroxide, brucite ($Mg(OH)_2$).

The chief magnesium ore mineral is magnesite, but it is principally used for the production of magnesia (MgO). Magnesium metal is mainly produced from natural magnesium-rich solutions such as seawater, saline lake waters and oil-well brines.

40% of the world's magnesium metal is used in making light *aluminum alloys, though magnesium alloys themselves are also used extensively where weight is at a

Three stages in the obtaining of aluminum from its oxide, alumina (Al_2O_3). The metal is usually extracted by means of the Hall-Héroult Process: alumina is dissolved in a smelter filled with a molten electrolyte, usually the aluminum mineral cryolite. Direct current is passed to the bottom of the smelter by means of carbon electrodes, and a crust forms on top of the liquid. Alumina is piled on top of the crust, which is periodically broken to allow the alumina into the mixture beneath. The metal accumulates on the bottom of the smelter, from where it is siphoned off into crucibles (top) to be poured into molds (top left). The final product may be in the form of ingots or "logs" (above).

premium. In the USA the principal consumer is the aerospace industry; in Europe magnesium alloys are used extensively in the manufacture of transmissions and crank-case housings for motor vehicles. Magnesia is much used in the manufacture of refractory bricks for lining the basic oxygen furnaces required to purify steels.

Magnesium has atomic number 12, atomic weight 24.32, and chemical symbol Mg. ETCS

Aluminum

Aluminum always occurs in combination with other elements, and is one of the major constituents of the predominant group of rock-forming minerals, the silicates. It is also contained in the minerals diaspore (AlO . OH), boehmite (AlO . OH), gibbsite ($Al(OH)_3$) and *corundum (Al_2O_3).

The principal sources of aluminum ore are *bauxites. These contain aluminum hydroxides and hydroxyoxides, and form as surface residua left from the leaching effects of groundwater in tropical climates.

The major uses of aluminum take advantage of its lightness, resistance to corrosion and good electrical conductivity. It is a major component of light alloys.

It has atomic number 13, atomic weight 26.98, and chemical symbol Al. ETCS

Yellow sulfur encrusting the edges of a vent inside the active crater of Gunung Papandayan volcano, Jawa. The economic importance of sulfur hinges primarily on the many uses of sulfuric acid (H_2SO_4).

Silicon

Silicon is the second most abundant element of the Earth's crust (27.72% by weight), the most abundant being *oxygen (46.60%). The compounds in which silicon occurs constitute the principal group of rock-forming minerals, the silicates and aluminosilicates. Most abundant of these is the simple oxide, *quartz (SiO_2).

High-purity silica sand is required for the manufacture of glass, and is used also as an abrasive. Moderate-purity silica sand is used for metallurgical molds and furnace linings; while impure sands are used as aggregate in cement and concrete. Silicon itself is used in ferrosilicon alloys, and for the manufacture of silicones, which are used as resins, adhesives and lubricants.

It has atomic number 14, atomic weight 28.09, and chemical symbol Si.   ETCS

Phosphorus

A reactive nonmetallic element, phosphorus is present in very many compounds – probably as many as is carbon – and is of great biological importance, phosphates playing a vital role in all energy-transfer processes. Nucleic acids, of which chromosomes consist, are phosphates.

Phosphorus is found in the phosphate minerals, a large group with over 150 members, all of which contain the orthophosphate ion (PO_4^{3-}). They fall into three classes. Primary phosphates are those which have crystallized from aqueous solution or from a melt. They include *apatite, by far the commonest and most important phosphate. Secondary phosphates are formed by low-temperature reaction of other minerals with phosphatic waters. Often beautifully colored, good specimens are much prized. Rock phosphates are formed by the action of water on phosphatic organic debris such as bones, shells and guano. They also include altered corals and phosphatic oolites.

Phosphate rock (impure fluorapatite) is mined from vast deposits found in the USA, the USSR, north Africa and elsewhere. Much is treated with sulfuric acid to give "superphosphate" (CaH_2PO_4), the main phosphate fertilizer. Another process is to heat the rock with silica and coke in a furnace, producing elementary phosphorus used in matches and incendiary bombs and in metallurgy. Most phosphorus thus made is converted to phosphoric acid. Phosphates are added to detergents as water softeners.

Phosphorus has atomic number 15, atomic weight 30.97, and chemical symbol P.   PCG

Sulfur

It is quite common for sulfur to occur in the native uncombined state. It is also a fairly common constituent of minerals, primarily sulfides such as *pyrite, *chalcopyrite, *galena, and *sphalerite; and sulfates, such as *anhydrite, *barite, and *gypsum. Large quantities also occur as the bad-egg-smelling hydrogen sulfide gas (H_2S) in "sour" *natural gas and crude oil.

Native sulfur is found principally in bedded deposits which formed by reduction of gypsum and anhydrite in *evaporites. It is worked by pumping hot water under pressure through the outer casing of a drill pipe into the sulfur-bearing formation (the Frasch process): this melts the sulfur (melting point, 113°C, 235°F) and forces it back up to the surface through an inner tube in the pipe. Sulfide *ore deposits are usually of hydrothermal origin.

Sulfur, as sulfuric acid (H_2SO_4), is mostly used in the manufacture of phosphatic fertilizers. Sulfuric acid is also used in the bleaching of fibers, in the vulcanization of rubber for motor vehicle tires, and in many other industrial and chemical applications.

Sulfur has atomic number 16, atomic weight 32.07, and chemical symbol S.   ETCS

Potassium

This is a common element in the major groups of rock-forming minerals, the silicates and aluminosilicates. It is found principally in *micas, *feldspars and the feldspathoid *leucite. The major ore minerals of potassium are, however, *sylvite and carnallite ($KCl.MgCl_2.6H_2O$). The potassium ore minerals occur in *evaporite deposits.

About 95% of the world's potassium is used in fertilizers, usually as potassium chloride, but also as the sulfate and in combination with nitrogen and phosphorus.

Potassium has atomic number 19, atomic weight 39.10, and chemical symbol K. ETCS

Calcium

A major element of the Earth's crust, calcium occurs combined in silicates, such as anorthite *feldspar; carbonates, such as *calcite, *aragonite and *dolomite; sulfates, such as *anhydrite and *gypsum; phosphates, such as *apatite; and the fluoride, *fluorite.

The principal calcium-containing materials of industrial importance are *limestone, dolomite, gypsum, anhydrite and fluorite. Limestone and dolomite occur as the major component of sedimentary carbonate rocks; gypsum and anhydrite occur in sedimentary rocks formed by the evaporation of seawater (see *evaporites); and fluorite occurs in hydrothermal/replacement deposits of carbonate rocks (see *ore deposits). Calcium metal is also obtained from calcium chloride produced from oil-well brines.

Calcium is used mainly as an additive to molten metals to remove oxygen, halogens, sulfur and phosphorus; and also as a reducing or dehydrating agent in organic chemistry. The calcium-containing materials have widespread use.

Calcium has atomic number 20, atomic weight 40.08 and chemical symbol Ca. ETCS

Rare Earth Elements

These elements are normally grouped together because they are produced from the same minerals and have similar chemical properties. They occur chiefly in *monazite, bastnaesite ($(Ce,La)CO_3F$, which accounts for 60% of rare-earth-element production) and xenotime (YPO_4).

Bastnaesite is chiefly obtained from pegmatites, whereas monazite is produced from sedimentary concentrates in beach sands and as by-products of *tin placer working and processing *uranium ores.

Scandium is used principally as a radioisotope and in high-intensity lamps, yttrium in color TV phosphors, neodymium in lasers, and the rare earth elements in general in special steels, ceramics, glasses and catalysts.

They have atomic numbers, atomic weights and chemical symbols as follows:

Element	Atomic number	Atomic weight	Symbol
Scandium:	21;	44.96;	Sc
Yttrium:	39;	88.91;	Y
Lanthanum:	57;	138.91;	La
Cerium:	58;	140.12;	Ce
Praseodymium:	59;	140.91;	Pr
Neodymium:	60;	144.24;	Nd
Promethium:	61;	~147;	Pm
Samarium:	62;	150.40;	Sm
Europium:	63;	151.96;	Eu
Gadolinium:	64;	157.25;	Gd

Although it is a rare mineral, vanadinite is exploited as an ore of the technologically important metal, vanadium. Color is frequently red, but may also be yellow, orange or brown. Vanadinite forms a solid-solution series with mimetite by replacement of the vanadium by arsenic.

Terbium:	65;	158.93;	Tb
Dysprosium:	66;	162.50;	Dy
Holmium:	67;	164.93;	Ho
Erbium:	68;	167.26;	Er
Thulium:	69;	168.93;	Tm
Ytterbium:	70;	173.04;	Yb
Lutetium:	71;	174.97;	Lu
			ETCS

Titanium

The principal occurrence of titanium is in the oxide minerals *ilmenite and *rutile. It is also found in the silicate mineral *sphene and as an accessory element in many other silicates. Workable titanium mineral deposits occur mainly as sedimentary concentrates in beach sands: such accumulations have been derived from the *erosion of plutonic *igneous rocks, in which ilmenite and rutile occur as accessories. In the case of ilmenite, occurrence in igneous rocks may provide workable magmatic concentrations (see *ore deposits).

Titanium metal has a higher strength-to-weight ratio than steel, is highly resistant to corrosion and can withstand high temperatures. Its dioxide (TiO_2) is the major source of the white pigment used in paints and plastics.

It has atomic number 22, atomic weight 47.90, and chemical symbol Ti. ETCS

Vanadium

The minerals in which vanadium is a principal element are uncommon. They include patronite (VS_4), carnotite (K_2O . $2U_2O_3$. V_2O_5 . $2H_2O$) and vanadinite ($Pb_5Cl(VO_4)_3$).

About 80% of the world's vanadium is used in the manufacture of high-strength steels which are widely used for gas and oil pipelines. A further 10% is used in nonferrous alloys (i.e., alloys which do not contain iron), in particular as an additive to *titanium. Of the remaining uses, the most important is as vanadium pentoxide (V_2O_5), a catalyst used in the manufacture of sulfuric acid.

Vanadium has atomic number 23, atomic weight 50.95, and chemical symbol V. ETCS

Chromium

In only one mineral does chromium occur in significant amounts, *chromite. Chromite occurs in layers formed by selective crystallization and settling in *igneous intrusions of basic composition. It is also found in large pod-shaped bodies in ophiolitic ultrabasic rocks.

There are two types, metallurgical chromite and refractory chromite. The former is used as a source of chromium metal for the manufacture of hard and stainless steels. Refractory chromite is used in combination with magnesia (MgO) and a binder to make bricks used in blast-furnace linings. Chromite is used also as a raw material for the manufacture of chromate and dichromate, from which other chromium chemicals may be derived.

Chromium has atomic number 24, atomic weight 52.01, and chemical symbol Cr. ETCS

Manganese

The principal minerals of manganese, which occurs only in combined form, are oxides such as pyrolusite (MnO_2) and braunite (Mn_2O_3); and hydroxyoxides, such as manganite (MnO . OH) and psilomelane (an amorphous hydrated oxide, $[Ba,H_2O]_2Mn_5O_{10}$). Manganese nodules, which occur on the deep *ocean floor, consist of two complex hydroxyoxides. Other minerals include *rhodocrosite and *rhodonite.

The three types of manganese *ore deposits are all of sedimentary origin: shallow marine chemical precipitates associated with *sandstones and *siltstones; residual deposits, similar to *bauxites and laterites, left as surface accumulations by deep *weathering in tropical climates; and manganese nodules we've mentioned (see *ore deposits).

About 95% of manganese output is used in making a variety of hard steels. Other uses include that of pyrolusite as a catalyst in the manufacture of chlorine, bromine and iodine; and of sodium and potassium permanganates ($NaMnO_4$ and $KMnO_4$) as disinfectants.

It has atomic number 25, atomic weight 54.94, and chemical symbol Mn. ETCS

Iron

5% by weight of the Earth's crust is iron. Native iron occurs, though rarely, in basaltic *igneous rocks which have been contaminated by coal or wood, the additional *carbon having acted as a reducing agent (i.e., having removed the *oxygen from previously existing iron oxides); and is also found alloyed with *nickel in iron *meteorites. Iron forms a large variety of minerals: silicates, such as *olivine; oxides, such as *magnetite and *hematite; hydroxyoxides, such as *goethite; carbonates, such as *siderite; and sulfides, such as *pyrite and *pyrrhotite.

Most iron ore is produced from sedimentary banded *ironstones of *Precambrian age, which are rich in hematite, magnetite and siderite. In France there are ores of *Jurassic age, consisting primarily of goethite, and these are also sedimentary in origin.

As the essential component of every variety of steel, iron is obviously the most important of all industrial metals. Also, of course, it has played a large part in the development of our modern civilization – one thinks of the Iron Age. But from a scientific point of view, iron's most important property is that it becomes magnetized: paleomagnetic studies have been vital in establishing the theory of *plate tectonics.

Iron has atomic number 26, atomic weight 55.85 and chemical symbol Fe. ETCS

Cobalt

The principal occurrences of cobalt are in the minerals smaltite ($CoAs_2$), cobaltite (CoAsS), linnaeite (Co_3S_4) and cobaltiferous *pyrrhotite. *Ore deposits can occur in hydrothermal veins, magmatic cobaltiferous pyrrhotite deposits, or nickeliferous laterites (see *bauxites and laterites).

Cobalt has three major uses: it is employed in a variety of alloys; in cobalt-molybdenum catalysts, which have been employed in the desulfurization of high-sulfur *coals and in hydrocracking crude oil shale; and for pigments, particularly blue, in the glass, enamel and pottery industries.

It has atomic number 27, atomic weight 58.94, and chemical symbol Co. ETCS

Nickel

Nickel never occurs as a native metal, but it is a subsidiary component of the *iron alloys present in iron *meteorites. Most commonly, it is found combined in sulfides such as millerite (NiS) and pentlandite ($[Fe,Ni]_9S_8$). In addition, *pyrrhotite may contain up to about 5% nickel. The metal is also found in the arsenide, niccolite (NiAs), and in a silicate, garnierite, similar to the mineral *serpentine.

There are two very different kinds of nickel *ore deposit: pyrrhotite-pentlandite deposits in both intrusive and extrusive basic and ultrabasic *igneous rocks; and sedimentary residual deposits, similar to *bauxites and laterites.

Nickel is used primarily to produce hard steels of high tensile strength. It is also employed in *NiFe* (nickel-iron) batteries used in the mining industry to power cap lamps.

Nickel has atomic number 28, atomic weight 58.71 and chemical symbol Ni. ETCS

Copper

Copper occurs quite commonly as the uncombined native metal; and also in a large variety of minerals, including sulfides, such as *chalcopyrite and *bornite; oxides, such as cuprite; and hydrated carbonates, such as malachite and azurite.

Porphyry copper deposits of hydrothermal origin are the main sources of copper ore. These are large, low-grade disseminations of copper minerals in altered rocks centered on high-level, normal porphyritic igneous stocks.

Copper is one of the most important industrial metals. Metallic copper is malleable and an excellent conductor of electricity. Copper alloys – such as bronze (copper with tin) and brass (copper with zinc) – are also important. Copper chemicals, too, find many applications: the sulfate is of particular use as a fungicide.

Copper has atomic number 29, atomic weight 63.54 and chemical symbol Cu. ETCS

Zinc

The element zinc does not occur in the uncombined state, but is found principally as the sulfide, *sphalerite, and the carbonate smithsonite ($ZnCO_3$). Zinc *ore deposits are of hydrothermal origin.

Zinc is resistant to corrosion since, on exposure to air, a thin surface coating of zinc oxide (ZnO) forms, and this protects the metal from further attack. For this reason, its major use is in galvanizing, the process whereby steel is coated with a thin layer of zinc to prevent rusting. It is also used for the manufacture of die castings, especially in the motor-vehicle industry.

Zinc has atomic number 30, atomic weight 65.38 and chemical symbol Zn. ETCS

Gallium

Although gallium is as common in the Earth's crust as, for example, *lead, it forms no important minerals. It does, however, occur as a trace element dispersed at low concentrations in other minerals. It is produced entirely as a by-product of processing aluminum (90%), zinc and phosphate (together 10%) ores – total world production is about 15 tonnes a year.

When alloyed with *arsenic and *phosphorus, gallium can be used in the manufacture of light-emitting diodes, whose main application is in the construction of colored numeric displays for electronic calculators.

Gallium has atomic number 31, atomic weight 69.72 and chemical symbol Ga. ETCS

Germanium

Like gallium, germanium is not found as a major constituent of any reasonably abundant minerals, but occurs as a trace element, at low concentrations, in many minerals.

Five stages in the processing of copper from the raw ore to the pure metal. *Above far left*, the arrival of the untreated ore. Larger pieces are separated from smaller in this case by the simplest of methods. The ore is then pulverized and concentrated, the latter being usually by use of water to "wash out" some of the unwanted material (*below far left*). Smelting commences with the production of "matte", a mixture of copper and iron sulfides: it is this matte that is smelted by addition of air to achieve iron oxide (which may be removed as a slag with silica) and copper sulfide. Further addition of air provides sulfur dioxide (gaseous) and metallic copper (*above left*). Refining removes trace impurities, the result being metallic copper that is more than 99.9% pure (*below left*). *Above*, the final product: copper whisky stills at Convalmore Distillery, Dufftown, Scotland.

The bulk of germanium supply is obtained from *zinc smelter residues.

Together with gallium, indium and silicon, germanium is one of the principal materials used for semiconductors in solid-state transistors, diodes and rectifiers.

It has atomic number 32, atomic weight 72.60, and chemical symbol Ge.   ETCS

Arsenic

Though it does occur as a native metalloid (metalloids are so named because they have some metallic and some non-metallic properties), arsenic is principally found in the mineral *arsenopyrite. It is also found in sulfides, such as *realgar and orpiment (As_2S_3); arsenides, such as smaltite ($CoAs_2$) and chloanthite ($NiAs_2$); and sulfo-salts, such as tennantite ($[Cu,Fe]_{12}As_4S_{13}$).

Arsenic minerals are frequently found in connection with hydrothermal sulfide deposits of many types (see *ore deposits). It is hardly surprising, then, that its production is almost entirely as a by-product of the smelting of arsenical sulfide ores.

Arsenic compounds are extremely toxic and are used principally in pesticides and insecticides (and detective stories).

Arsenic has atomic number 33, atomic weight 74.91 and chemical symbol As. ETCS

Strontium

The two minerals in which strontium occurs significantly are the sulfate *celestine and the carbonate *strontianite.

Strontium chemicals are used mainly in pyrotechnics – for fireworks, distress flares and tracer ammunition. They burn with a strong maroon-colored flame. Strontium chloride is used to produce the phosphors in the activated coating of fluorescent lights and color television screens. Strontium is also added to the glass used to manufacture the face-plates of color television sets in order to absorb the small quantities of X-radiation which would otherwise bombard the viewer.

It has atomic number 38, atomic weight 87.63, and chemical symbol Sr.   ETCS

Zirconium

The principal occurrence of zirconium is in the silicate *zircon: it is also found as the oxide, baddelyite (ZrO_2). Zircon is worked mainly from sedimentary placer concentrates in beach sands (see *ore deposits). It is found in association with *rutile, *ilmenite and *monazite.

Zircon's uses are, in order of decreasing importance, in foundry sands, in the manu-

facture of corrosion-resistant alloys, in refractories for metallurgical furnaces, and in ceramics.

It has atomic number 40, atomic weight 91.22 and chemical symbol Zr.   ETCS

Molybdenum

The two main minerals of molybdenum are the sulfide, molybdenite (MoS_2), and a molybdate, wulfenite ($PbMoO_4$). It is produced from large low-grade hydrothermal *ore deposits associated with calc-alkaline intrusive *igneous rocks; and is supplied partly as a by-product of some porphyry *copper deposits, but mainly from geologically comparable porphyry molybdenum deposits.

About 75% of the world's supply is used in the manufacture of high-strength corrosion-resistant steels employed in construction and for tools. Paint manufacturers are today turning toward non-toxic molybdate pigments to replace toxic *lead and *chromium compounds. Molybdenum disulfide (MoS_2), which has properties similar to those of *graphite, is used as an additive to greases and lubricating oils.

Molybdenum has atomic number 42, atomic weight 95.95, and chemical symbol Mo.   ETCS

Silver

Quite commonly, silver occurs as a native uncombined metal. In addition, it is found

in the sulfide, argentite (Ag_2S); in a variety of sulfo-salts which include the "ruby silvers" pyrargyrite (Ag_3SbS_3) and proustite (Ag_3AsS_3); and in the chloride, cerargyrite or "horn silver" (AgCl).

Apart from a small amount of silver found alloyed with *gold in sedimentary placer deposits, the major silver *ore deposits are of hydrothermal origin. These include vein deposits and disseminated deposits in *sedimentary rocks.

The principal applications of silver and silver compounds are industrial: they are mainly used in monochrome photography, in the electrical industry and for the manufacture of silverware. Silver is also used significantly in coinage, commemorative medals, and jewelry.

Silver has atomic number 47, atomic weight 107.88, and chemical symbol Ag.

ETCS

Cadmium

The sole significant occurrence of natural cadmium is in the form of the sulfide, greenockite (CdS). The metal is produced entirely as a by-product of the processing of *zinc sulfide ores, in which small grains of greenockite occur: such ores rarely contain more than 0.4% cadmium.

Around 40% of the world supply of cadmium is used in electroplating, to protect parts of automobiles, household app-

One of the principal uses of the metal strontium is to provide the reddish-maroon coloration in flares and fireworks – as in this catherine wheel.

A silver mine at Guanajuato, Mexico. While the best known use of silver is as a decorative metal, it has several more important uses: where better electrical conduction than even that of copper is required; in photography; and for silverware.

liances, electronic equipment and numerous fastening devices such as nuts and bolts. The second largest user of cadmium is the pigment industry: cadmium salts give strong red and yellow colors. The metal is also used in stabilizers for vinyl plastics, and for electrodes in nickel-cadmium batteries.

It has atomic number 48, atomic weight 112.41, and chemical symbol Cd.   ETCS

Tin

In nature, tin is found mainly in the form of its oxide, *cassiterite, though it is also found, less commonly, in the sulfide stannite (Cu_2FeSnS_4). Cassiterite occurs in hydrothermal vein deposits associated with intrusive granitic *igneous rocks, commonly in proximity to other minerals such as *tourmaline and *topaz. The most important concentrations from an economic point of view are, however, secondary sedimentary placer deposits derived from the *erosion of primary ore.

The main uses of tin are, in decreasing order of importance, for tinplate, solders, bearing alloys, bronze chemicals, and the manufacture of float glass. Molten tin is very fluid and runs over a surface easily, and has therefore been used to coat steel, forming a corrosion-resistant covering. The main use of tinplate is for tin cans.

Tin has atomic number 50, atomic weight 118.70, and chemical symbol Sn.   ETCS

Antimony

Only rarely does antimony occur in the uncombined state as a native metal. Its commonest mineral is *stibnite, though it is also found as a constituent of pyrargyrite (Ag_3SbS_3), bournonite ($CuPbSbS_3$), tetrahedrite ($[Cu,Fe]_{12}Sb_4S_{13}$) and other less common minerals.

Antimony oxide is used to flame-proof cable coverings and plastic upholstery in automobiles – in the USA it is now a legal requirement for seating in automobiles to be treated in this way. Antimonial lead is used for electrodes in batteries.

Antimony has atomic number 51, atomic weight 121.76, and chemical symbol Sb.
ETCS

Iodine

The heaviest naturally-occurring member of the halogen family of elements (the others being fluorine, chlorine and bromine), iodine is found worldwide but usually in very low concentrations. In trace quantities it is vital to animals, and it is also found in many plants, notably seaweeds. Elementary iodine is a volatile violet-black solid, too reactive to occur as such in nature. Most is produced from calcium iodate ($Ca(IO_3)_2$) found in Chile saltpeter; it is also extracted from oil-well brines containing sodium iodide (NaI).

The main uses of iodine and its compounds are as antiseptics and in pharmaceuticals. Several major dyes contain iodine. Silver iodide (AgI), being light-sensitive, is used in photographic film emulsions.

Iodine has atomic number 53, atomic weight 126.90, and chemical symbol I (J, for *jod*, in Germany).   PCG

Caesium

Caesium (or cesium) is not an abundant metal, and its normal occurrence is as a minor constituent of lithium- and potassium-rich minerals. It is, however, a major component of the rare mineral pollucite. Economic concentrations of caesium occur only in pollucite-bearing pegmatites.

Because light falling on its surface causes the emission of electrons, caesium is used in photoelectric cells, spectrophotometers, infrared radiation detectors, etc.

Caesium has atomic number 55, atomic weight 132.91, and chemical symbol Cs.
ETCS

Barium

Barium occurs combined in two principal minerals – *barite and *witherite. It is also found in an uncommon *feldspar, celsian ($BaAl_2Si_2O_8$).

The major ore mineral, barite, occurs in hydrothermal veins and in disseminated deposits which form strata-bound replacements of carbonate *sedimentary rocks.

Small amounts of barium chemicals are used in the glass industry, but this is insignificant compared with barite's use as an unmodified industrial mineral. It has a high density – $4.5g/cm^3$ ($290lb/ft^3$) – and this explains why 75% of the world's production is used as a weighting agent in drilling muds, to prevent gas or oil blowouts. It is also a strong absorber of X- and γ-radiation, and thus is used as an additive in concrete used to construct nuclear power plants. "Barium meals" are used for obtain-

Above, tin ore in slate. *Right*, panning for tin at Taraouadji, Niger. In the processing of cassiterite, tin's principal ore, smelting is initially performed using coke or coal as a reducing agent. The smelting is performed twice, the resulting crude tin being partially remelted to extract further impurities.

ing medical X-ray photographs, in which the additive shows up, so revealing details of the structures of internal organs.

Barium has atomic number 56, atomic weight 137.36, and chemical symbol Ba.

ETCS

Tungsten

There are two principal tungstate minerals, *wolframite and scheelite ($CaWO_4$). These two minerals are found mainly in hydrothermal vein deposits spatially associated with plutonic *igneous rocks of granitic composition. An association with *tin minerals is quite common.

Tungsten steels and tungsten carbide (WC) are extremely hard materials, and both are used in the manufacture of cutting and machining tools. Tungsten metal is used to make the filaments in electric light bulbs.

It has atomic number 74, atomic weight 183.86, and chemical symbol W (from its little-used alternative name, wolfram). ETCS

Platinum Group

The platinum group of metals contains, aside from platinum itself, palladium, rhodium, iridium, osmium and ruthenium. They are found largely in the uncombined native state; as platinum alloys; as separate native elements; and as alloys with each other. Platinum is also found combined as the arsenide, sperylite ($PtAs_2$).

Their main occurrence is in magmatic *ore deposits in seams in layered basic intrusions, though sperylite also occurs associated with iron-nickel sulfide deposits. Sedimentary placer deposits may also contain the platinoid elements.

Platinum is used principally as a catalyst in oil refining and in catalytic converters designed to reduce emissions of carbon monoxide and nitrogen oxides from motor vehicle exhausts. It is also used in jewelry and to manufacture crucibles used in chemical laboratories and for handling mol-

ten glass. Palladium is used to coat electrical contacts; and rhodium is used in the glass-fiber industry and, with platinum, for manufacturing thermocouples. Iridium is used as a catalyst in the petroleum industry. Osmium and ruthenium have few industrial applications.

The metals have atomic numbers, atomic weights and chemical symbols as follows:

Platinum:	78;	195.09;	Pt
Palladium:	46;	106.40;	Pd
Rhodium:	45;	102.41;	Rh
Iridium:	77;	192.20;	Ir
Osmium:	76;	190.20;	Os
Ruthenium:	44;	101.10;	Ru

ETCS

Gold

The main occurrence of gold is as an uncombined native metal alloy with *silver (white gold) or *copper (red gold): usually about 85–95% of the alloy is gold. It can also occur in combined form in tellurides such as sylvanite ($[Au,Ag]Te_2$).

Gold can be found either in hydrothermal gold-quartz veins or in placer deposits formed by mechanical sorting during sedimentation, particularly associated with rivers (see *ore deposits): such placers may be of either recent or ancient origin. The bulk of the world's gold is obtained

from a *conglomerate of *Precambrian age in the Rand of South Africa.

Investment demand for newly mined gold and gold coins accounts for, respectively, 45% and 20% of gold output. Industrial uses in dentistry and electronics, and of course its use in jewelry, account for most of the remainder.

Gold has atomic number 79, atomic weight 197.0 and chemical symbol Au. ETCS

Mercury

At ordinary temperatures, mercury is a liquid, its freezing point being $-39°C$ ($-38°F$). It occurs naturally as small fluid globules in the principal mercury mineral, *cinnabar. Both cinnabar and native mercury occur in shallow hydrothermal veins associated with volcanic rocks (see *ore deposits).

Mercury's major use is as a liquid electrode in the electrolytic production of chlorine and caustic soda (sodium hydroxide, NaOH). It is also employed in the electrical industry for automatic switches and blue mercury-vapor lamps. Its more familiar uses are in thermometers and barometers, and as a component of the amalgam used in filling dental cavities.

It has atomic number 80, atomic weight 200.61, and chemical symbol Hg. ETCS

Lead

It is exceptionally uncommon for uncombined lead to be found in nature, its most usual occurrences being as the sulfide *galena, the carbonate *cerussite and the sulfate *andesite. All lead *ore deposits are of hydrothermal origin.

40% of the world's lead output is used in the making of antimonial lead electrodes for batteries, used mainly for motor vehicles (see *antimony). Tetraethyl lead is used in large quantities as an additive in gasoline (petrol) to reduce "knocking". Lead has many other minor uses: for example, in low-melting-point alloys used for soldering, and in combination as fluxes in glass and glaze manufacture.

Lead has atomic number 82, atomic weight 207.21, and chemical symbol Pb.

ETCS

Bismuth

The uncombined native metal bismuth, which is silver-white with a faint tinge of red, occurs quite commonly. The most important naturally occurring bismuth compound is its sulfide, bismuthinite (Bi_2S_3). Both occur associated with other elements in hydrothermal vein deposits (see *ore deposits).

Bismuth is produced only as a by-product of the smelting and refining of lead and copper ores. In the USA, 38% of the supply is consumed by the pharmaceutical industry. In combination with metals such as *tin, *lead and *mercury, bismuth forms a series of "fusible alloys", which have low melting points and are used in casting; and it is used as an additive in various alloys.

Bismuth has atomic number 83, atomic weight 209.00 and chemical symbol Bi. ETCS

Radium

Radium occurs as a natural disintegration product of unstable *uranium isotopes: it is itself radioactive. It forms no minerals in its own right, but occurs as an accessory in the principal uranium minerals *uraninite and carnotite ($K_2O.2U_2O_3.V_2O_5.2H_2O$). Radium is thus found in the same ore deposits as is uranium, and processed solely as a by-product of the processing of uranium.

Radium produces X-rays, and therefore has applications in any technique that requires a source of X-radiation – for example, in the treatment of cancer. Small quantities are used in luminous paints.

It has atomic number 88, atomic weight 226.05, and chemical symbol Ra. ETCS

Thorium

A radioactive metal in the actinide series, thorium occurs in the minerals thorianite (ThO_2) and thorite ($ThSiO_4$), found chiefly in pegmatites. The major source, however, is *monazite sand, a phosphate of cerium and other rare earths containing up to 10% thorium.

Because of its occurrence, and the analogy between the actinides and the lanthanides, thorium is sometimes treated with the *rare earth elements. Its chemistry differs radically, however, as thorium always has a valence of four, and most

resembles *zirconium and hafnium. Thorium oxide is used to make incandescent gas-lamp mantles, and is added to the *tungsten filaments in electric light bulbs. Thorium metal gives strength to *magnesium alloys. Of great potential importance is the conversion of thorium in nuclear reactors to uranium-233, a nuclear fuel.

Thorium has atomic number 90, atomic weight 232.04 and chemical symbol Th. PCG

Uranium

The main occurrence of the radioactive metal uranium is in *uraninite and carnotite ($K_2O . 2U_2O_3 . V_2O_5 . 2H_2O$). In zones where water has percolated down through uraniferous deposits, torbernite ($Cu(UO_2)_2P_2O_8 . 12H_2O$) may occur.

Uranium minerals occur principally in strata-bound deposits in sandstones and conglomerates (see *ore deposits). The former were deposited hydrothermally at low temperatures, whereas the origin of the latter is not clear. Uranium minerals can also occur in hydrothermal vein deposits.

By far the most important of the world's producers of uranium is the USA. In the near future, however, the large new ore deposits which have recently been found in Australia may make that country a major producer.

The principal uses of uranium are in

nuclear fission reactors to produce electrical power, and in military weapons.

It has atomic number 92, atomic weight 238.07, and chemical symbol U. ETCS

Petroleum

Man's quest for petroleum began when God told Noah to coat the ark with pitch. In prehistoric and early historic times, petroleum from natural seepages on the surface of the Earth was used for many purposes, military and medical as well as nautical. But

The first production platform for the giant Brent oilfield in the North Sea on tow down Stavanger Fjord, Norway, to its destination some 400km (250mi) away. This platform, launched on August 4 1975 and to date the biggest, heaviest and most expensive offshore oil production unit ever built, towers some 130m (420ft) above the waterline – the height of a thirty-storey building – and has a towing draft of 80m (260ft).

An oil tank "farm" on Kharg Island in the Persian Gulf: the flames are from the burning off of unwanted natural gas. The yellow rectangle over toward the coast is a sulfur plant. Petroleum products from the Middle East have in recent years ensured the dominance of the Arab states in the world's economy.

it was only when petroleum products were required in large quantities, first paraffin for lamps, then petrol for cars, that Man's search became scientific instead of haphazard.

We now know that a petroleum accumulation requires five conditions: an organic-rich source rock, generally a fine-grained shale, from which the oil was generated; a porous reservoir rock to contain the accumulation; an impermeable cap rock above the reservoir to prevent its escape; and a configuration of the rocks such that the oil is trapped in the reservoir beneath the seal. Given a source, a reservoir, a seal and a trap, the fifth condition is that the source rock must have been heated sufficiently to expel the oil.

Types of Petroleum. Petroleum is the name given to solid, plastic (i.e., pliable) and liquid hydrocarbons which occur naturally within the Earth. Solid and plastic forms go under a variety of names, including elaterite, ozokerite and bitumen. With increasing fluidity these grade through tar into what is variously termed crude oil, rock oil, natural oil – or simply oil or "crude".

Crude oils are oily liquids whose color varies from dark brown to tan and yellow-green. Again, they are very variable in density and viscosity, ranging from thin, colorless, translucent light oil to heavy, dark, viscous asphalt. They consist of varying proportions of four main groups of organic compounds: the aromatics, the paraffins, the naphthenes and the asphalts.

The Origin of Petroleum. Petroleum is found in rocks of all types, igneous, metamorphic and sedimentary, of all ages ranging from the Precambrian to sands less than a million years old. The majority of petroleum geologists believe, however, that regardless of where it now occurs most oil was generated from marine muds of Cambrian or younger age. (Natural gas is generally believed to have formed either from land-derived plant material or from the breakdown of crude oil at high temperatures.)

Criteria cited in favour of an organic origin for petroleum include the fact that it commonly contains microscopic particles of identifiable matter and that it shows the property of levorotation, the ability to rotate polarized light leftward, a phenomenon peculiar to organic substances.

Facts which confirm the formation of petroleum from sedimentary rocks include that it occurs within basins of sedimentary rocks, and not within the shield areas of continents which are composed of igneous granites and metamorphic gneisses. Moreover, it is sometimes found within porous sand beds that are entirely surrounded by impermeable shales, demonstrating the improbability of migration from a distant igneous source. Where it does occur in igneous rocks these are generally found to be adjacent to organic-rich sediments from which it is reasonable to suppose it came.

The most favorable conditions for the extensive preservation of organic matter are underwater environments where current circulation is minimal and where the bot-

Name	Properties	Composition (weight %)		
		Carbon	Hydrogen	Sulfur, Nitrogen, Oxygen, etc.
Kerogen	Solid at normal temperatures and pressures. Insoluble in petroleum solvents.	75	10	15
Asphalt	Solid at normal temperatures and pressures. Soluble in petroleum solvents.	83	10	7
Crude Oil	Liquid at normal temperatures and pressures.	85	13	2
Natural Gas	Gaseous at normal temperatures and pressures.	70	20	10

tom is stagnant and anaerobic (without oxygen). Thus the most likely places for petroleum source beds to form are in fine-grained clays in shallow sheltered embayments and deep waters of restricted bottom circulation: these conditions are most commonly found in marine environments. In fact, most ancient identified petroleum source rocks are of marine origin, though there are several notable exceptions.

Rapid burial by continuous sedimentation is a further factor favoring the preservation of organic matter. This is typically found where the crust of the Earth is unstable and rapidly subsiding. As a mud is buried, it is compacted, due to the weight of overlying sediment, it loses porosity, and water is expelled. With increasing depth of burial, pressure and temperature reach the point at which petroleum forms. The physical conditions at which this occurs are a matter for debate, and probably vary considerably depending on the chemistry of the organic material and on the catalytic effect of the clays and pore fluids. $65°C$ ($150°F$) is generally accepted as a temperature requirement for the formation of significant quantities of petroleum.

For the incipient petroleum to emigrate from the source rocks they must be interstratified with permeable carrier beds – sands are ideal. Petroleum may migrate for considerable distances through such carrier beds in response to hydrodynamic pressure gradients within a sedimentary basin. There is debate as to whether these fluids are already free oil, or whether the hydrocarbons migrate as an emulsion in water.

In the ordinary course of events, oil will migrate up through permeable rocks to the surface of the Earth, where it is dissipated, oxidized and destroyed. But, in certain favorable circumstances, upward-migrating oil may be trapped beneath impermeable strata, through which no fluid can flow. In this way is a petroleum reservoir formed.

Petroleum Traps and Reservoirs. A situation in which oil is retained within porous rocks beneath an impermeable seal is termed a trap. Four main types are generally recognized, though variations are countless.

Commonest are *structural traps*. These are due to bending or upward arching of porous strata into domes or linear folds, termed anticlines. A second type of structural trap is caused by rocks moving along faults in such a way that a porous reservoir formation is juxtaposed with impermeable strata.

The second type of petroleum trap is a *stratigraphic trap*. In this variety the trapping situation is due to lateral variations of permeability in stratified sequences. Examples include channel and beach sands and porous reef limestones enclosed in impermeable shale.

The third group of traps are called *combination traps* because they contain elements of both structural and stratigraphic control. An example would be petroleum

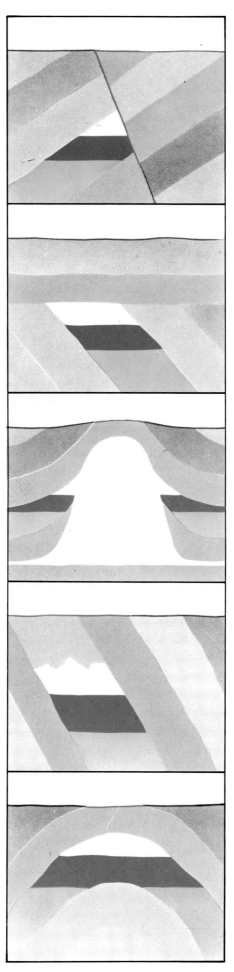

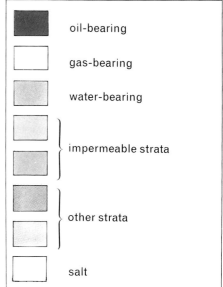

Various types of petroleum trap: from top to bottom, faulting, unconformity, salt dome, facies change and anticline.

accumulated within a sand which was erosionally truncated (an unconformity) on the crest of a faulted block.

The fourth type of trap is a *hydrodynamic trap*: these are very rare and are seldom specifically sought. They occur where the upward flow of oil due to buoyancy is offset by a fluid potential gradient such that pore water is moving downdip. In these rare instances petroleum may accumulate in uniformly dipping strata with no structural or stratigraphic trapping elements.

Another group of traps, sometimes classified on their own, are those due to *salt domes. Beds of salt often occur interbedded with other sedimentary rocks. Because of their low density and plastic properties, these salt beds often deform, creating domes and vertical pillars which, in moving upward, intrude into, fracture and dome the overlying sedimentary layers. When this occurs in petroleum-rich strata, many traps may be formed in folded and faulted porous beds adjacent to the salt dome.

Global Distribution of Petroleum. World petroleum reserves at the end of 1974 were as follows:

Area	Thousand Tonnes	%
Middle East	55,322,930	56.1
Sino-Soviet area	15,260,300	15.6
Africa	9,356,050	9.5
North America	6,625,000	6.7
Latin America	5,555,360	5.6
Western Europe	3,536,150	3.6
Western Far East & Australasia	2,883,270	2.9
Total:	98,539,060	100.0

These figures can in no way be regarded as indicating the true distribution of petroleum. The search for oil is governed by many factors, economic, geographic and political, and this means that the reserves discovered to date may give a false picture of the resources of different regions. In particular, only in recent years has it been technically possible to explore and exploit petroleum deposits beneath the deeper seas.

Methods of Exploration and Production. As we have seen, the first step in petroleum exploration is to find a basin of sedimentary rocks – more specifically, a sedimentary basin in which marine shales (potential source rocks) are interbedded with porous sandstones and limestones (potential reservoirs), and in which these have been so deposited and deformed as to form traps. Furthermore, the sediments must have been at some time sufficiently heated to generate oil.

Exploration methods range from surface geological mapping to subsurface geophysical techniques (such as seismic, gravitational and magnetic studies). Present exploration techniques can locate only *potential* traps. Each trap must be tested in turn by drilling bore holes.

This is done by means of a drilling rig or derrick. In modern rotary drilling a multitoothed bit is rotated on the end of a length of drill pipe. Drilling mud is pumped down the drill pipe and flows upward between the outside of the drillpipe and the side of the borehole. On reaching the surface the rock chips are collected so that the geologist can identify the rocks and so detect any signs of petroleum.

Petroleum is generally drawn from the reservoir through perforations in the steel casing lining the sides of the borehole. Where the reservoir pressure is sufficiently high, petroleum flows naturally to the surface, where it is fed through a system of valves (termed a Christmas Tree) to a pipeline. In low-pressure reservoirs it may be necessary to pump the oil to the surface.

Conclusion. Probably more geologists are currently employed in the hunt for petroleum than in all the other professional branches of the science put together. Conversely, the petroleum geologist is only one of many specialists, ranging from economists to engineers, who are involved in the exploration and production of petroleum.

RS

Natural Gas

Natural gases include the inorganic compounds of carbon and sulfur, the inert gases, and the gaseous hydrocarbons. They emanate from volcanoes – oxides of carbon and sulfur, water vapor, and other gases in minor quantities – but the natural gases we are most concerned with are those that are to be found in the pore spaces of many

different types of rock, of various ages, and in many different parts of the world.

Inert gases, including nitrogen and helium, are rare but can occur locally in significant quantities. The reasons for their origin are little understood. It is known that helium forms from the decay of radioactive minerals, and there is therefore a strong presumption that its presence in porous rocks indicates the existence of radioactive minerals, such as uranium, at greater depth. But very little is known of the origin of concentrations of nitrogen.

The most important gases from an economic point of view are the gaseous hydrocarbons. Of major importance are: methane (CH_4), ethane (C_2H_6), propane (C_3H_8), pentane (C_5H_{12}) and hexane (C_6H_{14}). The origin of gaseous hydrocarbons is intimately associated with that of liquid *petroleum. It is generally accepted that these gases are organic in origin, and can form in three ways:

(*i*) Biogenic gas is the name given to gaseous hydrocarbons, principally methane, formed by bacterial degradation of organic matter. To be within the environmental tolerance of the bacteria, this can occur only at or near the Earth's surface. For example, methane is extensively produced from rotting vegetation in swamps, where it is termed "marsh gas".

(*ii*) The second main origin of natural gas is from organic matter disseminated in sedimentary rocks. It is generally believed that oil (liquid hydrocarbon) forms from organic matter contained within muds of marine origin which have been buried and heated to over about 65°C (150°F). Conversely, it is thought that natural gas tends to be generated within continental sediments from plant-derived organic matter, which may be dispersed throughout sands and muds, or coal.

(*iii*) The third origin is from the breakdown of liquid hydrocarbons. This is believed to take place when temperatures exceed about 150–175°C (300–350°F). If the temperature becomes much greater than this, gaseous hydrocarbons are themselves destroyed to leave a residue of carbon.

The origin, migration and entrapment of hydrocarbon liquids and gases are intimately associated. Nevertheless, within a single sedimentary basin they often show some degree of positional segregation. Gas tends to be present in the deep traps in the center of the basin, oil in a peripheral ring around the gas zone, and water in the shallow traps of the basin margin.

An explanation for this arrangement is termed for its proposer Gussow's Principle. The idea is that, within a porous trap, gas will tend to rise to the roof above the oil, and the oil to float above the water. Fluids surplus to the capacity of each trap will flow out below the spill-point and migrate up into the next trap where the process is repeated. Thus gas fills the deepest traps, oil those of intermediate depth, and water the shallowest.

An alternative explanation for the observed zonation in a basin is that gas occurs in the center below the gas/oil maturation boundary, and oil occurs in the peripheral zone within the optimum thermal window for oil generation and preservation (between about 65°C and 175°C (150–300°F)). Traps around the basin margin are barren of hydrocarbons because they have been flushed out by meteoric water.

Whether these hypotheses are correct or not, it is clear that there is a close relationship between the origins of oil and natural gas.

Commercially, natural gas is not as valuable as oil. Gas deposits are generally worth developing only if they occur close to major centers of population or industry. Natural gas is often produced with oil – where this has happened far from any potential market, the gas has often been burned as a waste product or reinjected into the reservoir to maintain pressure and so productivity.

But, while our resources of geologic natural gas are limited and often unexploitable for economic reasons, biogenic gas manufactured on the surface of the Earth could be one of Man's most important energy sources in the centuries to come.   RS

Coal

Coals are organic rocks consisting mainly of altered accumulations of terrestrially derived plant materials, originally deposited as peats, with varying but generally small amounts of mineral matter. Ever since a land flora established itself on the Earth's surface some 400 million years ago, the potential for the formation of substantial coal deposits has existed. The development of a coal-forming peat, however, particularly its subsequent preservation, depends on a number of factors, understanding of which permits full exploitation.

The Origin of Coal and the Formation of Coalfields. Many of the world's coal seams originated from peats deposited in extensive coastal swamps that were probably in many ways similar to those forming on the coasts of Florida and New Guinea at the present time. These swamps are vast low-lying regions with virtually no relief and containing extensive deltaic spreads; and over them numerous wide, sluggish rivers flow, carrying large quantities of muds, silts and finely degraded organic matter.

The peats in the swamps accumulate in a wide belt so close to sea level that any slight change in the relative levels of land and sea will either cause flooding of extensive areas by the sea or, alternatively, expose previously submerged areas of the shallow sea bed that before had fringed the swamp. Thus the swamp belt, which may be tens of kilometres wide, does not usually maintain a stable position with the passage of time, but shifts backwards and forwards broadly parallel to the margins of the associated landmass. These movements allow all forms of lateral and vertical transition of sediments and peats to take place, but, at any point in time, a series of peat types can be expected to develop perpendicular to the coast.

Among the more important factors

Connecting detonator tapes to the firing line preparatory to blasting the coalface in the Mount Isa coalmines, Queensland, Australia. Because of the inherent dangers, alternative techniques to underground working are now increasingly being used.

An example of the type of swamp in which coal deposits are laid down: the Everglades in southern Florida. In swamps such as these dead and decaying vegetation accumulates in the form of peat which, on burial, becomes subject to compaction and a certain degree of heating. The degree of coalification depends upon the amount of heating, and represents a progressive increase in carbon content and decrease in volatile content.

governing the development of a coal-forming peat and its subsequent preservation are: the climate, which is probably most favorable when warm and oceanic with a high rainfall; the plant types present, their abundance and their degree of evolutionary development; and the rate of subsidence of the region in which the peat is accumulating – to ensure continuous peat formation: if subsidence is too slow, rapid destruction of the dead organic matter will occur by atmospheric oxidation; if it is too rapid, a premature submergence will occur under sediments transported by streams and rivers draining the land surrounding the swamps. The balance between these factors is critical and, if a major coalfield is to form, then the same sequence of inter-related conditions must be repeated time and time again over periods of many millions of years, the periods between peat accumulations being occupied by the deposition of muds, silts, sands and even limestone when increased subsidence carries the region below water level. This repetition of sedimentary environments in the vertical sense is described as "rhythmic" or "cyclothemic" deposition.

The seams formed from peats in coastal-swamp environments (paralic coals) are characteristically relatively thin after compaction – generally less than 3m (10ft) thick. The original peat may be compacted to as little as one twentieth of its original thick-

The world's major coalfields.

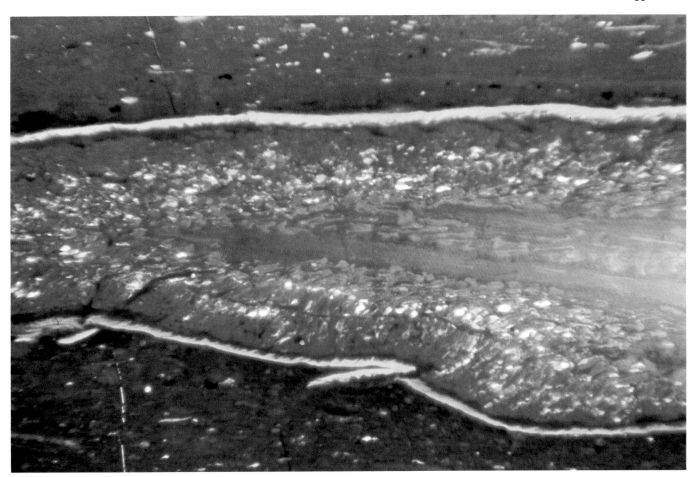

ness in the time bituminous coal takes to form. A major coalfield will contain many such seams.

In contrast to paralic coals, seams may develop within large inland continental basins (limnic coals). Typically, coal seams in these environments are extremely thick (up to 200m (650ft)) but few in number, and each results from subsidence that is just sufficient to allow continued growth of the peat: the *Carboniferous coal deposits of the French Massif Central are a good example of limnic development.

However, whether limnic or paralic, the vast majority of coal seams are regarded as forming *in situ*; the minority, in contrast, having developed from organic remains that were washed into, or drifted into, the position where the peat ultimately formed.

Coal Types. The plant remains in peats usually occur in two principal forms, still recognizable even in mature coals: as discrete and often sizeable parts of trees, bushes and smaller plants; and as a matrix composed of a wide size-range of tissue fragments, in varying stages of decay, from different plant organs and including seed and spore coats, leaf cuticles and highly resistant plant impregnations such as tannins and resins.

The different factors which control plant growth and the deposition and accumulation of coal-forming peats must also influence the chemical and physical compositions of the coals produced: the coal developed will reflect the plant components

from which it originated. Thus, a coal formed from a forest peat will be rich in modified wood and bark tissues, whereas a coal developed from a rich organic mud may be almost wholly composed of resistant spores, some heavily degraded organic matter and finely pulverized minerals; or, alternatively, have algae as its major organic component.

Peats which have contrasting plant compositions, having formed in different parts of the coal-forming swamp, lead to the concept of "type" in coals. Fundamental differences in coal type are obviously introduced very early in the history of the coals. Biochemical changes resulting from extensive fungal and bacterial degradation of the massive input of cellulose, lignin and proteins from the plants, with a consequent enrichment of the peat in more resistant waxy constituents, contribute further to these type differences. The depositional environment, however, may become sterile due to microbial overactivity, in which case further decay can occur. The particular depositional environment will govern the relative proportions of different plant constituents remaining in the peat: oxygen-bearing waters are important in modifying the proportions of organic constituents in peats – and so must influence type.

The technological implications of type variation may be considerable. For example, a coal rich in waxy spores will yield much greater amounts of gas and tar than a coal predominantly derived from the mass-

Cuticularized leaf in bituminous coal of Jurassic age. Due to the processes that act upon peat during its transformation into coal, such well-preserved remains are not altogether common: however, plant debris when found in or overlying coal seams can be of considerable value in studies of coal formation.

ive structural parts of trees. Similarly, differences in flora can produce type – and hence technological – variation in coals which in every other way have suffered the same histories.

Coal Rank. Later changes which modify peats into coal are purely physical and chemical in character and are mainly related to the length of time that the organic accumulations are exposed to high temperatures in the Earth's crust. Pressure from overlying rocks and stresses caused by Earth movements also influence the progress of these changes. In this way coals of different rank or "degree of coalification" are produced, these terms referring to the particular level of physical and/or chemical development attained by a fuel. Using one or several of a wide range of different properties, all fuels can be placed at some point within a continuous rank series that extends from peat to anthracite and beyond, towards a graphite-like end-product. The well-known terms brown-coal, lignite, bituminous coal and anthracite refer to specific subdivisions of increasing rank within this series.

Many coal properties display a systematic variation with rank increase and can be used to estimate rank level. The changes that occur are related to the composition of the original material, primarily carbon, hydrogen and oxygen, with smaller amounts of nitrogen, sulfur and other elements. As rank increases, the proportion of carbon relative to oxygen rises, nitrogen and sulfur remain approximately the same – as does hydrogen until, at higher rank levels, it too begins to fall with the development of highly aromatic products. These changes in the elementary composition of coals with rising rank are accompanied by the release of a variety of gases during the coalification process, predominantly a loss of differing proportions of water, carbon dioxide and methane.

All the organic constituents display changes, to a greater or lesser degree and depending on the original compositions, with rank increase. If possible, the principal coal constituent, vitrite (forming on average perhaps 70% of all coals), is used for rank estimation, first because the properties of vitrite vary in a relatively linear manner with rank increase, and second because the properties of a single constituent will give a more precise estimate of rank than will the properties of a whole coal, which is a mixture of constituents of differing composition.

Rank changes, then, reflect the response of the organic matter of fuels to geological conditions: as a consequence systematic lateral and vertical variations of rank can be observed within all coalfields. If a borehole penetrates a vertical succession containing numerous coal seams, the rank would be expected to vary systematically from seam to seam down the borehole. The carbon content and the calorific value would rise, and volatile-matter yield would fall.

The rate at which changes in individual properties take place in a number of boreholes over a coalfield area will not necessarily be the same, because the downhole variation is closely related to the rate at which past temperatures have increased with depth in the crust of different parts of the coalfield. Geothermal gradients of approximately $1C°/30m$ ($0.02F°/ft$) would be usual and so a temperature of $200°C$ (about $400°F$), regarded as a maximum for anthracite formation in regional coalification processes, would be reached in seams which were originally at the considerable depth of 6000m (about 20,000ft) below the surface. Variations in the geothermal gradient from this value may, however, be substantial and the gradient much higher.

Distribution and Age of Coals. There has been only one major coal-forming period that widely affected both hemispheres – during the Carboniferous and Permian periods, reaching its acme approximately 300 million years ago. There have been two more recent but less important periods of coal formation, in early Jurassic times and during the Tertiary era: this latter coal-forming period is probably still continuing in certain parts of the world, notably in the Far East. Many of the older coal occurrences today form isolated coalfields, but originally they were parts of much more extensive depositional areas that have become disrupted as a result of Earth movements and subsequent weathering and erosion.

Earlier discussion has shown that a number of factors control the development of coals of different type and rank. The same general types of coal can be recognized through the different eras, but clearly the opportunity for greater rank change exists the longer the coal has been in the crust. Consequently, the older a coal, the more valuable it usually will be as an energy source: thus, coals deposited during the Paleozoic and in the early part of the Mesozoic are generally of bituminous or higher rank, while late Mesozoic and Tertiary coals are predominantly brown coals, lignites and sub-bituminous coals.

World Coal Resources. The total amount of coal in the Earth's crust is acknowledged as being vastly greater than the total resources of *petroleum that may ultimately be discovered. The table gives a breakdown of world resources and reserves either by continent or by major country.

Universal acceptance of the precise definitions of "recoverable reserves", "total reserves" and "total resources" is probably not possible because of the different opinions that exist on how reserves and resources should be estimated. In general, however, "recoverable reserves" means those coals whose precise position in the crust is known through detailed prospecting and which could be removed economically by current mining methods. The re-

World Solid Fuel Resources (excluding Peats) in Megatonnes

Country or Continent	Recoverable Reserves	Total Reserves	Total Resources
USSR	136,000	273,200	5,713,600
China, Republic of	80,000	300,000	1,000,000
Asia, Remainder of	17,549	40,479	108,053
USA	181,781	363,562	2,924,503
Canada	5,537	9,034	108,777
Latin America	2,803	9,201	32,928
Europe	126,775	319,807	607,521
Africa	15,628	30,291	58,844
Oceania	24,518	74,699	199,654
World Total	591,191	1,402,274	10,753,880

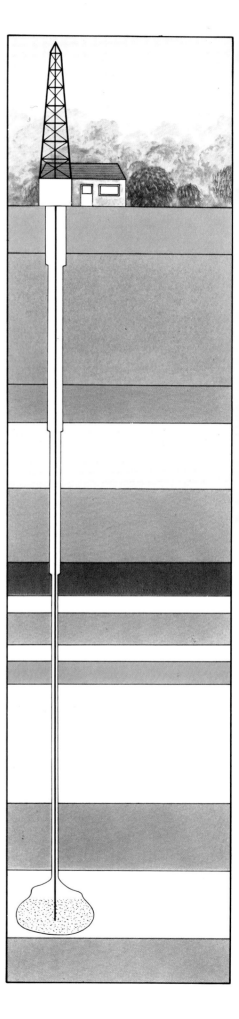

mainder of the reserves comprise coal which has been proved to be there by prospecting, but which could not be removed economically by mining technology at the present time. Quantifying "total resources" is more speculative and uncertain, but the term can be considered to describe estimates of coal that might reasonably be expected to occur in areas which have not been adequately prospected. The uncertainty in such figures lies in deciding what constitutes "adequate

prospecting". Clearly opinion on this point can range widely.   DGM

Salt

Salt has played a major part in the gastronomic and economic life of Man since the days of the Romans: our word "salary" is derived from the Latin *salarium*, the money given to Roman soldiers to buy salt, *sal*.

By "salt" we generally mean common salt, sodium chloride (NaCl), known to geologists as ★halite. It is one of a group of sedimentary rocks termed the ★evaporites.

The Origin of Salt. The classic theory for the origin of salt and other evaporites is the "evaporating-dish" model. This envisages an embayment of the sea in an arid climate, the mouth of the embayment being restricted by a shallow sill. Continuous evaporation in the embayment raises the concentration of the seawater so that dense brine flows to the embayment floor, but cannot escape to the sea because of the sill. When the brine is sufficiently concentrated, evaporite crystallization can take place. Once begun, this process may continue for some time as seawater intermittently flows over the sill to replace that which has been lost by evaporation and crystallization.

There is a second model for evaporite formation which explains those deposits which show evidence of a very shallow

Salt pans along the shore near Ruwenzori National Park, Uganda. Salt pans are small enclosed bodies of water which are permitted to evaporate so that concentrations of salts (evaporites) may build up. In some cases – as here – the salt pans are artificially created.

origin, perhaps not under water at all.

Many modern desert coasts are bordered by salt marshes, usually known as sabkhas. Here there are no rivers bringing sand and mud to the sea, and the shores consist largely of carbonate muds and skeletal debris sands. Seawater, already concentrated by evaporation in coastal lagoons, is drawn into the carbonate muds beyond the reach of the tides to replace pore water lost by evaporation at the surface. Thus a build-up of salinity occurs to the extent that evaporite minerals begin to form, not just by crystallization, but also by extensive replacement of the carbonate muds.

This process is well documented in modern sabkhas, which would suggest that the term "evaporite" may be inappropriate as a collective name for halite and its associated salts.

Economic Importance. There are three main reasons why salt is of economic importance. Apart from the fact that it is in itself a natural resource of great value, it is closely associated with the occurrence of petroleum, and there is strong evidence to indicate that it is connected with the occurrence of certain metallic ores.

Large salt deposits such as those of

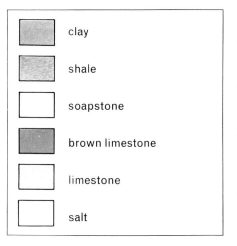

	clay
	shale
	soapstone
	brown limestone
	limestone
	salt

A typical salt well. Hot water is passed down the shaft to dissolve the salt: the resulting brine is then drawn up through the central pipe and the salt extracted from it by evaporation or otherwise.

Germany, England, and the Gulf Coast of North America are extensively mined – for more than just table salt. Halite and its associated salts are the basis for extensive chemical industrial complexes, the source for the world's requirements for rare earth elements, sodium and potassium, for the halogens, principally chlorine and bromine, and also for sulfur.

Salt deposits are closely related to *petroleum deposits for three reasons: structure, seal and source. Salt beds often deform plastically and domes of salt move up through denser sediments forming fold and fault traps in which *petroleum can be preserved. Secondly, because of its plasticity (pliability) and impermeability, salt makes an excellent seal to trap petroleum within porous reservoir beds. And, thirdly, salt is commonly interbedded with organic matter because they both require stagnant conditions for their preservation. For these reasons petroleum source beds are often found in intimate proximity to salt deposits.

The third important economic aspect of salt is the observed correlation between evaporites and metallic sulfide ores, such as those of lead, zinc and copper. These ores appear to be of low-temperature origin, and often replace reef limestones adjacent to evaporites. In this situation metallic ions, themselves concentrated from seawater, may have been carried in the residual brines and precipitated in carbonate reefs around the basin margin.

From all of this it can be seen that the common salt of our dinner table is, in terms of origin, occurrence and economic significance, a mineral whose importance is very, very far from common.   RS

Building Stone

The raising of stone for building has been going on since time immemorial, and certainly for centuries before the science of geology evolved. The particular stone selected depended on many factors, not least the type of building envisaged. The main priorities were always durability, local availability and nearness to the Earth's surface. Difficulty of working a stone seems never to have been a major constraint.

The selection of stone for building was thus inextricably bound up with the local geology. One has only to look at the domestic and older ecclesiastical architecture of a country to see this. The building-accounts for an English almshouse, built 1439–44, show that for special stone fetched a distance of only about twenty kilometres the cost of haulage was almost double that of the stone itself. In this light, one can understand why local geology was, until relatively recently, crucial.

As a result many materials that are not very durable had to be used, especially in domestic situations – though ecclesiastical building, on the other hand, could often afford the cost of haulage, which explains the widespread use of French Caen Stone in Southern England in the Middle Ages.

The petrology of building stone is immensely variable, and so a meaningful classification on scientific principles is impossible – as well as unnecessary. All rock types of sufficient durability have been used. Each group has its own unique qualities and appearance: to appreciate this one need only contrast the cold granitic exteriors of Aberdeen or Rennes with New York's brownstones or the warm limestone facades of Oxford.

Basic requirements determining the choice of stone are equally difficult to quantify. If a harmony with previous buildings has to be achieved then color is the prime consideration: otherwise texture may be important, and a stone with grain-size sufficiently small to allow smooth surfaces to be generated will be preferred. Smooth surfaces are needed because only accurately shaped blocks will withstand stress well and so allow the architect full rein. The durability of particular types of stone in the face of *weathering has been investigated, but if this factor is as vital as has been claimed one can hardly explain the widespread and successful use of Bath Stone, which in this respect is at best about four times weaker than Portland Stone.

The best test to find out whether a particular stone stands up to a particular environmental attack is always the simple if irreversible one of time. Again, environments change: the sulfation of limestones in many industrial cities, with the consequent

unsightly and damaging exfoliation, has been largely checked by recent legislation over use of fuels.

One vital property especially difficult to control is that the quality of each type of stone should be consistent. Even within a single quarry the stone can, and does, vary enormously. Much can be done at the time of quarrying by a skilled overseer, who can check quality as the stone is removed, and by marking the lie of the stone, if sedimentary, on the extracted blocks to allow this to be reproduced in the building.

The techniques of extraction of the stone were amazingly simple until very recently – the methods of the Egyptians of the Pharaonic era, who worked granitic rocks with extreme sophistication, remained essentially unchanged. Mechanical aid has come in more recent times. The few concessions to modern technology include large-scale wire and sand saws, compressed air tools and electric coal-cutting machines and scabbing hammers.

New techniques of building with stone have also helped the stone extraction industry. Precast reinforced concrete blocks have – for economic reasons – come to be used more and more in building. This has generated a need for crushed stone, enabling inferior material to be used up. Harmony with preexisting stone buildings has been achieved by use of new techniques of cladding thin slabs of the matching material to the outsides of the concrete structures: cladding can be similarly fixed to brickwork or steel.

A whole range of artificially created stone products have been made. Two examples are limestone chips set in a matrix, cut and polished, and used as interior marble flooring; and reconstituted stone, where the natural material is crushed and recast in blocks simulating the original in color and texture, with the advantage that the blocks are of uniform size.

Quarrying of stone has always brought benefits other than just the stone itself. The early development of paleontology was greatly aided by the fossils provided by quarrymen. The modern decline in the quarrying industry has frequently made new fossil material impossible to obtain from the original source.   HST

Selection of rock for use as building stone depends as much on factors such as local availability as on the inherent suitabilities of the rock concerned. Where the rocks are relatively soft, wire saws (endless wires driven at speed and pulled through rock bodies) or even, as in this quarry in Menorca, circular saws may be used for extraction.

The Rocks of the Earth

Rocks are divided into three classes or types: *igneous, *sedimentary and *metamorphic. A fourth type of "rock", *soil, consists of the eroded products of true rock. Major entries on these types of rock are included in this section in the order given above. The order in which more detailed entries are arranged reflects the ways in which rocks are commonly classified. Igneous rocks are arranged in order of decreasing acidity or quartz content and, after each rock type is described, its fine-grained or volcanic equivalent is dealt with. Sedimentary rocks are arranged in order of increasing particle size, with rocks of organic origin left to the end of the section. Finally, metamorphic rocks are arranged in a sequence such that rocks in whose formation pressure is dominant are dealt with before those in which heat is a progressively more important factor.

Igneous Rocks

Whenever we see that most spectacular of natural phenomena, a volcanic eruption, we obtain direct evidence of the existence of molten rock material issuing from within the Earth. The molten material at depth is termed *magma* and when it solidifies the product is an igneous rock.

Nearly all igneous rocks are composed of silicate materials, and the magmas from which they form are derived from partial melting of the Earth's mantle and crust, layers of the Earth which are normally in a solid state. The deepest zone of the Earth, the core, is thought to be permanently molten, but this is not the source of the igneous rocks we see at the surface today (geophysical and other evidences suggest strongly that the core is composed of iron and nickel rather than silicates). Nevertheless, igneous activity in the past – and continuing to the present day – may be responsible for the distinct compositional zones of the interior of the Earth, if we assume that the planet formed originally as a homogeneous body. To support this assertion we must consider first the way in which rocks melt.

It is generally thought that when existing solid rocks begin to melt, as a result of local thermal perturbation within the Earth, the melting will be only partially completed before the magma collects together and begins to migrate away from the source region under the influence of gravity. It is thus possible to make a distinction between the molten phase (the magma) and the remaining unmelted material (the refractory residue). If we assume the Earth originally had a composition like that of the chondritic *meteorites, it must have consisted of a mixture of metallic iron and nickel with silicate materials. Heating such a mixture first produces a melt rich in iron and nickel while leaving a refractory residue of silicates. The melt phase is in this case denser than the refractory residue and we can thus assume that it migrates downward, at the same time displacing the silicates upward. This is an attractive hypothesis to explain the formation of the core.

Subsequently, continued heating of the silicate layer induces further episodes of partial melting within it. In this case the melting behavior is more complex, but there is a general tendency for certain constituents (e.g., silicon, aluminum, sodium, potassium, water) to become concentrated in the melt while others, notably magnesium, are concentrated in the residue: thus silicate magmas tend to migrate upwards. This has resulted in the differentiation of the silicate-rich parts of the Earth into an upper layer known as the crust, rich in silica, aluminum and the alkali metals, and a lower zone, the mantle, rich in magnesium silicates, which are relatively silica-poor, and rather impoverished in aluminum and alkalis.

Hence igneous activity can be viewed as one of the most fundamental Earth processes, and probably all the rocks we see at the surface of the Earth today owe their ultimate origins to this activity.

Igneous rocks at the surface of the Earth are of course subsequently subjected to *erosion and *weathering and give rise to sedimentary materials such as sandstones and clays. Both igneous and these *sedimentary rocks can also become reconstituted as *metamorphic rocks under the influence of heat and pressure; for example, in episodes of mountain-building. Finally, igneous activity is also effective in expelling water and carbon dioxide from the interior of the Earth, thus making a substantial contribution to the oceans and the atmosphere. Some of the carbon dioxide may become fixed again in solid rock when it combines with calcium to form *limestones.

Classification. Igneous rocks are classified in a number of ways and one of the most important distinctions, based on mode of occurrence, is into the two main groups, *extrusive* and *intrusive*.

The extrusive rocks are those which reach the surface as magma and are ejected either as streams of molten material (lavas) or as explosive ejections of droplets and already solidified particles (see *vulcanicity). The explosively ejected rocks are known as *pyroclastics*, those composed of small particles being termed *ashes* or *tuffs*, while coarser material is termed *agglommerate*.

The intrusive rocks represent magma

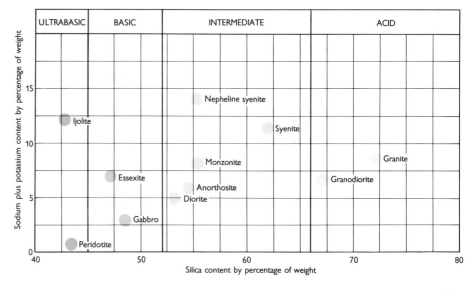

The classification of igneous rocks according to their silica and sodium-plus-potassium contents, with the positions on the table of several common plutonic igneous rocks shown (their fine-grained equivalents would be similarly placed).

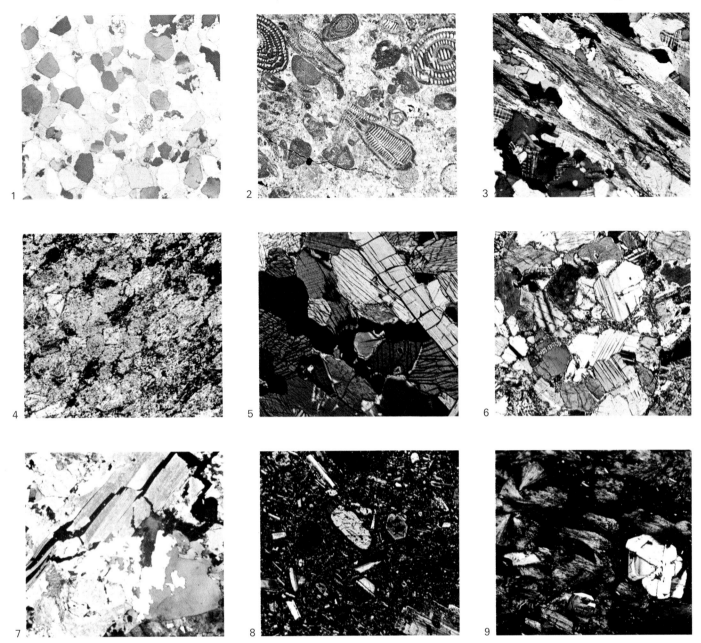

trapped below ground and solidifying there, and in order to see them we must wait until erosion has removed the overlying strata. The intrusive rocks generally cool more slowly than the extrusive and thus crystallize to a larger grain-size. The coarser-grained intrusive rocks (crystals generally greater than 5mm (0.2in) in diameter) are usually known as *plutonic* rocks.

Igneous rocks may also be classified in terms of their chemical compositions, one of the oldest divisions being based on their silica content (that is the weight per cent of SiO_2 determined by chemical analysis). The divisions are *acid rocks* ($SiO_2 > 64\%$), *intermediate rocks* ($SiO_2 = 53–64\%$), *basic rocks* ($SiO_2 = 45–53\%$), and *ultrabasic rocks* ($SiO_2 < 45\%$). The total content of the alkali metal oxides ($Na_2O + K_2O$) is also a useful indication of the nature of igneous rock types.

Apart from alkalis and silica, the other main chemical constituents of igneous rocks may be described quite simply. Firstly,

alumina, aluminum oxide (Al_2O_3), is a major constituent (10 to 20% by weight) of nearly all rock types, the major exception being *peridotite, which has much less than this. Consequently, those rocks such as granites and syenites which have a high total content of silica plus alkalis (75% or more), together with substantial quantities of Al_2O_3, contain only relatively small amounts of other constituents such as calcium oxide (CaO), iron oxides (FeO and Fe_2O_3) and magnesium oxide (MgO). Conversely, the content of these constituents is much higher in gabbros which have a low total of alkalis plus silica. Peridotites have relatively low silica and alumina as already noted and are very rich in magnesium, fairly rich in iron, but unlike the gabbros rather poor in calcium.

The chemical compositions discussed above have been expressed in terms of oxides, but this is a matter of convenience rather than reality. The rocks themselves consist of mixtures of minerals, most of

Thin sections of a selection of igneous, metamorphic and sedimentary rocks viewed under crossed polars: (1) and (2) are sedimentary, (3) and (4) metamorphic and the remainder igneous. (1) Quartz sandstone, a typical arenaceous rock, having particles principally of silica; (2) limestone, whose principal constituents are calcite and dolomite, shown here containing fossils; (3) gneiss, a rock formed during high-grade regional metamorphism; (4) hornfels, a result of metamorphism through heat; (5) peridotite, an igneous rock whose primary constituent is olivine; (6) gabbro, a typical basic rock; (7) granite, a typical alkaline rock; (8) basalt, the fine-grained equivalent of gabbro; and (9), rhyolite, the fine-grained equivalent of granite.

which are silicates, and it happens that silicates have compositions which can be expressed in oxide terms. For example, the chemical formula of magnesium orthosilicate (the mineral *olivine) is Mg_2SiO_4, and this can be expressed as $2MgO + SiO_2$ (magnesium oxide plus silica). In detail, rocks are classified by their mineral assemblages and by the relative proportions of the different minerals they contain. Several parameters are of importance:

Hound Tor, Dartmoor, UK, a striking formation of granite. Granite is described as a plutonic igneous rock, since it is intruded into the rocks of the crust at depth, so becoming visible at the surface only through the influence of erosion.

(*i*) The ratio of the dark-colored or ferromagnesian minerals (olivine, pyroxene, hornblende and biotite mica) to the pale-colored minerals (feldspars, feldspathoids and quartz). This ratio is often called the *color index*. A high color index implies a high content of magnesium, iron and often calcium.

(*ii*) The relative proportions of the alkali *feldspars (orthoclase, sanidine, microcline and perthites) to the calcium-bearing plagioclase feldspars (all *plagioclases that have more calcium than does albite). A high content of alkali feldspar implies richness in alkali metals, and conversely a high plagioclase content implies richness in calcium.

(*iii*) The presence or absence of *quartz or *feldspathoid minerals (nepheline, leucite, sodalite). This parameter is a reflection of the balance between alkalis and silica. An excess of silica means that the rocks contain free quartz. When alkalis and silica are balanced only feldspars will be present, while a silica deficiency is expressed by the presence of feldspathoids, compositionally similar to feldspars but poorer in silica.

A consideration of the chemical characteristics of the various groups allows some generalizations about mineralogy to be made. Granites, syenites and feldspathoidal syenites are all poor in ferromagnesian minerals and have color indices of 20 or less (i.e., less than 20% of the rock is made up of ferromagnesian minerals). Color index rises to about 50 in the gabbros and is between 20 and 50 in the diorites and syenodiorites. The peridotites have color indices approaching 100. Granites are the only rocks with enough silica to contain significant

amounts of free quartz, whereas the feldspathoidal syenites are sufficiently alkali-rich to contain significant quantities of feldspathoids, usually nepheline: most of the other igneous rocks contain neither quartz nor feldspathoids except in small quantities, their major light-colored constituent being a feldspar of one sort or another. Rocks poor in alkalis are characterized by plagioclase feldspar (except for the peridotites which contain little or no feldspar) – typical examples are gabbros and diorites. Increasing alkali content leads to rocks containing significant quantities of both plagioclase and alkali feldspar (syenodiorites, granites), while high-alkali rocks are characterized by alkali feldspar well in excess of plagioclase (e.g., syenites and feldspathoidal syenites).

Fine-grained extrusive rocks are frequently mineralogically difficult to identify, and it is often necessary to resort to chemical analysis for accurate identification. Nevertheless, many of the mineralogical criteria discussed above apply equally well to them.

Textures. Igneous rocks are normally examined by microscope in sections cut thin enough to transmit light. Polarized light is employed to produce interference colors, which make otherwise similar minerals easily distinguishable.

Plutonic rocks consist of large crystals intimately interlocking and showing a variety of textures as a result of the individual minerals beginning to crystallize in sequence, rather than simultaneously, as the magma is cooled. The typical texture of a rapidly cooled extrusive rock is rather different: here large crystals of feldspar and olivine, which formed as a response to slow cooling while the magma was still underground, are set in a fine-grained matrix which crystallized rapidly after eruption.

Magmatic Differentiation. Once magma is formed and begins its ascent towards the surface, several processes capable of modifying its composition *en route* may come into operation: these come under the general heading of magmatic differentiation.

One of the most important single processes is fractional crystallization. The textures seen in an extrusive rock illustrate how this may happen, because in general the large, early-formed crystals do not have the same overall composition as the magma from which they have crystallized. Like the refractory residues mentioned earlier, they tend to be poorer in silica and alkalis than the liquid fraction, though individual cases vary. Removal of the crystals from the liquid – for example, by gravitative settling – will give rise to a new magma richer in certain elements than was the original. Detailed studies have shown that this process has frequently operated to produce variation in magmas.

Environments of Igneous Rocks. Specific details of the occurrences of different rock types are given under the individual headings of those types. However, it is possible to make some generalizations arising from the above discussions of igneous differentiation processes.

*Peridotites, characteristic of the Earth's mantle, are highly refractory rocks having very high melting temperatures (e.g., 1500°C (2700°F)) and high densities (around 3.5g/cm^3 (225lb/ft^3)). Less refractory rocks, such as *basalts, with melting temperatures around 1000–1200°C (1800–2200°F) and densities of about 2.8g/cm^3 (180lb/ft^3), are characteristic of low-lying regions of crust, particularly the ocean basins. The least dense and least refractory rocks (melting temperatures 500–800°C (900–1450°F); density

2.5g/cm^3 (160lb/ft^3)) are granites and they are characteristically found in areas of continental crust.

Thus, as the broadest of generalizations, it is possible to view the Earth as a body firmly stratified with regard to density, and (excluding the non-silicate core) with a similar stratification in melting temperatures. All this has probably largely come about by igneous activity.

Granite

Probably the one rock name that is familiar to almost everyone is granite: it is a symbol of all that is hard and durable, it resists erosion and forms age-old hills and rugged cliffs.

Granite is one of a large class of rocks that solidify below ground as an *igneous intrusion, subsequently being exposed to view by the *erosion of the overlying strata. This general group, the plutonic igneous rocks, is of widespread abundance, particularly in deeply eroded terrains where ancient rocks are exposed, and granite, using the term in its broadest sense, is the most abundant type. Granites, in this sense, are coarse-grained rocks (crystals larger than 5mm (0.2in) in diameter) consisting predominantly of *feldspar and *quartz with subordinate amounts of dark minerals such as *biotite, *hornblende and, occasionally, *pyroxene.

In the narrow sense, granite is defined as a rock with more than 20% quartz and with an alkali feldspar (orthoclase, microcline or

Thin section of tourmaline granite under crossed polars at a magnification of about ×50. The large gray crystals are quartz and feldspars, the red-brown crystals biotite. The smaller, blue, crystals are of tourmaline.

perthite) as the dominant remaining constituent. Granitic rocks containing less quartz and a higher proportion of calcium-bearing plagioclase feldspar are termed granodiorites, and these in turn grade into diorites, which are rocks poor in quartz and alkali feldspar, but rich in *plagioclase.

Granites, diorites, and related rock types make up huge intrusive masses termed batholiths in the cores of eroded mountain ranges such as the Sierra Nevada. In the world's most ancient shield areas, granites make up the majority of the exposed rock. It is in these areas – for example, in northern Canada, Rhodesia, and parts of Greenland, India and Australia – that we see deep into the structure of the continental crust and perceive its dominantly granitic character. In contrast, granites are virtually absent from the ocean basins, in which areas *basalt is the almost exclusive rock type.

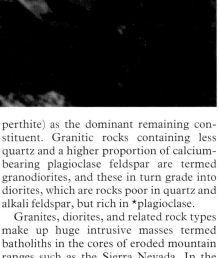

Granite seen in close-up. In this sample the groundmass is of potash feldspar.

Rhyolite

Rhyolite is the name given to lavas having the same general chemical composition as granite. Like the granites, they are often rich in visible quartz and alkali feldspars, and may contain these as larger crystals (phenocrysts) sparsely set in a much finer-grained matrix.

They form from granitic melts which are erupted onto the surface of the Earth and hence cool very quickly, as a mass of mainly very small, even submicroscopic, crystals: the lack of willingness of rhyolitic liquids to crystallize is a consequence of the very high viscosity of such melts when they reach the surface of the Earth and lose their dissolved water. Granites, in contrast, crystallize underground, retaining their water, and this, coupled with slow cooling, allows them to solidify as coarse-grained rocks.

A further consequence of the high viscosity of the de-watered rhyolitic melts is the tendency for eruptions to be explosive. Some rhyolites are erupted in the form of gas-charged clouds of molten droplets known as *nuées ardentes*. Such clouds travel down the slopes of volcanoes with immense speed and are highly destructive. The collapsed and solidified products of similar eruptions may reach 300m (1000ft) in thickness and cover many tens of square kilometres. When deposited hot enough for the individual fragments of glass to weld together, such deposits (termed ignimbrites) become massive and difficult to distinguish from lava flows.

Rhyolite eruptions are characteristic of areas of active mountain-building such as island arcs (e.g., Sumatra, Japan, West Indies), and *mountain chains such as the Andes.

Granodiorite

Granodiorite is the name given to one of the most abundant members of the granite family of plutonic igneous rocks. Strictly defined, a granodiorite contains more than 20% quartz, and of the feldspar present more than two thirds is a calcium-bearing *plagioclase. A granite, in contrast, contains a higher proportion of alkali feldspar and less plagioclase. Granodiorites are widespread and abundant rocks occurring in the same sorts of geological environments as granites.

Rhyodacites

Rhyodacites are volcanic rocks found as lava flows and ignimbrites, and have a slightly more basic composition (i.e., they are somewhat richer in calcium, magnesium and iron, and are slightly poorer in silica) than true rhyolites. In chemical composition they are broadly equivalent to the granodiorites. Most provinces of rhyolitic volcanics contain substantial amounts of rhyodacite as well as true *rhyolite.

Syenite

Syenite is a coarse-grained plutonic igneous rock type similar to granite but devoid of essential *quartz. It is very much rarer than granite and is usually found in the eroded roots of volcanoes which have erupted lavas much richer in alkalis than usual (such associated volcanic rocks include phonolites, trachytes, and alkali basalts). Since a richness in alkalis and a poorness in silica relative to granite is the characteristic feature of such provinces, many syenites contain not only abundant alkali *feldspar but also minerals of the *feldspathoid group such as nepheline or sodalite. The dark minerals of syenite are also often different from the *biotite and *hornblende so characteristic of granitic rocks. In syenites we see the comparatively rare sodium-bearing *amphibole arfvedsonite and the sodium-rich *pyroxene aegirine. Apart from the alkali metals, syenites are often relatively rich in rare elements such as rubidium, the rare earth elements, fluorine and chlorine. As a result many rare and unusual accessory minerals may be found in them.

The typical environment of the alkali-rich igneous activity which gives rise to syenites is that of the continental area subjected to intense *faulting and rift formation. One of the best known such areas is the Gardar province of West Greenland, where numerous syenite intrusions of *Precambrian age are found. The volcanism associated with the East African *rift valley also includes syenite, but here it is a rare type, probably because the activity is too young for many of the intrusive rocks yet to have been exposed by erosion.

Dacites

Dacites are volcanic rocks which correspond broadly in chemical composition to the plutonic rock type, quartz *diorite. They are similar to rhyodacites and rhyolites in general character, though less rich in silica. Like these other types, they are usually very fine-grained, though they may contain larger crystals of *plagioclase, *hornblende and *biotite.

Dacites are among the most abundant eruptive products of island-arc volcanoes. They are characteristically accompanied by andesites and more rhyolitic types. As with the latter, their eruptions are frequently explosive.

Phonolite

The name of this lava type is an adapted Greek version of the old name "clinkstone", given because these rocks supposedly emit a bell-like note when struck by the hammer. Nowadays, however, phonolite is defined as a fine-grained extrusive igneous rock characterized by the presence of abundant alkali *feldspar accompanied by the feldspathoid *nepheline. Thus phonolites are the volcanic equivalent of the plutonic rocks known as nepheline syenites. Phonolites are comparatively rare rocks but form extensive *plateaux of lava in parts of the East African *rift valley.

Trachyte

Trachytes are volcanic rocks rich in alkali *feldspar and carrying additionally either *nepheline or small amounts of *quartz. They thus grade with an increase of silica into rhyolites and with a decrease into phonolites. Like these other types, trachytes form flows of generally pale-colored lava because of their richness in feldspar, and this contrasts markedly with the darkness of their common associate, basalt.

It was Charles *Darwin who first speculated that trachytic lavas might be derived from basaltic liquids by a process of fractional crystallization in which minerals that crystallized early, forming while the melt was still underground, became separated from the remaining liquid. The separate eruption of the crystal-rich accumulation and the remaining liquid was supposed to give rise to the two lava types.

Darwin was wrong in detail but right in principle. The modern science of igneous petrology has followed this general idea in many detailed investigations of the origin of igneous rock types.

Carbonatite

For many years carbonatite was one of the most problematic of all igneous rock types.

Carbonatites are mostly found as small intrusions in association with alkali-rich igneous rocks – for example in East Africa – but they are unique in being composed dominantly or exclusively of carbonate material rather than the silicates characteristic of all other igneous rocks. Ancient carbonatite intrusions are composed of calcium, magnesium or iron carbonates; though at least one volcano, Oldonyo Lengai in Tanzania, has in recent times emitted ashes and flows of sodium carbonate: since this material is readily soluble in water it is not surprising that none is preserved among the products of older volcanic activity.

The origin of carbonatites was formerly obscure because carbonate rocks normally occur as sedimentary *limestones, and some time elapsed before it was conclusively shown that carbonatites were not simply large fragments of sedimentary strata accidentally incorporated in intrusive complexes but are themselves intrusive rocks. The discovery of the first carbonatite lava flows near Ruwenzori in Uganda and the subsequent eruptions of Oldonyo Lengai clinched the argument about the existence of molten carbonate liquids as genuine products of igneous activity.

Carbonatites have considerable economic as well as theoretical interest. They tend to be rich in rare elements such as cerium, niobium, thorium and phosphorus, and are extensively mined. One unique carbonatite intrusion, Phalaborwa in South Africa, is the site of a rich copper mine.

Diorite

Diorite is a fairly abundant type of plutonic rock: together with quartz diorites, granodiorites, and granites, it is characteristic of the intrusive igneous activity of mountain-building environments. The cores of eroded mountain chains such as the Sierra Nevada, western USA, contain large masses of all these rock types, commonly nested as separate intrusive bodies but collectively making up the huge plutonic complexes known as batholiths. Diorites themselves are coarse-grained rocks, consisting ideally of about 60% of the *plagioclase feldspar andesine, with *hornblende, *biotite, *magnetite and minor amounts of quartz and alkali feldspar. Compositionally speaking, they are thus halfway between granites and gabbros.

Andesite

Many of the loftiest peaks of the Andes are towering, steep conical volcanoes in which the lava-type andesite is a prominent constituent. Our everyday image of what a volcano ought to look like is based on such as these, for they are characteristic not only of active areas of mountain-building in continental areas such as the Andes and Central America, but also of the large island arcs like Japan, volcanic peninsulas such as Kamchatka, and the numerous arcs of smaller islands like the Aleutians and the West and East Indies (see *volcanic landscapes).

Steep andesite volcanoes are, of course, rather readily eroded once they have become extinct, and so most of our knowledge of them is derived from modern examples. In the past geological record, andesite is not prominent as such, but great volumes of sedimentary rock are composed of andesitic and rhyolitic debris eroded from former volcanoes and deposited in nearby areas of subsidence.

Andesites are frequently associated with lesser quantities of basalt (a more silica-poor lava-type), and of dacites, rhyodacites and rhyolites. The origin of andesitic melts is still a matter of considerable debate.

Most andesite volcanoes are situated above areas where the sea floor is being subducted into the underlying mantle: indeed, most of them border the Pacific Ocean, and this great collection of volcanoes has been dubbed the "Ring of Fire". The remelting of this material as it is taken down into hotter regions of the mantle is clearly one of the possible sources of andesite melts.

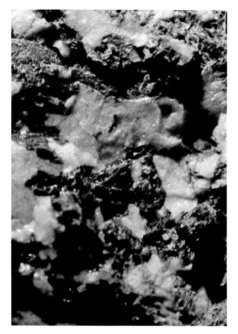

Gabbro is described as a basic rock, implying that it is free from quartz. This is in strong contrast with granite, one of whose primary constituents is quartz.

Gabbro

Gabbros are the plutonic equivalent of *basalt, the most abundant of all lava types. They characteristically have about 50–60% of the calcium-rich plagioclase feldspar labradorite, accompanied by *pyroxene and *magnetite with or without *olivine. Gabbros are considerably poorer in silica than are granites (about 50% by weight as opposed to about 70%) and considerably richer in calcium, magnesium and iron.

Although basaltic lavas are extremely common, gabbro is not itself particularly so; and among intrusive rocks it is much

Gabbro in thin section. The main constituents are plagioclase feldspars (calcium-aluminium-silicate), seen here as elongate gray crystals showing twinning (the alternating black, gray and white stripes represent parts of crystals in which the atoms are differently orientated). The other main constituents are olivine (blue) and pyroxene (here seen as one large purplish crystal enclosing many individual plagioclase crystals). The texture shows that the plagioclase crystallized relatively early, the pyroxene relatively late.

less common than granite. Probably the main reason for this lies in the low viscosity and consequent high mobility of basaltic melts, meaning that they tend to reach the surface before they solidify, and so appear as lava flows rather than as plutonic intrusions.

Nevertheless, gabbroic intrusions have played a highly important part in the study of igneous rocks and the processes by which they are formed. Among others, studies of the Skaergaard intrusion of East Greenland and the Stillwater intrusion in the USA have made it possible to determine the way in which magmas crystallize; that is, the way different minerals appear in sequence with falling temperature and the way in which the remaining liquid progressively changes in composition. If the crystals and liquids are separated from each other, usually under the influence of gravity, the progressive tapping off of the liquid fraction can give rise to a great variety of different rock types. This process, fractional crystallization, is the cause of much of the variation in composition seen in igneous rocks.

Basalt

Basalt is the most abundant of all lava types. Apart from wide occurrence in continental areas, it is the principal rock type of the *ocean floor. Basalts are fine-grained rocks consisting predominantly of small crystals of calcium-rich *plagioclase, *pyroxene and *magnetite and frequently carrying embedded crystals of *plagioclase, *pyroxene or *olivine. Dense and black, basalts frequently show small vesicles

A xenolith of andesite in adamellite from Cumberland, UK. Xenoliths may be fragments that have crystallized from the magma earlier than the surrounding rock, and hence having a slightly different composition; fragments derived from the country rock; or, in the case of those found in kimberlite, fragments brought up from considerable depths within the Earth. The term "xenolith" means, literally, "foreign rock".

Curtain of basalt at Skaftafell, Iceland. Some of the basalt columns have been dislodged by the stream, and their tumbled fragments are visible in the foreground: note their generally polygonal cross-section. Basalt is an extremely common igneous rock, comprising over 90% of all extrusive rocks. Beneath a sediment and lava veneer, basalt dike swarms form 90% or more of the Earth's ocean crust.

where bubbles of gas were trapped.

Basaltic magma is highly mobile, and individual lava flows are capable of covering hundreds of square kilometres: in some continental areas, lava *plateaux made mainly of basalt have in the past covered millions of square kilometres. Because of their fluidity, basaltic lavas are often erupted without significant explosive activity, and basaltic volcanoes are normally wide and of low angle compared with those formed from more viscous melts (see *volcanic landscapes; *vulcanicity).

One of the best studied basaltic volcanoes is the currently active island of Hawaii. Much of the volcanism of Iceland in the early 1970s was also basaltic in nature – as befits an area which is, geologically at least, little more than an elevated part of the sea floor.

The study of basalts has led to much information about the interior of the Earth, for basaltic melts are among those least modified by other processes on their way to the surface. It is evident from the wide occurrence of basalt that it must be one of the commonest melting-products of the Earth's mantle. This observation places significant constraints on hypotheses concerning the nature of mantle materials.

Peridotite
A comparatively rare rock type at the surface of the Earth, peridotite is, at depth, probably the most abundant of all the Earth's materials, being a major constituent of the mantle. Peridotites are variable rocks but are essentially characterized by a richness in *olivine and *pyroxene, the latter usually a magnesian rather than a calcitic variety. Peridotites are thus comparatively poor in silica (40–45% by weight) and very rich in magnesium oxide (up to 40% or more). Iron oxide is the remaining major constituent, accompanied by varying minor amounts of calcium, aluminum, sodium and other elements.

Peridotites are found in two important and contrasted environments. Firstly, they occur within fold *mountain chains, particularly where old oceans have closed up and disappeared between two advancing continental blocks. In these situations, parts of the old sea floor together with adjacent parts of the upper mantle are occasionally caught up and incorporated in the fold belts. The whole assemblage of sea-floor basalts and associated upper-mantle peridotites is known as the *ophiolite association*; and its study forms one of the most fascinating recent developments in the Earth sciences, for it is only in such areas that we can determine the nature of oceanic crust and observe the upper mantle directly.

Secondly, fragments of mantle peridotites are brought to the surface as blocks included in the volcanic vents known as *kimberlite pipes. The individual blocks are relatively small (up to about 1m (3.25ft) in diameter) and do not give as much structural information about the mantle as do the ophiolites. Nevertheless, it is possible to show that they come from considerable depths (at least 200km (125mi)) within the mantle, being thus the only samples obtainable from such relatively deep levels within the Earth. All peridotites are more or less susceptible to hydration, with the formation of rocks known as *serpentinites.

Kimberlite
Named after the town of Kimberley in South Africa, kimberlite is one of the world's rarest yet most fascinating igneous rock types, for the narrow pipe intrusions (see *igneous intrusion) of kimberlite bring *diamonds to the surface.

Kimberlite itself is not easy to define as a rock type for it is very variable and subject to much accidental contamination by other rocks torn off the sides of the pipe. Kimberlite intrusion seems to be associated with the emission of large quantities of water and carbon dioxide, so that when an intrusion comes near the surface the eruption becomes violently explosive; and the vents

Xenoliths are fragments of rock included within another rock. Here, xenoliths that are believed to be fragments from the earth's mantle are seen in kimberlite. They are important, not only because they indicate the possible constitution of the mantle, but because they sometimes contain diamonds.

therefore usually widen near the surface, evidently having acted as huge ball-mills as material caved in from the side and was ground to pieces and mixed with kimberlite coming from below by the action of escaping gases: the eruption must involve very rapid transport of material to the surface from great depth because large fragments of dense upper-mantle *peridotites are frequently brought to the surface.

However, when relatively uncontaminated kimberlite is found, it proves to be a fine-grained rock rich in *olivine and phlogopite *mica, usually with a matrix of *calcite and *serpentine. The diamonds are usually rather sparsely distributed within the matrix and it is necessary to process many tonnes of ore to extract even small quantities.

Kimberlites are common only in South Africa and adjacent areas, and in the Yakutia area of Siberia, though few are sufficiently rich in diamonds to be economically important.

Eclogite

An extremely rare though interesting rock type, eclogite is like *basalt in composition but has a completely different, high-pressure, mineralogy. Whereas a basalt crystallizing near the surface of the Earth usually consists essentially of normal *pyroxene and *plagioclase feldspar, the same chemical composition in eclogite is represented by *garnet plus a rare pyroxene called omphacite. Experiment shows that this mineralogy can be achieved by allowing basalt to recrystallize under a much higher pressure than that under which it originally crystallized. Thus some eclogites represent basalts transformed by intense pressure during, for example, episodes of mountain-building. However, eclogite is also a rare member of those rocks brought to the surface from deep within the mantle by *kimberlite pipes: it is thus a constituent of the mantle – albeit a rare one. Such eclogites may have originated by the trapping and crystallization of basaltic melts deep within the Earth, or may represent fragments of former *ocean-floor basalts which have been subducted and mixed into the upper mantle.

Whatever their origin, eclogites are handsome rocks, consisting as they do of green pyroxene and orange-red garnet. They are also probably the source of most of the diamonds in diamond pipes.

Obsidian

Obsidian is a naturally occurring volcanic glass of rhyolitic composition. Obsidians are usually black, with an intense glassy luster and a marked conchoidal (shell-shaped) fracture. They form by the rapid chilling of viscous, relatively dry rhyolitic magmas, and are mainly known from modern volcanoes since they devitrify rather rapidly. Famous examples of obsidian lava flows come from the Aeolian Islands north of Sicily. Obsidians were much used by primitive Man for the fashioning of arrow-heads and axes.   KGC

Sedimentary Rocks

Although sedimentary rocks form a very small proportion by volume of the rocks of the Earth's crust, the chances of encountering them on the Earth's surface are high: about three quarters of the Earth's surface rocks are of sedimentary origin.

Sedimentary rocks are formed by *diagenesis from sediments, accumulations of solid material formed on the Earth's surface as a result of the various processes that shape the surface of the Earth. Thus sedimentary rocks form a thin surface veneer to the Earth's crust, averaging about 2.5km (1.5mi) in thickness. The oldest sedimentary rocks known are some 3500 million years old.

As the processes which lead to the formation of sedimentary rocks are going on around us, the best way to begin to discover the clues as to how they form is to examine a present-day environment. In a typical coastal environment a variety of sediments are being deposited, and these may later be buried and converted into rocks. Very fine-grained sediments – muds – are accumulating in the salt marshes, whereas coarser grained sediments – sands – are being deposited on the dunes and beach.

What features of these sediments would the geologist expect to be preserved as fossil clues to the interpretation of ancient environments? The size of the sedimentary particles obviously has something to do with the environment of *deposition, for the marsh muds accumulate in relatively calm-water conditions, the dunes are deposited by wind, and the beach sand is being laid down in the area where the waves are breaking – in other words, in turbulent high-energy environments. The distributions of grain-sizes in the beach and dune sands are different, the dune sands showing better sorting (that is, one particular grain size predominates). The shape of the grains in each environment is also distinct, the dune sand grains being very well rounded. All the features just described (grain-size, sorting and shape) are collectively known as the *texture* of the sediment.

Thus sediments transported for long periods in "high-energy" environments and deposited in similar conditions are well-sorted and rounded, whereas lower-energy conditions are indicated by less rounding and sorting. Experiments passing water currents at various velocities over a variety of grain-sizes show (not surprisingly) that faster currents carry larger grains; and reveal that grains are not only carried in suspension in the moving fluid, but also by bouncing (saltation) and rolling along the sediment surface. Thus the muds of the salt-marsh environment were deposited from suspension, but the sands were largely transported by rolling and bouncing.

Sediments and sedimentary rocks may contain other clues to their origins. Ripple marks are a common sight on beaches and can indicate the direction of flow of the current that produced them. But, as well as surface features, there are structures *within* sediments that record the direction of current flow, as can often be seen when a sand dune is cut open. Cross-bedding is a common structure within sandstones, and is produced by deposition of sediment on the steeper sides of ripples or larger structures. Experiments in which the velocity of water flowing over sediment is increased reveal a succession of structures, both laminated sediments and small- and large-scale cross-bedding being present.

Organisms living within, or on, coastal sediments or in other environments are also potential fossil environmental indicators (see *paleoecology).

Examination of the different types of

Chalk cliffs at Seaford Head, Sussex, UK. The rocks of which these cliffs are composed were laid down during the late Cretaceous. Note the orientation of the strata: in the nearground these dip toward the sea while, immediately behind, the dip is toward the land.

coal and oil), calcium carbonate (as limestone), aluminum and iron oxides, and *evaporites.

The different types of sedimentary rock described here can be related not only to weathering processes, but also to the *climatic zone of the Earth in which they formed (for weathering is obviously greatly influenced by climate) as well as to the different parts of tectonic plates over which surface processes may be operating.

Argillaceous Rocks

By definition, argillaceous rocks contain particles less than 1/16mm in diameter and are usually even more finely grained, requiring electron-microscope techniques to study them. They form about 60% of all sedimentary rocks, and on average contain 60% *clay minerals (formed by the chemical weathering of alumino-silicate minerals), 30% *quartz, 5% *feldspar, and around 1% each of iron oxides and organic matter. Such finely disseminated organic matter can act as a source for oil – indeed it is estimated that 75% of all organic matter is buried within argillaceous deposits, rather than concentrated in materials such as *coal (see *petroleum).

When first deposited, argillaceous sediments contain considerable quantities of water, which may amount to as much as 80% of their volume. Such large amounts of water are held between the flakes of clay minerals, which become packed better together during burial by a process termed compaction. The amount of compaction plays a major role in determining the nature of the resultant argillaceous rock.

The types of clay mineral present in an argillaceous rock give some indication of the degree of chemical *weathering, and hence climatic conditions, experienced by the source region: the mineral *kaolinite, for example, indicates intense chemical weathering. The type of source rock is more difficult to interpret, save in the case of fuller's earth deposits that contain clay minerals resulting from the break-down of volcanic ash. Fuller's earth deposits may therefore result from the weathering *in situ* of falls of volcanic ash.

Argillaceous rocks are described as *siltstones, *clays, mudstones, *shales or *marls.

Clay and Mudstone

Clay is an *argillaceous rock resulting from the first stages of the compaction of mud. It is soft and easily absorbs more water, becoming plastic (see *clay minerals). Unlike clay, mudstone will not absorb water. It has lost much of the water present when the sediment was originally a mud, and breaks in a random fashion, reflecting its homogeneous internal structure.

Shale

Shales result from the compaction of mud.

mineral grains present in a sedimentary rock enables conclusions to be made about both the nature of the source rock (which contributed the rock fragments) and the *weathering processes that caused its disintegration. Such an interpretation is relatively easy for a coarse-grained sediment, such as beach pebbles, but is more difficult with a sand, in which a constituent grain may be only a part of a single crystal: thus pebbles could be matched with known exposures of probable source rocks, but a sand grain consisting wholly of the mineral quartz could have been derived from a multitude of rock types. A sandstone composed entirely of *quartz contains virtually no information about its source, but is

testament to the fact that the various surface processes (weathering, *erosion, *deposition) fractionate original crustal materials into pure deposits of, in this example, silica. The way in which the source rock has been fractionated into its various components by surface processes determines to which of the major categories the eventual sedimentary rock belongs. (Most sedimentary rocks, classified as either detrital or chemical-organic, are also classified according to their grain-size as *rudaceous rocks, *arenaceous rocks or *argillaceous rocks.) The degree of fractionation accomplished by surface processes has tremendous economic importance, producing concentrates of carbon (as

Eroded chalk in Kansas. Chalks are made up of millions of plates of calcium carbonate ($CaCO_3$) called coccoliths. These plates were secreted by certain unicellular golden-brown algae (called coccolithophores). Strictly speaking, all rocks termed chalks date from the Cretaceous, though the name is sometimes applied to similar rocks of Tertiary age.

During compaction, the constituent mineral flakes (*clay minerals, and often *micas) become oriented parallel to the original bedding, with the result that the shale fractures easily along one plane; this property, which characterizes shales, is known as *fissility*.

Black shales contain small amounts of iron *pyrites (iron sulfide), which renders the whole rock black in color. The iron sulfide is precipitated during *deposition in conditions lacking in oxygen, usually in deep-water areas, or regions without any oceanic circulation.

Oil shales are a variety of shale containing over 30% of organic material, derived either from planktonic remains or from the decomposition of vegetable debris washed in from nearby land.

Marls

These are *argillaceous rocks which contain a mixture of quartz, clay minerals and calcium carbonate, the latter being present in amounts between 35 and 65%.

Argillite

Argillite is a little used term to describe rocks that have characteristics mid-way between *shale (a sedimentary rock) and *slate (a metamorphic rock).

Siltstone

The term siltstone describes *argillaceous rocks that contain grains of diameter 1/16–1/256mm. These rocks, bordering as they do on being *arenaceous rocks, contain more *quartz than other argillaceous rocks.

Siltstones are commonly laminated, the laminations (thin sheets) being due to variations in grain-size, organic content, or amounts of calcium carbonate. When the laminations are due to seasonal changes (such as meltwater during spring, which deposits silts that contrast with clay sedimentation during the rest of the year), they are termed *varves*.

Arenaceous Rocks

Together with *rudaceous rocks, the arenaceous rocks, often collectively known as sandstones, comprise about one quarter of all *sedimentary rocks. Their grain-size is 0.0625–2.00mm, and they include *quartz with smaller amounts of *clay minerals.

Arenaceous rocks are divided into three groups according to the composition of their clasts, the fragments of which they are made up. The *quartz sandstones contain only small amounts of *feldspars and other minerals, the *arkoses contain appreciable amounts of feldspar, and the *graywackes show rock fragments set in a muddy matrix.

Quartz Sandstones

Quartz sandstones are the result of a considerable amount of fractionation of rock debris released by *weathering processes, as shown by their being deficient in minerals unable to withstand chemical weathering. They are usually moderate to well sorted, and the clasts are similarly rounded. These textural features result in quartz sandstones often exhibiting high porosity and permeability, at least immediately after deposition: later the spaces between the grains may be filled by mineral cement, and such a diagenetic modification destroys the porosity of the rock (the same process affects *conglomerates). Mineral cements are most commonly composed of *quartz or *calcite, but iron minerals such as limonite may also form cementing media. Sediment porosity may also be reduced by compaction, due both to the reshuffling of the grains and to the minerals being dissolved at grain contacts. The porosity and permeability of sandstones has great economic significance, as reservoirs of oil, gas and water may form within rocks in which the pore spaces have not been lost.

Two varieties of colored sandstones deserve special mention. *Greensands* are quartz sandstones which contain a few per cent of the green mineral *glauconite, which forms only in marine conditions. Another iron mineral, *hematite, characterizes the sandstones and shales that make up an association of sedimentary rocks termed *red beds*. The red coloration is due to the presence of coatings over the sand grains of hematite derived by the oxidation of iron-rich minerals swept in from the source area; and indicates a degree of

Oil-bearing shale. This fine-grained sedimentary rock is formed by cementation of particles of silt, usually containing also tiny fragments of other materials. Rich in fossils, shales can be readily split into separate layers, or laminae. Oil shales contain kerogen, a solid material that must be treated before yielding hydrocarbons.

Sandstone with a cement of ferric oxide. About one third of all the sedimentary rocks exposed on the Earth's land surfaces are sandstones; and it is probably owing to this frequency of occurrence rather than to any inherent favorable qualities that sandstone has been so much used as a building stone.

Carboniferous sandstone showing cross-bedding. Much can be learned about a sedimentary rock's environment of deposition by examining its internal structures. Cross-bedding arises when sediment has been deposited on the steeper side of ripples (or larger structures): the scale and nature of the cross-bedding depends upon the velocity of the current.

aridity of the depositional environment, where the water table has remained low, so permitting iron minerals originally in the sediment to be oxidized immediately after burial. Most *arkoses exhibit a red color for the same reason.

Quartz-rich sandstones are common in shallow marine and coastal plain environments

Arkose

Arkoses are *arenaceous rocks that contain not only *quartz but also notable amounts (of the order of 25%) of *feldspar. Feldspar will withstand almost as much mechanical degradation as quartz during transport, but is much more susceptible to chemical breakdown. Thus the occurrence of the mineral in proportions greater than a few per cent is indicative of fairly arid conditions in which transport of detritus from its source area has been relatively rapid after the essentially granitic source rocks have been mechanically weathered. And so most arkose sandstones were laid down not only in fairly arid conditions but also in proximity to source lands experiencing rapid erosion. Arkoses often characterize periods of deposition immediately following mountain-building episodes, and are almost exclusively confined to continental environments. The high content of orthoclase feldspar gives arkoses their characteristic pink color, which may be enhanced by staining by iron oxides.

A graywacke of Silurian age. Graywackes are made up of rock fragments that vary from fine to coarse in quality, sometimes being even pebbly. The fragments are generally poorly sorted. Graywackes are formed in regions of rapid erosion and varied rock, and are commonly found in geosynclinal areas.

Graywacke

The term graywacke is derived from the German *grauwacke*, meaning "gray and hard", a word used to describe the color and texture of a rock commonly found in the Harz Mountains, and containing a mixture of the *weathering products of *igneous and *metamorphic rocks. These products include both particulate material resulting from the mechanical break-up of the source rocks and *clay minerals resulting from chemical weathering. Thus graywackes result from incomplete fractionation of weathering products, reflected in the characteristic poor sorting and rounding of the clasts. Indeed the sand-sized particles often "float" within a matrix of clay minerals (see *conglomerates).

Graywackes are usually dark and extremely hard, and the finer-grained varieties can easily be confused with *basalts. They commonly occur in association with black *shales, forming an alternating sequence of shales and sandstones, the latter often showing evidence of having been deposited by turbidity or density currents. A turbidity current (see *deposition) is a slurry of sediment and water which, being denser than water, will flow downhill within a body of water and, as it does so, deposit a sand that has a mud matrix with sand-sized particles becoming progressively finer upwards; hence the term graded bed is applied to such sandstone units. In addition, graded beds show an upward change in sedimentary structures reflecting the decrease in current velocity as the turbidity current passes by. The base of graded beds may also show erosional features (sole marks), either as flute casts caused by the erosional effect of the turbulent base of the current, or groove casts produced by pebbles or organisms being dragged along and cutting linear marks in the underlying sediment.

Rudaceous Rocks

Being coarse-grained (grain-size greater than 2mm), rudaceous rocks are generally deposited in high-energy environments, and the large rock fragments of which they are formed (pebbles, boulders, etc.) give good indications of the composition of their source rocks. On the basis of grain shape, this rock group is divided into *breccias and *conglomerates.

Breccias

Breccia is Italian for rubble, and, as the word suggests, breccias consist of angular fragments of rock. Such angularity indicates a minimal amount of transport (since otherwise the corners would be knocked off), so we usually find breccias relatively close to their source. They are generally associated with modern and fossil coral reefs. The parts of the reef exposed to wave action are broken up, and the resultant debris accumulates as a submarine scree slope on the open-sea side of the reef.

Pulpit conglomerate near Coalville, Utah, from the top of the Cretaceous. Conglomerates are made up of rounded fragments whose size may vary from small pebbles to large boulders. They have generally been deposited fairly close to their place of origin, with the exception of tillites.

Breccia deposits may form when mechanical *weathering breaks off rock fragments from cliff faces, beneath which talus or scree deposits form. In this case, the breccias will be banked up against their source rocks. They may also be deposited by sporadically flowing rivers in arid regions. Rock debris loosened by prolonged mechanical weathering is washed away during rainstorms, and carried relatively short distances by flash floods. Such deposits are common on alluvial fans and may contain imbricate structures, in which flat pebbles are stacked together and inclined in the direction of current flow.

Diagenetic processes can result in the formation of solution breccias. For example, *evaporite deposits may be dissolved away or partially replaced by limestones, leaving behind a breccia.

Conglomerates

Conglomerates are distinguished from *breccias on the basis of the rounded nature of their clasts.

The clasts may be packed together in two ways. If they all touch each other the conglomerate is said to show *grain support*, and has been deposited in very high-energy conditions which persisted long enough for the sediment to become well sorted. Conglomerates with grain support are likely to have formed in a beach environment, or possibly on the floodplain of a large river not subject to periods of drying up. In contrast, some conglomerates show *matrix support*, in which the larger clasts are held apart by a finer matrix of sand and clay: this indicates that the sedimentary material was transported and/or deposited relatively rapidly, so that little grain-sorting could occur. Conglomerates with matrix support were commonly deposited on alluvial fans as a result of flash floods; in this case the matrix is composed of sand-sized particles.

Conglomerates may be divided further on the basis of the origin of their clasts. *Extraformational conglomerates* are composed of clasts from outside the area of deposition, whereas *intraformational conglomerates* are derived through the erosion of local and recently deposited sediments – for example, from an adjoining river bank. The variety of different rock types making up the clasts again provides clues to the origin of the conglomerate. Pebbles and boulders composed of only one rock type either derive from a source area exposing only one rock type or have resulted from the elimination of all but the most stable components during prolonged weathering, erosion and transport.

Tills and Tillites

Till, or boulder clay, is the sediment deposited directly from the melting ice of a glacier. Tillite is till converted to rock.

The formation of glacial ice involves the incorporation of rock debris, often as scree material. Armored with this rock material, the ice flows over rock surfaces and grinds them away, producing a rock flour which also becomes impregnated in the ice. When the ice melts, all the rock material is released and may be deposited: it is characterized by a huge range of grain-sizes, from clay grains up to boulders. Tills and tillites are therefore distinguished by being extremely poorly sorted, and by matrix support (see *conglomerates) of the larger clasts.

Similar deposits may be produced by mud flows on alluvial fans and submarine slopes. These deposits are termed tilloids when their non-glacial origin can be proved, but this is often hard; clues may be provided by the pebbles in a tillite being striated by glacial erosion and the deposit itself resting on a striated rock surface.

Calcareous Rocks

Calcareous rocks, *limestones and *dolomites, contain at least 50% of either *calcite ($CaCO_3$) or *dolomite ($CaMg(CO_3)_2$). Both minerals may be formed as direct precipitates from seawater, but by far the most important way the calcium and magnesium (originally liberated by chemical weathering) become "fixed" is through secretion of carbonate minerals by animals and plants. At the present time, calcareous sediments (apart from deep-sea oozes) are commonly found only in tropical and subtropical environments where carbonate-secreting organisms flourish. Thus the occurrence of ancient calcareous rocks can be used as a paleoclimatic indicator.

Limestone

The majority of limestones formed in water only a few metres deep. If conditions were turbulent, then the grains within the limestone show grain support (see *conglomerates), with the pore spaces filled with a *calcite cement. However, if conditions were calmer, the pore spaces are filled with a carbonate mud; that is, clay-sized particles composed of calcium carbonate. The mud may even form the bulk of the limestone. There are four main types of grain:

Erosion of thin-bedded limestone at Osmington, Dorset, UK. The two most important constituents of limestones are calcite (calcium carbonate, $CaCO_3$) and dolomite (calcium magnesium carbonate, $CaMg(CO_3)_2$): on occasion, they may include small quantities of iron-bearing carbonates.

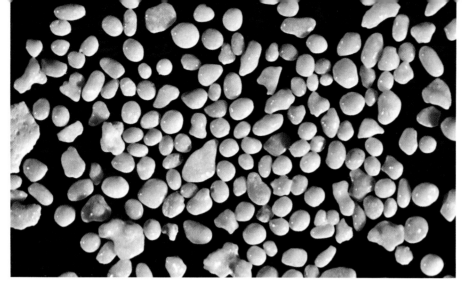

Oolites. These are spherical grains formed by the addition of successive envelopes of carbonate material as they roll around in turbulent environments. The oolitic material is coated around a nucleus, which may be a quartz grain or a fragment of a shell, and the particles reach a maximum size of 1mm (0.04in).

Pellets. These elongate grains are of carbonate mud excreted by sediment-feeding organisms.

Fragments of recently deposited limestone. Such grains, produced by local erosion, look rather lumpy. Examples forming today are termed grapestone.

Shell material. This is a common constituent of many limestones, ranging from undamaged fossils to an accumulation of sand-sized particles produced by the break-up of shells in turbulent conditions.

As the components of limestones form entirely within the area of deposition, rather than having been transported from elsewhere, they provide excellent clues to the nature of the depositional environment. Thus a well-sorted oolitic limestone was probably deposited in a warm, shallow, turbulent tide-influenced sea, for that is where oolites are forming today. In contrast, a carbonate mud containing pellets indicates calm-water conditions.

Reef limestones consist of a framework of organisms, usually corals and algae, filled with broken coral and shell material eroded by wave action in the surf zone of the original reef.

Algal limestones are laminated rocks in which the laminae (sheets) were produced by the growth of algal colonies, the organisms both secreting and, on their sticky surfaces, trapping carbonate material. The laminations exhibit a variety of patterns which may aid environmental interpretation. Today, such algal sediments are forming as "mats" around very flat coastal areas. *Pisolites* are a variety of algal colony in which the algae grow around a nucleus of shell material: they may grow to become several centimetres in diameter.

Deep-sea calcareous oozes are formed in areas of ocean where there is very little deposition of land-derived sedimentary material. Sedimentation is very slow and dominated by the accumulation of the calcareous remains of plankton raining from the surface waters. Such deposits are uncommon in the geological record; but the northwest European Cretaceous, though not a deep-sea deposit, is an accumulation of plankton (see *coccolithophores).

Nearly all limestones forming at the present time are of calcium carbonate in the form *aragonite, which is unstable and so during *diagenesis changes to *calcite. Thus nearly all ancient limestones are of the latter mineral.

Dolomite

Dolomite is a term used to describe both a mineral which is a mixed calcium-magnesium carbonate $(CaMg(CO_3)_2)$ and the *calcareous rock composed of it. Dolomite rocks have two origins, both involving the replacement of *limestone.

One type occurs within the basin of deposition, only a few tens of centimetres below the surface and very soon after the deposition of the limestone; this process is associated with the formation of *evaporites.

The second type takes place at depth some considerable time after deposition, and produces a coarse-grained dolomite rock. The replacement is caused by magnesium-rich solutions percolating through the limestone, the magnesium itself being derived from other limestones which generally contain a few per cent of magnesium carbonate – particularly in fossil algae and some animal shells. The dolomitization process occurs by magnesium atoms substituting for calcium "one for one", resulting in a volume reduction of up to 13% which increases the pore spaces in the rock. Thus dolomites are often porous, and can form reservoir rocks for oil and gas.

Certain limestones are relatively more susceptible to dolomitization, especially those, such as reef limestones, with an initially high magnesium content.

Ironstone

There are three principal types of sedimentary ironstones: banded ironstones, oolitic ironstones and clay ironstones. The banded ironstone formations date from the early *Precambrian, 1700 to 3200 million years ago, when the Earth's atmosphere contained no oxygen. The other types formed in the geologically recent past when the atmosphere *did* contain oxygen.

Banded ironstone formations consist of alternating laminae (sheets) of iron oxide, iron carbonate or iron sulfide, and silica (*chert), and contain various sedimentary structures, including ripple marks and mud cracks, suggesting that they were deposited

Ironstones are sedimentary rocks containing a high proportion of iron minerals. This sandstone is rich in limonite, once thought to be a single mineral with a fixed chemical composition but now known to be a mixture of minerals, notably goethite, in amorphous, colloidal or cryptocrystalline form.

in shallow water. It is suggested that the Earth's atmosphere at this time contained considerable amounts of carbon dioxide, the presence of which would produce rainwater and river water much more acid than that of today, thus enabling significant quantities of iron compounds to be transported in solution.

Oolitic sedimentary iron ores have formed during the last 600 million years or so (i.e., the Phanerozoic) and possess all the features of *limestones: oolites, shell fragments and mud matrices. But they are composed not of calcium carbonate but of iron minerals, including iron carbonate (*siderite), and iron-alumino-silicates (*chamosite) that can form only in conditions lacking oxygen. Probably these minerals were precipitated just beneath the sediment surface, to be reworked for only short periods into oolites and other particles. Chamosite is known to be forming today in sediments deposited in deeper waters in front of equatorial deltas. This type of iron ore is the only common sedimentary iron ore of northwest Europe.

Clay ironstones, the third type of sedimentary *iron ore, are quantitatively insignificant today but were the origin of the steel industry associated with a number of coalfields. They consist of rounded accumulations of iron carbonate that have replaced the *shales of many coal-bearing strata, especially overlying *coal seams.

Evaporites

Evaporites are sedimentary rocks produced by the evaporation of seawater, and so their presence in the geological record indicates an arid, hot climate.

In an experimental situation, evaporation of seawater results in the formation of *calcium carbonate, then *gypsum (calcium sulfate), and finally the most soluble salts, including *halite (common salt). However, evaporites cannot be accounted for by invoking simple evaporation, for 1.5m (4.9ft) of halite requires the drying up of a sea 100m (325ft) deep – and there are many salt deposits that are hundreds of metres thick. Moreover, the proportion of different minerals in evaporites is not the same as that produced by the simple evaporation of seawater. So, clearly, evaporites must be formed by some kind of recycling process, whereby water evaporated is constantly replaced.

Before modern examples of evaporite deposition were known, ancient deposits were explained using the barred-basin model, in which a land-locked sea has only one outlet to the open ocean (like the Mediterranean), across which lies a bar of material. However, evaporites are now known to form today in coastal plain areas, sabkhas, in which the *limestone sediments are replaced by evaporite minerals formed by evaporation of water from within the pore spaces of the sediment. The result of this process is that the original limestone is replaced by fine-grained *dolomite, and that calcium sulfate (*anhydrite) grows within the sediment and deforms it.

Phosphorites

The average phosphorus content of *igneous rocks is about 0.1%, but workable deposits (phosphorites) of *apatite – the ore of phosphorus – contain around 30% of this mineral.

The bulk of the world's *phosphorus reserves are in the form of marine phosphorites, found particularly in Morocco, the Spanish Sahara and the western USA. These deposits are of large area, and appear to have been formed by the replacement of *limestones in regions where cold currents welled up from deeper parts of oceans. These phosphorus-rich currents resulted in sea-bottom reactions whereby *calcium carbonate was replaced by apatite and related minerals. The phosphatization produced nodules and pellets up to 2cm (0.8in) in diameter: such nodules are known on present-day ocean floors. Apatite may be formed also by direct precipitation. Another type of phosphate deposit is *pebble-phosphate*, produced as a residual deposit of weathered phosphatic limestone, the best known deposit being in Florida.

The only type of phosphate deposit whose origin is beyond doubt is *guano*, produced by accumulation of bird excreta.

Cherts

Chert is composed of almost pure silica, either extremely finely crystalline (crystals visible only using high-powered microscopes) or crypto-crystalline, showing no evidence of a regular crystal structure. A number of other rock names familiar to the layman are in fact varieties of chert: *jasper is red chert, the coloration being due to iron oxide; flint, which is black, is commonly found in the Chalk of Northwest Europe; and *opal too can be found in chert deposits.

There are two distinct types of chert: cherts replacing *limestones (as chert nodules or chert bands) and truly bedded cherts associated with either *shales or banded *ironstone formations.

The replacement origin of cherts in limestones can easily be proved by direct observation, for often it can be clearly seen that fossils and limestone particles have been replaced by silica. The source of the silica is likely to be within the limestones themselves, in the form of siliceous microfossils such as *sponge spicules, or plankton such as *radiolarians. These remains are distributed throughout the limestone and after burial are dissolved, if the pore waters become slightly alkaline, and concentrated by replacing certain parts of the rock.

Bedded cherts are known from deep-sea sediments of a variety of ages: they have been sampled by deep-sea drilling of both the Atlantic and Pacific oceans. The latter examples appear to be organic in origin, for they contain remains of siliceous planktonic organisms. Cherts are also associated with volcanic rocks, such as submarine lavas, and it is probable that volcanic ash falling into ocean water stimulates "blooms" of siliceous plankton.

However, some cherts may be inorganic in origin, for precipitates of opaline silica have been observed in some Australian lakes which have very alkaline waters. Again, an organic origin for chert in the *Precambrian banded ironstone formations seems unlikely for there are no known siliceous organisms of this age. Perhaps the absence of living organisms resulted in higher concentrations of silica dissolved in seawater.

At present, there is no infallible method to determine whether cherts that do not contain siliceous fossils are organic or inorganic in origin. Indeed, cherts older than 100 million years have recrystallized, so that any fossils that they might once have contained would now have disappeared.

Bauxites and Laterites

Laterites and bauxites are the residual products of chemical *weathering, the material that has not been dissolved after even the most intense attack by acidic groundwater. As well as such chemical attack (largely induced by percolating water becoming acid due to decaying vegetation), conditions in which mechanical *erosion and removal of material is virtually zero are necessary for laterites and particularly for bauxites to form.

Such deposits are therefore found in tropical climates in lowland or flat-lying areas with little surface drainage. The residual cap of this weathering profile, composed of hydroxides of iron and aluminum, is termed a laterite. When most of the iron compounds are leached from a laterite, it becomes bauxite – the ore of *aluminum. Both types of deposit are usually colored in deep hues of red, brown and orange. RCLW

Bauxite, the major ore of aluminum. Bauxite consists of hydrated aluminum oxide, usually with iron oxide as impurity. It is a claylike, amorphous material formed by the weathering of silicate rocks, especially under tropical conditions. High-grade bauxite, being extremely refractory, is used as a lining for furnaces. Synthetic corundum is prepared from bauxite; and the ore is also an ingredient in some quick-setting cements. Leading bauxite-producing countries include Jamaica, Australia, the USSR, Surinam, Guyana, France, Guinea and the USA (especially Arkansas, Alabama and Georgia).

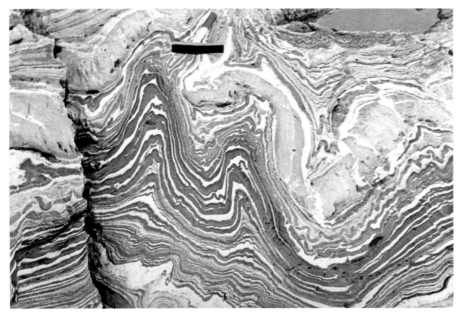

Marble, one of the best known metamorphic rocks, is much used for building and decorative stonework. Marble results from the thermal metamorphism of limestone: if the limestone was pure the marble consists of recrystallized calcite, but original impurities give rise to the rock's characteristic "marbled" appearance. (The term marble is often applied in building to stones that are not, in fact, marble.)

Metamorphic Rocks

Metamorphism is a term that implies change of form. We use it for a set of processes that change the form of preexisting rocks, either without altering their chemical composition or, much more commonly, producing rocks whose compositions are slightly or even wholly new. In general the changes occur in the solid state, although a vapor or liquid phase may also be involved.

Metamorphism. The most common environment in which metamorphic rocks are available to us is the *mountain chain, where erosion of a temporarily thickened part of the continental crust exposes once deeply buried sedimentary and igneous rocks that have been deformed and have undergone mineralogical changes in response to increased pressures and temperatures. Why then do they not change back, on returning towards the surface, into their premetamorphic forms?

As far as metamorphosed *sedimentary rocks are concerned, that is precisely the change effected by *weathering, which alters the chemical composition of rocks by addition of water to produce hydrous silicates, and by addition of carbon dioxide to produce carbonates. The metamorphism of sedimentary rocks, on the other hand, involves the production of water vapor, carbon dioxide and other gaseous substances, which are much less dense than the volatile-depleted silicate minerals and therefore move upward and away during the metamorphic episode, so altering the composition: this is irreversible until the rocks once more reach contact with the airs and waters of the Earth. This type of metamorphism is called *prograde* metamorphism, and takes place principally in response to increasing temperature. However, since the Earth is hotter inside than outside, there tends to be a correlation between pressure and temperature within a particular mountain chain.

*Igneous rocks start off as rather dry mineral assemblages, because they form at very high temperatures from silicate melts. Their metamorphism commonly involves the *retrograde* absorption of volatiles (water, carbon dioxide, etc.), which are borrowed from adjacent metamorphosing sedimentary rock masses. Older metamorphic rocks, whether ultimately of sedimentary or igneous derivation, may be involved for a second time in metamorphic episodes within mountain belts, and their behavior then is more like that of igneous rocks than that of wet sediments.

Temperatures at very great depths within the crust may be so high that the rocks begin to melt. This process starts at about 600°C (1100°F) in rocks of favorable composition and becomes important at 700–900°C (1300–1650°F). Because rocks are complex chemical compositions they melt incongruently, the liquid coexisting with solid crystals – in the same way that salty water and ice can coexist – over a range of temperatures. In some cases, the liquid separates from the solid crystals, and two types of rock, one igneous and the other metamorphic, are generated. Clearly neither is likely to retain the chemical composition of the initial rock.

A further environment in which the studies of igneous and of metamorphic rocks come very close to one another is in the mantle, where temperatures are high enough for recrystallization to occur entirely within the solid state, without the participation of volatiles, and where temperature perturbations may result in the production of liquids (magmas) together with a solid residuum. As the mantle is not easily accessible, most of the information about metamorphic changes within it comes from laboratory experiment, coupled with seismic studies and direct inspection of those rare mantle fragments emplaced in the crust by unusual processes.

Classification. Classification is essentially a three-part exercise, involving: a de-scription of the rock in terms of its 1cm–1m texture and chemistry, which may be a guide to the premetamorphic nature of the rock; a description of its mineralogy, which reflects the temperature and pressure of metamorphism; and a description of its fabric, which reflects the type of deformation of the rock. All three tend to be applied genetically though often insufficient information is recorded in the rock for unique classification.

It is sometimes possible to decide whether a metamorphic rock is of sedimentary or igneous parentage by use of exactly the criterion which a sedimentary or igneous petrologist would use: texture. In many weakly deformed sequences of metasediments, sedimentary structures such as grading, bedding, cross-bedding, ripple marks and even fossils are preserved – even though the minerals comprising the original rock have reacted to give new minerals. Similarly, in meta-igneous rocks, the coarse grain-size of plutonic intrusions (see *igneous intrusion) may be inferred, even though the grains themselves are now aggregates of other minerals, as may the former presence of insets in meta-volcanics (see *igneous rocks).

More usefully, the chemical characteristics of sedimentary rocks are frequently apparent, despite the fabric of the rock bearing no relation to the sedimentary environment. The extreme products of physical weathering, *quartz sandstones and quartzites, may be identified in their metamorphosed state by the abundance in them of *quartz, which is stable throughout the crustal pressure-temperature range. (Meta-sandstones are sometimes called psammites.) The extreme products of chemical weathering are aluminous *clays, in which the ratio Al:(Na + K) – that is, aluminum:sodium-plus-potassium – is much greater than 1:1, a feature which distinguishes them from acid igneous rocks whose mineralogy is dominated by alkali feldspar ($NaAlSi_3O_8$ or $KAlSi_3O_8$). The aluminous character of metaclays, often called pelites, is revealed by their high content of the aluminous mica *muscovite at low and medium grades of metamorphism and by the presence of *aluminum silicate minerals (Al_2SiO_5) at medium and high grade. Limestones and dolomites retain *calcite and *dolomite to high grade as marbles; and metamorphosed sandy or muddy carbonates may often be distinguished as calcsilicates from metamorphosed intermediate igneous rocks (which they resemble, chemically) by their having a pale colored magnesium-rich *amphibole or *pyroxene, rather than the darker, iron-richer silicates of the meta-igneous group.

Basic and intermediate igneous rocks form metabasites (epidiorites or prasinites to European workers; greenstones to North Americans) on metamorphism, characterized by amphibole with *plagioclase at medium grade and *pyroxene(s) with plagioclase at high grade: at very high pressure such rocks lose plagioclase, and *garnet becomes as abundant as pyroxene.

Ultrabasic rocks appear dark and silky with *serpentine as low-medium grade *serpentinites* and, at high grade, contain abundant *olivine and are termed *peridotites*.

Classification: Mineral Facies of Metamorphic Rocks.

Information about the conditions of temperature and pressure under which metamorphism has taken place comes from considering the relative stabilities of minerals or assemblages of minerals. By way of comparison, the existence of pure water in the form of ice implies temperatures below 0°C (32°F); in the liquid form, temperatures between 0° and 100°C (32–212°F); and as steam, above 100°C (212°F). In the terminology of metamorphism ice, water and steam are three *mineral facies* (sometimes termed metamorphic facies). In the example of water, the temperatures quoted above are correct only at a pressure of 1 atmosphere: at 10 atmospheres, ice is in equilibrium with water at −0.1°C (31.82°F) and water with steam at 180°C (356°F); so that mineral facies are pressure- as well as temperature-sensitive. If any composition such as a rock can exist as two distinct mineral assemblages (though some minerals may be common to both assemblages) then for a given temperature the pressure is fixed at some definite value. Most rock compositions can exist as numerous different mineral assemblages over the whole range of pressure and temperature encountered within the Earth, so for a given composition a pressure-temperature graph can be divided up into a series of spaces, representing the stability of particular mineral assemblages, separated by lines representing reactions to other assemblages. As most rocks within a broad chemical grouping – such as pelites or metabasites – have a number of minerals in common, reactions between the common set of minerals can be identified in many different individual rocks. By choosing the common mineral set large or small, the number of mineral facies may be increased or decreased to suit particular problems.

Some mineralogies are much less informative about conditions of metamorphism than others. For example, the mineralogy of quartzite (quartz) is practically without information content, while hornblende-plagioclase-metabasite restricts the temperature to less than about 300°C (572°F) without placing a useful limit on the depth of formation. The occurrence of a sillimanite-pelite close by the metabasite would limit the depth of formation to within 15–25km (roughly 9–16mi) while further limiting the temperature.

It is common to designate the mineral facies of a metamorphic rock by prefixing

the rock-composition name with as much mineralogical detail as desired. However, metabasites are often called by a range of facies-specific names of which the commonest are *blueschist* (glaucophane-lawsonite-metabasite), *greenschist* (actinolite-chlorite-epidote-metabasite), **amphibolite* (hornblende-plagioclase-metabasite), *granulite* (clinopyroxene-orthopyroxene-garnet-plagioclase-metabasite) and *eclogite* (clinopyroxene-garnet-metabasite).

Classification: Metamorphic Fabrics.

The texture of metamorphic rocks provides a valuable range of techniques for finding not merely the conditions of metamorphism but also parts of the path (in terms of temperature-pressure and time) by which the rock arrived at these conditions. The basis for these techniques is the principle that a volume, plane or line must be later than any feature which it cuts.

On the microscopic scale, many minerals or metamorphic rocks characteristically grow as large crystals or *porphyroblasts*, often to as much as 10 times the dimensions of *groundmass* crystals. On the mesoscopic scale, sedimentary or volcanic bedding may be folded and crystals of flaky or tabular minerals such as micas, amphiboles and pyroxenes may be produced in reactions, may rotate, or may recrystallize so as to be aligned with their shortest dimensions perpendicular to the axial planes of the folds in a *cleavage*. Where the oriented crystals are visible to the naked eye (bigger than about

A slightly metamorphosed shale, the clearly defined planes along which the rock has split marking the original bedding planes. The product of the regional metamorphism of shale is slate or schist.

0.1mm) the cleavage is called a *schistosity* (see *schist) and in rocks where schistosity is the most prominent planar feature the texture is often appended to the rock composition name, as in "peliticschist" or "hornblende-schist". Deformation and the development of schistosity may result in the complete obliteration of primary planar features such as sedimentary bedding.

In strongly deformed rocks bands of minerals with different physical properties are often segregated. Thus, in many strongly deformed peliticschists, a compositional banding develops, parallel to the schistosity, defined by alternations of mica-rich and quartz/feldspar-rich sheets (*foliae*). Such a banding may or may not be parallel to earlier compositional bandings – mostly it is quite impossible to tell. It is called *gneissose* banding (sometimes foliation in European literature) and the rocks *gneisses. A rock can be schistose and gneissose at the same time, although the one term does not necessarily imply the other.

Linear fabrics which occur in metamorphic rocks include densely-packed small (1–5mm (0.04–0.20in)) folds affecting bedding, cleavage or schistosity, called microfolds; as well as the alignment of tabular or needlelike crystals so that their long axes lie parallel.

Environments of Metamorphism.

The main occurrence of metamorphic rocks is in mountain chains, but there are great differences between the types of metamorphic rocks found in different types and ages of mountain chains.

One of the most restricted types of metamorphic rock association is found in, for example, the Californian Coast Ranges, the Sanbagawa of Honshu (the main island of Japan) and in some parts of the Western Alps of Switzerland. Here the occurrence of glaucophane-metabasites, low-grade pelites and serpentinites implies a depth of around 20–30km (12.5–18.5mi) and a temperature of about 300°C (572°F). Almost all occurrences of such associations are younger than 500 million years, and it is likely that such rocks were rapidly and deeply buried in subduction zones (see *plate tectonics), from which they emerged by erosion before they had a chance to heat up to the temperatures characteristic of such depths.

At the other extreme are mountain chains such as the Sierra Nevada of California, the Ryoke-Abakuma of Japan and parts of the fragmental Hercynian chain of Europe (Western Spain and Portugal, the Central Pyrenees, the Massif Central of France, the Black Forest and the Bohemian Massif), where heating to temperatures around 600°C (1100°F) at depths of less than 10km (6.2mi) is inferred from the occurrence of cordierite-andalusite pelites. Fabric studies show this heating to have taken place after much of the deformation history; and the occurrence of voluminous intrusive and extrusive calc-alkaline *igneous rocks suggests that such regions may be eroded analogues of modern island arcs or Andean mountain chains.

Intermediate between these extremes – low temperature at great depth and high temperature at shallow depth – is possibly the most abundant type of metamorphic mountain chain, exemplified by much of the Alpine chain, the Caledonian-Appalachian mountain belt and perhaps many of the older metamorphic regions comprising the big continental landmasses. Such regions are characterized by mineral facies representing largely post-deformational recrystallization along geothermal gradients of around 20–30C°/km (58–87F°/mi). A search for modern analogues has proved controversial.

Slate

A low-grade metamorphic rock, slate is derived from sand-free *clay with or without a component of volcanic ash. The grain-size is too fine (less than 0.1mm) for individual crystals to be distinguished with the unaided eye; but, under the microscope, slates can be seen to contain abundancies of minerals such as *muscovite and *chlorite and frequently albite (NaAlSi$_3$O$_8$) and *quartz. There are numerous other minerals (many present in only minor quantities), of which graphite (a form of *carbon) in black slates and *hematite in red slates are notable.

Frost shattering of slates of Precambrian age from Finnmark, Norway. Slate results from the regional metamorphism of argillaceous sedimentary rocks, especially shale: little recrystallization has taken place. Slate's primary importance has been as a roofing material, though it is less used for this today.

All the major minerals in slates – except quartz, albite and hematite – have sheet-like structures and sheet-like form, and are aligned to give one direction of almost perfect "splitability". This is known as slaty cleavage or flow cleavage.

Slates form by water loss (which helps to rotate the flaky minerals into the cleavage plane), compaction and recrystallization at temperatures up to about 450°C (about 850°F) in situations where the principal stress is acting in a direction perpendicular to the cleavage plane. Slates are used in many parts of the world as a roofing material and for decorative stonework.

Phyllite

The low-grade metamorphosed sediments called phyllites are comparable with *slates but are not restricted in premetamorphic composition to very pure *clays. Thus the sheet silicates such as *muscovite and *chlorite are usually less abundant, and quartz and albite usually more abundant, than in slates: for this reason the almost perfect cleavage of slates is only to a lesser extent present in phyllites.

Although it is sometimes said that there is a steady progression in properties from slates to phyllites to *schists with increasing metamorphic grade or temperature, slates and phyllites form alongside each other in the same metamorphic situations, and it seems that compositional controls are more important than temperature. Because of its imprecision, the term phyllite has, outside Europe, fallen somewhat into disfavor.

Schist

Metamorphic rocks characterized by the presence of visible (as opposed to microscopic) flaky or tabular minerals aligned in a cleavage are called schists.

Metamorphosed clayey rocks known as pelites contain the sheet-like micas *muscovite and *biotite in the middle grades of metamorphism, and the crystallization of micas in a cleavage in such rocks gives them the name mica-schists. Metamorphosed intermediate and basic *igneous rocks and some muddy dolomitic *limestones contain green amphibole (*hornblende) at moderate metamorphic grade, and alignment of amphibole in these rocks gives them the name hornblende-schists.

Thus schistosity is a term applying to texture and is not specifically restricted to any one compositional group of rocks. The texture is thought to develop in several different ways, including: concentration of the aligned mineral phases in distinct shear planes through the rock; crystallization or recrystallization of the aligned minerals

Schists split easily along well-defined cleavage planes due to the presence in them of flaky or platy minerals such as the micas and amphiboles. They have undergone metamorphism at comparatively low temperatures and pressures.

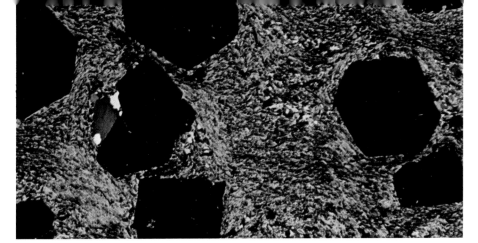

Garnet-muscovite-quartz-schist viewed under crossed polars. Garnet porphyroblasts have grown in a matrix of muscovite (colored), quartz (gray and white) and iron oxide. Note the schistose texture and the quartz segregations. In general, schists contain rather less quartz and feldspars than do gneisses: similarly, though mica schists usually contain muscovite rather than biotite, the latter mineral is also common.

with their smallest dimension parallel to the direction of principal stress; and rotation of the aligned minerals toward the plane of the cleavage by compressive shear deformation. The degree of alignment tends to be less perfect than in fine-grained *slates.

Gneiss

A metamorphic rock characterized by compositional banding of metamorphic origin is known as a gneiss. Many moderate- to high-grade metamorphic rocks, thought to comprise much of the lower continental crust, are gneisses consisting of layers richer in *quartz and *feldspar and layers richer in *amphibole or *pyroxene. Gneisses may be either meta-sediments or meta-igneous rocks. Gneissose banding also occurs, though less commonly, in low- and moderate-grade metamorphism, where mica- or amphibole-rich layers segregate from quartz-feldspar-rich layers.

Many different modes of formation of gneissose banding have been suggested, and it probably can originate by diverse means: differences in the properties of individual minerals can lead to the recrystallization of some away from sites of high stress; at high temperatures silicate liquid may form and separate from refractory crystals (this is one way in which the gneisses called migmatites have been shown to form); and original compositional differences may be preserved through intense deformation which has totally destroyed the original geometry (kneading neapolitan ice-cream has the same effect!). *Augengneisses* are metamorphosed rocks of granitic composition containing large lens-shaped crystals (eyes or augen) of *feldspar in a banded matrix of quartz, feldspar and micas.

Gneisses are hard, massive and decorative rocks and are used in many parts of the world as *building stones.

Hornfels

Hard, fine-grained and often dark *metamorphic rocks, hornfelses lack planar or linear structure and so splinter rather than cleave under the hammer. They originate by rapid metamorphism – leading to

The term gneiss is applied to a broad class of coarse-grained rocks with a banded, foliated structure and poor cleavage that have been subjected to, usually, high-grade metamorphism. The thickness of the bands varies from about 1mm (0.04in) to several centimetres. This picture shows the boundary between two bands, the one to the right being quartz.

A gneiss from the Lugard Falls, Athi River, Kenya, showing characteristic gneissose banding. The rounded shapes are a result of erosion by the waters of the river.

fine grain-size – in the absence of a stress field, so that platy or elongate minerals grow in a criss-cross texture. Such conditions are found close to bodies of *igneous rock intruded at shallow depth (usually not more than about 5km (3mi)); *aureoles* of metamorphism surround such intrusions.

As sediments tend to be the commonest rocks at shallow levels, hornfelses are often meta-sediments, but the term hornfels really describes texture rather than composition. It is easy to crush hornfels into pieces, and so this material is much used for road metal and concrete aggregate.

Mylonite

The grain-size of metamorphic rocks whose constituent minerals have reached chemical equilibrium results from competition between two processes: deformation, which tends to reduce grain-size, and crystal growth, which increases grain-size. Rarely in prograde metamorphic situations, but not unusually in rigid dry bodies of igneous or previously metamorphosed rock, deformation becomes concentrated into thin planar shear zones where its effect is particularly marked. The grain-size of rocks in such zones becomes microscopic or even submicroscopic, and such rocks are described as mylonites.

Because of their fine grain-size, most mylonites are dark gray or black, but usually some banding parallel to the plane of shearing can be picked out through subtle color or reflectivity differences.

Considerable heat is dissipated from mylonitic shear zones and the temperature may rise so that the growth of, for example, *feldspars and *quartz outlasts grinding during deformation. Porphyroblastic mylonites thus produced are called *blastomylonites*. Very rarely temperature may rise enough for melting, and such frictionally-produced melts are called *pseudotachylites*.

Mylonites probably form over only a restricted depth range (possibly about 10–30km (6–19mi)): at greater depths temperatures are high enough for rock masses to deform much more generally; and at lesser depths the stress is insufficient to cause intense grinding and instead a clear fault-plane results (see *faulting).

Amphibolite

These are of moderate grade, of chemical composition similar to basic or intermediate *igneous rocks, and consist mainly of the blue-green or green-brown amphibole *hornblende: thus amphibolite is both a term implying metabasite composition and also a mineral facies term applicable to rocks formed within a restricted range of temperature and pressure conditions.

Hornblende and *plagioclase are stable together during metamorphism in the temperature range 450–700°C (840–1300°F). The association is limited to depths of occurrence probably shallower than about 50km (30mi). Accessory minerals in amphibolite include: *quartz, *epidote and *garnet, confined to amphibolites formed at depths greater than about 10km (6mi); *pyroxene, which appears at lower pressures and higher temperatures; and *sphene and *apatite.

The lower part of the basaltic oceanic crust is probably converted to amphibolite shortly after its formation, and remains thus until taken back into the mantle in a subduction zone. Amphibolite is also an important constituent of *gneisses.   SWR

Soils

For the most obvious of reasons, soil is to Man one of the most important features of the Earth: it supports the plants whose existence we depend on. Only slightly less obviously, soil has determined the course of the evolution of our civilization; for had it not been for the early emergence of the great agricultural civilizations our modern society would indeed be very different.

There are many definitions of soil, but the simplest is that it is the medium in which land plants grow. Plants need an aqueous solution carrying certain minerals, from which the roots absorb their food, and air so that the roots can respire. Soil is, therefore, a complex of air, water, organic matter, plants and burrowing animals, in addition to weathered rock residues. The smallest volume that can be called a soil is the pedon (Greek *pedon*, ground), and so soil science is known as pedology.

The science goes back to earliest times, but Vasily Dokuchaiev (1846–1903) laid the foundations of pedology in the late 1900s. He concluded that soil formation was the result of the following factors: the climate, the nature of the parent rock, the mass and character of the vegetation, the age of the soil, and the relief of the locality.

Though modern research has modified his work, two main generalizations have resulted from the Russian school: first, that a soil is characterized by a distinct series of layers (horizons) forming a soil profile; and second that, in areas of fairly uniform climate, there exist soil belts corresponding to the *climatic zones.

The soil profile is divided into A, B, C and D horizons, of which D is the bedrock.

In wet climates, leaching of minerals is important in the A horizon, which is then known as the eluvial layer, and the minerals accumulate in the B horizon, the illuvial layer: this is what takes place in a *podzol*. In a climate with a strong dry season, accumulation of salts may take place in the A horizon, so that eluvial and illuvial layers are reversed, as occurs in a tropical laterite.

The *texture* and *structure* of a soil refer respectively to the size and shape of its particles and to whether the soil is blocky, platy, crumbly, etc. These properties are partly determined by the nature of the parent rock but more often by the vegetation. Texture and structure are also affected by the amount of humus, the end-product of organic activity of plants and

Sections through different soils showing how they are composed of layers having distinctive characteristics. *Below left,* a podzol in a sandy regolith. Here the uppermost horizon (A_1) is a thin dark layer rich in humus. This overlies the gray leached horizon (A_2) for which the podzols are named. Immediately below this is the dark horizon (B_{2h}) charged with humus illuviated from the A_2 horizon. Deeper down (B_{22} and B_3), the proportion of humus decreases until the zone of unaltered regolith (C) is reached. *Below right,* waterlogged soil showing gleying and reflecting the influence of Man in pedogenesis. Here the parent material is a 5000-year-old marine

animals in the soil. Soil fertility is determined by the amount of organic activity, and humus, particularly alkaline humus, greatly improves the structure of the soil. The amount present varies greatly: peaty soils may contain 100% organic matter, whereas in desert or Arctic climates soils contain very few species of plants or animals, and hence less than 1% organic matter.

A soil developed over a long period *in situ* usually has a well developed profile with conspicuous horizons, and is described as a mature soil. In areas of recent alluviation or glacial deposition, young soils, with poorly developed profiles, occur; these may also be called residual and transported soils, depending on their origin.

Because of their close association with

vegetation, and hence climate, soils were classified by Dokuchaiev into zonal belts paralleling the climatic belts – for instance, podzols are associated with cold humid climates and *chernozems* with sub-humid to semi-arid steppe. Thus soils may be zonal, in equilibrium with the climatic (weathering) regime; intrazonal, modified and with anomalous constituents; or azonal, not yet in equilibrium.

These ideas were accepted in the USA, and the concepts of pedalfers and pedocals added: *pedalfers* are leached soils in humid areas where aluminum and iron accumulate in the B horizon, and *pedocals* occur in more arid regions where calcium and other minerals accumulate in the A horizon. Tropical soils are characterized by the leaching out and elimination of silica and alkaline earths, and the concentration of hydrated iron and aluminum oxides. Such soils are often lateritic, but tropical soils can also be marly and black, then being known as margalitic. The most important factor in the distribution of margalitic soils, it has been suggested, is not overhead climate but actual soil climate.

Climate-oriented classifications are useful under conditions of uniform climate, but certain rock types exert a profound influence on soil type. This is seen particularly in limestone soils which contain much more lime than do normal zonal soils. Relief also affects soil development: soils on the hilltops are, for instance, better drained than those in the valleys, where *gley* soils occur. A topography of alternating hills and valleys produces alternating belts of soils, a feature known as a *catena* (Latin, chain). In western Europe the brown forest soils have been subjected to long-continued cultivation by Man and are now essentially manmade soils.

Soils develop over a long period, and may reach a steady state or equilibrium. However, if slopes are being actively eroded, the material removed is deposited on lower slopes or in the valleys. Such activity produces an erosion/deposition phase or cycle in the process of soil formation. Soils may also become buried, in which case they are known as fossil soils – examples of such soils are rare.

Both the classification and the mapping of soils are a subject of some controversy because there is no sharp borderline between one soil-type and another. Detailed work shows that there is probably no single factor predominant in soil formation, and this is reflected in the different systems of classification. The US, European and UK systems all use as a basis for classification different criteria which reflect national landuse policies (or lack of policy, as in the UK). In the USA soils are classified according to their visible characteristics in the field, and not on a genetic basis: since 1960 the official classification has been known as the Seventh Approximation. In the UK the classification and mapping of Soil Series and Soil Types reflect local conditions of humus, drainage and texture.   MMS

clay deposit. Under changed conditions a layer of peat many metres thick was built up on the surface of the clay. With the exception of a thin band, this was all dug away by Man, the peat workings being subsequently flooded. A silty ooze was deposited on the bed of the resulting mere; this forms the A horizon of the present soil. When the mere was drained, the upper surface of the clay cracked, allowing air to penetrate and oxidize the iron salts present, giving rise to the yellow coloration along the cracks (C_{1g1}). Further down (CG and G), the iron remains in a lower oxidation state, the waterlogged clay retaining its blue-green "gleyed" appearance.

The Geological History of the Earth

The Age of the Earth

If there is a single quantity that appears to us limitless, unplumbable, infinite, then that quantity is time. Yet modern scientific investigations assure us that, although there is a great deal of it, time is of strictly finite extent. In a more restricted sense we talk of "geological time", meaning the timespan encompassing all matters connected with the origin, evolution and structure of the *Earth. It is now known with some certainty that the Earth formed about 4600 million years ago.

The incredible trial-by-error process of *evolution has produced Man from microbe, and microbe from primeval aggregation of complex organic molecules, over a period of some 3500 million years. Early geologists obtained a historical perspective of the Earth from the succession of sedimentary strata and from the progressive evolutionary sequence shown by the fossils which they contain. Fossils may be used for stratigraphic correlation of sedimentary rocks in different parts of the Earth. A *relative* time-scale based on stratigraphical and paleontological evidence was first produced early in the last century and has been used, with improvements and refinements, ever since. However, until quite recently, there was no adequate way of deciding the true ages (sometimes called *absolute* ages) to be attached to the relative time-scale. Nor was it possible to correlate from one region to another the vast thickness of largely unfossiliferous Precambrian rocks.

A momentous scientific discovery, made in 1896 by Henri Becquerel, was the radioactivity of the element *uranium. Much later, it was found by other workers that *thorium and certain varieties (isotopes) of *potassium and *rubidium are also radioactive, and that they all decay by atomic transmutations into decay (or daughter) products at characteristic rates totally unaffected by any known physical or chemical conditions. By the early years of this century, Rutherford and co-workers were attempting to determine the absolute ages of certain uranium minerals by measuring their contents of radioactive uranium and radiogenic lead: given the approximately known rate of radioactive decay of uranium to lead, the ages could be calculated. For various reasons, these early attempts were rather inaccurate, but they sufficed to show that the history of the Earth could be measured in terms of thousands of millions of years.

Only within the last few decades has it become possible to measure the absolute age of rocks (and their constituent minerals) of almost any geological age with a high degree of precision and accuracy. The basic geological events which can be dated are the crystallization of an *igneous rock from a molten magma, and the recrystallization of an igneous or *sedimentary rock to form a *metamorphic rock. Other geological events sometimes amenable to dating are the deposition of a sedimentary rock and the uplift and cooling of a mountain belt.

Absolute age measurements based on radioactive decay are usually termed "isotopic", "radiometric", "radioactive" or "radioisotope" ages. That branch of the Earth sciences concerned with the interpretation of the ages of rocks is geochronology, while the more specialized technical and methodological aspects related to the measurement of absolute ages are termed geochronometry.

Principles of Radioactive Decay and Isotopic Dating. A specific kind of atom characterized by a particular atomic number and mass number is termed a nuclide: the atomic number is the number of protons in the nucleus, while the mass number is the sum of the number of protons and neutrons in the nucleus. Isotopes of a given element have the same atomic number, but different mass numbers.

Three types of radioactive decay are of direct interest to the geochronologist. In *alpha-decay* the nucleus of the parent atom emits two protons and two neutrons, that is to say a helium nucleus, so that the mass number decreases by four and the atomic number by two. In *beta-decay* the nucleus emits an electron so that one of its neutrons turns into a proton, and the atomic number increases by one. In *electron-capture*, a proton picks up an orbital electron and changes into a neutron, so that the atomic number decreases by one. The last two processes leave the mass number unchanged.

The radioactivity of a given radionuclide decreases by an exponential process. The fundamental law states that the number of atoms disintegrating in unit time is proportional to the total number of radioactive atoms present, with a proportionality factor λ (lambda, the decay constant) which has a characteristic value for each radionuclide: it represents the probability that an atom will disintegrate in a particular unit of time, which is mostly in years for longlived radionuclides.

An alternative constant is more commonly used to characterize a radionuclide. This is the *half-life* and represents the time required for the number of radioactive atoms in any given initial amount of radionuclide to decrease by half. It can be shown that the half-life equals $0.693/\lambda$.

Once λ has been evaluated for a particular radioactive element, we have a basis for calculating absolute geological ages from laboratory data.

To be of any real use to the geochronologist, a radionuclide must have a half-life within one or two orders of magnitude of the age of the Earth. If the half-life were much longer than this, the accumulated amount of decay nuclide would be too small to measure, while if the half-life were very much shorter the radionuclide would have decayed completely.

Several assumptions are necessary for calculating the age of a rock from the measured parent-to-daughter ratio and the appropriate half-life. The calculated age will be the time since crystallization of the rock or mineral only if there has been no loss or gain of either parent or daughter nuclide by processes other than the radioactive decay of the parent, if the half-life of the parent nuclide is accurately known, and if accurate correction can be made to allow for any of the daughter nuclide having been incorporated into the rock or mineral from the environment or from the source region of the rock at the time of crystallization.

In very many cases, the validity of the above assumptions can be demonstrated beyond any doubt. In other cases there may be various complicating factors – which can nonetheless yield highly useful geological information. For example, some minerals only begin to retain atoms of the daughter nuclide within their crystal lattice at a temperature much below that of crystallization. In this case, the calculated age may be significantly lower than the true age of crystallization, but may yield the age of uplift and cooling of a mountain belt.

Direct measurements of half-lives are, of course, basic to geochronology. Much patient effort has gone into the exact measurement of the radioactivity of such nuclides as uranium-238, uranium-235, thorium-232, potassium-40 and rubidium-87.

Precise measurement of the radioactivity of a rock can be extremely difficult because of the low disintegration rate and/or the low energy of the radiation of some long-lived radionuclides, and because they generally make up only a tiny proportion of the rock

A rock sample is taken prior to its being dated using the principle of radioactive decay.

in which they occur.

In practice, the amounts of both parent and daughter nuclides in a rock are precisely measured by means of a mass-spectrometer; and, by isotopic analysis with the same instrument, both parent and daughter nuclides can be distinguished from respectively non-radioactive and non-radiogenic nuclides.

The uranium-to-lead (U-Pb) method is historically the most important of the radioactive rock-dating methods. However, the potassium-argon (K-Ar) and rubidium-strontium (Rb-Sr) methods are nowadays used far more widely because they are applicable to such a very wide range of commonly occurring igneous, metamorphic and sedimentary rocks and minerals. In contrast, uranium and thorium minerals are relatively scarce although when they do occur they can yield very valuable data.

Whenever possible, more than one of these methods is applied to a given rock. All common potassium minerals also contain small traces of rubidium, because these two elements are chemically very closely related: consequently, both the potassium-argon and rubidium-strontium methods may frequently be carried out on the same sample. Furthermore, a given rock body may contain a sufficiently wide range of minerals for *all* the major dating methods to be used. A typical case would be a *granite or a *gneiss.

Many cases are known where two or more of the dating methods yield the same age (within analytical error) for different minerals separated from a single rock sample. In such cases the age of a particular geological event is obviously well established; and the rock is said to possess a "concordant age pattern". However, many cases are also known where different minerals from the same rock sample, or rock unit, yield different apparent ages – where either the same dating method or different dating methods have been used: in such cases the rock is said to possess a "discordant age pattern". Such patterns may give much fundamental information on the history of a rock, particularly should it come from a terrain that has been subjected to several igneous and/or metamorphic events at widely separated times.

The Carbon-14 Method. The dating methods we have discussed are applicable to rocks mostly in the age range of millions of years, although the potassium-argon method can sometimes be used for dating rocks significantly younger than one million years. Another important method, and perhaps the most familiar, is the carbon-14 method used for dating archaeological material. This method works on somewhat different principles to those described above, although radioactivity is still the fundamental process.

Natural carbon consists of the stable isotopes carbon-12 and carbon-13 in the ratio 89:1. In addition, minute amounts (of the order of one part in one million million) of radioactive carbon-14 are continuously generated in the upper atmosphere by the action of cosmic-ray-produced neutrons on the stable isotope nitrogen-14. After combining with oxygen to form carbon dioxide and mixing with the lower atmosphere, the rate of decay of carbon-14 (it has a half-life of 5570 years) reaches equilibrium with the rate of production of new carbon-14.

The carbon-14 dating method depends on the following assumptions: the rate of carbon-14 production in the upper atmosphere over the period of usefulness of the method has been constant; the rate of mixing of carbon-14 in the atmosphere-biosphere-surface ocean reservoir is rapid relative to the rate of decay; when carbonaceous material is completely removed from the reservoir by, say, the death of an organism, no further carbon-14 is added, while that which is present decays with a half-life of 5570 years; the concentration of carbon-14 incorporated into living organisms over the period of usefulness of the carbon-14 method has remained constant.

It follows that if the carbon-14 concentration in the carbon from a piece of fossil wood of unknown age were measured and found to be one-half that in a living piece of wood, the age of the fossil wood would be 5570 years. To generalize, the period since the time of isolation from the active reservoir can be measured by determining the amount of remaining carbon-14 and comparing it with the original (i.e., equivalent to present) amount of carbon-14. Materials which can be dated include wood, charcoal, peat, leaves, flesh, hair, horn, bone, hide, rope, parchment and carbonate shells.

Although the overall concentration of carbon-14 is very small, it can be quantitatively measured by its radioactivity. The low disintegration rate of carbon-14 (about 16 atomic disintegrations per minute per gram of modern carbon), as well as the low energy of carbon-14 beta-particles, necessitates the use of high-sensitivity, low-

The Age of the Earth

Cenozoic

Life on the land in the Cenozoic has been characterized by the radiations of the mammals and the insects in the animal kingdom and of the angiosperms (flowering plants) in the plant kingdom. There has been a continuation of sea-floor spreading, perhaps the major results of which have been the separation of Australasia from Antarctica and the convergence of Africa–Arabia on Eurasia. Climate has deteriorated progressively throughout the era, the climax of the deterioration being the ice age, still continuing, of Quaternary times.

Mesozoic

On land, life in the Mesozoic was dominated by the dinosaurs and the gymnosperms (naked seed plants): dominant in the sea were the marine reptiles among the vertebrates and the ammonites among the invertebrates. The Mesozoic saw the progressive breakup of Pangaea, and late in the era, the creation of most of our modern continents. The climate was warm and equable throughout.

Paleozoic

Most invertebrate groups evolved early in this era, the faunas being characterized by, for example, the trilobites, graptolites, primitive mollusks, arthropods and brachiopods. With respect to the floras, the first vascular plants emerged early in the era, and plants first colonized the land during the Silurian. The first vertebrates (the jawless fish) evolved in the Ordovician, and primitive amphibians took to the land during the late Silurian/early Devonian. Towards the end of the era the amphibians were replaced by the reptiles as the dominant land animals. Climate alternated between warm equable periods and rather shorter ice ages. The seas were comparatively widespread, especially in the early part of the era. Successive closures of ancient oceans resulted in the welding together of continental masses to form Pangaea.

Precambrian

This is a long, complex and comparatively poorly understood phase of Earth history. There were a number of important mountain-building episodes and at least one major ice age, late in the Proterozoic. In the early Archaean occurred the differentiation of the lithosphere, hydrosphere and atmosphere, and that of the crust, mantle and core. Later, possibly in the late Archaean, free oxygen became available in the atmosphere for the first time, almost certainly as a result of photosynthesis on the part of primitive plants (algae). The first multicellular organisms appeared close to the end of the Precambrian.

Period	Epoch	Age
Quaternary 2	Holocene	0.01
	Pleistocene	2
Tertiary	Pliocene	7
	Miocene	26
	Oligocene	38
	Eocene	54
	Paleocene	65
	65	
Cretaceous		136
Jurassic		190
Triassic		225
Permian		280
Carboniferous 345	Pennsylvanian	315
	Mississippian	345
Devonian		395
Silurian		440
Ordovician		500
Cambrian		570
Proterozoic		2500
Archaean		4600

- Early civilizations
- Emergence of Man
- Start of main Himalayan folding
- Extinction of dinosaurs
- Main fragmentation of Pangaea; transgression of sea over land
- Start of fragmentation of Pangaea
- Worldwide regression of sea from land
- Formation of Pangaea
- Animal life takes to the land
- First land plants
- Major transgression of sea over continents
- First multicellular organisms
- ? 3000: Free oxygen in atmosphere
- ? 3500: First unicellular organisms
- 3780: Age of oldest known terrestrial rocks
- 4600: Formation of Earth
- ? 10,000: "Big Bang"

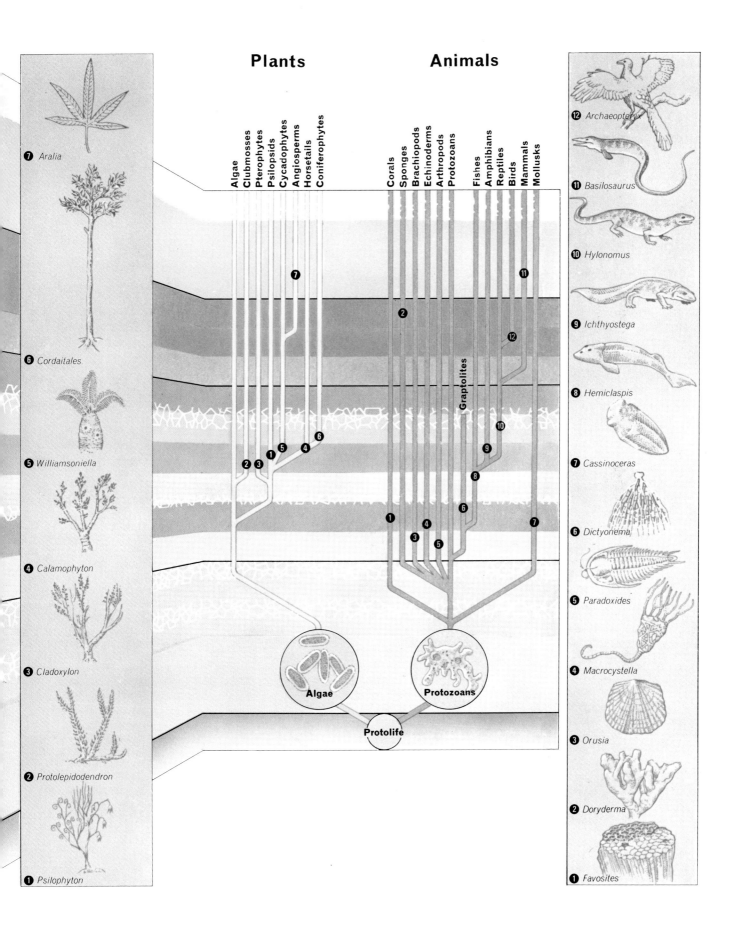

The geological timescale. In the left-hand column of this diagram are the **eras** of Earth history. In the next column of this diagram are the **periods** into which the eras are sub-divided: the Archaean and Proterozoic are frequently con-sidered as eras in their own right; the Carboniferous period is often considered to comprise two periods, the Mississippian and the Pennsylvanian; and the Tertiary and Quaternary are further subdivided into **epochs.** In the third column are shown the more notable events in Earth history: ice ages are shown by a white ice symbol (the ice age early in the Proterozoic is uncertain). On the opposite page are shown the evolution of and possible relationships between the major plant (green) and animal (brown) groups: each is illustrated by an early member. (All dates are in millions of years before the present.)

Plants

Animals

Algae
Clubmosses
Pterophytes
Psilopsids
Cycadophytes
Angiosperms
Horsetails
Coniferophytes

Corals
Sponges
Brachiopods
Echinoderms
Arthropods
Protozoans

Fishes
Amphibians
Reptiles
Birds
Mammals
Mollusks

Graptolites

Algae

Protozoans

Protolife

⑦ *Aralia*

⑥ *Cordaitales*

⑤ *Williamsoniella*

④ *Calamophyton*

③ *Cladoxylon*

② *Protolepidodendron*

① *Psilophyton*

⑫ *Archaeopteryx*

⑪ *Basilosaurus*

⑩ *Hylonomus*

⑨ *Ichthyostega*

⑧ *Hemiclaspis*

⑦ *Cassinoceras*

⑥ *Dictyonema*

⑤ *Paradoxides*

④ *Macrocystella*

③ *Orusia*

② *Doryderma*

① *Favosites*

background-radioactivity counters. (In contrast to the other dating methods we have discussed, carbon-14 cannot be measured by mass-spectrometry.)

The maximum practical age limit of the most sensitive counting methods now in use is about 50,000–60,000 years. Above about 40,000 years only about one per cent of the original amount of carbon-14 remains and measurement becomes correspondingly difficult. Five to ten grams of a fairly young carbon sample can be measured routinely to about 0.5 per cent, corresponding to an error of as little as ±30 years on the date. In practice, however, a single carbon-14 date in the range of about 0–5000 years may have an actual error of about ±150 to 200 years because of uncertainties arising from various factors associated with variations in atmospheric carbon-14 concentrations.

The reliability of the carbon-14 method has been tested by dating materials of independently known age. Back to about 5000 years ago this is possible by using known archaeological, historical and tree-ring dates. In fact, evidence from tree rings has shown that the rate of carbon-14 production in the upper atmosphere has *not* been constant in the past, and our techniques of calculation have been adjusted accordingly.

Application of the carbon-14 method has shed light on numerous aspects of late Pleistocene geology and paleontology, the study of vegetation development (in combination with pollen analysis) and, above all, the dating of archaeological specimens, which so dramatically illustrate the rise of mankind.

Age of the Oldest Terrestrial Rocks. On every continent there are large Precambrian "shield" areas composed of igneous and sedimentary rocks, ranging from weakly metamorphosed to strongly metamorphosed, with isotopic ages of about 2600–2800 million years. Areas yielding this age make up about 5% of the Earth's land surface. Younger shield areas can sometimes be shown from the study of discordant age patterns to be reheated and reworked 2600–2800 million-year-old rocks. It is now thought that about 50% of the area and volume of the Earth's continents may already have been in existence about 2600 million years ago.

Recently there have been discovered a few areas on Earth which have isotopic ages greater than 3000 million years. The oldest reliably dated rocks come from West Greenland. They comprise a surprising variety of metamorphosed igneous and sedimentary rocks, including granitic gneisses, volcanic lavas, banded ironstones, schists, quartzites, etc., mostly characteristic of the well-known early Precambrian assemblage termed the "granite-greenstone" associ-

ation. Rubidium-strontium and uranium-lead measurements conclusively show that all these rocks were formed between about 3700 and 3800 million years ago. Rocks of similar age and character are currently beginning to be found on the North American and African continents.

Contrary to many earlier ideas, it is now evident that by about 3800 million years ago, and possibly well before this, surface cooling and geochemical differentiation of the Earth had progressed sufficiently to produce continental crust not too greatly different from that in evidence today, although it was probably much smaller in area and considerably thinner. From the sedimentary rock types present it can be concluded that free water already existed on the Earth's surface.

The Age of the Earth. As we have already said, the Earth is considered to be about 4600 million years old, although no terrestrial rocks approaching this age have yet been found – indeed, it is rather unlikely that they ever will be. The evidence from recent studies for this estimate of the Earth's age is circumstantial and is based upon the following indications:

Isotopic age measurements by the uranium-lead and rubidium-strontium methods on most *meteorites have yielded a solidification age close to 4600 million years. Meteorites are of interest in that they solidified very early in the history of the Solar System.

The oldest rocks and soils from the *Moon yield isotopic ages close to 4600 million years. This has therefore been interpreted as the age of solidification and chemical differentiation of the Moon.

The growth on the Earth of the two radiogenic lead isotopes, lead-206 and lead-

207, approximates closely to a very simple pattern that would show that 4600 million years ago the isotopic abundance ratios of lead were identical in the parent material of the meteorites and of the Earth. The overall growth curve is based on several major lead ore deposits of different geological ages, independently dated by other isotopic age methods. If the growth curve is extrapolated back through time from the oldest lead ore deposit, using the simple radioactive decay equation described earlier, it passes through the measured lead isotope ratio of 4600-million-year-old iron meteorites. The latter contain no uranium, so that no radiogenic lead has had a chance to form within them since they solidified. Lead-204 is a non-radiogenic isotope which has always been constant in amount, and forms a reference against which the quantitatively varying radiogenic isotopes can be compared.

Thus, although it is highly probable that the Earth attained something like its present form and overall consistency 4600 million years ago the term "age of the Earth" is used in the current scientific literature to signify that point of time, 4600 million years ago, when the isotopic composition of lead was the same in the Earth and in the parent material of the meteorites. It seems very likely, nonetheless, that the entire Solar System condensed within a relatively short period of time from disaggregated primeval matter about 4600 million years ago. SM

Principles of Stratigraphy

Stratigraphy is essentially the study of the history of the Earth as preserved for us in the stratified rocks. That history includes great episodes of mountain building, the intrusion and extrusion of vast quantities of molten rock, and the alteration of huge areas of older rocks by heat and pressure.

The early history of stratigraphy centers around two great battles between wealthy intellectuals, the "neptunists" versus the "plutonists" and the "catastrophists" versus the "uniformitarians", plus the sound common sense of one simple working man – William *Smith, who is well named the "Father of Stratigraphy".

The "neptunists" were those, mainly German, thinkers who back in the 18th century maintained that all the rocks had been laid down as chemical precipitates in water. The great Goethe, himself an amateur geologist, adhered to this school. The "plutonists", on the other hand, recognized (symbolically at least) the hand of the fiery king of the underworld in rocks which were so like the lavas seen erupting from modern volcanoes that they must have formed in a hot molten state. Both schools were right, of course, each in their own way. The battle is long over, though geologists still find details to argue about.

And most geologists would probably say that the second battle, that between the "catastrophists" and the "uniformitarians", has also ended – in favor of the latter. The great catastrophe of Noah's flood was long seen as the explanation for

most of the phenomena of the rocks, especially the fossils they contain. But as knowledge of the stratigraphical record grew, more and more catastrophes and new creations had to be dreamed up to explain the long succession of faunas and floras in the stratified rocks.

The uniformitarians, on the other hand, maintained that everything in the rocks could be explained by the processes seen going on in the world at the present day. Given enough time (and geologists were provided with more and more of it as the 19th and 20th centuries rolled by!), the ordinary processes of *erosion, transport and *deposition were sufficient to interpret everything that we see in the sedimentary and the activity of modern volcanoes explained the volcanic rocks.

The great battle-cry of the uniformitarians was the dictum "The present is the key to the past!", and the man who shouted this battle-cry most loudly was *Lyell, whose book *Principles of Geology* was the foundation stone of all modern geology, and of stratigraphy in particular. The frontispiece of that book and the clearest demonstration of the principle of uniformitarianism was a picture of the Roman pillars in the Temple of Serapis near Naples. These bear, in the boring made in them by marine organisms, an obvious record of the rise and fall of the sea since Roman times.

While the educated men fought about theories, more practical men were recognizing the principles of stratigraphy without really realizing what they were doing. As far

back as 1719, John Strachey recognized the meaning of the succession of strata in the Forest of Dean coalfield of western England. And at the end of the 18th century William Smith, son of a Cotswold blacksmith, was seeing the realities of stratigraphy as he surveyed coal mines and the routes of canals in the brief heyday of that form of transport. He was not an educated or wealthy man, he did not succeed as a businessman (in fact he went to prison for debt), and he was not an intellectual giant, but he recognized two simple principles of stratigraphy – practically the only two principles the science has, even today.

He recognized firstly the Law of Superposition, which states simply that in normal circumstances the younger deposits will rest on the older, that the succession will follow bed upon bed in chronological order. This may seem so obvious that it hardly needs spelling out – and indeed the principle had been recognized back in the 17th century by *Steno – yet it had been overlooked by most of the learned armchair geologists of the day.

Of course, it is not always as simple as this. With lateral movements of the Earth's crust, the strata may be tilted; and greater dislocations can even lead to older strata being pushed over younger – in major mountain belts, where compressive movements have reached their greatest intensity, the strata may be completely inverted. But all these complications can be recognized for what they are, and the simple truth remains that, when comparatively undisturbed, younger rocks must always rest

An early view of the origin of the rocks of the Earth, that held by the "catastrophists", illustrated in a painting by John Martin.

on older.

William Smith's second great principle was that layers of sediment can be recognized by means of the fossils they contain. In other words, rocks in different places can be recognized as having been formed at the same time because they contain the same sorts of fossils. They can be *correlated*. Thus fossils embedded in a block of sandstone on the north coast of France are very similar to those found in the Portland Stone of southern England: they can therefore be recognized as of approximately the same age (in fact, both were formed towards the end of the *Jurassic).

As the work of unravelling the geological history of the Earth went on, it was soon realized that, at any one place, the record was far from complete. Thus, on the south coast of England near Seaton, Cretaceous rocks rest directly on Triassic rocks, with no Jurassic sediments preserved in between. This is a very simple example of an *unconformity*. Such breaks in the record had, in fact, been recognized long before by *Hutton, the first protagonist of uniformitarianism.

The older rocks may have been folded and altered considerably before the younger

An unconformity between Ordovician and Cambrian rocks in Wales. The older Cambrian rocks were first deposited and tilted then, after a period of erosion, Ordovician rocks were deposited on the Cambrian surface. The unconformity represents a period of many millions of years during which there was no deposition.

The catastrophic theory of the origin of rocks was replaced by the views of the "uniformitarians" who maintained that different layers of rock represent different periods in a long history of deposition, younger rocks therefore appearing above older rocks. This principle is demonstrated here in an illustration from Sir Charles Lyell's *Elements of Geology* (1838).

rocks were laid down across their eroded surface. Thus in the Grand Canyon of the Colorado River, where Cambrian marine sediments rest on the dark, much altered rocks of the early Precambrian near the bottom of the gorge, we find that much of the later part of the Precambrian record is missing, the unconformity being a very obvious one. Far less obvious is the break *above* the Cambrian, where the next, similarly horizontal, sediments are of Carboniferous age (in some places, Devonian) – there are no sediments from the Ordovician or Silurian. Then follow Permian sediments, with here and there along the rim of

the canyon some patches of Triassic strata. But there's nothing from the last 200 million years or so of the Earth's history.

The stratigraphical record is therefore full of gaps, even in the best of sections, and nowhere on Earth do we appear to have anything like a *complete* pile of sediment.

Just as the rocks may be present in one place and absent in another, obviously they may change in character from place to place. The sum total of the characteristics of a rock is what we call its *facies*. We say that one set of strata are in a "sandy facies", another of approximately the same age in a "limestone facies". Or we may go further

and interpret the actual environments that the rocks represent. We talk of a "shallow marine facies", or a "sand-dune facies", and so on.

The next stage in working out the stratigraphy of an area is therefore to interpret the successive environments recorded in the sediments. In this way we build up paleogeographical maps which show the supposed geography of a particular region at some specific time in the past.

Obviously the fact that strata vary laterally in their facies makes correlation all the more difficult (and all the more fun). What is more, it is not only the rocks that vary but also the fossils they contain, since clearly one can't expect to find the remains of a marine creature in a desert sand or the inhabitants of deep muddy seas in limestones that were laid down in shallow water. Therefore some fossils are much more useful to us than others. A free-swimming, rapidly-evolving creature is more useful for correlation than a slowly-evolving form that only lived on a particular type of sea floor. Microscopic organisms are often more useful than larger forms because there are far more of them and one can find large numbers in a very small sample of rock (such as is recovered from a borehole). In recent years, for example, the fossilized remains of plant spores and pollen have proved extremely valuable, not only because they are abundant and easily preserved, but also because they tend to be scattered everywhere, both in the sea and on land, and can be said to be relatively independent of facies.

Since rocks of a specific age vary in facies from place to place, it also follows that rocks of the same facies may differ in age from place to place. This is known as *diachronism*, where the rocks, as it were, cut across the time planes.

Apart from fossils there are several other ways of correlating rocks. These may be listed as follows:

(*i*) *Radiometric methods:* In these the known rates of breakdown of the radioactive isotopes of particular elements (such as rubidium, potassium and carbon) are used to provide an estimate in years of the age of the rock concerned (see *age of the Earth).

(*ii*) *Tectonic methods:* The use of major events, particularly mountain-building episodes, to elucidate the history of the Earth is full of difficulties. There is no real evidence that such major events were synchronous over the whole of the Earth's surface or even parts of it (and this method is anyway reminiscent of old-fashioned catastrophism). Nevertheless, within particular areas, the effects of major Earth movements must have been widespread.

(*iii*) *Paleomagnetic methods:* The periodic reversals of the Earth's magnetic field, as recorded in certain rocks, particularly those of the *ocean floor, have provided a very useful tool for dating the later part of the stratigraphical record, and especially for unravelling the story of the relative movements of the continents.

(*iv*) *Paleoclimatic methods:* Marked changes in climate, often accompanied by marked changes in the relative height of land and sea, provide another obvious means of correlation (though used mainly for the *Quaternary).

(*v*) *Volcanic methods:* Volcanic events are often sudden, short-lived and widespread in their effects. Lavas and ashes which are erupted at the surface (or on the bottom of the sea) will obviously take their place in the regular succession of strata, and they can therefore be used as marker horizons.

However, we have to recognize that intrusive rocks, often emplaced far below the Earth's surface and only exposed as a result of deep erosion, do not follow the rules: they commonly cut across strata of different ages. This does not mean that they, and rocks altered out of recognition by heat and pressure (see *metamorphic rocks), are not part of stratigraphy. They are just a more difficult part of the story.

It has been suggested that the catastrophist school of thought was demolished by the uniformitarian school. It must be said, however, that in recent years there has been something of a swing back, if not to catastrophism, then to what may be called "neocatastrophism". This involves the recognition that Man's experience has not been long enough to recognize all the processes that are going on at the present day, and that the stratigraphical record (and equally the *fossil record) reveals a certain periodicity that involves both sudden violent happenings as well as long periods of "nothing much in particular". In these terms, it can be argued that the theory of *plate tectonics itself is somewhat catastrophic in theme, since long periods of "quiet" subduction of ocean floor are followed by short periods of "violent" continental collision. The argument is hardly over; perhaps it has barely begun.   DVA

Precambrian

Most of the Earth's history belongs to that period of time known as the Precambrian. Problems such as where the Earth originated, how it evolved, how the continents and oceans took shape, and what conditions were like on the proto-Earth are a constant source of speculation – and hypotheses relating to the early development of the Earth abound. A prime difficulty with a more objective geological approach is to construct some sort of meaningful time-scale. Just how difficult this is can be judged from the way that the Earth's "age" has increased a million-fold (from 4004BC to 4,600,000,000BC) over the last few centuries as we have gradually begun to appreciate the enormity of geological time. Progress in understanding the Precambrian has gone hand-in-hand with the construction of reliable timescales.

The development of stratigraphic principles during the 19th century led to great advances in our knowledge of the fossiliferous sediments deposited on the Earth's surface since the beginning of the *Cambrian some 570 million years ago. Fossils allowed sedimentary rocks from widely different geographical areas to be arranged in a definite time sequence. But the lack of easily recognizable fossils in rocks that were clearly older than the Cambrian sediments posed great problems to the early stratigraphers. They were content, therefore, to group these rocks into one great geological period, the "Pre-Cambrian", and to concentrate their efforts on the fossiliferous strata of the Phanerozoic (post-Precambrian). Nevertheless, as the science of geology has advanced, so new techniques have been developed for dealing with unfossiliferous Precambrian rocks in order to place them in some sort of time sequence and thus relate them to the events that were taking place on Earth. This has been made possible through detailed studies of:

(*i*) the igneous history of the Precambrian: periods when magmas were formed, volcanoes erupted and dike swarms injected;

(*ii*) the metamorphic history: periods when rocks were heated as a result of igneous activity or deep burial, and as a consequence changed their mineralogy;

(*iii*) the structural history: when the rocks were severely strained, folded and faulted through Earth movements;

(*iv*) the sedimentary history: periods when erosion took place, basins were formed and the eroded debris dumped in them.

By examining the ways these processes interrelate it has been possible to build up a *relative* time-scale of events in the Precambrian in many parts of the world. Moreover, with the development and application of methods for dating rocks using radioactive isotopes and their decay products, it is now possible to set these events against an *absolute* time-scale (see *age of the Earth). Of course, much more work needs to be done before a detailed picture emerges.

What, then, do we know about the earliest stages of the Earth's history? The answer is: very little. The *Earth is known to be about 4600 million years old. Rocks formed more than 4000 million years ago have been recovered from the *Moon – indeed there are few lunar rocks younger than 3200 million years – yet on Earth the oldest crustal rocks so far recognized are no more than 3780 million years old, and there are only a few areas of Precambrian rocks where ages greater than 3200 million years have been recorded. Does this mean that during the first third of its history the Earth was so hot and mobile that the rocks then formed have only rarely been preserved? or does it mean that the continental crust as we know it did not exist in any great volume prior to 3800 million years ago? The solution of this problem is fundamental to an understanding of early Precambrian geology.

Whatever the explanation for the absent record of the first 700 million years or more of the Earth's history, it is clear that our planet's surface must have been very different 4000 million years ago. Some reflection of this difference is found in Precambrian terminology: terms such as "greenstone belt", "komatiite", "anorthosite" and "granulite" rarely appear in the literature of Phanerozoic rocks, but are part of the basic vocabulary of the Precambrian geologist.

The Precambrian, though several times longer than all the other geological periods combined, has not suffered repeated subdivision into smaller and smaller time intervals. This is partly because such detailed evidence is usually lacking. Nevertheless, due recognition of the changing character of crustal processes during the Precambrian is provided by division into the (earlier) Archaean and the (later) Proterozoic, the dividing line lying approximately 2500 million years ago. The Archaean seems to have been primarily a period of crustal formation, while the Proterozoic is characterized by extensive "reworking" of this older crust in several periods of deformation, metamorphism and associated magmatic activity. However, this reworking was not entirely pervasive, since many areas of essentially unmodified Archaean crust still survive on all continents, being commonly known as Archaean kratons. Areas of the Earth's crust which have remained relatively stable since the end of the Proterozoic are known as shield areas, the best example being the Canadian Shield.

It is worth reiterating that Phanerozoic principles of stratigraphy are difficult to apply to the Precambrian, especially the Archaean. Although, locally, vast thicknesses of Precambrian sedimentary rocks do occur, episodes of sedimentation are in fact subsidiary to periods of magmatism and crustal generation, metamorphism and deformation. Hence it is necessary to treat Precambrian history in terms of such events. Most geological events, whether episodes of magmatism, metamorphism, deformation or erosion and sedimentation, are related in one way or another to convection in the Earth's mantle, and ultimately to the operating thermal regime. Since sedimentation is directly linked with *erosion, and erosion is dependent on relief above sea level, the thinner sedimentary record in the Archaean might just be a reflection of a thinner continental crust.

Archaean. Most geologists recognize two different types of Archaean terrain – low-grade and high-grade respectively – which tend to be separated geographically. Low-grade terrains include the greenstone belts, which have generally been deformed, but have suffered metamorphism at relatively low temperatures. The high-grade terrains, on the other hand, are composed largely of granitic *gneisses which have suffered strong deformation and recrystallization at moderate to very high temperatures. The relationship between the low- and high-grade terrains in terms of Precambrian crustal processes is not known.

Greenstone Belts. A typical greenstone belt consists of a thick (several kilometres) sequence of basalt lavas and overlying sediments which have been folded into an upright syncline and invaded by granite. Individual greenstone belts may be separated by intervening regions of granite *gneiss.

Many of the lavas are pillowed flows, the pillows having formed as the hot lavas were extruded under water. Similar pillow structures characterize basaltic lavas erupted on the present *ocean floors at mid-ocean ridges and in marginal-basin spreading centers behind island arcs. Indeed some geologists regard the greenstone belt lavas as preserved relics of the basaltic crust of Archaean oceans or marginal basins.

The lavas are rather unusual, however, in that many are much richer in magnesium than modern ocean-floor lavas (their special name, komatiite, taking account of this and other chemical differences). The differences imply much more extensive melting of the Earth's magnesium-rich mantle than is possible at present spreading centers, and some geologists have speculated that meteoritic impacts (as on the Moon) may have triggered off this extensive melting. However, at higher levels in many greenstone-belt sequences, lavas such as *basalt, *andesite and *dacite, with close compositional similarities to lavas of present-day volcanic island arcs, are found: this would suggest a marginal basin model.

Above the lava sequence there is normally a succession of immature sediments (*graywackes), probably derived by rapid erosion of the adjacent pile of volcanic lavas and granite gneiss and dumped into the basins with very little chemical *weathering. Other types of sediment, such as silica-rich *chert and banded *ironstones, may be present but are subordinate to the graywackes.

The greenstone-belt cycle seems to have been terminated in most cases by folding and deformation of the lavas and sediments. At the same time, the lavas were altered and metamorphosed so that the original igneous minerals were replaced by green-colored secondary products such as chlorite and hornblende (hence the name "greenstone"). At the same time the belts were invaded by granitic magmas, which were themselves often highly deformed.

Greenstone belts appear to have been formed over a period of almost 1000 million years. Those in southern Africa are older

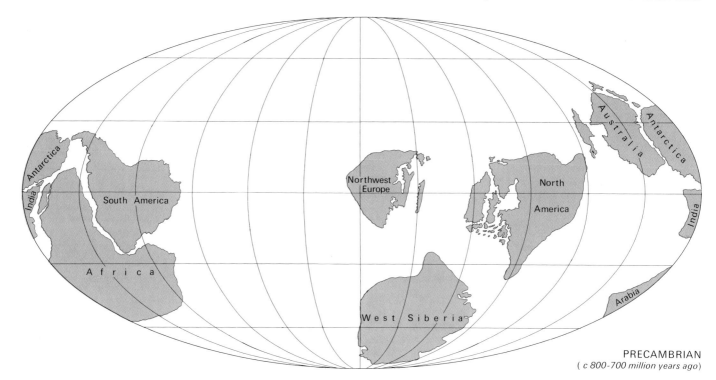

PRECAMBRIAN
(c 800-700 million years ago)

(3400–3000 million years) than those in Canada and western Australia (2900–2500 million years), although belts of different ages may occur in adjacent areas. The younger greenstones tend to occur in a series of roughly parallel belts, while the older ones are more irregular and cuspate in form. Nevertheless, the general characteristics of individual greenstone belts are essentially similar through this period. Why they apparently ceased to form after the close of the Archaean is not known.

Many major economic *ore deposits (nickel, chromium, copper, gold, silver) are closely associated with greenstone belts.

High-grade Gneiss Terrains. The best example of a high-grade gneiss terrain is the North Atlantic kraton, extending from Labrador through Greenland to northwestern Scotland, though equivalent rocks are found on other continents. The dominant rock type in the high-grade terrains is a "gray gneiss" composed essentially of the minerals quartz, plagioclase, hornblende and biotite. Chemically these gneisses are equivalent to the huge batholiths of granite, granodiorite and tonalite which dominate the Andean cordillera and western North America; and it is probable that most gray gneisses are just plutonic igneous rocks of this type but strongly deformed.

Sediments do occur in these high-grade terrains, but are shelf-type sediments such as sandstones, shales and limestones – obviously of different character from the ill-sorted graywacke sediments of the greenstone belts. Moreover, they form smaller, thinner units and have, like the gneisses, almost everywhere suffered high-grade metamorphism so that they are now represented by quartzites, kyanite schists and marbles (see *metamorphic rocks). Interbedded or associated with these sediments are considerable proportions of basic and ultrabasic *igneous rocks, but like the sediments these are so strongly deformed and metamorphosed that it is impossible to guess at their original nature. They could represent lava flows or sills of basic magma intruded into the sediments. It is usual to refer to this assemblage of rocks as supracrustal, the supposition being that they were deposited on a gneissic continental crust.

Two other rock types are commonly found in the high-grade terrains: granulites and anorthosites, and there is still considerable uncertainty as to their origin. Granulites are often very dry, having clearly been metamorphosed at such high temperatures that the water normally present in minerals such as biotite and hornblende has been expelled, so that anhydrous minerals such as pyroxene have crystallized in their place. Moreover, most granulite terrains have a higher proportion of basic rocks than the normal gray gneisses. This would concur with the suggestion that granulites may be sections of the deeper continental crust that have been uplifted during Earth movements.

Anorthosites appear to be restricted to the Precambrian, but they occur in both

Precambrian phyllites showing the tight folds typical of ancient rocks.

Archaean and Proterozoic. They are usually found in the same areas as granulites, suggesting that they may be a rock type characteristic of the deeper parts of the Earth's crust.

They are coarse-grained *igneous rocks similar to gabbros, but much richer in calcium-rich plagioclase (anorthite). A point of great interest is that similar anorthosites are abundant on the Moon, and make up a large proportion of the primitive lunar crust.

Just as greenstone belts seem to have formed over almost 1000 million years of Archaean time, so the high-grade gneiss terrains give a range of ages, and these are interpreted as times of crustal generation. During this period the gneisses were often strongly deformed and complexly folded. There is evidence in many areas of high-temperature metamorphism producing granulites about 2800–2600 million years ago. This seems to have coincided with attainment of some degree of crustal stability and the change from Archaean to Proterozoic types of crustal evolution.

Proterozoic. The Proterozoic was, compared with the Archaean, dominantly a period of crustal reworking rather than of crustal generation. Some idea of what this means can be seen by comparing the Andean and the Alpine mountain belts of the present day. The (Archaean) Andean belt is characterized by extensive crustal generation, in the form of huge batholiths, with subsidiary deformation; while the (Proterozoic) Alpine belt is characterized by strong deformation producing abundant thrusts and folds, but with very little granite. Obviously there are gradations between these two extremes.

In the northern hemisphere in particular there are two main episodes of Proterozoic activity: the first closely followed the Archaean itself, but peaked about 1800 million years ago; the second peaked about 1000 million years ago. Equivalent Proterozoic activity is known from elsewhere in the world, though not necessarily peaking at the same time.

There are two other ways in which the Proterozoic differs from the Archaean. Firstly, over large parts of the Archaean which were later to become active Proterozoic belts there were extensive swarms of roughly parallel basic dikes. In these areas the crust was fractured and extended and basic magma from the mantle was injected in large volumes, in some cases over a period of more than 100 million years. Curiously, the main trends of the later episodes of deformation run parallel to these dikes.

Secondly, the pattern of sedimentation in the Proterozoic bears more resemblance to that in later (Phanerozoic) mountain belts. Geosynclinal sedimentary basins have been recognized in northwestern and northeastern Canada and equivalent sedimentary

basins are known on other continents, all about 2200–2000 million years old. An interesting feature of these basins is the presence of considerable thicknesses of banded ironstones formed largely of magnetite and quartz: such sediments are not found from the later Proterozoic or Phanerozoic. It has been suggested that they are mainly chemical precipitates associated with a change in the character of the Earth's atmosphere from reducing to oxidizing. This may bear some relation to the rapid evolution of photoautotrophic organisms such as blue-green *algae.

The event some 1800 million years ago was a major period of crustal shortening. This produced major thrusts and folds, particularly near the margins of the mobile belts with the stable Archaean kratons. Some idea of the amount of deformation which took place can be gained by examining the basic dike swarms. These show clear cross-cutting relationships on the kratons but, traced into the Proterozoic mobile belts, are deformed and flattened almost into parallelism with the enclosing gneisses. At the same time, the original igneous minerals in the dikes were replaced by hydrous minerals such as hornblende, so that the whole rock has taken on a new planar fabric. The gneisses and associated sediments were strongly deformed and recrystallized in a similar manner, so that effectively the whole crust developed a new grain parallel to the margins of the mobile belt. It is clear that there was considerable ingress of water – presumably from the Earth's mantle – during the development of these fold belts.

The later Proterozoic Grenville belt of southern Canada, and its probable equivalent in southern Scandinavia, is a linear belt with more similarities to the younger mountain belts of the Phanerozoic. Certainly there are associated sedimentary rocks and considerable deformation. But a characteristic of the Grenville belt is the presence of very large bodies of anorthosites, some over 100km (60mi) across. Anorthosites are also common in southern Scandinavia. The Grenville belt has been interpreted as a continent-continent collision, like some of the Phanerozoic orogenic belts (e.g., the Himalayas), but this was accompanied by very high-temperature metamorphism with extensive melting of the lower crust or mantle to produce the anorthosites and dehydration of the sur-

Some of the earliest fossils occur in Precambrian limestones. These algal mats called stromatolites are up to a metre in diameter.

rounding rocks to produce granulites. Why this particular belt should be characterized by such high-temperature metamorphism is not known.

Conclusion. In summary, it is possible to see a continual change in tectonic processes through the Precambrian. The Archaean was dominantly a period of crustal generation and very mobile tectonic conditions, so that most of the rocks were deformed and folded and generally "gneissified": sediments played a relatively minor role. Toward the end of the Archaean, parallel linear series of greenstone belts began to develop. By the end of the Proterozoic, however, the pattern of tectonic processes was much closer to that seen in the Phanerozoic. Mountain belts were much more linear and sediments were relatively more important.

It is tempting to link these changes to an increasing influence of *plate tectonics as we know it today – perhaps with much smaller lithospheric plates or even a different mechanism of convection. This change would be expected if thermal gradients were higher in the Archaean, gradually falling as short-lived radioisotopes decayed away and other heat-producing elements such as uranium, thorium, potassium and rubidium were gradually trans-

ferred to the continental crust through magmatic processes.   JT

Paleozoic

The Paleozoic, the era of ancient life, lasted roughly 345 million years, from about 570 to about 225 million years ago. It encompassed the Cambrian, Ordovician and Silurian in the lower Paleozoic, and the Devonian, Carboniferous (Mississippian and Pennsylvanian) and Permian in the upper Paleozoic.   AI

Cambrian

In 1831 Adam *Sedgwick, professor of Geology at Cambridge University, traveled to north Wales to try to unravel the geological succession of the deformed and poorly fossiliferous rocks of this area (his earliest researches were aided by a young student, Charles *Darwin). His labors culminated in his defining the Cambrian, naming the system for the Latin name for Wales, Cambria.

The scope of the Cambrian has been rather reduced since Sedgwick's original definition. Sir Roderick *Murchison's

work in southeast Wales resulted in his defining the *Silurian system: each worker included as many strata as possible in their respective systems as they worked up and down the geological column. Before long it was realized that there was a major overlap which was only resolved later, after much acrimony, by Charles *Lapworth – who established the *Ordovician.

The importance of the Cambrian lies in the fact that, although the Earth is about 4600 million years old and life at least 3500 million years old, the beginning of the Cambrian heralds the appearance of metazoa (multicellular animals) with mineralized skeletons capable of being preserved as fossils. Furthermore, these fossils occur in relative abundance. In general, Precambrian rocks have only a sparsely preserved flora of microscopic *algae and fungi, as well as more prominent algal mounds called stromatolites. A notable exception is a late Precambrian soft-bodied fauna of coelenterates and worms.

Unfortunately, in only a few areas of the world was deposition of sediments continuous during the Precambrian-Cambrian transition: more usually there is a prominent unconformity. The lowermost Cambrian in many places is characterized by *conglomerates and coarse *sandstones that represent ancient beach deposits formed as the sea transgressed over the Precambrian continents. Local topographic features of these continents can often be recognized where the conglomerates have accumulated in hollows and depressions. The continents as a whole were, however, almost completely worn down to pediplains (see *plateaux). The worldwide transgression of the sea continued, with some halts and reversals, for much of the Cambrian at a modest rate of about 16km (10mi) per million years. By the end of the period the land area had been greatly reduced and there were vast areas of shallow, warm sunlit seas.

Although the base of the Cambrian is recognized by the advent of fossils with hard parts, debate on the exact horizon has been prolonged, and an International Commission is now working toward an acceptable solution. The boundary may be drawn within what is now the lowermost Cambrian, which is zoned, in the absence of *trilobites, by use of *archaeocyathids and small shelly fossils. This procedure will automatically place below the Cambrian some fossils with hard parts, but these are restricted to the uppermost Precambrian.

Radiometric dating determines the base of the Cambrian as being about 570 million years old and the top as being about 500 million years old. Worldwide subdivision of the Cambrian has not proceeded apace with other systems, and only the broad divisions of lower, middle and upper are used. Each division, with the exception of the lowermost Cambrian, is zoned by trilobites. The regional distribution of trilobites, combined with our imperfect knowledge, means that the Cambrian of each major area

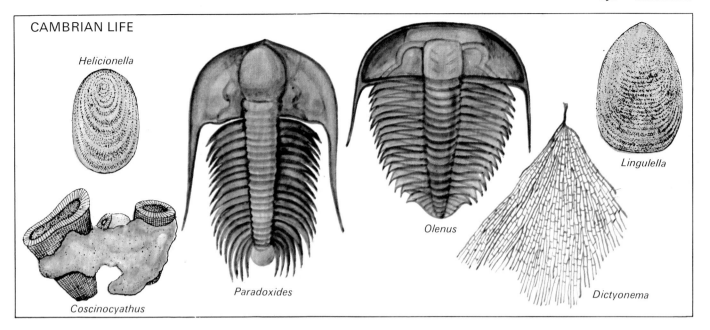

CAMBRIAN LIFE

Helicionella

Lingulella

Coscinocyathus

Paradoxides

Olenus

Dictyonema

has its own subdivisions which are yet to be united in a worldwide system.

The Origin of Hard Parts. The appearance of the Cambrian faunas is often regarded as sudden or abrupt. In terms of the Earth's age this is true, but the development of fossils with hard parts took at least 20 million years. Why did fossils appear in appreciable numbers when underlying Precambrian rocks which have suffered no subsequent metamorphism fail to yield fossils despite careful and widespread searches? The possible reasons for the development of hard parts remain a subject of lively debate, and no single idea has yet won general acceptance. Many representatives of the fauna must have had Precambrian soft-bodied ancestors that failed to fossilize: it is often argued that animals akin to trilobites and *brachiopods could not exist

without fossilizable hard parts, but in fact the exoskeleton need only be chitinous and weakly mineralized to function efficiently. Furthermore, the earliest Cambrian trilobites have noticeably thin exoskeletons.

Factors that have been invoked to explain skeletalization include the amount of free oxygen, the development of new ecological niches and food chains, and the late Precambrian ice age. The primitive atmosphere of the Earth was oxygen-free, the build-up to the present atmospheric level of 21% being due to the photosynthetic activity of plants, in particular marine algae. It has been proposed that, by the beginning of the Cambrian, oxygen had reached about 1% of modern levels. This value would, by allowing the synthesis of skeletal tissues such as collagen, enable the metazoa to diversify rapidly. Increased oxygen may

have also permitted more sophisticated behavioral activities, so promoting further skeletalization. Many of the earliest Cambrian fossils are very small, and the subsequent increase in size could again be connected with greater amounts of oxygen. The addition of oxygen to the atmosphere would shield the Earth from lethal ultraviolet radiation and allow organisms to live in very shallow water and near the surface of the oceans where algae, the primary producer in food chains, flourish.

Another theory suggests that at the end of the Precambrian a supercontinent broke up. The consequent development of mid-ocean ridges (see *ocean floor) would displace sea-water and account for the observed transgression. The increase in the area of shallow marine environments caused by continental fragmentation and this trans-

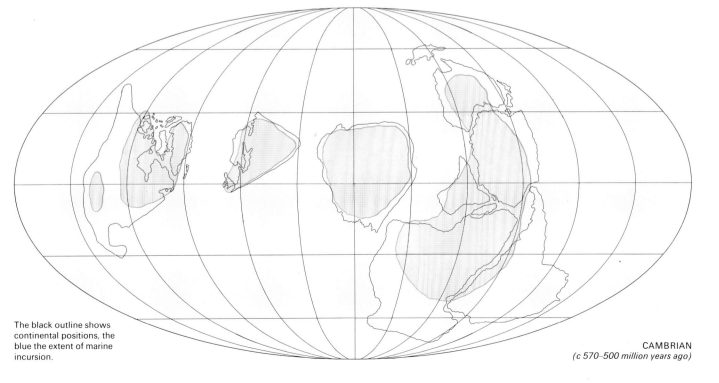

The black outline shows continental positions, the blue the extent of marine incursion.

CAMBRIAN
(c 570–500 million years ago)

gression would have offered many ecological niches that in turn encouraged diversification.

It has also been argued that hard parts evolved as protection against predators. The notion that the Cambrian was largely free from predators is being discredited, but the acquisition of an exoskeleton was probably connected more with the support of walking and feeding organs than with defense.

Cambrian Faunas. Not only is the Cambrian uniquely interesting because of the appearance of fossils, but the faunas, which were all marine, underwent fascinating developments in character and diversity. Many of the earliest Cambrian creatures disappeared after a relatively short time to be replaced by more long-lasting groups. Early Cambrian faunas are notable for their paucity of species, but diversity thereafter rapidly increased. By the end of the period every skeletonized metazoan phylum, with the exception of the *Bryozoa, had arisen, although the individual members are often very different from their modern descendants.

Phosphatic skeletons characterize many of the earliest fossils, which include inarticulate *brachiopods and *mollusks such as gastropods and the "cap shells" or monoplacophora. Several groups of small tubular and cone-shaped shells are of uncertain classification, and some may have been skeletal plates of larger animals. The distinctive hyolithids first occur in the lowermost Cambrian: they had a conical calcareous shell with a plate to fit over its opening, and two curved struts that may have been involved in locomotion. The archaeocyathids also had a calcareous skeleton that formed a single- or double-walled cup perforated by holes for feeding from water currents. They have affinities with the *corals and *sponges, but are regarded as a distinct phylum. It may be significant that archaeocyathids and sponges are rarely found together. Although the former flourished in the lower Cambrian, they were extinct by the end of the period, perhaps due to their inability to compete against the sponges.

Phosphatic skeletons were superseded by the widespread adoption of calcareous ones. The reasons for this change may be attributable to a decrease in the magnesium:calcium ratio of seawater, which would have enabled rigid skeletons to form. The argument that the Cambrian seas were less acidic, so allowing secretion of calcareous skeletons, is negated by the existence of Precambrian limestones and calcareous *algae. The biochemical pathways that led to calcification are still being studied.

Slightly later, the *trilobites arose and rapidly dominated the Cambrian seas. Their main evolutionary trends in the Cambrian were the fusion of the posterior segments into a solid plate (pygidium) and lateral migration of the eyes. Other mollusks include tiny bivalves, but the cephalopods did not appear until the uppermost Cambrian. The *echinoderms are especially interesting. One significant development in the eocrinoids and crinoids was a protrusion of the lower body as a stem to lift it clear of the sediment. Some echinoderms, such as the crinoids and edrioasteroids, continued successfully into post-Cambrian times: a conspicuous number, however, had a brief heyday but died out before the end of the Cambrian. Pentaradial symmetry, so characteristic of later *echinoderms, was by no means prevalent and different groups ranged from asymmetrical to radially symmetrical. These odd echinoderms are best regarded as "experimental" forms that despite their novelty could not compete against their better adapted relatives. Similar features may be seen amongst other Cambrian phyla – indeed, in some cases the entire phylum became extinct.

*Graptolites, which achieved a great diversity in the Ordovician and Silurian, first appeared in the middle Cambrian. Unlike the great majority of later graptolites, which floated, those in the Cambrian seas were sessile.

Conodonts have been recovered in sparse numbers. A few have peculiar shapes that were presumably "experiments", but the great majority are simple cones. The apparatus of the conodont animal must have been much simpler than in the later Paleozoic, when the shapes of conodonts became more diverse and complex.

Some of the varying activities of the fauna are preserved as *trace fossils: these are invaluable because they are often found in rocks, like sandstone, which lack body fossils. Several distinctive varieties are attributed to trilobites ploughing through the sediment; but it is curious that some of these trace fossils occur before the appearance of trilobites, and so may represent the activities of other arthropods. The growing diversity of the faunas is reflected in the increasing number and type of trace fossils above the basal Cambrian. Many trace fossils were produced by soft-bodied animals; and the presence of worms, possibly phoronid-like, and naked mollusks has been identified.

The Burgess Shale. Our knowledge of Cambrian faunas and the early evolution of metazoa without any hard parts would be greatly impoverished were it not for the celebrated middle Cambrian Burgess Shale, discovered accidentally by the American geologist C. D. Walcott in the mountains of southern British Columbia. He collected thousands of superbly preserved specimens from a small quarry which he opened above the town of Field. The fossils include numerous arthropods which are only distantly related to trilobites: it is significant that their exoskeletons are often thin and would not in normal circumstances fossilize. Trilobites with their limbs preserved are also known.

The completely soft-bodied worms show the most remarkable preservation. The gut, sometimes with food inside it, muscles and other organs can be identified. Some of the worms can be placed in groups which exist today, while others seem to be primitive chordates that by the beginning of the Ordovician had evolved into true fish.

A number of worms cannot, however, be placed in any known group. They had unusual or even bizarre morphologies: their approaches to problems like walking on or burrowing in sediment were rather unorthodox and not, in the long run, successful.

The Burgess Shale also contains a large number of sponges and some coelenterates.

Cambrian floras. Little is known about the Cambrian floras. The Burgess Shale has many macroscopic algae (seaweeds), which no doubt reflects their general abundance. Calcareous algae were sometimes sufficiently prolific to form reefs. Microscopic algae were also present but scant information is available. The continents, devoid of vegetation and subject to severe erosion, were deserts.

Paleogeography. The mechanism of *plate tectonics can explain many hitherto unconnected geological facts, although the concept will need some refining in due course. One consequence is that it allows us to work backwards, using present-day plate movements combined with a knowledge of the rates of sea-floor spreading, to reassemble the continents as they were in the past. Following this procedure back as far as the *Permian, we can deduce the presence of a supercontinent (Pangaea). Pre-Permian continental configurations remain rather speculative, as it is not easy to determine the size and shape of the continents that went to form Pangaea.

Several avenues of approach lead to a possible Cambrian paleogeography, but the paucity of data leaves many conclusions rather unfirm. It is argued that mountain-building results from the collision between two continental plates previously separated by an ocean, and so the identification of post-Cambrian/pre-Permian mountain ranges may indicate the sites of Cambrian oceans. The best documented of these is the proto-Atlantic, of approximately the same shape as the modern North Atlantic, whose closure later resulted in the Caledonian orogenesis. Paleomagnetic data gives information on the paleolatitudes of continents, but the scatter of results is often embarrassingly wide (see *geophysics).

Despite the uncertainty of the available information, Cambrian continental or kratonic areas were probably centered on North America, the Siberian shield, China-Korea, and the Baltic shield. In addition, Africa, Australia, South America and Antarctica formed a very large landmass called Gondwanaland. This supercontinent may have been a single whole, but some geologists believe that it was split by a central sea so that Africa-South America and Australia-Antarctica formed separate blocks – the critical evidence may be buried beneath the Antarctic icecap.

The Main Range of the Canadian Rockies. This section of the Rockies was formed by overthrusting of blocks of Cambrian limestone originally deposited in marine environments.

The relative positioning of the Cambrian continents across the globe depends largely on scanty paleomagnetic data, and considerable refinements and rearrangements will probably be necessary in the future.

Faunal Provinces. Only for trilobites have faunal provinces been recognized. Other groups are either insufficiently known or do not appear to show any systematic geographical variation capable of being resolved into provinces.

The distribution of trilobites was only partially controlled by sedimentary facies, and biogeographical factors are more important. The swimming or floating agnostid trilobites, however, ignored most provincial barriers that constrained the other bottom-living trilobites and had a widespread distribution.

It has been shown that some of the Cambrian continents were encircled by broad sedimentary belts that changed systematically from the shallow water of the continental shelf to the deeper water of the open sea. First identified in North America, these consist of an inner detrital belt of sand and mud derived from weathering of the continent, a median carbonate belt of limestone and dolomite, and an outer detrital belt of dark muds, silts, impure limestones and sometimes cherts, that were deposited in deeper and often oxygen-poor water on the continental slope. Occasionally carbonate conglomerates slumped over the shelf edge into this outer belt. The same threefold division of sedimentary belts has been recognized off the Siberian and Chinese-Korean continents, but in the Baltic region the median limestone belt is generally absent and so it is believed that the open ocean came close to the Baltic landmass.

In the lower Cambrian, two main trilobite provinces have been identified. The Olenellid province consists of two subprovinces centered on North America and northwest Europe/maritime North America. The Redlichiid province includes Australia, southeast Asia, Africa and Antarctica. The affinity, for instance, between the trilobites of northwest Europe and maritime North America is taken as evidence of their proximity on one side of the proto-Atlantic. When the Atlantic reopened the province was split into two sections that now lie thousands of kilometres apart.

Better knowledge of middle and upper Cambrian faunas gives a more complex pattern. The trilobites of the inner detrital belt and most of the carbonate belt are largely endemic. Four endemic trilobite faunas can be recognized in Siberia, North America, China-Australia and central Europe. Endemism arose because access by more cosmopolitan trilobites, including the otherwise ubiquitous agnostids, was restricted, sometimes to the extent of making correlation with other areas difficult. In some places the carbonate belt acted as a faunal barrier. Elsewhere, no obvious barrier is present and factors like temperature and salinity, which left no trace in the sediments, must be invoked. In contrast, the outer detrital belt was an area of open access to migrating trilobites because it lay on the edges of the open sea. The consequent mixing makes separate provinces difficult to identify. The lower Cambrian provinces seem, however, to persist and three provinces centered on western Europe, North America and southeast Asia/Australia have been defined.

Climate. The end of the late Precambrian glaciation must have heralded a distinct amelioration of climate that persisted during the Cambrian. The widespread distribution of carbonates and *evaporites suggests that much of the Earth's surface was warm. Archaeocyathids could form patch reefs, and it has been suggested that, like modern hermatypic corals, they could only live in tropical and subtropical conditions. The continents are generally assumed to have been dry for all or much of the year. The poles may have been in open seas, icecaps being apparently absent. Latitudinal temperature gradients do not appear to have been extreme.

Conclusion. The Cambrian is one of the most exciting geological periods to study, the great length of time that separates us from it making the intellectual challenge all the greater. Future research should reveal exactly why fossils with hard parts appeared relatively quickly, and what steps were involved in this process. Cooperation with other scientists and the development of new analytical techniques will help to clarify our notions on Cambrian faunas and their distribution.   SCM

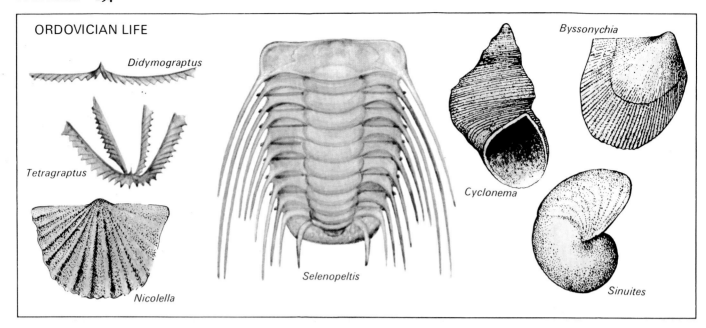

ORDOVICIAN LIFE

Didymograptus

Tetragraptus

Nicolella

Selenopeltis

Cyclonema

Byssonychia

Sinuites

Ordovician

The term "Ordovician" was introduced by Charles *Lapworth in 1879 to include those rocks which had been described as upper Cambrian by Adam *Sedgwick and as lower Silurian by Roderick *Murchison. When the *Cambrian and Silurian systems were first proposed it was believed that the Cambrian included rocks entirely older than the Silurian, but as investigations proceeded it became apparent that the two systems overlapped, and that the upper part of Sedgwick's Cambrian was essentially the same as the lower part of Murchison's Silurian. In an attempt to resolve this conflict of opinion Lapworth introduced the name Ordovician for the disputed beds. The name is for the Ordovices, a tribe of Celts from Wales.

On the radiometric timescale, the Ordovician period is accorded a duration of approximately 60 million years, from around 500 to about 440 million years ago.

Ordovician Fossils. Ordovician strata yield representatives of all the major invertebrate phyla; and it is from this period that abundant fish remains first occur.

The *trilobites, which were the dominant organisms in the Cambrian, are present also in great abundance in Ordovician strata; and many genera, and even families and higher taxonomic categories, appear for the first time in this period. Likewise, trilobites persist in great strength into the Silurian, but with the appearance of many new forms replacing earlier stocks.

In the Ordovician, and for the first time in the geological record, *brachiopods, *bryozoans, *gastropods, *bivalves,

*nautiloid cephalopods, *crinoids, *echinoids, and rugose and tabulate *corals all become locally common. The Ordovician representatives of these groups, together with the trilobites, were primarily, but not exclusively, inhabitants of the sea floor, and collectively they constitute the so-called shelly facies. Their abundant fossil representation is generally indicative of the fact that the sediments containing them accumulated in shallow seas, on or bordering the continental areas.

Ordovician sediments which accumulated in the deeper waters, beyond the epicontinental seas, yield abundant remains of pelagic organisms. Chief among these are the *graptolites; appreciably less common are the remains of inarticulate ("horny") brachiopods and phyllocarid *crustaceans. This association of graptolites with the

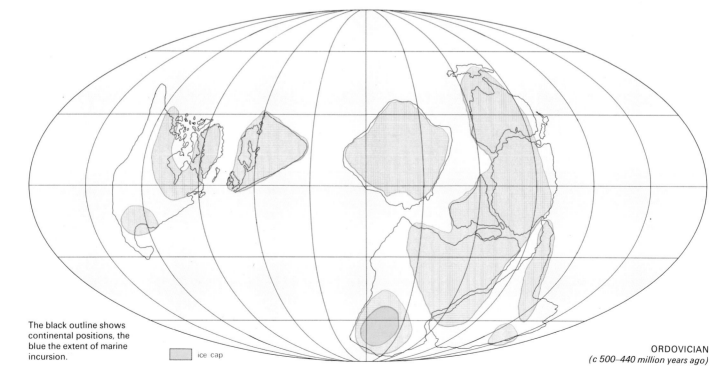

The black outline shows continental positions, the blue the extent of marine incursion.

ice cap

ORDOVICIAN
(c 500–440 million years ago)

Ordovician black shales exposed on a beach at Abereiddy Bay, Wales. These shales contain graptolites and indicate deposition in deep water during early Ordovician times.

remains of other pelagic organisms, and the virtual absence of shelly fossils, characterizes the Ordovician graptolitic facies.

Correlation. Fossils provide the primary evidence enabling the recognition of Ordovician strata throughout the world and their detailed correlation with the standard British Ordovician succession. The problems attending the classification and correlation of Ordovician rocks merely within the confines of the British Isles, arising from the diversity of their contained faunas, are considerable: on a world scale, such problems are increased, and the issue is still further complicated by the increasing appreciation that most, if not all, Ordovician faunas were differentiated into provinces, just as are organisms of the present day.

The outcome of this is that different classifications have had to be erected in different areas, because the fossil faunas, the bases of these classifications, are different. The sea-floor-dwelling shelly faunas appear to have suffered a greater degree of provincialism than the pelagic graptolites. For this reason, the shelly faunas play a restricted role in world-wide correlation, while the graptolites offer the most useful means of establishing contemporaneity. Interchange of faunas between provinces occurred from time to time

and instances of complete merging have been documented: this has enabled the correlation of the several schemes of classification to be made with a fair degree of confidence.

Ordovician Rocks. For descriptive purposes, present-day occurrences of Ordovician rocks can be grouped into two broad categories based on tectonic and lithological criteria: those preserved on the relatively stable *Precambrian shield areas (kratons); and those found in long, linear fold belts bordering the kratons.

The kratons are composed of Precambrian *igneous rocks and *metamorphic rocks covered with a veneer of younger sediments. Several such kratons, of continental proportions and varying relative elevation, existed during the Ordovician.

The European kraton was centered about Finland and western Russia; the presently exposed portion is referred to as the Baltic Shield. The Ordovician North American kraton included the present Canadian Shield, while the Siberian kraton extended over most of what is now northeast Asia. The immense Gondwanaland kraton of the Ordovician Period was subsequently dismembered by continental rifting, and the separated fragments are now widely distributed throughout the southern hemisphere: they include the eastern half of South America (the Brazilian Shield), much of Africa and Antarctica, and the western half of Australia (the Yilgarn and

Pilbara Blocks).

The Ordovician sediments which accumulated on these kratons are principally of shallow-water origin. Variations in thickness reflect differential warping of the kratons, such that the successions may be grouped generally into thinner carbonate platform facies and thicker basinal facies of rocks made up of eroded and weathered particles (clastic rocks). The relative stability of the kratons has ensured little, if any, structural disturbance of the Ordovician rocks there since the time of their formation.

However, Ordovician rocks exposed in the fold belts bordering the kratons have been intensely deformed, and generally exhibit metamorphism – to varying degrees. The successions attain a great thickness, usually of the order of thousands of metres, and the different rock types can be seen to have accumulated in a wide range of depositional environments. Fossiliferous limestones, together with sandstones, siltstones and shales, were laid down in the relatively shallow waters of continental shelf areas. Other sedimentary rocks (e.g., *graywackes) give evidence of their having been transported by turbidity currents and having accumulated in waters deeper than those in which they were initially deposited. Deeper water, abyssal environments are represented by cherty and argillaceous *limestones and black, often pyritic, shales with graptolites.

For descriptive purposes, it is convenient where possible to distinguish between eugeosynclinal and miogeosynclinal Ordovician successions within the fold belts (see *geosynclines): the former are thick sequences of sediments with associated volcanic rocks which formed some distance from the kratons; the latter are adjacent to the kratons and lack volcanic rocks.

Acceptance of the theory of *plate tectonics has necessarily modified earlier interpretations of the origin of linear fold belts. They are now considered to have arisen at ancient destructive, or compressional, plate margins and to define the sites of former oceans. Stages in ocean closure, leading to the ultimate collision of the bordering continents, are reflected in successive episodes of deformation and metamorphism in the rocks of the fold belts. The distinctive characteristics of the miogeosynclinal and eugeosynclinal successions within these belts can readily be interpreted in the context of this dynamic process.

The non-volcanic miogeosynclinal facies incorporates those sediments which were deposited on the submerged margins of the continental portions of adjacent plates: their major structural characteristics were acquired only at a late stage in ocean contraction, when opposed continental margins collided. Sediments of the eugeosynclinal facies, on the other hand, accumulated on the oceanward sides of the continental margins and include ocean-floor (or abyssal) deposits together with continental-slope and -rise turbidites. The abundant volcanic rocks incorporated within the eugeosynclinal successions, the evidence of successive phases of deformation and metamorphism which they portray, and their association with upthrust wedges of very dark rocks, all point to their intimate involvement in the contraction process. The scale and intensity of deformation may have been such as to force slices of the eugeosynclinal facies along shallow-dipping thrust planes into the miogeosynclinal or even the kratonic regions. Hence, whereas the relationship between these three facies was essentially gradational at the time the sediments were deposited, the boundaries of the eugeosynclinal facies are now usually tectonically defined in the ancient fold belts.

The proto-Atlantic. The Paleozoic Caledonian and Appalachian fold belts, or orogens, portray a classic series of ancient plate-tectonic processes. A combination of paleomagnetic, tectonic, petrological and sedimentary evidence points to the existence of what has been termed a proto-Atlantic Ocean occupying approximately, but not precisely, the site of the present Atlantic. The spreading process which initiated the proto-Atlantic commenced as far back in time as the late Precambrian and continued through the succeeding Cambrian, when, paleomagnetic evidence suggests, the opposing continental margins achieved a maximum separation of the

order of 2000km (1250mi). Contraction of the proto-Atlantic commenced at the start of the Ordovician and finally led to the collision of the opposing continents in Silurian and Devonian times. In terms of present-day geography, the resulting mountain belt, the Caledonian-Appalachian orogen, extends from Spitzbergen through Scandinavia and into Britain and Ireland; and it is traceable on the western side of the North Atlantic from Newfoundland southward to Alabama.

The geosynclinal belt in Europe extends from western Ireland, through Wales, the English Lake District and Scotland to Norway. The Ordovician successions within this belt vary considerably in thickness, owing partly to the different environments in which the sediments were deposited and partly to the extent of the development of volcanic rocks. Thicker successions, totaling some thousands of metres, are present in north Wales and the English Lake District; in both areas, volcanic *andesites and *rhyolites are extensively developed. The successions contrast markedly with, for example, the 34 metres (110ft) of Caradoc and Ashgill black graptolitic shales, with thin volcanic ashes, exposed at Moffat in the Southern Uplands of Scotland.

There is local evidence of plate-tectonic processes. Thus at Ballantrae in southwest Scotland an association of early Ordovician black graptolitic shales with ocean-floor volcanic lavas and wedges of black rocks is interpreted as the site of a subduction zone along which the floor of the contracting proto-Atlantic was consumed.

The separate recognition of eugeosynclinal and miogeosynclinal facies within the geosynclinal belt is most readily achieved in Scandinavia, on the southeast side of the proto-Atlantic Ocean. The contraction is now tectonically defined: slices (nappes) of eugeosynclinal rocks resting on, and partially masking, the miogeosynclinal successions were carried eastward into their present positions along low-lying thrust planes as the margins of the proto-Atlantic converged in mid-Paleozoic time. Limited exposure, coupled with major faulting in post-Ordovician times, obscures facies relationships on the northwest side of the proto-Atlantic Ocean. However, in northwest Scotland, early Ordovician platform carbonates are preserved which compare readily with those of similar age in North America; it is concluded that these carbonates accumulated on the Ordovician North American kraton and that their present location on the eastern side of the North Atlantic is a consequence of continental rifting and sea-floor spreading from the Mesozoic onward.

A complete section through the Appalachian orogen is best seen and documented in Newfoundland. Ordovician strata in east Newfoundland compare with those which accumulated on the margin of the European kraton and thus reflect the conditions in Ordovician time on the southeast side of the proto-Atlantic Ocean. Westward, the Or-

dovician of central Newfoundland is contained within a belt of volcanic-bearing rocks of great thickness, analogous to, and originally continuous with, that extending from Ireland to Norway. The Ordovician of western Newfoundland includes platform carbonates like those in northwest Scotland; moreover, these rocks are over-ridden by thrust sheets composed of rocks originating in the volcanic geosynclinal belt and carried westward during the contraction phases of the proto-Atlantic Ocean.

Ordovician Geography and Climate. To the extent that ancient linear orogenic belts define the sites of former oceans, it is believed that four major continental areas existed during the Ordovician Period. These were: (a) North America and Europe west of the Caledonian-Appalachian orogen (the North American kraton); (b) Europe and North America east of the Caledonian-Appalachian orogen and north of the Hercynian fold belt (the European kraton); (c) Asia east of the Urals (the Siberian kraton); and (d) Gondwanaland, comprising South America, Africa, India, Australia and Antarctica.

The precise outlines of these continents and the widths of the intervening oceans are still matters for speculation. Paleomagnetic data provides evidence for the latitudinal positioning of the continents, but they are of course unable to provide information on their longitudinal separation.

In terms of present-day geography, paleomagnetic evidence suggests that the Ordovician south pole was located in northern Africa, and therefore the Ordovician equator must have extended from California, west of Hudson's Bay, north of Greenland and Scandinavia, across northeast Asia, to western Australia. This configuration of lines of latitude places the bulk of the Ordovician continental areas in the southern hemisphere. Only southeast Asia, the eastern portion of Australia and northwestern North America were located in the Ordovician northern hemisphere, and all of these were within approximately 30° of the Ordovician equator.

Confirmation of this latitudinal positioning of Ordovician continental areas has been provided from other sources. Intensive postwar exploration in the Sahara under the auspices of the oil industry has revealed decisive evidence, including glacial sediments (tillites), glaciated surfaces and glacial striations, pointing to a late Ordovician ice advance, with the ice sheets radiating from a polar region in the continental area of northern Africa. The undoubted presence of late Ordovician tillites in the Saharan region has supported the theory that deposits with similar characteristics exposed elsewhere are likewise of glacial origin.

The composition and distribution of Ordovician marine faunas, coupled with the nature and distribution of sediment types (particularly *evaporites), support the location of the Ordovician equatorial zone in the present Arctic and sub-Arctic regions of

Volcanic rocks are commonly found in the Ordovician succession. These mountains in Borrowdale, England, are formed of Mid Ordovician volcanic rocks extruded subaerially. They were exposed as a result of uplift and erosion some 500 million years later during the Tertiary.

northern Europe and North America. In addition, the faunas and sediments provide some evidence for a progressive strengthening of *climatic zones during the Ordovician, lowest temperatures being achieved late in the period.

We have already mentioned the marked provincialism displayed by Ordovician marine faunas, and available evidence strongly supports the view that the overriding influence in the development and distribution of the provinces was climatic – more specifically, latitudinal variation in water temperature. In the North Atlantic region, where Ordovician faunas have been studied in greatest detail, two major provinces – named American and European – can be deduced from brachiopod, conodont, graptolite and trilobite fossil remains. If the geographic distributions of the two provinces are plotted on the Ordovician map, the American province can be seen to straddle the paleoequator, extending from about 30°N to about 30°S, while the European province occupies middle and high southern latitudes.

The two provinces became increasingly differentiated during the early Ordovician, reached a maximum disparity in the middle of the period, and then merged in the late Ordovician, giving virtually cosmopolitan faunas in the North Atlantic region. This breakdown of provincialism is generally attributed to the contraction of the proto-Atlantic Ocean, which opened up new migration routes and led to a merging of hitherto distinct faunas.

In the case of the graptolites, however, it seems at least equally likely that climatic deterioration in higher latitudes, as evidenced by the late Ordovician glacial advance, played a primary role. All late Ordovician graptolite faunas are confined to within a few tens of degrees either side of the Ordovician equator – that is, within the limits of the early and middle Ordovician American province. It would appear that the European province no longer existed in the late Ordovician, presumably because water temperatures in middle and high latitudes fell below the tolerance limits of the graptolites.

Conclusion. We have seen that the Caledonian-Appalachian orogen and the bordering North American and European kratons preserve contrasting facies of Ordovician rocks. From their nature and their present-day distribution it is concluded that an ancient ocean – the proto-Atlantic – occupied the site of the orogen in late Precambrian and early Paleozoic times.

Loss of ocean crust along subduction zones on the margins of this ocean was accompanied by the convergence of the North American and European kratons; and their ultimate collision and joining later in the Paleozoic resulted in the disappearance of the proto-Atlantic.

At the opening of the Ordovician, the ocean had reached its maximum width, which may have been as great as 2000km (1250mi). The period was to witness the first stages in the contraction of this ocean. Viewed in this context, the lithological characteristics of Ordovician strata become meaningful, and the present distribution on either side of the modern Atlantic is readily explained.   DS

Silurian

The Silurian system was named by *Murchison in 1835 after an ancient British tribe which inhabited South Wales and parts of the Welsh Borderland, where Murchison did much of his early fieldwork. In these areas, Murchison described sequences of Silurian sediments which were rich in marine fossils, mostly *brachiopods, *trilobites and *corals.

*Sedgwick, who named the Cambrian system, did not describe his faunas in the same detail as Murchison, leaving most of the paleontology to others, and by 1852

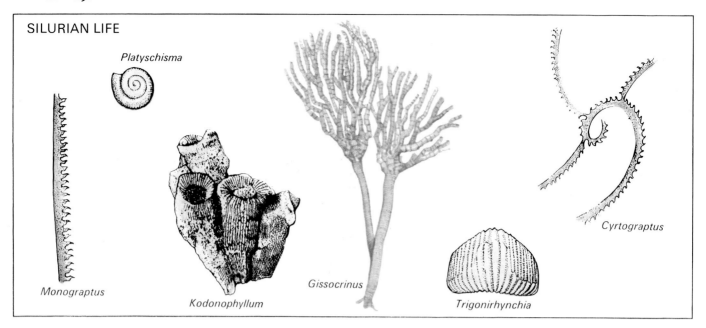

SILURIAN LIFE

Platyschisma

Monograptus

Kodonophyllum

Gissocrinus

Trigonirhynchia

Cyrtograptus

Murchison was claiming most pre-Devonian rocks for his own Silurian system. It was not until after Murchison's death in 1871 that Charles *Lapworth redefined the Silurian.

The four series in the Silurian are, from youngest to oldest: the Pridoli (named after a town in Czechoslovakia); the Ludlow (a town in Shropshire); the Wenlock (Wenlock Edge is a scarp formed by the Wenlock Limestone in Shropshire); and the Llandovery (a town in South Wales).

At the moment, it is not possible to be precise about the relative lengths of time represented by each series, but both graptolites and brachiopods show more evolutionary changes during the Llandovery than during any of the other series, so it may be that it represents rather more than a quarter of the 30 million years or so of the

Silurian, whose approximate span is from 440 to 410 million years ago.

Life in the Silurian. The later parts of the Silurian probably saw the first colonization of the land by primitive plants, though life's real conquest of the land had to wait until the Devonian. Some freshwater fish were present in rivers and lakes, but the vast majority of Silurian life was marine.

The marine faunas in the Silurian have two main components: those animals, such as the brachiopods, that lived on the sea floor (benthos) and those, like the graptolites, that inhabited the water above. Although trilobites occur, they are far less important than in the preceding Ordovician. Mollusks, corals, bryozoans, ostracods, crinoids and fish are also present, but are only common in a few out of the many marine environments.

The proto-Atlantic, which in the Ordovician separated the Canadian Shield (including Scotland) from the Baltic Shield (to which England was attached), became progressively narrower during the Silurian. As a result the free-swimming (pelagic) larval stages of brachiopods and trilobites were able to cross the ocean freely, so that these faunas were the same on both sides of the ocean. In the early Silurian most brachiopods and trilobites were worldwide in their distribution – no ocean was wide enough to act as a permanent barrier to their pelagic larvae. Many *ostracods, however, do not have a pelagic larval state – their eggs hatch out on the sea floor – and as a result even the narrow Silurian proto-Atlantic was a barrier to their migration. The freshwater fish were also restricted until the late Silurian (when Norway may have collided with

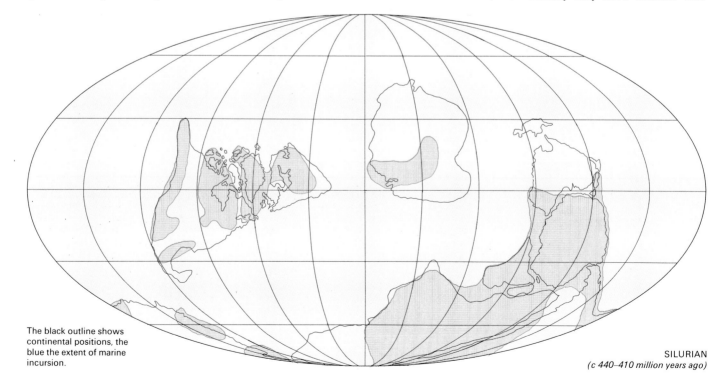

The black outline shows continental positions, the blue the extent of marine incursion.

SILURIAN
(c 440–410 million years ago)

Greenland).

The early Silurian brachiopods of Britain were among the first group of fossils in which communities have been described. These communities appear to be more closely related to varying depths of water than to varying types of sediment. Three lines of evidence for this are: firstly, that maps of community distribution at any one time show a consistent community sequence from the shore to deep water; secondly, in the transgressive lower Silurian of the Welsh Borderland different local successions show the same vertical sequence of communities (from shallow to deep); and, finally, in modern marine-bottom communities there is a progressive increase in the diversity of animal species with depth, and similar increases in diversity with depth are seen in the Silurian brachiopod communities.

Brachiopods make up as much as 80% of the preserved Silurian benthos. They are all suspension feeders; and, apart from competition for food, there was probably no interaction between one brachiopod and its neighbors on the sea floor. Each brachiopod community, therefore, did not form a discrete unit; there is little correlation between the ecological distribution of any two species, and there is a continuous gradation of changes in the brachiopod distribution from the shore to deep water, with no natural breaks.

Stratigraphical classification. Murchison defined the Wenlock and Ludlow as formations (that is, as mappable sedimentary units) which he could trace across the Welsh Borderland and South Wales. But as stratigraphy developed it became necessary to have a term which referred to rocks of the same age as these formations but in distant areas. Hence the concept of "series" developed: this is a time-stratigraphic unit, which embraces all rocks formed during a defined interval of time. At present, series can be correlated only by means of fossils; in time, radioisotope methods may become sufficiently accurate, but errors of 10 million years (perhaps up to 10 graptolite zones) are still present in the lower Paleozoic timescale.

The Llandovery, Wenlock and Ludlow series were established on the basis of brachiopods, trilobites and corals, which occur in the shelf sediments of the type areas. Subsequently, Lapworth and others recognized graptolite zones in deeper-water sediments of Wales, northern England and southern Scotland. These deep-water sediments include very few brachiopods and trilobites, and those that are present are usually of different species than are the shallow-water shelly faunas, so it was not easy to correlate the graptolite zones with the established series. It is only in recent years that this correlation between the deep-water graptolites and the shallow-water shelly faunas has been achieved.

There are two main snags in correlating the graptolite zones with the shelly faunas: one is that graptolites are not very common in shallow-water deposits, and the other is that it is not very easy to use many shallow-water fossils as accurate time indicators. In contrast with the graptolites, most brachiopod lineages do not show progressive evolutionary changes through time: most alterations in brachiopod assemblages are the direct result of environmental changes, and so cannot be used for correlation with any degree of certainty, as the environmental changes are unlikely to occur simultaneously over wide areas. Most

Silurian brachiopod species appear suddenly over much of the world, survive for 10 or 20 million years without much change, and then become extinct. However, a few lineages – such as *Eocoelia* and *Stricklandia* – show progressive evolutionary changes with time, and it is these that are most useful in correlation.

Both *Eocoelia* and *Stricklandia* occur in the lower Silurian beds near Llandovery in South Wales, so that the presence of members of either lineage in other places enables a correlation to be made with the sequence at Llandovery. A few graptolites are also known from the Llandovery area, but many more are known from other areas where *Eocoelia* and *Stricklandia* occur. It is thus possible to correlate most fossiliferous lower Silurian beds with the type sequence at Llandovery.

In 1925, O. T. Jones mapped the Llandovery area, and instead of giving formal names to the formations there, he gave them letters and numbers. In correlating shelly facies it has now become common usage to equate the brachiopod faunas to these formations at Llandovery; for example, some sandstones in Scotland can be referred to "C$_1$" if they contain an *Eocoelia* with strong ribs (*E. hemisphaerica*) similar to that occurring in the C$_1$ mudstones at Llandovery. It has been accepted by many stratigraphers that stratigraphic zones should be based on a type section, but it is only in the Llandovery that this concept is carried through to its logical conclusion: shelly faunas all over the world can be dated by direct reference to the type area.

Sufficient links are now known between these key Llandovery brachiopod lineages and zonal graptolites for the two systems to be fairly well integrated. This integration has allowed stages to be erected for the Llandovery (a stage is a time-stratigraphic term intermediate between a zone and a series and it, too, should be based on a type section). The four Llandovery stages have been defined most specifically by the level of their *base*. As the basal Llandovery formation (A$_1$) at Llandovery rests unconformably on the Ordovician, and as it is very sparsely fossiliferous, the basal stage (Rhuddanian) has its type section in the south of Scotland where deposition was continuous from the upper Ordovician into the Silurian. The other three Llandovery stages all have their type sections near Llandovery.

In the Wenlock and Ludlow, there are no described lineages in shallow-water brachiopods, so graptolites present the only reliable means of correlation over large distances. Fortunately, in the type areas (Wenlock Edge and Ludlow are both in Shropshire), recent work has provided many new graptolitic horizons, and the Wenlock and lower Ludlow graptolitic beds over most of the world can be correlated with the type sections.

The graptolites and brachiopods are the main groups of fossils which have so far proved most useful in correlation, but other groups are also likely to be valuable. *Conodonts can be used to distinguish most of the Llandovery and Wenlock stages (but not the finer divisions recognized by graptolites and brachiopods). Ostracods, acritarchs (see * dinoflagellates) and also spore assemblages are beginning to prove useful in the upper Silurian, but again no very fine stratigraphic intervals have so far been shown to be correlatable over large distances. Conodonts and acritarchs can occur in most types of marine facies, while spores and ostracods are more abundant in shallow-water environments. All these groups have great potential for stratigraphic zonation of the upper Silurian, but before they can be of much use many more detailed studies will be necessary.

Geography, Sediments and Climate.

During the Silurian, the continents were distributed very differently from the way that they are today. Their positions can be deduced (with varying degrees of certainty) partly from paleomagnetic data, partly from faunal distributions (especially Ordovician faunas, which were much more provincial than those in the Silurian), and partly from distributions of igneous rocks (which indicate ocean floor or continental margins above areas where ocean floor was being subducted). (See *plate tectonics.)

The continents were never too far apart for the pelagic larval stages of brachiopods and trilobites to cross the relatively narrow oceans, and few continents acted as barriers to migration because, for much of the time, they were covered by sea. Thus most of the Silurian marine faunas have a worldwide distribution. The principal exception was with Gondwanaland, the large southern-hemisphere continent which split up in the Mesozoic to later form South America, Africa, Antarctica, India and Australia. Gondwanaland was covered by sea only around its margins and in a few other localized areas. Argentina was probably close to the south pole in the Silurian and the regions near this pole show a peculiar cold-water marine fauna containing a limited number of brachiopod and trilobite species (the *Clarkeia* fauna).

At the end of the Ordovician the south pole was situated in what is now the Sahara Desert (it was not until later that Gondwanaland had moved such that the pole had migrated to Argentina). Glacial deposits and erosive features are now exposed clearly in the Sahara, and glacially derived sediments may have extended as far east as Arabia, as far north as Normandy, and as far west as Argentina. When large masses of ice are present on continents (as they are today), there is a corresponding drop in sealevel. In the very late Ordovician the seas were restricted in area, and a non-sequence is present in most shallow marine sedimentary deposits at the base of the Silurian, probably as a result of the late Ordovician ice cap. During the Llandovery, sea levels rose, and most Llandovery marine sequences show a progressive deepening (and spread) of the sea, which may be linked to melting of the ice.

Shropshire, Wales, Western Ireland, New England and New York State all lay on one or other margin of the proto-Atlantic, and all these areas show the effects of this rise in sea-level. In North America the gradients were much less than in Britain, and the sea spread rapidly over a much greater area of the continent, but by the end of Llandovery time, even in Britain, there were very few land areas left outstanding. As a result, the supply of land-derived sand was often reduced, and limestones and muds are the characteristic deposits in shelf areas. In those parts of North America and Europe (notably England and Sweden) where limestones and dolomites occur, corals, bryozoans and algae often flourished, as these areas were not far from the equator, and the shallow seas were warm enough to support a rich abundance of life.

In the late Wenlock, many seas in North America and western Europe became shallower, and it may yet be shown that some of the changes in sea-level are eustatic (that is, world-wide), perhaps once again connected with the development of ice sheets, though the eustatic changes could equally well have been caused by uplift (or subsidence) of an oceanic ridge. *Tillites of Silurian age occur in Argentina and Bolivia: they appear to be early Silurian, but tillites do not contain fossils, and it is quite possible that some parts of Gondwanaland were covered by ice throughout the Silurian.

Limestones or dolomites covered parts of Australia and a strip of Asia extending from Malaysia to Afghanistan as well as the central parts of northern Europe and North America. Over other parts of the continents there was extensive deposition of mud, especially to the east of the carbonate belt in Asia and across central Europe.

On the margins of many continents, sites of former subduction zones can be detected by the presence of *andesites and other calc-alkaline *igneous rocks. In these belts, where oceanic crust was descending beneath the continents, the sites of ocean trenches may be marked by great thicknesses of slumped deposits; or, in other areas, large quantities of graywackes originally deposited on the ocean floor may have been scraped off above the descending ocean plate and plastered onto the margins of the continents. Belts of deep-water sediments occur along both margins of the proto-Atlantic (notably in southern Scotland, Ireland and Newfoundland) and also in Nevada and California (on the western margin of North America), and in eastern Australia. In each of these examples, considerable amounts of material were added onto the margins of the continents. The presence of calc-alkaline igneous rocks in these areas confirms their association with subduction zones active during the Silurian.

In the Ludlow of many areas (including North America, Britain and North Africa) there was a gradual retreat of the sea. In Michigan, Ontario and New York, rock salt and other *evaporite deposits indicate high temperatures with evaporation of the shallow seas during the latest Silurian time. But in other areas there was a gradual change in the environment from marine sediments to river sediments, as in the Old Red Sandstone deposits of northwest Europe and northeast North America. This Old Red Sandstone continent developed after the closure of the proto-Atlantic. It is likely that collision took place between Norway and Greenland in the late Silurian, but the ocean did not finally close, in the northern Appalachians, until late in the early Devonian.

At one time all Old Red Sandstone deposits were thought to be of Devonian age, but we now know that river floodplain environments appeared in many areas well before the end of the Silurian. The red color is not diagnostic of desert environments: all the Old Red Sandstone sediments were laid down in water, either on floodplains or in lakes, the red coloring being due to the ferric oxide *hematite, which coats the sand and mud particles. Many Silurian marine sequences also contain red beds associated with local transgressions of the sea over nearby land. As the land surface was oxidized at this time, more hematite was transported and deposited without reduction at these times of transgression than would otherwise have been the case. Though red sequences can occur in marine sediments which were deposited in a variety of water depths, they are more common in thicker sequences that have been deposited more rapidly.

Conclusion. Plate-tectonic theory shows us that, compared with today, the continents and oceans were arranged very differently in the Silurian. But looking at the Silurian world in more detail we find that the nature of an ocean would not have been very different from that of a modern ocean; the only obvious distinctions would be in the marine organisms – for example, graptolites would be there instead of modern *coccoliths and *diatoms (though the *radiolaria would be much the same). By contrast, apart from the ice caps, the Silurian continents were very different from those of today: the areas where rocks were exposed above sea-level were bright red and lacked soils. These land areas were much smaller than are the extensive land areas of today, most of the continents being covered by shallow seas. The bottom faunas of these seas were dominated by brachiopods, with the occasional patch of coral or calcareous algae in some of the warm shallow areas. In the rivers and lakes, there were a few primitive fish, but most of the land areas were almost barren of life, except for some rare rootless plants that were just starting to colonize the soil-less landscapes. During the Silurian, living was undoubtedly much easier in the sea than on the land.   WSM

Devonian

The Devonian lasted for approximately 65 million years, from about 410 to 345 million years ago. It is the first period from which fossil remains of animals are common, with extensive preservation of continental sediments. The highlights of the period were the establishment of land plants, the appearance of abundant freshwater life, fish and arthropods, and, with the first tetrapods around the end of the period, the initial stages of the vertebrates' conquest of the land. The period also probably witnessed the evolution of insect flight, a development associated with the evolution of larger plants, though the first fossil winged insect is not known until the upper

Carboniferous. In the marine facies, the period witnessed a most remarkable surge of reef growth, largely due to an enigmatic group of fossils, the *stromatoporoids.

Devonian sediments are exciting in their diversity. The shorelines of the continents advanced and retreated during the period; and, in areas of subsidence, the sedimentary record shows evidence of extensive deltas, coastal plains and beaches.

On the other hand, the economic potential of Devonian rocks is limited. No substantial *coals were formed and there are only minor amounts of *iron ore. Devonian fossil reefs (in the form of limestones) are extensively used in the marble industry. The best-known oil-bearing rocks of Devonian age are the Alberta reefs and sandstones of the Moscow area on the Russian Platform.

The name Devonian was proposed by Adam *Sedgwick and Sir Roderick *Murchison in 1839 when it was realized that the fossils that could be found in south Devon, UK, were the marine equivalents of the Old Red Sandstone continental sediments, farther north in the UK, which were unquestionably stratigraphically below the *Carboniferous yet above the *Silurian. The relationships were confirmed in Germany, Belgium and Russia, where marine and continental facies are interbedded. In North America the same type of alternation between marine and nonmarine sediments is superbly displayed in the Catskill Mountains of New York and in gorges and creeks westwards.

The Devonian system is now divided into seven stages, and during the last few years it has become possible to begin to define these stages in an internationally acceptable scheme. But there are still difficulties over the precise interpretation and limits of some of the stages. *Graptolites, which prove so useful for the correlation of strata in the Ordovician and Silurian, became extinct quite early in the Devonian, but the *ammonoids, which replaced the graptolites in terms of stratigraphic usefulness, are not sufficiently abundant until the middle and upper Devonian to form a satisfactory basis for correlation. Thus any scheme of zones and correlations is particularly uncertain for the lower Devonian. A series of *conodont zones is used for the correlation of marine strata.

Paleogeography. In a reconstruction of Devonian world geography the large land mass comprising what are now northern North America, Greenland and northern Europe, was situated astride the equator. Australia was also in the tropics and formed the northern part of the southern continent of Gondwanaland.

For the most part Devonian sedimentation follows sequentially that of the Silurian. There are breaks, particularly where continental Old Red Sandstone facies rests on older rocks of the Caledonian fold belt and, for instance, where sediments of mixed facies transgress the *Precambrian rocks of the Russian Platform. From an inspection

of the paleogeographic map, and from all other available evidence, we find that the Euro-American Old Red Sandstone continent and the sea surrounding it enjoyed a tropical climate. Evidence for glaciation is more limited than in the Silurian or Carboniferous. The south pole was situated in the area of Buenos Aires, and glacial sediments of Devonian age are known from adjacent areas of South America and from South Africa.

The global structure during the Devonian and the distribution of plates and continents are still very uncertain. For instance, the geosynclinal belt across Europe, with its associated zones of spilitic volcanics and serpentines, has been interpreted as site of an old ocean; and the thin oceanic-type deposits of the central European Devonian may also be associated with this ocean.

But detailed examination does not justify such assertions. There is no doubt that the Devonian was a time of global mobility, but the major tectonic events in Europe and North America appear to be those already initiated in the early Paleozoic and represent merely the continuation of Caledonian plate movements and the welding of the Old Red Sandstone continent. The lower to upper Devonian Acadian orogeny which affected the Appalachian region, with extensive metamorphism and granitic intrusion, probably represents a further episode in continental growth.

Life. The Devonian is often referred to as the "Age of Ferns" or the "Age of Fishes". Although nonvascular plants must earlier have been present on the land, probably since Precambrian times, it was only in the late Silurian that there evolved vascular plants with a waxy cuticle and stomata so that they could live out of water; and the first well-preserved vascular plant fossils are known from the lower Devonian, those petrified in silica in the Rhynie Chert of Scotland being especially famous. Starting from small leafless and rootless *psilopsids vascular plants evolved to *ferns, seed ferns, calamites and lycopods by the middle Devonian, many reaching considerable size and growing in forests such as that described from the upper Devonian of New York. In contrast with later Paleozoic floras, those of the Devonian are remarkably cosmopolitan.

The lower Devonian fish faunas are dominated by pteraspid and cephalaspid ostracoderms (see *jawless fishes), and mostly bottom-dwelling forms with a flattened body and extensive armor: the only survivors beyond the Devonian of such jawless fish are the lampreys and hagfish.

The origin of jaws is still not fully understood, but it is now clear that the small, jawed *acanthodians had evolved towards the end of the Silurian. The *placoderms, which are related to the cartilaginous-skeletoned sharks, evolved jaws independently and appeared in the early Devonian. A group of the placoderms, the arthrodires, often reached large size and

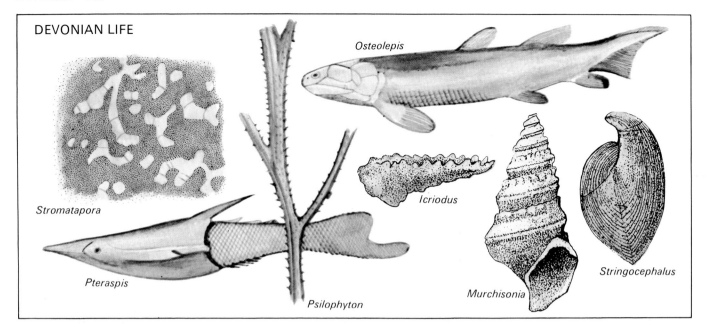

DEVONIAN LIFE

Osteolepis

Stromatapora

Icriodus

Pteraspis

Murchisonia

Stringocephalus

Psilophyton

were the dominant carnivores. (It is likely that complex food webs – plant, invertebrate, vertebrate – evolved about this time.) True sharks are known from the middle Devonian, as are the more advanced lungfish and coelacanths.

In addition, Devonian rocks contain the first well-preserved specimens of *bony fish. There are a few scales preserved from earlier rocks that indicate that this group had its origins in the *Silurian, but upper Devonian sandstones from Australia have recently been discovered containing almost perfectly preserved remains of a bony fish, *Moythomasia*, which is probably the ancestor of the group.

The extensive preservation of fossil fish in the Devonian is probably because of the combination of the relatively good preservation potential for their bony skeletons

in the many estuary and lake environments, and the evolutionary burst of the ostracoderms at the beginning of the period. The evolution of the lungfishes, which can breathe not only with gills but also with an adapted air bladder, is consistent with the periodic drying-out characteristic of tropical areas of low rainfall or periodic rainfall. In some respects, it is surprising that the emergence of the tetrapods from the water was delayed as long as it was, but presumably this event must have had to await the necessary evolutionary combination of extra-aquatic support and air respiration. In respect of support, the ostracoderms, with their bony external skeleton of large plates, were better adapted than the group which, though they had internal bony skeletons, actually made the evolutionary advance.

The marine faunas, in many respects, continue the general scene set during the Silurian, and there are relatively few Devonian innovations. Orthocone *nautiloids were already on the wane, but goniatitic *ammonoids became important. The establishment of the terebratulid brachiopods in the early Devonian completed the evolution of all the orders of the phylum *Brachiopoda.

The rugose and tabulate *corals are represented by essentially the same forms as in the Silurian, but there is an evolutionary burst of diversity and abundance and in size of the *stromatoporoidea: the biological affinities of this group are still somewhat puzzling, but it is possible that they are related to the *sponges. Reefs in which stromatoporoids are the principal skeletal elements are widespread, particularly from

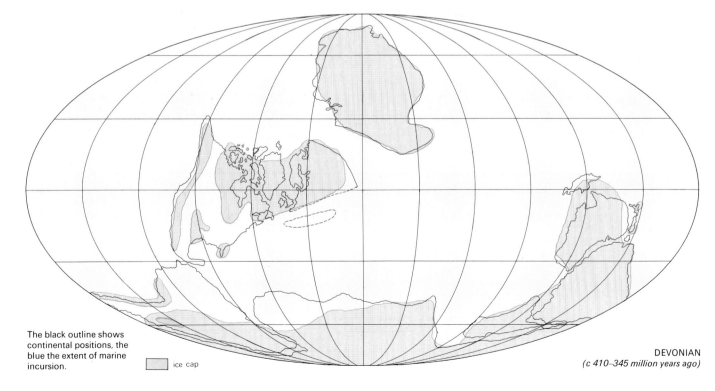

The black outline shows continental positions, the blue the extent of marine incursion.

ice cap

DEVONIAN
(c 410–345 million years ago)

Reef deposits forming high ground in the Atlas Mountains of Morocco. These limestones were deposited near the shoreline of the Devonian continent.

the middle Devonian: those in Alberta are important oil and gas reservoir rocks. (The role of *algae in the Devonian reefs is somewhat uncertain, but certainly it was no greater than it had been in the Silurian.) The reefs frequently exhibit vertical and lateral facies changes, and much prominence has been given to the proposed analogues between this lateral distribution of sediments and modern tropical reefs: reef, forereef and backreef facies have been recognized in the Devonian. However, there is little evidence of any actual reef zone that reached above wave base, and in general the reefs were constructed in quiet-water environments – though subject to storm damage. Many of the stromatoporoids grew in very shallow and extensive lagoonal areas associated with *stromatolites. In the late Devonian the stromatoporoid-dominated reefs suffered drastic reduction, and the communities and facies of the uppermost Devonian and the following Carboniferous are for the most part relatively poor.

The *trilobites of the Devonian, although represented by fewer taxa than in the lower Paleozoic, exhibit much skeletal elaboration. Loss of visual areas and blindness are characteristic of the trilobites

found in deeper-water facies. Devonian *bivalves continue stock already present in the lower Paleozoic.

The Devonian in Europe. There are several well known areas of Devonian sediments.

In the British Isles, north of the line of the Bristol Channel and Thames estuary, which corresponds with the trend of the Devonian shoreline, the sediments are mainly of continental Old Red Sandstone facies: estuary, river and lake environments. Soil horizons similar to present-day caliches (crusts of calcium carbonate typical of subtropical areas) are frequent and are locally referred to as cornstones. No wind-blown dunes are known, although the climate was likely to have been dry over northern and northwest Europe. The river sediments are frequently in cycles in each of which the finer particles are toward the top: the conglomeratic bottom part of the cycles has been interpreted as river channel deposits, the finer upper part probably representing overbank and floodplain deposits. The conglomerates frequently yield fish and plant debris, though of a fragmentary nature.

Sedimentation in southern Britain and Ireland was interrupted by a major tectonic phase of the Caledonian orogeny during the middle Devonian, so that there is regional uncomformity between the lower and upper Devonian, with the middle Devonian

missing.

The shoreline which ran across southern Britain continued eastward through Belgium and northern Germany into Russia. But this shoreline oscillated, with transgressions and regressions of the sea, and at times lay much farther to the south. South of the general line of the shore the thickness of Devonian sediments rapidly increases; this essentially east-west belt of mainly marine sediments across northern Europe is referred to as the Armorican geosyncline. In the south of Belgium, around Couvin, the thickness exceeds 5000m (16,000ft) and south of the Ruhr it exceeds 7000m (22,500ft).

The facies about the shoreline are highly diverse – as is to be expected. Relatively little work has yet been done but beach, estuarine and intertidal facies are well represented. Carbonate sedimentation is well-developed in the Torquay, Brixham and Plymouth areas of southern England, the Eifel area of Germany and the Namur district of Belgium. On the east side of the Rhine the Rhenish Slate Mountains are formed of mainly clastic sediments with apparently greater deltaic influence.

The sea floor did not constantly deepen southward. A number of submarine rises are known, and these have a general trend parallel to the shoreline. Sedimentation on these relatively positive areas is often quite thin. Some of the rises were volcanic, and marine sedimentation was interrupted by lava flows and ashes. Several atolls have been described from northern Germany where corals and stromatoporoids grew on extinct volcanic piles in a manner remarkably like those of the Pacific ocean today. Land-derived sediment scarcely extended to such areas and so the sediments are often almost completely formed of the remains of marine organisms: goniatite and clymenid *cephalopods, styliolinid *mollusks and *conodonts.

The extensive Devonian sediments that outcrop on the Russian Platform between Minsk, Leningrad and the Baltic are much thinner – only a few hundred metres – and because of this the general facies picture differs markedly from that of western Europe. There is great facies diversity but sedimentation is the result of local and irregular subsidence of the basement allowing lenses or wedges of river, evaporitic and shallow marine sediments to be preserved. Shallow marine carbonates are persistent during the early upper Devonian, but there are repeated and widespread discontinuities and no extensive reef formations. In contrast, the Devonian in the Ural Mountains is more complete, geosynclinal and akin to that of western Europe.

The Devonian in North America. In many respects the Devonian sediments of North America are a reflection of those of Europe. In recent years, knowledge of the North American Devonian has been greatly extended beyond the classic areas of the Appalachians and eastern Canada to include the Canadian Arctic Islands, Alaska and western North America.

The Rocky Mountains were the site of a large *geosyncline, with sedimentation continuing from the Silurian. Thinner sequences are present on the kraton, including the famous Alberta reefs and evaporites. In East Greenland, Bear Island and Spitzbergen, sedimentation was continental. The Devonian of the Canadian Arctic Islands has many similarities with that on the opposite side of the Old Red Sandstone continent in northwest Europe, with alternations of thick marine and nonmarine sediments. Similarly, the Devonian of the North West Territories, Alberta and Saskatchewan, are the mirror image of the Devonian of the Russian Platform and the Urals, though evaporites and reefs are more prominent. In British Columbia the stratigraphy is less clear in the tectonized zone of the Rockies, and sedimentation appears to have been oceanic, with volcanics, cherts and shales. The thick clastics and turbidites of the upper Devonian of the Yukon seem to be derived from an oceanic zone of uplift rather than from the Old Red Sandstone continent.

In the Eastern and Central United States of America the Devonian can be traced in great detail from the thick continental facies of the Catskill Mountains westward, as the thickness decreases and marine sediments, at first clastics and then limestones and shales, predominate. Repeated marine transgressions result in an alternating succession of shales and limestones, and this is most prominent in western New York between Buffalo and the Finger Lakes and in the corresponding area to the south in northern Pennsylvania. Relatively little modern work has been done on the shoreline and continental sediments of the Catskills, where around 3000m (10,000ft) of mainly upper Devonian sediments can be inferred to have been deposited.

Elsewhere. The structural relationships of the Devonian of northwest Africa are not properly understood, but the stable African kraton did not supply much clastic sediment. However, the proximity of this area to Europe, and more especially to North America, is clear from repeated faunal similarities. The Devonian of northwestern Australia is somewhat analogous to that of north Africa with prominent carbonates and reefs in the early upper Devonian. The separated Devonian deposits of eastern Australia are geosynclinal in character.

Conclusion. We have already described the Devonian as both the "Age of Ferns" and the "Age of Fishes", but we could well add to these honorific titles a third, the "Age of the Conquest of the Land". By the end of the period both plants and animals were well established on land, in striking contrast to the situation in the preceding Silurian; and in a geologically rather short extent of time, by the end of the succeeding Carboniferous, the world would see the rise of such comparatively advanced forms as the reptiles. Only a little reflection is necessary to realize the immense importance of the revolution that took place in the Devonian.   RG

Carboniferous (Mississippian and Pennsylvanian)

The Carboniferous world was dominated by three major continents: Laurasia, made up of North America, Greenland and Europe west of the Urals; Angaraland, formed of Siberia with China and Korea; and Gondwanaland, extending over India, Africa, South America, Australia and Antarctica. The reconstruction of these continents depends on detailed evidence from common rock sequences, common floras and common faunas.

Since the end of the Carboniferous these continents have moved in relation to each other, and, in the case of Laurasia and Gondwanaland, the continental regions have divided. Detailed comparison of the geology and fossils of the separated continental fragments can give proof of their original contiguity and, in some cases, rough and relative estimates of the time when the fragments separated.

Paleogeography. The positions of the continents during Carboniferous time can be found by determining paleomagnetic properties of their rocks. Certain igneous and sedimentary rocks contain magnetic minerals that may, at the time of the rocks' formation, become aligned to the magnetic lines of force around the world. Delicate experiments on oriented rock samples can determine the direction of the Earth's magnetic field at that time, giving a figure for the latitude of the sample region. In this way the position and orientation of the continents can be deduced (though their positions relative to the lines of longitude cannot be determined).

From paleomagnetic observations an interpretation of world paleogeography during the Carboniferous can be made, the data on the positions of Laurasia and Angaraland being very much more precise than those for Gondwanaland. This is a field of geology where much progress is being made at the present time and revision and improvements in our knowledge of Carboniferous geography are to be expected in the future.

Even during the Carboniferous the three great continents were moving relative to each other: Laurasia was rotating in a clockwise manner, and Angaraland was closing with Laurasia on a line now represented by the Ural Mountains. These two continents collided later, during the Permian, causing orogeny along the Ural line. Gondwanaland was also drifting during the period, but our knowledge of this is less precise.

The paleogeographical maps show that the three continents spread across most of the climatic zones of the Carboniferous world. Angaraland was mainly boreal, stretching almost to the Arctic; Laurasia lay in the equatorial and boreal zones; and Gondwanaland, though partly equatorial and partly austral, was largely centered over the South Pole. Carboniferous rocks were therefore deposited under widely varying climatic conditions, allowing glacial deposits to be laid down over a wide region of Gondwanaland, while tropical coal swamps spread over large areas of Laurasia.

Stratigraphy. No one method of stratigraphical division and correlation can be used throughout the widely different sedimentary environments found in the various climatic zones of the Carboniferous continents: the equatorial zone is the only region where the Carboniferous rocks can be divided and correlated on a truly international basis, though even here three major methods have been established in, respectively, northwestern Europe, the USSR and North America. No comparable series of divisions exists for either the boreal Angaraland region or the austral and glacial Gondwanaland area.

The absolute age of the Carboniferous system has been determined radiometrically. Results suggest that the period began 345 million years ago and ended about 280 million years ago. Absolute dating work is still being perfected and the present results will doubtless be subject to revision in the future. Nevertheless, the Carboniferous dates are well founded and the period is

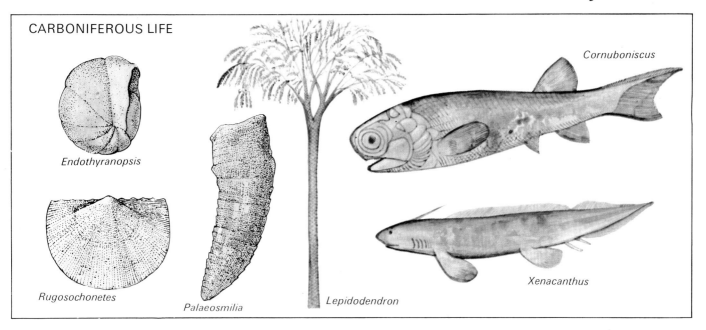

CARBONIFEROUS LIFE

Endothyranopsis

Rugosochonetes

Palaeosmilia

Lepidodendron

Cornuboniscus

Xenacanthus

dated more precisely than most of the others in the stratigraphical column.

The Carboniferous is divided into two parts, lower and upper Carboniferous, that are recognized under various names in many parts of the world. Originally the system was divided in Great Britain during the early part of the 19th century into lower Carboniferous or Carboniferous Limestone, a dominantly marine succession, and upper Carboniferous or Millstone Grit and Coal Measures: the Millstone Grit is a marine deltaic sequence and the Coal Measures a succession formed under dominantly terrestrial delta swamp conditions. No one group of fossils can be used to divide and correlate these rocks, formed as they were under different environmental conditions.

Classically corals and brachiopod shells are used as zonal fossils to divide the marine sediments of the Carboniferous equatorial belt. To a certain extent these groups have now been augmented by fossil *foraminifera as well as conodonts, which are minute toothlike fossils probably formed by lowly fishes. The marine deltaic sequences are divided into zones using conodonts and goniatites (coiled and chambered mollusk shells) and the delta swamp environment by mussel shells, plants and fossil plant spores. The microfossils foraminifera, conodonts and plant spores are particularly important for division and correlation of the Carboniferous because of their great variety and number and their occurrence in many different types of sedimentary rock. Some of these groups are sufficiently widely distributed to make possible correlations between distantly separated regions.

In the USA the Carboniferous is often divided into the Mississippian, beginning some 345 million years ago, and the Pennsylvanian, beginning around 320 million years ago.

Carboniferous Equatorial Belt. Over much of the equatorial and warmer boreal belts, the *Devonian ended with a rise in sea level and the corresponding flooding of low coastal plains about the continents by transgressive seas. Wide shallow shelf seas were formed, and here marine life found conditions suitable for rapid establishment and development. The change from Devonian red sandstones and siltstones, formed in arid desert conditions, to Carboniferous gray and blue-gray marine limestones and shales is often striking.

This change in sedimentation, however, did not take place at the same moment of

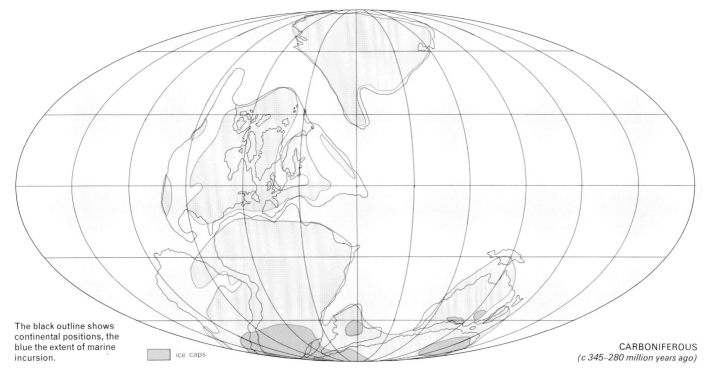

The black outline shows continental positions, the blue the extent of marine incursion.

ice caps

CARBONIFEROUS
(c 345–280 million years ago)

During late Carboniferous times, many coals were formed in swamp environments. Plant remains are common and, in some localities, even large trees have been preserved.

time around the world. The Devonian land surface was irregular in some regions, with uplands and intervening basins: the basins were first flooded by the Carboniferous seas, the uplands standing out to form islands. Elsewhere, in places where coastal lowlands were absent, desert conditions persisted into the Carboniferous. In Devonian marine basins sedimentation continued into the Carboniferous without break.

Along the margins of the seas, quantities of detritus from eroding mountain chains were carried down by rivers to form wide deltas and delta-swamps. The deltas were formed of thick wedges of sand separated by bands of silt and clay: deep burial and cementation have changed the sequences into sandstone, siltstone and shale. The deltas often built up above the sea level and exposed sands and muds to form a land surface. Under tropical conditions this low-lying land was rapidly colonized by terrestrial vegetation to form luxuriant swamp forests only just above the sea. The forests produced quantities of woody plant debris which fell below the water in the swamps to form thick layers of peat, and subsequent burial of the peat beneath later sediments caused the slow transformation of the peat into *coal.

Further out to sea, beyond the swamps and deltas, muds were laid down; and beyond this, in clear-water marine conditions, limestones were formed on the sea bed. Under favorable conditions of sinking sea floor and warm climate, coral and algal reefs and shell banks were formed. Where a restricted embayment of the sea developed in a hot climate, high evaporation resulted in the crystallization of *evaporites on the sea bottom.

All these marine sediments occur in the shallow shelf regions around the Carboniferous continents. In the deeper regions of the ocean, marine trenches gathered thick sequences of sands and muds which can now be recognized by successions, often folded, of dominantly gray and black sandstones and shales.

Life was abundant on land and in the tropical and subtropical seas. After a significant crisis among the marine invertebrates that had taken place toward the end of the Devonian, members of many old lines becoming extinct, a new period of evolution and diversification began with the opening of the Carboniferous. The wide shallow shelf seas around the continents gave abundant secure habitats for creatures with calcareous shells and skeletons. *Corals, *brachiopods (lamp shells) and *mollusks abounded and dense masses of *crinoids (sea lilies) formed groves on the sea floor in clear-water conditions: many limestones, known as crinoidal limestones, are largely composed of the separated fragments of

sea-lily stems, cups and arms. *Crustaceans were also plentiful, as were the rapidly evolving *insects on land. Among the vertebrates, fish were present in both marine and lake environments. The first evolutionary radiation of the tetrapods (four-legged animals) was in progress, with numerous true *amphibians and the first *reptiles appearing before the close of the period.

The flora was equally diverse and luxuriant in the equatorial belt. The seaweeds were widespread, with some marine forms forming thick calcareous crusts that are now algal limestones. On land, plant life had developed to a rich flora of giant tree-ferns, horsetails and seed-bearing ferns that grew in extensive swamps.

Western Laurasia: North America. The western extremity of the Laurasian continent comprised what is now North America. Here, a twofold division of the Carboniferous has been conventional since the last century, the divisions being in 1906 raised to the rank of systems, the Mississippian and Pennsylvanian, because of significant differences in their lithology and their separation, over wide areas, by a major hiatus. This division of the Carboniferous is roughly in agreement with the divisions in Western Europe (but not with the Russian system, where a threefold division is used).

The Carboniferous was a time of crustal unrest in North America, with downwarping of wide basins and uplift of broad

Limestone is the rock type typically found in lower Carboniferous deposits in many parts of the world. This hollow was partly filled with sand during a period of uplift, when shallow seas covered these limestones.

highlands. In Canada to the north lay an ancient landmass, and surrounding it a wide flat-lying shelf that was progressively covered by the Carboniferous sea. This shallow sea spread completely across southern North America and in its clear, warm waters vast spreads of limestones were deposited: these thick, pure limestones characterize the Mississippian succession. Earth movements caused a progressive reduction in the shelf sea towards the end of the Mississippian, and at the same time uplift produced the new highlands of the Oklahoma ranges, the ancestral Rockies and others. Uplift and erosion caused a widespread unconformity at the top of the Mississippian.

The earliest Pennsylvanian seas were restricted to a narrow belt between the ancient landmass to the north and an encroaching landmass, Llanoria, to the south. During the early and middle Pennsylvanian the sea spread widely over the shelf, covering much of what is now the USA, and newly formed uplands stood out of the sea as extensive elevated islands. The Pennsylvanian sea, like its Mississippian predecessor, was not to persist for long and, before the end of the period, the seas were once more receding.

The best section of the Mississippian is along the Mississippi River in Iowa, west Illinois and west Kentucky. Most of the lower half of the succession is richly fossiliferous limestone, including the Burlington Limestone and the oolitic Salem Limestone. The latter is extensively quarried in southern Indiana and is used as *building stone in many public buildings in the eastern USA. The lower Mississippian limestones stretch over the central and western part of the USA and into Canada.

In the Appalachian region to the east a sinking basin received sandy, non-marine sediments, but sediments of more marine aspect were deposited further to the south, where the basin was sinking more rapidly. These sand and clay deposits represent sediments carried in from land areas, including Llanoria, in the region of the eastern seaboard. The upper part of the Mississippian is not so well developed and is missing over most of the states west of the Mississippi River. Where found, it tends to be composed of alternating bands of sandstone and limestone.

The Pennsylvanian started with a new transgression of the sea over the low land of the central North American continent. At this time the eastern and southern seaboard was rimmed with actively rising mountain chains from Nova Scotia through to Texas. Canada was largely a vast emergent land surface and, between it and the marginal mountain chains, a broad basin developed that quickly involved much of the USA. In this basin, marine conditions encroached

from the west and east and great quantities of sediment from rapidly eroding mountains spread in wide deltas over the basin floor. East of Kansas and Oklahoma, both deltaic and marine sediments were laid down in cycles of deposition, each cycle being of the general sequence: limestone, shale, sandstone, fossil soil and coal. In Illinois, where the cycles are well developed, there are about one hundred cycles in the Pennsylvanian succession, each representing a period when the sea spread over the region from the west and allowed limestone to be deposited; followed by deltaic sedimentation from the east, first muds then sands, building up to sea level and allowing terrestrial swamp forests to develop. Peats that formed below such forests were later changed to *coal after burial below further sediments. In general,

subsidence of the basin kept up the sedimentation – when subsidence was greater than deposition, marine conditions from the west advanced over the region; when it was less, the deltas advanced over the area from the east. Some of the coal beds are thin and impersistent, indicating localized and temporary swamps, while others, such as the Pittsburg coal, are continuous over hundreds of square kilometres and may be more than 1.5m (5ft) thick.

The sea did not reach parts of the east side of the Pennsylvanian basin, and here thick sequences of non-marine strata were laid down. In Nova Scotia some 4000m (13,000ft) of non-marine strata were deposited in mountaingirt basins on the edge of the Canadian Shield. At Joggins on the Bay of Fundy these beds are exposed in sea cliffs that show the stems of trees preserved

A coal seam interbedded with sandstones and shales exposed in a quarry face. Coal deposits make upper Carboniferous rocks the most economically important rocks in North America, Europe and Russia.

vertically in their position of growth, an indication of rapid deposition, since the trees were buried before they had had time to fall and decay.

The deltas spreading from the east dwindled and failed in the mid-continent region west of Kansas and Oklahoma. Here the marine influence is of increased importance, and so limestones form a larger portion of the succession. Clastic sediment derived from various land areas caused breaks in the limestone deposition, and in some regions thick conglomerate and sandstone sequences were laid down. In the western cordillera, complex Carboniferous sequences indicate that fragments of continental crust may have migrated from the south and west to become fused with western North America, possibly during Mesozoic times. Both Mississippian and Pennsylvanian rocks are present, and consist of limestone with wedges of clastic sediment and volcanic lavas, non-marine strata and coal seams being rare.

Central Laurasia: Western Europe. In Carboniferous times, western Europe lay adjacent to maritime Canada, with the central Laurasian uplands of the Canadian shield, Greenland and Scandinavia forming a continuous whole to the north. On the southern margin of the uplands, the Carboniferous shoreline can be traced across the UK on the flanks of the Scottish Highlands, continuing beneath the North Sea, where it is known from subsurface data, under Denmark and beneath the North German plain. Only in Britain is the shoreline exposed at the surface, elsewhere being deeply buried by younger rocks.

A wide continental shelf developed seawards from the shoreline. This can be seen in Britain stretching from southern Scotland across England to the southwest peninsula of Devon and Cornwall, where it terminates against a deep marine trench. The shelf can be traced eastward across Europe, and the seaward trench continues as far as Germany.

The base of the Carboniferous is marked by a change from continental to marine conditions over the shelf region. This was a major marine transgression which started in the south, adjacent to the trench, and progressively moved northwards during early Carboniferous times. The shelf region was a series of uplands and basins, so that the basins were invaded by the sea first with the uplands standing out as islands for much of the early Carboniferous. In many cases it has been shown that the uplands were positioned over Devonian *igneous intrusions of granitic rock (the low-density granite tends to rise and give upland topography). The whole shelf region was unstable and tended to sink, so the local movements of basins and uplands are superimposed on a broad pattern of re-gional subsidence. During the Carboniferous the emergent islands were progressively inundated by the sea, though some may have persisted as islands throughout the period. Because of their tendency to rise, the islands have no, or only a relatively thin, cover of Carboniferous rocks. In contrast the basins sank more rapidly and continuously and so contain thick sequences of Carboniferous strata.

Southern Europe. South of a line through southwest England, Belgium and Germany, Carboniferous geography, structure and history are quite different from those of the shelf-region to the north. Here, after a period of early Carboniferous deposition of limestone, shale and sandstone about emergent islands, strong Earth movements and volcanic activity produced a new series of mountain ranges across Europe. These mountains were formed during the early late Carboniferous, while sediments were being laid down on the floor of the more stable northern shelf. At this time, an arm of the sea is believed to have separated the northern shelf from the new mountains of southern Europe, but throughout the period the floor of this sea was being consumed along lines of subduction on the south side, adjacent to the new mountains. In fact, the new mountain ranges seem to have been caused by uplift, folding and igneous activity associated with the subducting ocean floor. (The present day Andes formed on the margin of the Pacific in the same way.)

During the latter part of the late Carboniferous, deep mountain-girt lake basins developed in southern Europe. Here thick coal seams represent periods when dense vegetation grew on swamps associated with the margins of the lakes, and these are quite distinct from the similar coal seams formed on swamps at the edge of the Carboniferous sea in Britain, Belgium and Germany.

Towards the end of the Carboniferous, the seaway between the northern shelf and southern Europe narrowed and finally closed. This collision caused a folding episode on the margin of the northern shelf, particularly well seen in southwest England. Uplift and folding ended Carboniferous deposition on the shelf; and the emergent land area began to undergo an extended period of erosion. In the landlocked lake basins of southern Europe, sedimentation was continuous from the Carboniferous through into the Permian.

Eastern Laurasia: The Russian Platform. The wide Russian Platform which forms Eastern Europe was a stable shelf area stretching east from the Baltic landmass. In pre-Carboniferous times it had only gentle vertical movements of uplift and subsidence. Shallow-water marine, deltaic and terrestrial deposits were laid down over the region and these conditions persisted throughout the Carboniferous period. After marine late Devonian conditions, the Carboniferous started with a limestone sequence followed by sandstones and shales with productive coal seams. The upper Carboniferous returns to a bedded limestone sequence particularly rich in fossils. This marine succession continues into the Permian without break though, during the early Permian, continental conditions again became established over the shelf.

To the west, the Russian Platform is continuous with the north-Europe/British shelf, with a common shoreline developed round the Baltic landmass. In the southwest of the platform the Ukraine Shield may also have been emergent during the Carboniferous, but between this region and the main Russian Platform a deep sedimentary basin developed.

This is the Donetz Basin, a deep trough in which the full Carboniferous succession was laid down with only minor breaks. There are some 10,000m (32,800ft) of sediments in the basin compared with about 500m (1600ft) on the platform. The initial lower Carboniferous limestones were followed by cyclic marine, deltaic and continental deposits, including thick coal seams. This sequence continues through the upper Carboniferous and into the Permian. During deposition, subsidence kept pace with sedimentation so that a continuous sequence of shallow-water deposits was laid down.

Angaraland. Angaraland comprised what are now the USSR east of the Urals and China: it is a vast area, now divided into several geological regions. Our knowledge of Angaraland is more limited than is the case with Laurasia and only tentative reconstructions of the geography can be made at the present time.

A central landmass appears to have provided sediment to surrounding wide continental and marine shelves to the west, east and south. Central Siberia (west and central Angaraland) was a shallow shelf region over which alternately spread marine and continental conditions. Only thin sedimentary deposits, composed of marine limestones with deltaic and continental strata up to a total of 200m (650ft), are present. Coal-bearing sequences associated with limestone continue to the top of the Carboniferous.

A deep basin, the Angara Trough, developed in the south and southwest, and much of it is filled with continental sandstones and shales, sediments derived from the central landmass. Marine limestones are found only in the center and south of the basin. A thick coal-bearing series is developed in the upper part of the lower Carboniferous.

At the beginning of the early Carboniferous, land was emergent over the Angara Trough as continental deposits of red sandstones spread south from the upland region. A later terrestrial coal-bearing series has wide distribution and is of latest Carboniferous and Permian age. The region was structurally unstable with volcanic activity present from the Devonian right through into Permian times.

The geology of eastern Angara, including the eastern USSR and China, is not known in detail, but the broad picture is generally similar to that for the western region.

Gondwanaland. The vast Gondwanaland continent lay in the southern hemisphere during Carboniferous times, stretching from the austral to the antarctic climatic zones. The southern part of the continent was centered over the South Pole and during the upper Carboniferous suffered a major period of continental glaciation. Ice sheets and subsequent meltwater laid down typical stratigraphical sequences composed of boulder clay, gravels and sands. Over large parts of Gondwanaland the glaciers invaded a previous land surface and initiated a thick succession of mainly unfossiliferous continental deposits that extends from the late Carboniferous through early *Cretaceous times.

Outside the area of continental glaciation, wide shelf seas and sinking basins allowed Carboniferous deposits to accumulate round the margins of Gondwanaland. In eastern Australia, glacial deposits follow marine lower Carboniferous limestones: in the west these fossiliferous marine limestones are again well developed, but the upper Carboniferous is thin. A break in succession occurs widely at the top of the lower Carboniferous, and this was possibly caused by a lowering of the sea level at the beginning of the Gondwana glaciation (a lowering resulting from the extent of the *glaciation).

Other fragments of Gondwanaland – such as India – show the continental margin with glacial deposits and marine shelf sedi-ments. In peninsular India, glacial tillite is developed at the base of a thick sequence of sediments laid down on a land surface. Marine conditions are found in the Salt Range to the north, where fossiliferous late Carboniferous or Permian limestones overlie the tillite beds. Similar deposits are found in South Africa and South America.

In northern Gondwanaland, now North Africa, unstable shelf conditions at the margin of the continent allowed marine and continental deposits to be laid down. In the west, lower Carboniferous marine limestones are followed by continental deposits, but in the east early alternating continental coal-bearing deposits, with marine intervals are followed by a dominantly marine upper Carboniferous.

Conclusion. From the economist's viewpoint, the Carboniferous is the most important period of the Earth's past, for it was during the Carboniferous that much of the world's important coal-bearing deposits were laid down. To the paleontologist it is also a time of importance, since the period saw the spread of the amphibians and of the shark-like fishes, and, most important of all, the beginnings of the rise of the reptiles. The insects too, were evolving rapidly. All these evolutionary trends were to be continued and accentuated in the succeeding Permian.   GALJ

Permian

The Phanerozoic phase of Earth's history is beginning to emerge as an era encompassing one full cycle of global activity – commencing with a number of continental masses more or less widely scattered across the face of the globe, their gradual convergence to form a united supercontinent (Pangaea), followed by rebound and drifting apart of differently delineated continents to occupy their present positions. The individuality of the Permian lies in its having occupied the pivotal position during this accordion-like cycle.

The Permian, named for the Perm province of the USSR, was instituted and described by the British geologist Sir Roderick *Murchison in 1841 while carrying out a geological survey of the Russian Empire: he based the system on his studies of portions of a thick pile of strata exposed along the western flanks of the Ural mountains. The stratigraphic confines of the Permian have been intermittently stretched throughout the past century, and even now no precise, internationally accepted definitions of its upper and lower limits have been reached. However, we can consider its start to have been around 280 million years ago, its end around 225 million years ago.

Stratigraphic Definition. Standard reference sequences for the preceding *Carboniferous and succeeding *Triassic systems may be based on more or less uninterrupted *fossil records, but there is no such simple solution for the Permian – small wonder in view of the unsettled

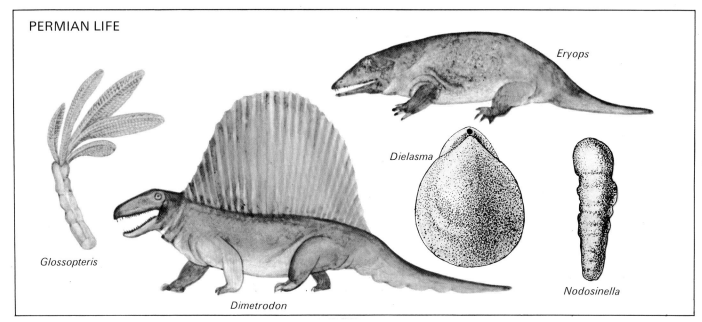

PERMIAN LIFE

Glossopteris

Dimetrodon

Dielasma

Eryops

Nodosinella

history of the Earth's crust during the period's course. A composite reference sequence has therefore been established, incorporating the most complete, suitably exposed, comprehensively studied marine sections from widely scattered areas. The lower Permian is thus represented by the southern Uralian Geosyncline, the middle Permian by the Texan Sea and the upper Permian by the Iranian Sea.

The lower stratigraphic limit of the Permian is generally taken at the contact between the Orenburgian Stage (uppermost Carboniferous) and the Asselian Stage (lowermost Permian) of the southern Uralian Geosyncline. An apparently uninterrupted marine succession here spans the boundary between the two systems. Marine strata of the uppermost Permian (Ali-Bashian Stage) are known only from the Iranian Sea

and the South China Sea and in each area an apparent interruption (possibly relatively brief) occurs between these strata and the overlying marine beds of the Griesbachian Stage (lowermost Triassic). For this reason, no mutual boundary can be defined between the Permian and Triassic systems.

There are four main phases of the Permian: early lower, late lower, middle and upper; though exact definitions of the boundaries between them are a subject for debate.

Tectonic Events. The major tectonic and physiographic features of the Permian world can be most effectively portrayed in the light of the theory of ★plate tectonics. We may consider five major continental plates being actively thrust up against one another through the persistent inward push of a series of encircling oceanic plates.

By the end of the middle Permian the union of these five plates had been well established. The continued subduction of the leading edges of the oceanic plates beneath the continental margin of the supercontinent so formed, Pangaea, now caused widespread uplift and the draining of virtually all major seas. In the late Permian we thus encounter a single sea-less supercontinent, a phenomenon probably unique in the history of the Earth.

Climate. With the virtual closure of the ocean between Laurasia and Gondwanaland toward the end of the Carboniferous, oceanic and atmospheric circulation patterns were radically altered, and this had far-reaching effects on global climatic conditions. Strongly differentiated climatic belts emerged on the continents, which were now strung out to form a broad arc

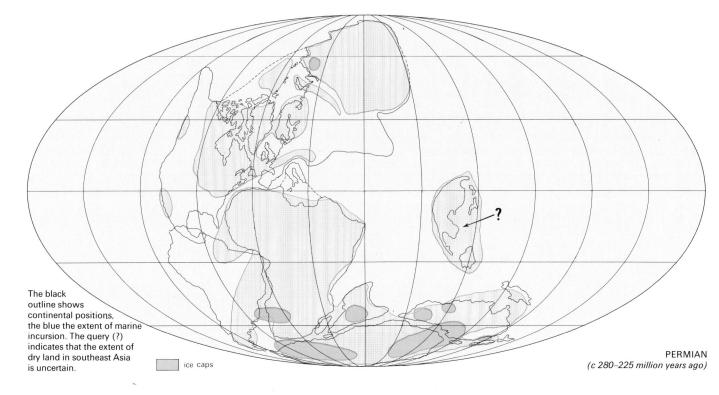

The black outline shows continental positions, the blue the extent of marine incursion. The query (?) indicates that the extent of dry land in southeast Asia is uncertain.

ice caps

PERMIAN
(c 280–225 million years ago)

Permian Kaibab limestones seen in the Grand Canyon. The limestones cap Paleozoic and Precambrian rocks and are seen on the horizon in this view. They are lighter in color than the older, redder rocks.

spanning one face of the globe from the south to the north poles.

During the opening phase of the Permian this configuration of the continents favored the extensive spread of ⋆glaciation in Gondwanaland, matching that which was to ravage the northern continents during the ⋆Pleistocene. The westward-flowing south equatorial warm-water current was deflected southward into the oceanic gulf between Africa and India, encouraging steady evaporation of water from the surface and subsequent snowfalls in the southern reaches of Gondwanaland. A continental ice sheet developed over Antarctica, which then as now straddled the South Pole, while a number of smaller dislocated ice caps spread from the peripheral upland regions.

The north equatorial current was deflected northward into the seas of the Cathaysian kingdom where moist, warm conditions were encountered in latitudes considerably farther north than might otherwise be expected. Hot dry conditions, with the development of extensive deserts, prevailed over much of the Euramerican kingdom and northern Gondwanaland. The southern part of Angaraland (roughly equivalent to today's Asian Russia) fell in the hot dry equatorial belt, while its northern parts lay in a warm-temperate humid zone.

The global climatic framework altered through the Permian in harmony with the migration of the united continents northward through some 15° of latitude: as one would expect, the climatic belts are generally seen to have shifted southward relative to the continents. In Gondwanaland, glaciation soon subsided and conditions became progressively warmer, while in Angara and northern Cathaysia they correspondingly became steadily cooler. The desert environment of Euramerica and northern Gondwanaland persisted throughout the Permian but shifted south with the Equator.

Flora. Global floral distribution patterns since the appearance of the first primitive land plants followed a trend of general homogeneity during the ⋆Devonian, a gradual development of provincialism through the Carboniferous which rapidly reached a peak in the middle phases of the Permian, followed by a reversal to greater uniformity in the ⋆Triassic and ⋆Jurassic. The considerable pressures brought to bear on the vegetation by the development of sharply differentiated climatic belts during the early lower Permian may be invoked to explain this diversification. Four floral kingdoms (Gondwana, Cathaysia, Euramerica, An-

gara), broadly reflecting the climatic zonation, can be distinguished.

Euramerican Kingdom. The earliest Permian floras are not markedly different from those of the late Carboniferous, the primary change being the decline of the arborescent lycopods and arthrophytes (⋆horsetails) which dominated the earlier coal swamps. The most notable newcomer was the widespread seed fern *Callipteris*, particularly *C. conferta*, whose first appearance is usually regarded as marking the start of the Permian. The ⋆conifers, represented by *Walchia*, a forest tree uncommon earlier, became locally common.

These early floras characterized the basins associated with the Appalachian ranges and the areas bordering the Texan and Donetz seas. Inland from these vegetated areas nothing is known of the flora, which was possibly very sparse owing to the desert conditions.

The sole preserved late lower Permian material derives from the region of the Texan Sea, surrounding which there were three distinctive floral provinces: one along the western slopes of the ancestral Rockies, the other two along the southeast and northeast shores of the sea. Each was dominated by a particular genus of seed fern (*Supaia*, *Gigantopteris* and *Glenopteris* respectively). The conifer *Walchia* was still prominent in the region; while the other representatives of the floras were not dissimilar from those of the early lower Permian.

Middle Permian floras are even more scanty, being known only from fragmentary material carried into the Zechstein Sea. The flora, clearly adapted to hot dry conditions, consisted chiefly of coniferous forest with the new genus *Ulmannia* dominating.

Upper Permian floras are unknown.

Cathaysian Kingdom. The flora of Cathaysia, similar to that of Euramerica during the earliest Permian, assumed an independent character thereafter, particularly through the appearance, diversification and dominance of the gymnosperm (see *plants) *Gigantopteris* and its relatives. This group of plants clearly favored warm moist conditions. Endemic species of fern and seed fern appeared, while the conifers were virtually absent.

Cathaysian plants are known from widespread localities through all four phases of the Permian. However, the western extension of the kingdom into the Tethyan region is based on very conjectural plant occurrences.

Angaran Kingdom. This kingdom shows progressive expansion and differentiation into regional floras during the early and middle Permian, by which stage four distinctive provinces are discernible: a Siberian province forms the core, and lies surrounded by the three peripheral buffer provinces of Petchora, Ural-Kazachstan and Far East. The warm temperate Siberian province appears to have been particularly characterized through most of the Permian by forests of *Cordaites*. Endemic species of fern, seed fern and horsetails were also much in evidence. The buffer provinces displayed features transitional in nature with Euramerica and Cathaysia.

Due to lack of information it is difficult to ascertain the situation in the upper Permian.

Gondwana Kingdom. The Gondwana flora was markedly distinct from those of the northern continents. The newly evolved division of plants, the *Glossopteridophyta*, unknown in the north, dominated the scene in the south. These plants owed their establishment to the spread of glaciation, which had almost annihilated the earlier vegetation. For a brief spell after the retreat of the glaciers the genus *Gangamopteris* stood as the chief representative of the division, but it was soon supplanted by the closely related *Glossopteris* which by the late Permian was the truly dominant element of the flora throughout the kingdom. Lycopods were not uncommon in certain areas during the earlier phases of the Permian, but became extremely rare later. The horsetails generally became more – rather than less – common, often forming dense bamboo-like stands in the swamps of the late Permian. The ferns and seed ferns were nowhere as significant as in the northern floras. The cycads and ginkgos were rare, as in the north, and only began to show their true mettle in the early *Mesozoic. The true conifers were likewise uncommon, but the closely related Cordaitales, with large parallel veined leaves, flourished around the coal swamps during the earlier phases of the Permian. *Noeggerathiopsis* was the southern representative of this latter plant group, and has at times been considered as identical with *Cordaites*, its northern counterpart.

All woody plants of the Gondwana Kingdom exhibit annual growth rings, and this reflects the seasonal nature of the prevailing climate.

Tetrapod Vertebrates. As the curtain rose on the Permian, the *reptiles were gaining in the war for supremacy over the *amphibians; and the scene was restricted to the equatorial belt traversing the southern part of the Euramerican continent. The theme that was to unfold during the Permian was of reptilian ascendancy and of successive waves of colonization until, by the close of the period, most of Pangaea was inhabited.

Throughout the lower Permian the tetrapods (*tetra*, four; *pod-*, foot) were confined exclusively to the paleotropics of Euramerica. The aquatic fauna was probably still dominated by the amphibians, some of which were of giant proportions, though life on the land was now securely in the thrall of the reptiles.

The varied reptile faunas fall very largely into two orders; the *cotylosaurs (stem reptiles descended directly from the amphibians) and the pelycosaurs (the forerunners of the *mammal-like reptiles). A bizarre characteristic of a number of the pelycosaurs was the development of a huge sail-like structure adorning the back. One such reptile, *Dimetrodon*, was over 3m (10ft) long and had a massive skull: an active predator, it held sway over its contemporaries. *Edaphosaurus*, somewhat smaller and with a less massive head, was the herbivorous counterpart.

The only early Permian tetrapod known outside Euramerica is *Mesosaurus*, a small, long-snouted amphibious reptile that fed on small *crustaceans. This enigmatic interloper, which put in an appearance in Gondwanaland for only a brief spell, was spawned by unknown stock.

Beginning near the early/middle Permian boundary, extensive colonization and adaptation occurred for several million years. Only two assemblages of middle Permian tetrapods are known, one from the Kazanian stage of the Russian Platform and the other from the slightly younger *Tapinocephalus* zone of the Karroo Basin, South Africa. Both represent diverse faunas flourishing on broad, low-lying swampy plains adjacent to extensive inland seas. The two faunas were similar in general though quite different in detail; and they reveal very marked advances over those of the early Permian. The reptiles had by this time achieved complete supremacy, amphibians being low in both numbers and diversity. The cotylosaurian and pelycosaurian orders of reptiles were still present, but poorly represented: the *mammal-like reptiles had assumed dominance.

Colonization beyond the Euramerican and western Gondwana continents was in all probability barred during the middle Permian by the sea and mountain tracts of the Uralian and Cape/Natal geosynclinal belts.

In the upper Permian the principal fossil tetrapod assemblages derive once again from the Russian Platform and the Karroo Basin. The faunas are still superficially similar. Now, however, with the final retreat of the Uralian and Cape seas, sporadic migration occurred for the first time into Angara, Cathaysia and east Gondwana, where (very sparse) remains have been encountered.

*Mammal-like reptiles continued to completely dominate the fauna but within the order significant evolutionary advances had taken place. The Dinocephalia had died out, while the dicynodonts had flourished and diversified to fill the available herbivorous niches. Among the carnivores the most significant newcomers were the cynodonts (dog-tooths), from which progressive stock, late in the Triassic, arose the *mammals.

Marine Invertebrates. Three major faunal realms, controlled by water temperatures, can be recognized in the Permian seas – paleotropical Tethyan, northern cool-water Boreal and southern cool-water Gondwana.

Mixing of Boreal and Tethyan faunas occurred in the southern reaches of the Uralian Geosyncline during the early lower Permian. A narrow passage enabled warm Tethyan waters to enter the essentially cool-water Uralian Sea. Thereafter the passage was blocked. The Tethyan realm in the east extended far into northern latitudes indicating the circulation there of warm waters. Recognition of the three faunal realms, and correlation of strata within each or from one to the next, rests primarily on the study of three very different groups of animals – the fusulinids, ammonites and brachiopods.

The *fusulinids, a superfamily of the order Foraminifera, phylum Protozoa, appeared in the early Carboniferous and became extinct at the close of the Permian. They were abundant and widespread, evolved rapidly and attained a remarkable degree of diversity. They were small (measured in millimetres) inhabitants of the sea bottom, and flourished in relatively shallow clear water away from the shores. They dwelt principally in warm-water seas.

During the early lower Permian the fusulinids were present, often in great numbers, throughout the Tethyan and Boreal realms. They were conspicuously absent in the glacial waters of the Gondwana realm (other than for an isolated occurrence along what is now the Chilean coast of South America).

In the late lower Permian, presumably due in large measure to the northward drift of the continents, the fusuline faunas of the Boreal realm became impoverished and included only cosmopolitan genera. The

Tethyan faunas, on the other hand, became further diversified, with a larger number of endemic genera. The fusulinids still enjoyed no success in the penetration of the Gondwana realm where the waters were still unfavorably cold.

By the middle Permian, with continued northward drift, they had become extinct in the Boreal realm. They remained absent in the Gondwana realm (except for occurrences in what is now New Zealand); but continued to be abundant and still further diversified in the Tethyan realm. The Iranian and Southern China seas were the only seas of significance to persist after the widespread regressions of the middle/late Permian transition. In these last remaining oases the fusulinids became progressively less diverse and abundant, and this continued until their extinction by the end of the Permian.

The ubiquitous retreat of the seas played havoc in all walks of marine life, so that the faunas that appear in the widespread early Triassic transgressive seas were radically altered in aspect. This is in marked contrast to the situation at the onset of the Permian, when marine life (and indeed terrestrial life) continued through from the Carboniferous essentially unchanged.

Some sixteen families and seventy genera of *ammonites are recorded from the Permian. This degree of diversity is relatively low when compared to their much greater differentiation in the Mesozoic. They are presumed to have been highly mobile, free-swimming animals which inhabited primarily shallow waters in paleotropical latitudes, though a number of cosmopolitan Permian genera, of great value in global correlations, occurred.

The ammonites showed greatest profusion in the Tethyan realm. The best exposed sections, yielding abundant ammonite assemblages throughout the lower Permian, occur in the southern Uralian Sea. At the close of the early Permian the Uralian Geosyncline ceased to subside and the sea began to recede, with the resultant development of extensive *evaporite deposits. Ammonites are thereafter rarely encountered in this area.

The best developed ammonite-bearing sections of the middle Permian are to be found around the mouth of the Texan Sea, which in turn began to dry out at the close of this phase. From here we hop to the northern margins of the Iranian Sea for well exposed, well developed, ammonite-yielding strata representing the late Permian. As we have already seen, these three sections have been chosen as reference for the lower, middle and upper Permian respectively.

Ammonites occur sparsely and in low diversity in the cool-water realms. A few cosmopolitan genera are occasionally encountered at scattered levels through the lower Permian (and lowest middle Permian) in the marginal seas of eastern and western Australia. These comprise the most tangible props for establishing correlations

between the sections of the Gondwana and Tethyan realms, the remaining bulk of the Gondwana marine invertebrates being essentially endemic to the realm. In the middle Permian, rare ammonite finds are made in the geosynclinal seas lining the northwestern and northern margins of Euramerica and in the Himalayan, Madagascar, and New Caledonian regions of Gondwanaland.

The *brachiopods, which constitute an independent phylum, have persisted from the *Cambrian to the present, but are now on the verge of extinction. During the Permian they occurred in great abundance and diversity in both paleotropical and cool-temperate seas, where they generally lived firmly attached to irregular rocky bottoms in the off-shore reaches of shelf seas. They are of considerable value in correlations within any particular realm but, since few forms appear to have displayed a sufficiently wide tolerance of different water temperatures to enable them to colonize beyond their parent realms, they are of little value in broader correlation.

At the end of the Permian there was a massive extinction of the brachiopods: of over 125 genera only two survived the start of the Triassic.

Conclusion. The dramatic chain reaction of change seen to reverberate through the Permian, sparked by the meeting of the northern and southern continents, brought the Paleozoic to an unequivocal close and ushered in the new Mesozoic era, during which the world was to see the beginnings of the mammals' slow rise to dominance. The true significance of the Permian's end can perhaps be seen from the name of the era which ended with it: the Paleozoic, or era of ancient life. JMA

Mesozoic

The Mesozoic, the era of "middle life", lasted some 160 million years, from about 225 to about 64 million years ago. It encompassed the Triassic, Jurassic and Cretaceous periods. Early in the era the first mammals appeared, but faunas as a whole were dominated by the reptiles, some of which achieved massive proportions. AI

Triassic

The Triassic period lasted about 35 million years, beginning about 225 million years ago and ending approximately 190 million years ago. It was the first period of the Mesozoic, an era characterized by the appearance of faunas and floras strikingly different from those of the preceding Paleozoic era, a change particularly marked among the marine invertebrate faunas, because it coincided with the widespread extinctions that took place during the late Paleozoic. These late Paleozoic extinctions and the subsequent expansion and diversification of the new marine faunas in the

Triassic are generally attributed to the reduction, during the late Paleozoic, in the area of the seas covering continental margins, the principal habitat for these faunas, and the subsequent expansion of the shelves during the Mesozoic.

The changes in the faunas were also reflected in the floras. The late Triassic marked the appearance and predominance of new groups. These changes, taken together with the extinction of much of the late Paleozoic flora, were probably associated with a considerable change in climate.

Paleomagnetic and facies evidence suggests that the position and configuration of the continents changed very little from the *Permian to the Triassic period. One difficulty in proving this is that few complete late Permian to early Triassic marine successions are known, and consequently very little evidence is available on which to reconstruct the geographical details of early Triassic seas. However, two Triassic world-wide marine margins can be identified, namely the circum-Pacific and Tethys.

In absolute contrast to the marine shelves and basins around the one supercontinent, Pangaea, were the huge landmasses of its two components, called Laurasia and Gondwanaland, which were joined along the lines of the incipient mid-Atlantic ridge and western Mediterranean. These two continental masses were characterized by extensive terrestrial deposits in which the remains of reptiles and other vertebrates, and of plants, were preserved.

Biostratigraphy. In the Alpine region the marine Triassic rocks are traditionally divided, on the basis of their *ammonoid faunas, into five stages.

One of the main problems of Triassic biostratigraphers has been the lack of a marine rock succession in which there is unambiguous fossil evidence of an unbroken sequence of rocks from Permian through to Triassic. Various Tethyan localities have been investigated, especially in the Himalayas, the Salt Range (Pakistan) and eastern Iran, as well as circum-Pacific localities in northeastern Siberia, North America and the Arctic. One result of these studies has been the introduction in recent years of four new stage names, to replace the single Scythian Stage representing the lower Triassic rocks, although so far these new stages have been mainly applied to the North American and Arctic successions where they are defined on the basis of their faunal sequences.

Subdivision of Triassic rocks is further complicated by the presence of very extensive sedimentary basins in Laurasia and Gondwanaland filled with terrestrial deposits and lacking a marine fauna. The presence of land plants offers the possibility of correlating between the marine and non-marine deposits by means of their fossil microflora, but many of these terrestrial deposits have been so oxidized that suitable microfossils are rare or absent. Furthermore, the task of establishing a useful

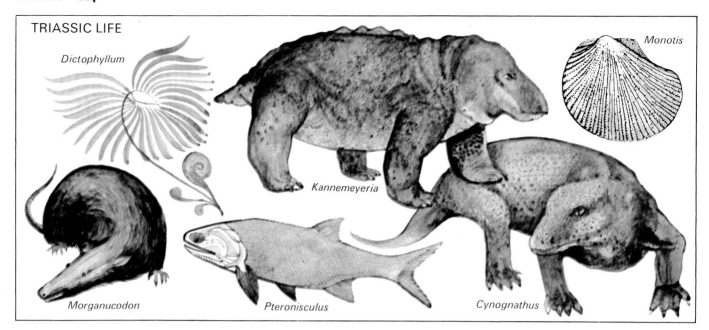

TRIASSIC LIFE

Dictophyllum

Monotis

Kannemeyeria

Morganucodon

Pteronisculus

Cynognathus

correlation scheme between Triassic microfossils and ammonoids has not yet been satisfactorily achieved: it remains a task for future biostratigraphers.

Fossils. Triassic fossil organisms valuable in providing evidence of the environment in which they lived are often much closer related to modern forms than are the Paleozoic fossils, so they can be used to interpret the environment with greater confidence.

During the Triassic the marine faunas expanded. ★Mollusks were the dominant invertebrates, of which the ammonoids are undoubtedly the most useful because they became widespread and abundant and displayed great evolutionary changes in their easily recognized and preserved shell morphology. The ammonoids had become almost extinct in the late Permian, but during the Triassic they rapidly evolved and diversified before approaching extinction again in the late Triassic.

After the ammonoids the lamellibranchs, a class of ★bivalve mollusks, were the most widely distributed and diverse group of Triassic invertebrates. Some, such as *Monotis*, are useful as guide fossils and are widely used for correlation in the upper Triassic rocks of both circum-Pacific and Tethyan marine deposits.

The special value of colonial ★corals is their toleration of fairly narrow limits of water depth, salinity and temperature; and consequently they are used to interpret the environment in much greater detail than most fossil groups permit. The Triassic corals are of particular use in this respect because during the middle Triassic there appeared in the Alps and the Mediterranean region the new order of scleractinian corals, to which modern reef-building corals belong. The main Paleozoic coral groups became extinct at the end of that era, but curiously there is an apparent absence of any record of early Triassic corals. This may be associated with the absence of hard parts in the ancestors of the scleractinian corals, but it remains a field for future investigation. By the end of the Triassic these new corals had become widespread throughout the world.

The main terrestrial and aquatic vertebrates were ★reptiles and ★amphibians, and these do not appear to have undergone important evolution in passing from the Permian to the Triassic period. The reptiles, especially the ★dinosaurs, increased in importance during the Triassic, becoming more diverse and numerous. ★Mammal-like reptiles, whose ancestors

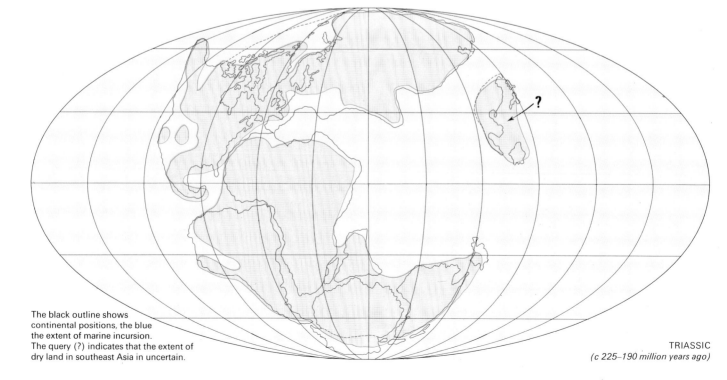

The black outline shows continental positions, the blue the extent of marine incursion. The query (?) indicates that the extent of dry land in southeast Asia in uncertain.

TRIASSIC
(c 225–190 million years ago)

Salt pseudomorphs in marls indicate the hot, dry conditions that existed in Triassic deserts. Pseudomorphs are the result of the filling of spaces left by crystals after they have been dissolved away.

can be traced back to the *Carboniferous, became widespread and common during the Triassic: they, like the ammonites, became nearly extinct by the end of the Triassic but survived to become the ancestors of the modern *mammals.

From the late Carboniferous up until the beginning of the Triassic, four floral provinces can be distinguished: the European, Angaran and Cathaysian warm provinces to the north of Tethys, and the Glossopteris cold-temperate province to its south.

The early Triassic floras appear to have been sparse and poor, and this is usually interpreted as a result of the unfavorable climate. Furthermore, they were mainly survivals from the Permian period. As many early Triassic terrestrial deposits are red beds with *evaporites, indicating arid or semi-arid hot conditions, the early and middle Triassic floras were probably concentrated around lakes and rivers where *Neuropteridium* ferns grew with conifers like *Voltzia*, while horsetails such as *Schizoneura* grew in the shallow waters. In contrast to this warm flora of Laurasia the southern continent of Gondwanaland was characterized by the cold-temperate *Glossopteris* flora.

The late Triassic floras reflect an evolutionary change by becoming more abundant and varied, with many new groups appearing: the main plant groups were conifers, ferns, cycads and ginkgos.

Paleogeographical Maps. During the Triassic period there were only two main continental margins, the circum-Pacific, which could be regarded as an external margin, and the Tethys, which appears as an internal margin, the two being separated by the huge land areas of Laurasia and Gondwanaland.

The available paleomagnetic data (see *geophysics) suggest that throughout the Triassic period there was no significant movement of Pangaea. This phase of only very slight plate-tectonic activity appears to have lasted from about 250 million years ago to about 170 million years ago.

The configuration of Pangaea during the Triassic, with Gondwanaland occupying a very southern and hence cool position, had a major influence on the world's climate. This climatic influence is recognizable in the types of rocks deposited – for example, desert dune sands and evaporites – and in the distributions of fossil animals and plants. Another major paleogeographical influence of this supercontinent was the greatly reduced length of coastline throughout the world: this had the effect of restricting the total area and distribution of marine shelves along the coastlines of the Pacific and Tethys, and influenced the distribution of faunas because marine shelves were the principal habitat of the great majority of the creatures that have come down to us as fossils.

One of the complicating factors that have affected the interpretation of Triassic paleogeography is the widespread effects of the late *Mesozoic and *Cenozoic orogenies, resulting from the convergence and in places the eventual collision of parts of the north and south margins of the Tethys ocean. This collision resulted in the Alpine, Himalayan, IndoBurman and Indonesian *mountain chains, where most of the evidence of the Tethyan marine Triassic rocks is exposed.

Triassic Paleogeography, Tectonics and Igneous Activity. Although most of the circum-Pacific continental margin of Pangaea remained mobile throughout the late Paleozoic and Mesozoic, the Triassic was a particularly quiet period. Much of the Tethys margin remained passive throughout the period. On a world-wide basis the tectonically quiet Triassic separates the late Paleozoic orogenic events that characterize the Pacific margin of the continents from the next major tectonic phase of the middle and late Mesozoic, when the conglomerated continent of Gondwanaland was split up by the opening of the new Atlantic and Indian Oceans.

There were three Triassic major paleogeographical provinces, namely the huge land area of Laurasia and Gondwanaland, and the two ocean margins of circum-Pacific and Tethys.

The Circum-Pacific Margin. In North America the geosynclinal belts that con-

Buttes formed by the remains of the Jurassic Navajo sandstone deposits overlying Triassic red beds in Utah, USA.

stitute the Triassic margin pass into the Canadian Arctic region where the Svedrup Basin contains thick Triassic miogeosynclinal facies (see *geosyncline): the same is true of Triassic deposits on the eastern side of the Siberian platform. The Cordilleran belt of North America must have represented an active continental margin during the Triassic, as indicated by the volcanic rocks, which were important in the central sector, where *andesite, *rhyolite, *dacite and pyroclastic deposits are preserved. Some small Triassic *granite intrusions represent an early stage of plutonism.

In the Antarctic peninsula (West Antarctica) there appears to have been strong *folding and some metamorphism (see *metamorphic rocks) during the Triassic. This orogenic phase may be related to the history through the Triassic of eugeosynclinal deposition and volcanism in New Zealand, when it was adjacent to West Antarctica: during the Mesozoic New Zealand was much closer to eastern Australia and the Tasman Sea had not yet formed.

The circum-Pacific margin of Gondwanaland extended around eastern Australia into New Guinea where it connected with the Tethys ocean margin of Gondwanaland. The continental margin of New Guinea seems to have been, like all of the southern margin of Tethys, a tectonically passive feature throughout the Triassic.

The Northern Margin of Tethys. The Alpine cycle, representing the accumulation of sediments and igneous rocks and resulting in their deformation and uplift into the Alpine mountain chain, began in the Triassic period. It started with marine transgressions over the eroded Hercynian mountain complexes and with the development of marginal troughs within and adjacent to the continental margin. The marine deposits are mainly *limestones, *dolomites, *sandstones and *shales. The transgressions which characterized the Alpine Triassic marine deposits occasionally invaded the continental hinterland of Laurasia, where the "Germanic Trias" facies of red beds and evaporites accumulated. By the end of the Triassic the Tethys sea had extended over the adjacent forelands covering the older Triassic red-bed facies.

In the Himalayas there appears to be an important gap in the sedimentary sequence at the Permian-Triassic boundary. Extensive carbonate and shale deposits, with some volcanics, represent shelf and deeper basinal sediments that accumulated on the northern margin of Tethys.

Unlike the quiet shelf and basin deposits that characterize much of the northern margin of the Alpine-Himalayan Tethys, the Triassic deposits of the IndoBurman ranges, Indonesia and the Philippines contain unmistakable evidence of an active continental plate margin (see *plate tectonics). This Triassic convergent plate margin extended around southeast Asia from Burma through the Philippines and the Ryuku arc to Korea and Japan.

It is in the Japan region that the northern margin of Tethys met the circum-Pacific margin.

The Southern Margin of Tethys. On the southern side of the present Mediterranean, in the Atlas mountains of Morocco and Algeria, are found Triassic rocks very similar to those of the Alpine mountain chain. As on the northern margin of Tethys, Triassic marine transgressions also invaded the older "Germanic Trias" red-bed facies, which was deposited on both the African and European continental blocks.

The correlation of the Atlas mountains to the east is uncertain. They may belong to the facies belts that pass through the Apennines, Southern Alps, Dinarides and Hellenides. It is not easy to determine what elements belong to the southern part of Tethys. In Libya, Egypt, northern Israel and on the Arabian peninsula Triassic marine-shelf deposits accumulated, and the sea probably extended onto the Ethiopian shield.

The southern margin of Tethys can also be identified in western Australia and on the present northern Australian shelf, where land-formed deposits lie in alternating layers with open marine-shelf sediments that accumulated around this part of Australia and New Guinea.

Laurasia. Much of the North American continent was above sea level, and suffered active *erosion during the Triassic. In late Triassic times, fault-bounded troughs formed in eastern North America in which great thicknesses of terrestrial sediments accumulated together with some basic volcanic rocks. The interior lowlands also preserve some records of the terrestrial deposits that accumulated thinly over large areas during the Triassic. In the west, along the margins of the present Cordillera, thin sequences of non-marine red beds with evaporites are found.

The European landmass, like that of North America, consisted of the eroded late Paleozoic mountain ranges as well as extensive lowlands, such as the North Sea Basin, where the detritus accumulated mainly as red beds with evaporites. Many, but not all, of the basins of *deposition were fault-bounded troughs. There is evidence that these Triassic sediments accumulated under hot arid and semi-arid conditions, which in northwest Europe gradually diminished in aridity during the Triassic, a waning perhaps associated with the encroachment of Tethys from the south.

The Siberian continental block, unlike the European region, was the site of an important igneous episode during the Permian and Triassic. *Plateau basalts and associated basic dikes and sills were erupted, forming a layer up to 1km (0.6mi) thick and covering about 1,500,000km² (575,000mi²). These Siberian "Traps" may be compared with the roughly contemporaneous plateau basalts of southern Africa, which then formed part of Gondwanaland.

Gondwanaland. Thick Triassic Gondwanaland deposits are found in large basins and in thinner widespread sequences. Large basins of terrestrial facies were developed in the late Triassic in northeastern and eastern South America, and are present in southern Africa where they consist mainly of sandstones and shales. Late Triassic plateau basalts have also been identified in South Africa. In India extensive continental red beds with a fauna of reptiles and amphibians have been dated as Triassic. In Antarctica, similar facies of terrestrial sandstones with reptile and amphibian fossils have been dated as Triassic on the basis of the plant remains that they contain.

Very similar facies were deposited in Australia, especially along the eastern margin, where sandstones of river or stream origin are known with *coals. Some volcanic activity of Triassic age is reported from the hinterland of eastern Australia. In early Triassic times small marine transgressions reached into the western and eastern margins of Australia. The present wide northwest Australian shelf region was, during most of the Triassic, the site of Tethyan marine infiltrations with the typical "Gondwana" facies.

Summary. Much of the circum-Pacific margin of Pangaea was an active plate margin during the Triassic, and consequently the site of tectonic and igneous activity, as was the northern margin of Tethys, east of the Himalayas and extending from Burma to Japan. The southern margin of Tethys and most of the western part of the northern margin of Tethys appear to have been a passive margin during the period, and so this region was tectonically quiet, with only a very little in the way of volcanic activity.

Within the continental blocks of Laurasia and Gondwanaland, very little igneous activity has been dated as Triassic. One exception is provided by the late Triassic volcanic rocks and dikes of the fault troughs of eastern North America. Parts of some dike swarms in west Africa and northern South America may also be of Triassic age and are possibly associated with the activity in eastern North America; and all this may be related to doming in the incipient mid-Atlantic ridge that was to form in the late Mesozoic. Other igneous activity on the continental blocks was concentrated in the two widely separated areas of southern Africa and Siberia, where plateau basalts were erupted.   MGA-C

Jurassic

The Jurassic, named for the Jura Mountains of Switzerland, began about 190 million years ago and ended about 136 million years ago. There were then only two major continents, Laurasia in the north and Gondwanaland in the south. During the early part of the period these were united at their western end and separated by a major equatorial ocean called Tethys, which widened towards the east.

Late in the early Jurassic, that section which we now know as northwestern Africa moved away from North America, so creating a narrow sea, the precursor of the modern Atlantic, which continued to widen throughout the remainder of the period. It was not until the Cretaceous that the North and South Atlantic and Indian Oceans came into existence.

Sea-floor spreading (see *plate tectonics) almost certainly took place in the Pacific region, but the evidence is largely lost today because the present Pacific floor is underlain by *basalt generated later.

Sea-level was comparatively high in the Jurassic: large areas of the present continents were inundated by shallow seas, and the land appears in general to have been fairly low-lying. Early in the period sea-level was at its lowest: subsequently the sea advanced more or less progressively to flood more and more of the lower-lying continental areas, reaching a maximum toward the end of the period when something like 25% of the present continental area was covered. However, in the very final stages of the Jurassic, the sea withdrew markedly from the continents, a process which continued into the Cretaceous. There are no reliable estimates of the precise range of sea-level differences during the Jurassic, but it was probably no more than about 200m (660 ft). The basic cause of these changes was probably vertical movements of the mid-ocean ridge systems, resulting in the displacement of seawater (see *ocean floor).

Most of what is now Western Europe was covered by sea for almost the entire period, and the same is true for what are today the margins of the Pacific. However, much of Gondwanaland was never flooded, the sea being restricted to shallow bays or straits. Most of North America also remained uncovered, but a shallow sea spread from the Pacific into the Western Interior of the United States and Canada during the middle and early part of the late Jurassic. Another area of persistent land was what we now know as Eastern Asia.

The climate of the Jurassic was appreciably more equable than that of the present day. There were probably no polar icecaps, temperate conditions extending as far as the Arctic and Antarctic. Similarly, the climatic conditions which today characterize the tropics extended well north and south spreading as far as, for example, Western Europe.

However, it should be borne in mind that details of climate zones, or for that matter of wind and ocean current distributions, have not yet been adequately worked out.

Vertebrate Fauna. The vertebrate terrestrial life of the Jurassic period was dominated by the reptiles. The *dinosaurs had first appeared late in the Triassic from a *thecodont stock which also gave rise to *pterosaurs and, later, birds. From small bipedal animals such as Coelophysis there evolved the huge, spectacular creatures familiar to us all. These include the herbivorous Brontosaurus, Brachiosaurus, Diplodocus and Stegosaurus as well as the carnivorous, bipedal Allosaurus.

Only two rich dinosaur faunas are known from Jurassic deposits, the Morrison Formation of the American Western Interior and the approximately contemporary Tendaguru Beds of Tanzania. The two faunas are strikingly similar at family and generic level, which would suggest rather strongly that free land communications existed between western North America and East Africa until quite late in the period – a fact that is not easy to reconcile with some modern paleogeographic reconstructions.

Flying animals include the truly reptilian pterosaurs and the first animals that could be called *birds as distinct from reptiles, as represented by the pigeon-sized Archaeopteryx.

There were two important groups of reptiles which lived in the sea, the dolphin-like *ichthyosaurs and the long-necked *plesiosaurs. Both of these groups, specimens of which are displayed in many natural history museums, had streamlined bodies and limbs beautifully adapted to marine life. *Turtles and *crocodiles are also found as fossils in Jurassic deposits.

Jurassic mammals, known mainly from their teeth alone, were small and obviously did not compete directly with the dinosaurs. They included a number of biologi-

JURASSIC LIFE

Rhamphorhynchus

Archaeopteryx

Gryphaea

Williamsonia

Arnioceras

Antrodemus

Stegosaurus

Macroplata

Cuspiteuthis

cally primitive groups such as the triconodonts, docodonts and multituberculates (see *primitive mammals). The fish faunas were dominated by the holosteans, characterized by heavy rhombic scales. Their evolutionary successors, the teleosts, probably appeared shortly before the end of the period (see *actinopterygians).

Invertebrate Fauna. Because they are far more abundant, the invertebrate fossil faunas of the sea are of more importance to stratigraphers and paleoecologists than are the vertebrates. By far the most useful for stratigraphic correlations are the *ammonites, a group of fossil *mollusks. They were swimmers that lived in the open sea, only rarely braving the fluctuating salinity and temperature of inshore waters. They are characteristically most abundant in marine *shales and associated fine-grained *limestones.

From a solitary family that recovered from near extinction at the close of the Triassic, there radiated an enormous diversity of genera. Many of these were worldwide in distribution, but increasingly throughout the period there was a geographical differentiation into two major realms. The Boreal Realm occupied a northern region embracing the Arctic, northern Europe and northern North America. The Tethyan Realm, with more diverse faunas, occupied the rest of the world.

Taking into account the evidence of the whole invertebrate fauna, together with the paleogeographic picture that can be deduced from the sedimentary rocks, it looks as though the differentiation into Tethyan and Boreal realms most probably developed as a result of the operation of a number of independent factors, of which possibly the most significant was irregular fluctuations in the salinity and temperature of the shallow seas close around the shores of the continents. In other words, the Tethyan Realm was marked by more environmentally stable open-sea conditions than was the Boreal Realm.

During the Jurassic, the ammonites evolved rapidly and had a high rate of extinction. No fewer than 65 zones of fossil ammonites are recognized in northwest Europe, which has become the classic region for study of this period. As the Jurassic lasted for some 65 million years, it would seem that each of these zones represents, on average, a duration of the order of a million years.

Where ammonites are absent, as in all

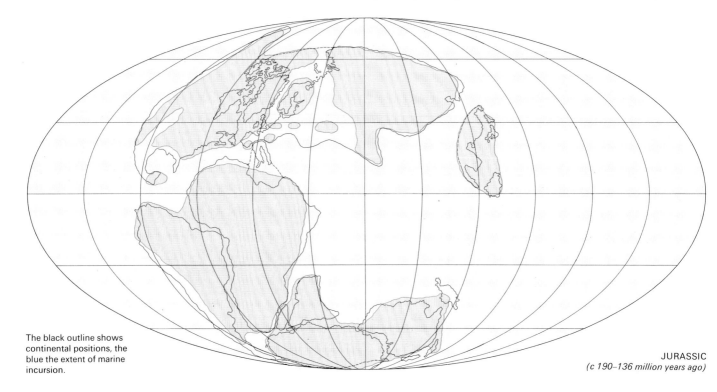

The black outline shows continental positions, the blue the extent of marine incursion.

JURASSIC
(c 190–136 million years ago)

Limestone concretions in shales deposited during early Jurassic (Liassic) times on the coast of Yorkshire, England.

non-marine and certain shallow inshore marine deposits, or where the rocks are known only from boreholes, a variety of other fossils have had to be used for correlation: of these, the most useful have proved to be *foraminifera, *ostracods, plant spores and *dinoflagellates.

In most facies the *bivalves, which flourished in shallow, muddy sea bottoms, are the most abundant and diverse of the microfauna. They included many cemented forms such as *Ostrea*, recliners like *Gryphaea*, swimmers such as the pectinids and limids and rock borers such as *Lithophaga*. However, the majority were burrowers; either relatively mobile, shallow burrowers or forms occupying deep, permanent burrows and normally found still in their positions of growth.

*Brachiopods were much more abundant and diverse than they are today. The range of depths below the surface which they occupied is far wider than for the bivalves, and a definite depth zonation can be established in Europe, just as with the ammonites.

*Echinoderms are best represented as fossils by the crinoids and echinoids, and were all inhabitants of shallow seas, unlike some of the modern representatives of this class. The echinoids include both primitive regular forms, like the cidaroids, and irregular forms, such as *Clypeus* and *Pygaster*.

*Corals belonged to the still extant Scleractinia group, and included reef builders such as *Isastrea* and *Thamnasteria*. Calcareous and siliceous *sponges are also quite common locally, even forming reefs. It seems likely that these sponge facies developed in somewhat deeper water than did the corals.

The invertebrate microfaunas are represented by abundant foraminifera, ostracods and radiolaria – foraminifera and ostracods are of great value to oil companies in correlation studies.

Not all Jurassic invertebrates lived in the sea. Some lived in continental environments such as lakes and rivers; they include a few genera of bivalves, gastropods and arthropods. These faunas are far less diverse than their marine counterparts.

Flora. With regard to the plant kingdom, the Jurassic might well be called the age of the gymnosperms, the non-flowering *plants, forests of which covered much of the land. They included the conifers, gingkos and their relatives, the cycads. *Ferns and *horsetails made up the remainder of the land flora. These and others of the Jurassic flora are still extant today in much the same forms.

Remains of calcareous *algae are widely preserved in limestone. Besides the laminated sedimentary structures produced by blue-green algae and known as oncolites and stromatolites, there are skeletal secretions of other groups. Many pelagic limestones have revealed themselves under the electron microscope to be composed largely of tiny chalky plates, known as coccoliths, which are secreted by certain planktonic algae called *coccolithophores.

It seems highly likely that the late Jurassic saw the emergence of the *angiosperms, the flowering plants, since well-developed flowering plants existed in the very early Cretaceous. However, it is not yet understood quite how they emerged, and a satisfactory direct evolutionary ancestor has yet to be identified with any degree of certainty.

Stratigraphy. Most of the Jurassic sedimentary rocks that can be observed today were laid down in the extensive shallow seas that flooded parts of the continents.

Western Europe has become classic ground for Jurassic stratigraphers because of the widespread development of marine and non-marine strata that are extremely rich in fossils. The succession in England is probably the most celebrated. This succession consists of an alternation of soft-weathering *clays or *shales with more

Reefs form in warm conditions. The reef deposits in northern European Jurassic strata show that equatorial conditions prevailed there.

resistant formations of limestones and sandstones. This alternation gives rise to a characteristic scarp and vale topography.

Only in the south of England is the Jurassic succession complete, with an unbroken transition to the Cretaceous, but even here the boundary beds are nonmarine, a local expression of the worldwide regression, mentioned earlier, of sea from land in the late Jurassic.

Looked at geographically, the principal facies change in Western Europe and Britain is from sandstone and sandy shale with subordinate bands of *ironstone in the north (Scotland, Scandinavia, northern England, northern Germany) to limestone and marls further south. This implies that the major source of land-produced sediment was in the north – probably one or more land masses including the Scandinavian Shield. Islands further south – such as the old Paleozoic horsts of the London-Brabant Platform, the Massif Central and the Brittany Massif – do not appear to have been major sediment sources.

The vertical sequence of alternations of clays with sandstones and limestones signifies changing depths of sea. The clays were laid down in relatively deep (around 100m (325ft)), quiet water and include thin layers of shale, rich in organic matter, which were deposited in stagnant or near-stagnant water. Sandstones and limestones

were laid down in shallower water.

The extent to which the inferred changes in depth of water signify local tectonic subsidence or worldwide sea-level changes is still in dispute. However, evidence is accumulating to suggest that some of the major changes in the classic European sequences were controlled primarily by eustatic uplift.

For the countries around the Mediterranean an interesting history of tensional tectonics has recently been worked out. In the late Triassic an extensive carbonate platform had developed, somewhat resembling the Great Bahama Bank. On this were laid down several thousand metres of extremely shallow-water deposits of limestone. These conditions persisted – everywhere from southern Spain and Morocco to the southern Alps, Austrian Calcareous Alps and Apennines to Greece – until late in the early Jurassic. A widespread collapse then took place and extensive sectors subsided considerably, resulting in the formation of deeper-water deposits. The thicker, more basinal deposits are marly limestones with *trace-fossil mottling; the thinner deposits were laid down very slowly on structural highs (probably seamounts). These deposits were from deeper water than any known further north in Europe, and probably were laid down in depths of several hundred metres. They are overlain by even deeper-water deposits of middle to late Jurassic age, thought by many to have been laid down at depths of several thousand metres. The youngest Jurassic de-

posits are fine-grained pelagic limestones composed largely of coccolith debris. They have hardly any benthonic fossils but contain the aptychi of *ammonites. Traces of vulcanicity and sedimentary fissure fillings, together with dramatic lateral changes in sedimentary thickness, provide additional support for a general, self-consistent interpretation of Jurassic geological history.

The zone of Jurassic limestone facies continues eastward into the Middle East – as exemplified by the Zagros ranges of Iran and their continuation into Saudi Arabia. Here, however, the whole sequence is of shallow-water carbonate-platform type. Thick upper Jurassic limestones are also known in the southern part of European Russia and around the Gulf of Mexico. Elsewhere in the world, limestone facies are subordinate to sandstone and shales.

Another well-studied region of epicontinental sea deposits is the American Western Interior, embracing the Rocky Mountain states. Here the lower Jurassic is represented by several hundred metres of flat-lying red or yellow *sandstones, known in different states as the Navajo or Nugget Sandstone, that was laid down in hot desert and forms some of the more spectacular landscape features of that part of the world, such as Monument Valley. These are overlain by a series of shallow marine sandstones, shales and limestones with layers of gypsum and salt, of middle and early late Jurassic age, and mark the influx of the sea from the northwest. As with the lower Jurassic, deposits were much thicker in the

Badlands composed of Jurassic deposits in Arizona, USA.

west and were derived at least partly from tectonically rising land still further west.

The youngest Jurassic deposits of the Western Interior are a group of non-marine multicolored shales, siltstones, sandstones and subordinate limestones known as the Morrison Formation, celebrated for the rich dinosaur faunas it has yielded.

The fauna of freshwater *gastropods and bivalves and the flora of charophyte *algae (stoneworts) indicate that these are mainly the deposits of lakes. As in Europe, therefore, the late Jurassic was marked here by a withdrawal of the sea from the land, in this area beginning perhaps rather earlier.

The strata described so far were laid down mainly in areas of tectonic stability. However, around the margins of the Pacific – for instance in parts of California, Japan, New Zealand, Chile and Argentina and in the Caucasus and Balkan Mountains – the deposits contain a fair deal of volcanic material and the rocks are frequently folded, sheared or even metamorphosed. All this points to regimes of great tectonic instability and probably signifies the proximity of these areas to Jurassic subduction zones (see *plate tectonics). It seems likely that much of the old Tethys ocean floor was consumed along the line including the present Caucasus ranges.

One other part of the world that warrants mention in a regional survey is the Karroo Plateau of South Africa. Here there are thick basaltic and associated *igneous intrusions. This part of Gondwanaland witnessed huge outpourings of lava from late Triassic into middle Jurassic times. (The Tasmanian dolerites and Ferrar dolerites of Victoria Land, Antarctica, are essentially contemporaneous.) Extensive outpourings of plateau basalt indicate the onset of a tensional regime which anticipated the breakup, during the Cretaceous, of Gondwanaland.

Conclusion. The Jurassic was a period of relevance to us in several ways. Economically important deposits such as coal and ironstone were laid down. In the animal kingdom, the precursor of modern birds appeared, and creatures familiar to us, such as turtles and crocodiles, coexisted with the giant reptiles, doomed to become extinct late in the Cretaceous. It was also a period of notable plate tectonic activity, as the land continued its slow evolution toward our modern continents.   AH

Cretaceous

The word "Cretaceous" is derived from the Latin word *creta*, chalk, and indeed it is the Chalk Formations of Western Europe that are the best known part of the Cretaceous stratigraphic sequence. The sequence was christened *Térrain Crétacé* by J. J. Omalius d'Halloy in 1822 on the basis of his study of the chalky sediments of the Paris Basin and adjacent areas; and our name for it is an anglicization of this.

The Cretaceous is the third and last period of the *Mesozoic era and lasted from 136 to 64 million years ago, a span of 72 million years. It followed the Jurassic and was succeeded by the Paleocene, the first period of the *Tertiary era. From our point of view, the most important Cretaceous products are coal, oil and gas, but many Cretaceous rocks are also of economic significance: they are used in, for example, ceramics, building and in the making of cement. From the paleontologist's point of view, the Cretaceous marks an important transitional stage in the evolution of life, from the ancient forms of the Jurassic to the "modern" forms of the Paleocene, the ancestors of the lifeforms we know today.

The Cretaceous World. The distribution of continents and oceans during the early Cretaceous resembled that during the Jurassic, but sea-floor spreading during the Cretaceous produced fundamental changes (see *plate tectonics). The North Atlantic opened progressively throughout the period, with rifting extending to the north of Britain. To the south, during the middle Cretaceous, rifting along the margins of what was to become the South Atlantic produced extensive salt basins and a seaway (if not actual oceanic crust) appeared. The principal breakup of Gondwanaland, with the northward movement of India and the

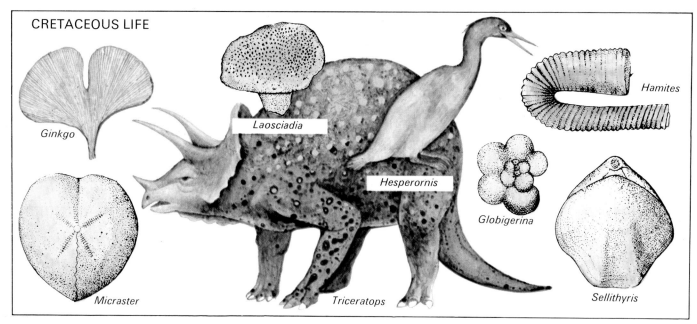

CRETACEOUS LIFE

Ginkgo

Laosciadia

Hesperornis

Hamites

Globigerina

Micraster

Triceratops

Sellithyris

separation of Antarctica, began during the period, while the floor of wide areas of the Pacific is also of Cretaceous date. Towards the close of the period the old seaway of Tethys was progressively eliminated as the African plate moved northwards to impinge upon the Asian plate, while the anticlockwise rotation of the Iberian peninsula produced the precursor of the Bay of Biscay.

Regression, the withdrawal of sea from the land, had begun in the late Jurassic, and it continued into the early Cretaceous. In many parts of the world, therefore, this is a time marked by *erosion or non-marine sedimentation. But renewed and progressive transgressive pulses, beginning during the late early Cretaceous, caused a cumulative rise in sea level of 600–700m (2000–2300ft) at minimum: this was a result of the construction of extensive mid-

ocean ridge systems and a surge in sea-floor spreading. By the end of the period wide areas of continental crust, including some which had not been submerged since the Precambrian, were flooded in the greatest marine transgression since the Ordovician – perhaps the greatest at any time of the Phanerozoic. Most of what is now Western Europe was covered by epicontinental sea, as were the margins of all of the other continental blocks: a seaway extended across the Sahara, central Europe was an archipelago with a gulf extending north to the Arctic ocean, while a major seaway extended the length of the Interior of the USA and Canada.

Cretaceous climates were equable and warm compared with those of the present. There appear to have been no polar icecaps – indeed, the poles probably experienced

warm, temperate conditions. Large reptiles were present close to the north pole, according to *fossil evidence in Alaska, while marine fossil faunas include forms which range from the then north pole through the tropics to high southern latitudes, suggesting mild climatic gradients. Reef-building *corals extended far to the north of their present ranges, as did giant *foraminifera, rudist *bivalves, and other warm-water indicators. Floras from Alaska include remains of cycads, palms and figs.

Flora. The land floras of the earliest Cretaceous, like those of the Jurassic, are dominated by gymnosperms – conifers, cycads and the like, resembling closely their living relatives. Late in the early Cretaceous, however, *angiosperms (flowering plants) became prominent and, by the close of the Cretaceous, forests in many areas were

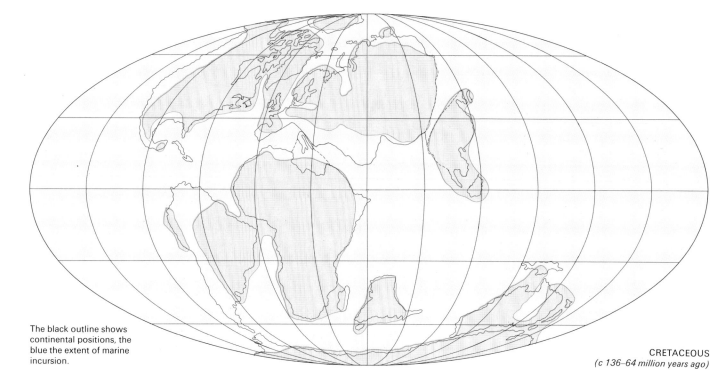

The black outline shows continental positions, the blue the extent of marine incursion.

CRETACEOUS
(c 136–64 million years ago)

Cliff dwellings built in Upper Cretaceous sandstones, Mesa Verde, New Mexico, USA.

dominated by deciduous, broad-leaved trees. By the middle Cretaceous up to 90% of floras consist of angiosperms with leaves which resembled modern beech, oak, maple, walnut, plane and magnolia, along with broad-leaved shrubs and the like. Cycads remained prominent in some upland floras, as did conifers. Various groups of marine calcareous *algae are abundant in shallow-water limestones. The chalk-producing algae (*coccolithophores) reached their peak in the late Cretaceous. Although known from the Jurassic onwards, it is only during the Cretaceous that coccolith sediments come to dominate deep-sea carbonate sequences, and only in the upper Cretaceous do they spread onto shelf areas. Chalk sedimentation spread over wide areas of Europe and North America, while chalks are known also from South America and Australia, indicating an extraordinary abundance of these microscopic organisms.

The late Cretaceous saw also the acme of the *dinoflagellates, another group of single-celled plants.

Fauna. The broad composition of Cretaceous invertebrate faunas resembles that of the *Jurassic. Protozoans are represented by *radiolarians and *foraminifera, the late Cretaceous seeing the diversification of the planktonic *globigerinids. *Sponges are a common element of the fauna of much of the European Cretaceous, and *brachiopods are in many places abundant. Coelenterates are also widespread, and the coelenterate faunas began to approximate to their modern appearance by the end of the period. *Gastropods spread during the late Cretaceous. Cretaceous *ammonite faunas were dominated by desmoceratids, acanthoceratids and diverse heteromorphs. The largest ammonite known is a late Cretaceous *Pachydiscus*. The first octopus (*Paleoctopus newboldi*) comes from the upper Cretaceous of the Lebanon.

Amongst the arthropods, many groups of crustaceans and many insects appear in much their modern forms; while the interdependence of insects and flowering plants probably developed early in the period. *Echinoderms are diverse and widely varied, with many irregular echinoids, asteroids and crinoids. Triconodont and symmetrodont *mammals died out during the Cretaceous, as did several groups of *birds. Fish faunas include chondrosteans, holosteans and cartilagenous sharks, while the teleosts undergo a major radiation during the period.

However, the most spectacular Cretaceous vertebrates are without doubt the *reptiles. Marine forms include *plesiosaurs up to 16m (50ft) long, and short-necked forms such as *Kronosaurus*, whose skull was some 4m (13ft) long. *Ichthyosaurs occur, but are far less significant than they were in the Jurassic; while a new group appeared, the mosasaurs, rather similar to the popular conception of sea-serpents, with paddle-like limbs and an array of sharp teeth. They reached sizes in excess of 10m (32ft). In addition, giant turtles (*Archelon*) with a carapace up to 4m (13ft) long have been found.

Also occurring in marine sediments, and best known from the chalks of Kansas in the US Western Interior, are the last of the flying reptiles, the *pterosaurs. One of the largest of these was *Pteranodon*, some species of which had wing-spans over 10m (33ft), the largest flying animals of all time. On land, the *dinosaurs dominated, with giant herbivores such as *Iguanodon*, the armor-plated *Ankylosaurus* and the horned *Triceratops*; and *Tyrannosaurus*, the largest known terrestrial carnivore, which reached heights of 15m (48ft).

The end of the Cretaceous, like the end of the Permian over 150 million years before, was a period of widespread extinction. At least sixteen higher taxa of animals died out by the close of the Cretaceous, including the rudistid *bivalves, the euomphalacean, trochonemalacean and nerineacean *gastropods, all *ammonoids, the *ichthyosaurs, *pterosaurs and sauropod dinosaurs, while the belemnitids died out early in the Paleocene. At a lower taxonomic

A chalk cliff on the Dorset coast, England. The chalk was deposited in a shallow water environment, and the flints, seen in regular horizons throughout the succession, formed later as a result of water dissolving and redepositing silica.

belt are major thrusts, some up to 400km (250mi) long, with lateral displacements of up to 35km (22mi), and a total crustal shortening of up to 110km (70mi). More important is its continuation, the Laramide orogeny, first named from the Laramie region in Wyoming, which affected the area west of the Sevier orogenic belt. This orogeny was responsible for the major structure of the Rocky Mountains, although their topographic relief is largely the result of post-Cretaceous uplift and erosion. Structures include major folds, with total vertical movement in some areas of over 10km (6mi); and in places Precambrian basement has been exposed.

This great orogenic belt extends for the length of North America, and there is major associated igneous activity (see *igneous intrusion), including the intrusion of enormous granite batholiths. These include the Baja, California, Sierra Nevada, Idaho, Boulder and Coast Range batholiths, the last-named having a total length of over 1500km (930mi). Evidence of *vulcanicity can also be seen in the USA: volcanic ash bands occur throughout the whole of the Interior, and are the principal source of radiometric dating of the late Cretaceous.

In South America, there was a similar phase of mountain-building during the late Cretaceous, again accompanied by major igneous intrusive and extrusive activity, along the line of the Andes and the Antillean Arc of Central America. Extrusive igneous rocks associated with the opening of the South Atlantic occur widely in the Paraná Basin of Brazil, covering an area of a million km^2 (400,000mi^2), with a maximum thickness of 1.5km (0.93mi). Smaller areas of similar, contemporaneous volcanics occur in Namibia.

In the Middle East, there were widespread movements in the early Cretaceous, and some granitic intrusion in Iran, with late Cretaceous ophiolites in the same area. In the Himalayan region, there was major submarine *vulcanicity at this time. Late Cretaceous volcanics are known from western Australia, while there were widespread volcanism and major granite intrusion in the Japanese islands. In the eastern Antarctic, there is evidence of extensive Cretaceous vulcanicity and intrusive activity.

In contrast, the deformation affecting western Europe north of the Alps was generally on a small scale. *Faulting affected the rift systems of the North Sea, while minor *folding can also be recognized. Igneous activity was also on a small scale. Bentonites (altered volcanic ashes) are known from southern England, while much of the *montmorillonite in the chalk may have a similar origin.

In the Alpine region, the initial phases of compressional deformation associated with

level, many families of forams, bivalves, echinoderms and other invertebrates went into decline or disappeared, while there was a major reduction in the diversity of many planktonic micro-organisms, including *dinoflagellates, *coccolithophores and *radiolaria.

The cause of the late Cretaceous extinctions is not fully understood, and many suggested mechanisms – such as variations in solar radiation, or epidemic disease – are of course untestable. It is, however, obvious that whatever mechanism was responsible, it affected not only marine organisms but also terrestrial forms – from coccolithophores to dinosaurs. Among recently advanced hypotheses are: reversals

in the Earth's magnetic field, which might temporarily reduce protection from cosmic radiation; fluctuating climatic conditions associated with global sea-level changes; or even phases of mountain building.

Tectonic and Igneous Activity. The best known Cretaceous orogenic (mountain-building) areas are in the Americas. The first of these in North America was the Sevier orogeny, which takes its name from the Sevier arch in southwestern Utah. Beginning in the early Cretaceous, and extending to the close of the period, it affected a belt running from southern Nevada across Utah into the southwestern corner of Wyoming and southeastern Idaho. The most notable features of this the closure of Tethys, precursors of the main Tertiary deformation phase, appear during the Cretaceous.

Northwestern Europe. The earliest Cretaceous paleogeography of northwestern Europe, the type area for the period, closely follows that of the late Jurassic. As a result of continued regression, non-marine deposition occurred in areas such as southern England, northern France and northwest Germany. Marine deposits of the Boreal Sea extend over northern England to the Soviet Union, while to the south, in southern France and parts of the Alps, are marine sediments laid down on the northern margins of Tethys. Land areas at this time included the Paleozoic blocks of the Anglo-Brabant Massif, Massif Central, Armorica, Vosges, Hartz, Baltic Shield and much of Wales, Scotland and Ireland.

Uplift during the early Cretaceous ended deposition of fine-grained clays and limestones in areas of non-marine deposition, and these are succeeded by a thick (up to 800m (2700ft)) sequence of fluvial and deltaic *clays and sandstones, best known from the Weald of southeastern England.

The English Lower Greensand, dating from a little later, marks a transgression by the sea, and takes its name from the widespread occurrence of the mineral *glauconite; and although most of it is neither green nor a sand, its name stresses one of the features of the northwest European Cretaceous, the wide occurrence of glauconite sediments. This, together with the widespread occurrence of *chert and sedimentary phosphates, is believed to be a reflection of changing patterns of ocean currents, and upwelling associated with the initial opening of the North Atlantic.

By the early late Cretaceous, a combination of transgression, reduced relief, and perhaps low water run-off had resulted in a great reduction in supply of land-derived sedimentary material. This, combined with a peak in diversity and abundance of the *coccolithophores, led to the widespread deposition of chalks, which dominate upper Cretaceous sedimentation in an area extending from Ireland to the Caspian. Many of the land areas were greatly reduced in size by this time. Non-chalk facies are chiefly known from areas such as the southern Paris Basin, Ireland, and parts of Germany: they included cross-bedded sandstones, greensands and calcarenites.

The European Chalk was originally regarded as a deep-sea deposit, comparable to *globigerina ooze. It is now known to have accumulated at comparatively shallow depths, between 50 and 300m (160–975ft), although the closest modern counterparts are deep-sea oozes (see *ocean floor). One of the most striking features of the sequence are flints, horned nodules of black silica with white rinds of partially siliceous material. Once regarded as some sort of a primary precipitate, they are now known to be diagenetic in origin, the silica being derived from sources such as *sponge skeletons, and possible *diatoms and *radiolaria (see *diagenesis).

In the North Sea area, drilling has shown that chalk sedimentation extended into the early Tertiary. In Denmark also some Paleocene chalks occur.

Conclusion. The Cretaceous was a period of evolution rather than of revolution. At its end, which signified the beginning of the Cenozoic and the end of the Mesozoic, dramatic changes took place. It was the end of the age of the giant reptiles, the beginning of the era of modern life. The dinosaurs and the ammonites were extinguished, but the first true birds appeared. The mammals, one day to give birth to man, had remained small and shrew-like throughout the Cretaceous, but their inexorable rise was to come with the beginning of the Cenozoic.

Perhaps it would not be too much to say that the most significant element of the Cretaceous was its ending.   WJK

Cenozoic

The Cenozoic is commonly described as "The Age of Mammals", and as one might expect from this nickname its most important characteristic has been the evolution of modern forms of life.

The Cenozoic has lasted for around 64 million years, and is commonly subdivided into two periods, the *Tertiary and the *Quaternary, each of which is further subdivided. The latter period, the Quaternary, is of special interest as it has seen the rise of modern Man.

Many of the major mountain ranges of the world have formed during the Cenozoic. Early in the era the Indian subcontinent was pushed against Asia, the crumpling of the land resulting in the Himalayas; and rather later the similar thrust of Italy, then part of Africa, against Europe resulted in the formation of the Alps. The later part of the era has been characterized by a series of Ice Ages from which, there is good reason to believe, we may not yet have emerged. AI

Tertiary

The name "Tertiary" was introduced by Arduino in the middle of the 18th century to denote the various rock types younger than the Mesozoic, the "Secondary". It became an internationally recognized time-stratigraphic unit when Desnoyers, in 1829, distinguished the *Quaternary as a separate period. Together Tertiary and Quaternary comprise the Cenozoic Era.

The Tertiary had a duration of about 60 million years, beginning around 64 million years ago and ending only 2 million years ago. *Lyell in 1833 subdivided the period into four epochs, the Eocene, Miocene and the Older and Newer Pliocene, later substituting the name Pleistocene for the last of these. (The *Pleistocene is now more usually considered as part of the Quaternary.)

Further subdivisions, the Paleocene for the earlier part of Lyell's Eocene, and the Oligocene for the earlier part of Lyell's Miocene, were contributed by later workers to complete the modern subdivision of the period.

The first part of the Tertiary, up to the end of the Oligocene, is sometimes called the Paleogene; and the later part, the Miocene and Pliocene (and sometimes the Pleistocene) are by some workers referred to as the Neogene.

The Tertiary can be thought of as the period characterized by the rise of modern *mammals. At the beginning of the period archaic forms predominated, but rapidly these were superseded by more and more modern forms. By the end of the Tertiary there were certainly man-apes in Africa. But the end of the period was characterized by cooling climates and a massive extinction: the emergence of modern Man himself was a matter retained for the Quaternary.

Paleocene

The Paleocene was named by W. P. Schimper in 1874 for continental and brackish-water beds in the Paris Basin. The beds chosen as typical of the Paleocene are, in actual fact, of early *Eocene age, and this has led to some problems concerning the real meaning of the term Paleocene. The Paleocene is now accepted as spanning the time between 64 and 54 million years ago, approximately.

The epoch is generally agreed to have two stages: the earlier Danian, named in 1847, the type area being Denmark; and the Thanetian, named later in 1873, the type area being the Isle of Wight, England. The Danian type area consists of limestones developed in a variety of facies, such as coral reefs, *bryozoan mounds (bioherms) and coccolith chalks (see *coccolithophores). The Thanetian type area is in contrast developed in a silty, glauconitic facies. (The Montian and Landenian stages of Belgium, used in parts of Europe as standard stages, may be considered chronologic equivalents of the Danian and Thanetian, respectively.)

A major floral and faunal discontinuity occurred between the end of the *Cretaceous and the beginning of the Paleocene: important elements of oceanic plankton disappeared, as did many ammonites, belemnites, mollusks, marine reptiles, and so on.

Significant paleogeographic changes occurred in the world during the Paleocene. The Norwegian Sea began to form as a result of the separation of Greenland and Scandinavia along the axis of the Reykjanes Ridge, a northern extension of the Mid-Atlantic Ridge. During the Cretaceous, the Atlantic Ocean, its borderland margins and large parts of the surrounding low-lying continents which were covered by shallow inland seas, had been characterized by widespread deposition of calcium carbonate under climatically equitable conditions; but, in the Paleocene, local mountain building accompanying the birth of the northeast

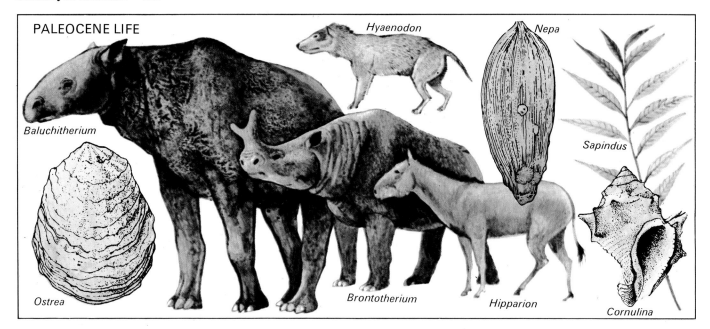

PALEOCENE LIFE

Hyaenodon

Nepa

Baluchitherium

Sapindus

Ostrea

Brontotherium

Hipparion

Cornulina

Atlantic changed the sedimentation patterns considerably, particularly in the marginal areas, with land-derived sediments becoming an important factor.

The North and South Atlantic had respectively opened to about 95% and 75% of their present-day width through sea-floor spreading (see *plate tectonics). The Atlantic and Arctic oceans were separated from each other, and a northward-flowing, warm "proto-Gulf Stream" brought subtropical faunal and floral elements as far north as the Labrador Sea and Rockall Bank. The Atlantic and Pacific were linked by the westward-flowing North Equatorial Current through the straits of Panama, which separated North and South America; it contributed to circumglobal transportation of tropical marine faunal and floral elements. In the southern hemisphere, Aus-

tralia separated from Antarctica and began its northward flight.

Climate was rather equitable all over the Earth, with evidence of subtropical floras and faunas as far north as the present-day latitude of London. Evidence of climatically controlled cycles is seen, however, in the deep-sea record of the Atlantic Ocean, with alternate equatorward and poleward expansion and retreat of high-latitude and low-latitude oceanic planktonic floras and faunas. There is no evidence of polar ice-caps.

Significant changes in the organic realm distinguish the Paleocene from the Cretaceous. Paleocene mammalian faunas were characterized by multituberculates, creodonts, condylarths (see *primitive mammals) and the earliest prosimians (ancestral forms of the *primates). Migration of mam-

malian faunas between Europe and North America occurred *via* a polar route between Spitzbergen and Greenland, and this resulted in virtually identical faunas in these two areas. However, trans-Eurasian migration was inhibited as a result of a north-south interior seaway – the Uralian Sea – which connected the polar regions with the equatorially situated Tethys Sea.

The marine realm was characterized by a rapid radiation of calcareous plankton following their virtual extinction at the close of the Cretaceous. Large (5–10mm (0.2–0.4in) diameter) tropical shallow water *foraminifera – known as *Nummulites* – evolved within the Paleocene and continued to flourish throughout the *Eocene; they were to decline in the early *Oligocene and become extinct at the end of the middle Oligocene.

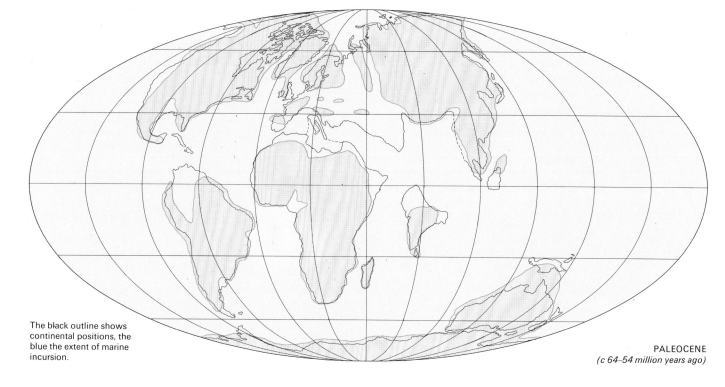

The black outline shows continental positions, the blue the extent of marine incursion.

PALEOCENE
(c 64–54 million years ago)

Eocene

*Lyell, in 1833, originally introduced the term Eocene to describe the lowest subdivision of the Tertiary, including the rocks between the *Cretaceous and *Miocene in which only a small (less than 5%) number of living molluscan species were believed to occur.

Numerous stage terms have been introduced over the years to correspond to a threefold subdivision of the Eocene. The most commonly accepted stages include (from oldest to youngest) the Ypresian, Lutetian and Bartonian. The Cuisian and Priabonian are commonly used synonymously with upper Ypresian and Bartonian, respectively.

The sediments of the type areas of the Eocene stages of northern Europe are deposited in shallow-water environments of the continental shelf. These sediments are generally characterized by abundant molluscan faunas, using which local stratigraphic correlations are possible. Faunas of nummulites, an extinct group of *foraminifera, are locally abundant at certain stratigraphic levels and enable regional correlation to be made.

The stratigraphic position of the Paleocene/Eocene boundary is controversial because of confusion surrounding Schimper's original characterization of the former epoch (see *Paleocene). He appears to have considered parts of the Ypresian as Paleocene in age in one instance, whereas in another he included them in the Eocene. As a result the Paleocene/Eocene boundary is drawn by some stratigraphers at the base of the Cuisian. This results in the stratigraphically undesirable practice of having the Ypresian stage straddle the boundary, its lower part being of Paleocene age.

In his original concept of the Eocene, Lyell included in its upper part the marine sands of Fontainebleau (Paris Basin). He subsequently extracted these rocks from the upper Eocene and included them in the lower *Miocene. In 1854 the term "Oligocene" was introduced for the time-stratigraphic interval between the Eocene (as modified by Lyell) and the Miocene and this term naturally included the sands of Fontainebleau and their equivalents.

The stratigraphic position of the Eocene/Oligocene boundary is also controversial. This stems from the recent, somewhat contradictory, evidence that the sands of Lattorf (East Germany), commonly assigned to the early *Oligocene, may in actual fact be of late Eocene age.

With these qualifications the Eocene is generally accepted as spanning the time between approximately 54 and 38 million years ago.

The final fragmentation of the Eurasian continent occurred about 50 million years ago with the separation of Greenland and Scandinavia near Spitzbergen. It is still uncertain whether there was communication between the Atlantic and Arctic oceans at this time, or whether a land barrier extended transversely from Greenland to Scandinavia in the vicinity of present-day Iceland and the Faeroe Islands.

One of the outstanding characteristics of the Eocene record in the oceans is the formation of extensive deposits of organically-derived silica, which implies the presence of waters rich in nutrients and a high marine population at this time. The distribution of these deposits in the Atlantic is primarily along the continental margins and across the Atlantic and Pacific along the (paleo)latitudes of the (paleo)equatorial belt, implying transportation by the warm-water oceanic current systems of the time.

Extensive areas of the continental margins – particularly in the Tethys Sea, which extended from the Indo-Pacific region to the Atlantic Ocean – were developed. The extraction of large amounts of silica by the oceanic plankton was balanced by the carbonate-rich environment developed in the shallow, marginal areas of Tethys, in which the developing nummulite faunas flourished (see *foraminifera): over 5000 years ago the Egyptians quarried the extensive deposits of these rock-forming fossils, which are exposed in the vicinity of Cairo, to build the pyramids.

During the middle and late Eocene the Pyreneean orogeny led to the formation of the east-west mountain range, the Pyrenees, between Spain and France. Although subtropical conditions existed in Europe as far north as the Paris and London Basins, the first evidence of (at least) minor *glaciation in the southern hemisphere is seen in the oceanic record near Antarctica. A significant worldwide cooling event occurred about 38 million years ago (Eocene/Oligocene boundary) with the formation of significant sea-level glaciation on Antarctica. The temperatures of the deeper-water layers (below 1000m (3300ft)) were lowered by several degrees, and as a result the thermal structure of the oceans as we know it today was essentially formed at this time. This played a significant role in climatic modification, as the movement of water masses over the surface of the globe is one of the major factors affecting climates.

In addition to the proliferation of the shallow-water, tropical faunas of nummulites, the Eocene is characterized in its early part by a great diversification of marine planktonic faunas and floras, followed by a gradual but inexorable decline in diversity, and the extinction of various lineages in the middle and later parts.

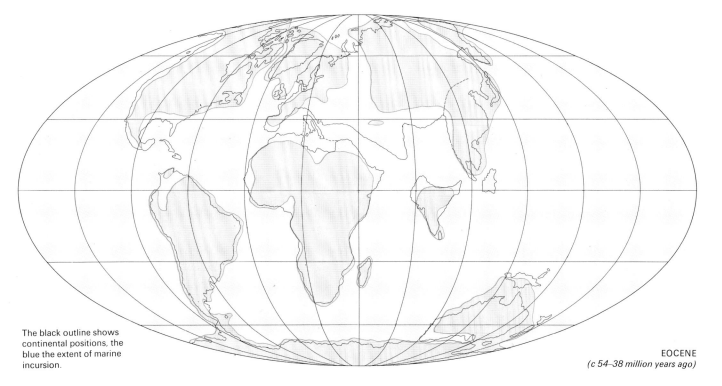

The black outline shows continental positions, the blue the extent of marine incursion.

EOCENE
(c 54–38 million years ago)

Flysch deposits consisting of shales and sandstones at San Sebastian, Spain. These rocks were deposited in the Tethys Sea during Eocene times.

Among the *mammals the Eocene is characterized by the initial appearance and rapid radiation of such groups as *rodents, *artiodactyls, *perissodactyls, *carnivores and certain *subungulates (elephants and related forms), *edentates, and *whales; and the extinction of amblypods, multituberculates, creodonts and condylarths (see *primitive mammals). Early Eocene mammalian faunas in Europe and North America were virtually identical; but from the middle Eocene onward the two faunas were totally isolated, a reflection of the disruption of the polar communication route by the separation of Greenland and Scandinavia. The continued presence of the north-south Uralian Sea in Eurasia, running from the polar regions to the equatorially situated Tethys Sea, resulted in continued separation of European and Asian land-mammal faunas. Current studies suggest that limited exchange of land mammals between the southern and northern shores of Tethys took place.

Oligocene

The term Oligocene was introduced in 1854 to describe strata intermediate in age and position between Lyell's *Eocene and *Miocene. These Oligocene strata were originally believed to be the result of a marine transgression which had covered a large part of northern Europe, but in fact the early Oligocene is characterized by world-wide regression – probably due, at least in part, to extensive glaciation on Antarctica – and much of the Oligocene is missing, or present only in marginally deposited shallow-water facies over a large part of the Earth.

The Oligocene, which covers a span of roughly 14 million years, from 38 to 26 million years ago, was a time of significant geographic, climatic and faunal changes. Isotopic studies indicate that major climatic cooling occurred at the Eocene/Oligocene boundary, 38 million years ago, and again during the mid-Oligocene, from 32 to 27 million years ago; and these events are reflected in significant invasions of equatorial latitudes by normally high-latitude calcareous plankton floras. The Oligocene was a time when the diversity of oceanic planktonic faunas and floras was generally low, and it represents a transitional period during which early Tertiary faunas and floras were gradually dying out or evolving into the late Tertiary, modern faunas and floras. The once prolific nummulitic fauna (see *foraminifera) became extinct at the end of the early Oligocene, and its ecologic niche was occupied later in the epoch by another group of "larger foraminifera" – the miogypsinids – which were to flourish in the Miocene.

In the southern hemisphere the separation of Tasmania and the South Tasman Rise from Antarctica, about 30 million years ago, removed the last barrier to what is now one of the major factors governing global oceanic circulation and climatic conditions because of the way that it mixes the waters of all the oceans, the circum-Antarctic Current. It appears to have remained relatively stable since its inception in the mid-Oligocene.

The epoch was characterized by the continued rise of the *mammals. There were, for example, the appearances of *Anthracotherium*, an *artiodactyl similar to the wild pig, with strong incisors and large canines; the rhinoceros (*Aceratherium*); and the tapir. Also in the Oligocene occurred the expansion of the prosimians and the disappearance, at the close of the epoch, of the titanotheres, a group of large *perissodactyls with concave skulls and prominent horny protuberances above the nose. The closure of the Uralian Sea in Eurasia permitted trans-Eurasian migration of mammalian faunas; and, at least intermittently, migration to and from North America *via* the Bering landbridge was possible.

Miocene

A time-stratigraphic term created by *Lyell in 1833, Miocene describes those strata in Europe and equivalents of the same age elsewhere which were believed to contain 20–40% of the species of *mollusks that are still extant. Numerous stage names have been applied within the epoch, but the following (from oldest to youngest) are these in most widespread use at present: Aquitanian, Burdigalian, Langhian, Serravallian, Tortonian and Messinian. The Miocene lasted some 19 million years, from about 26 to about 7 million years ago. Most of the major aspects of contemporary Earth history can be traced to this epoch.

Major Miocene Paleogeographic Events. The junction of Eurasia and Africa about 18 million years ago resulted in the interruption of the east-west Tethyan seaway and the origin of the Mediterranean Sea as we know it today. The immigration into Eurasia of African elephants, bovids and pigs in the early Miocene was the result of this dramatic closure of the Tethys seaway. The eastern part of Tethys evolved into what is now the Indian Ocean as India collided with Asia and the Himalayas were formed. In Europe this was the period of major Alpine orogeny, leading to the present-day Alps and related mountain chains to the east.

A major East Antarctic Ice Sheet formed in the late Miocene about 10 million years ago, and subsequently expanded, about 6 million years ago, to a size considerably in

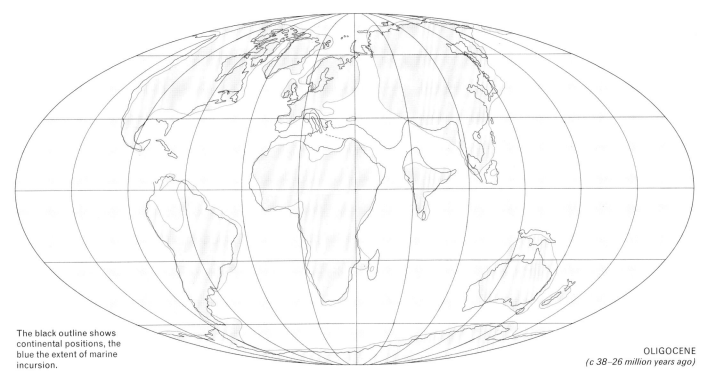

The black outline shows continental positions, the blue the extent of marine incursion.

OLIGOCENE
(c 38–26 million years ago)

excess of its present-day dimensions. This latter event was linked with the isolation of the world's oceans and the evaporation and desiccation of the Mediterranean Sea in what must be considered one of the most dramatic events in the history of the Earth: the opening of the Straits of Gibraltar 5 million years ago marked the end of the "salinity crisis", as Atlantic waters were able to flow once more into the Mediter-

ranean Basin. However, the once rich marine fauna and flora of the Mediterranean Sea, which had migrated southward along the west African Coast after having been expelled as the Mediterranean grew ever saltier during the late Miocene, never completely succeeded in reestablishing itself, and the Mediterranean today contains only an impoverished representation of the luxuriant life that thrived there during the

An eroded volcanic landscape in South Yemen. The peaks seen here are the remains of volcanoes that were active during late Tertiary times.

Miocene.

The initial appearances of the planktonic globigerina (see *foraminifera) and the larger benthonic foraminiferal group of miogypsinids which replaced ecologically the *Nummulites*, occurred rather before the

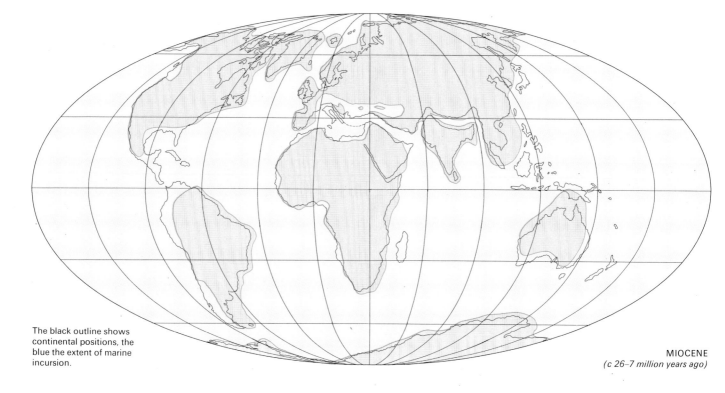

The black outline shows continental positions, the blue the extent of marine incursion.

MIOCENE
(c 26–7 million years ago)

beginning of the epoch. The upper limit of the Miocene is characterized by the extinction of a number of marine microfaunal taxa, but it is more significantly denoted as a major paleogeographic boundary: in the Mediterranean area the youngest Miocene is characterized by *evaporite deposits (anhydrite, gypsum and salt) which were formed in deep basins isolated from the world ocean. This evaporative phase, which lasted less than 1 million years, is known as the "salinity crisis" because of the extreme effect which it had upon the marine fauna and flora present in the late Miocene Mediterranean. A return to normal marine conditions denotes the base of the Pliocene in this area.

Life. The Miocene is characterized by the appearance and radiation, or continued development, of faunal and floral elements that were gradually to become our contemporary fauna and flora. Calcareous planktonic microfaunas and microfloras became important once again following their decline in the Oligocene, and their diversity has been shown to bear a close relationship to the temperature of the oceans.

Among the plants, sequoia, taxodium, cypress, poplars, oaks, laurels, camphor and palms thrived in central Europe. *Mammals reached their height during the Miocene, in particular the elephants, which underwent a remarkable diversification: the epoch witnessed the appearance of mastodons with four tusks (*Tetrabelodon*), and of *Deinotherium*, a gigantic, 5m-high aberrant elephant with lower incisors recurved towards the rear, which persisted in Africa up to the Pleistocene. Deer, anthracotheres and pigs frequented the oak-forest environment while tapirs and rhinoceros wallowed in low-lying swamp areas. Ruminants with twisted horns (such as antelopes) appeared among the *artiodactyls and occurred together with the simple-horned deer and giraffids. Monkeys appeared by mid-to-late Miocene time and the early separation of the basic primate stock towards the hominoid branch, which was eventually to lead to the *hominid lineage and Man himself, can be traced to within the epoch. Hyaenids, civets and cats lived in the forests and prairies. Crocodiles, tortoises, large salamanders and a variety of fish (e.g., perch, eels) inhabited inland waterways.

The horses continued their evolutionary trend towards modern forms, gradual reduction of the lateral toes occurring along with the development of teeth with a high crown and deep socket. The early Miocene *Merychyppus* and its late middle Miocene descendant *Hipparion* appeared in North America and emigrated to Eurasia *via* the Bering landbridge. The latter form appeared about 12 million years ago, and its appearance is one of the most useful and widespread stratigraphic indicators in mammalian biostratigraphy.

Pliocene

The name Pliocene was introduced by Sir Charles *Lyell in 1833 to denote those European strata (and, by extension, their equivalents elsewhere) in which more than half the species of *mollusks are still to be seen alive today. He changed his mind in 1839, and created the term *Pleistocene for the "Newer Pliocene", in which 90–95% of the various species of mollusks are still extant. The name Pleistocene is still with us, even although Lyell himself later tried to suppress the term in favor of "post-Pliocene" or "post-Tertiary".

In terms of geologic time, the Pliocene was an extremely short epoch. Its start can be considered as around 7 million years ago, and its end, and the beginning of the Pleistocene, as around 2 million years ago; so its total duration was only some 5 million years – little longer than some of the stages of other Cenozoic epochs. But, despite this, it was an epoch of marked, and comparatively rapid, change.

There were several major paleogeographic events during the Pliocene, each of which affected the global distribution of both plants and animals.

The Gibraltar Straits opened up, and so marine connections between the Mediterranean and the Atlantic were resumed.

The uplift of the Isthmus of Panama some 3½ million years ago resulted in a cessation of the exchange of marine fauna and flora between the Atlantic and the Pacific, an interruption of the connection that had continued, essentially undisturbed, between the two oceans for some 125 million years. The Isthmus of Panama provided a major migration route for North and South American mammals, allowing interchange between two mammal faunas that had been isolated from each other throughout the Cenozoic. This migration was predominantly southward from North America, with forms such as equids, mastodons, tapirs and llamas successfully colonizing the South American continent. In contrast to this general trend, such forms as sloths and armadillos made their way northward into North America.

About 3 million years ago polar glaciation began to develop in the northern hemisphere, and this resulted in the appearance of the Labrador Current, which displaced the Gulf Stream to its present position, south of latitude 45°. For over 100 million years before this the Gulf Stream had flowed into the Labrador Sea carrying tropical-water elements from the Caribbean. The glaciation also caused the formation of a Polar faunal province, completing the process of faunal localization which

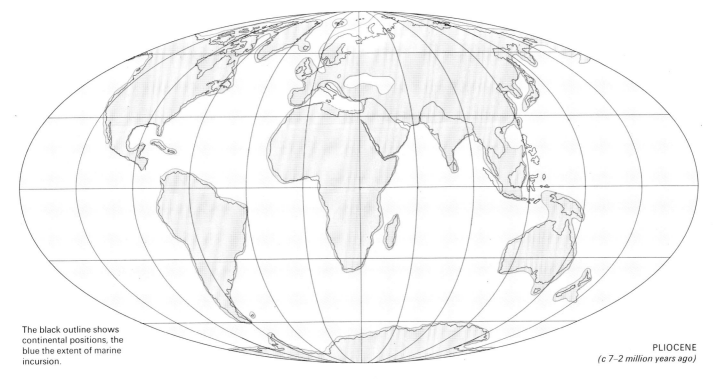

The black outline shows continental positions, the blue the extent of marine incursion.

PLIOCENE
(c 7–2 million years ago)

had begun millions of years before in the Paleocene and Eocene.

This marked deterioration in climate which characterized the Pliocene resulted in the extinction of numerous marine *plankton. In fact, the species of plankton alive today represent essentially no more than a relict fauna and flora that survived – and adapted to – the "ecologic trauma" that they experienced during this brief epoch.

The Pliocene also witnessed the demise of numerous elements of the prolific, if somewhat bizarre, mammalian fauna of the Miocene. Small mammals like the rodents continued to proliferate, and their fossils are extremely useful in making correlations over large distances.

The evolution of the elephant has been traced back as far as the early Pliocene in Africa, and mastodons and mammoths made their initial appearance in the late Pliocene and the early Pleistocene.

But among the mammals of the Pliocene, the most interesting to us is surely *Australopithecus*, Man's earliest definite ancestor. Fossil australopithecines dating back at least four million years have been found in East Africa. This comparatively recent discovery has, effectively, doubled the age of Man because, before it, the oldest known fossil men were those found in the Olduvai Gorge, dating back only two million years. (See *hominids.)

Major tectonic activity during the Pliocene occurred within the Mediterranean Basin: there was a significant downwarping of the sea floor, particularly in the Tyrrhenian Sea. There were also major orogenies, notably a continuation of the building up of the Andes and the Himalayas, as well as along the west coast of North America, giving rise to the features we now know as the Cascades, Olympic Ranges and California Coast Ranges.

The Pliocene can be viewed as the last epoch before the world became very much as we know it today, the end of the past and the beginning of the present. The Pleistocene would see the emergence of modern Man and most of the mammals that we are accustomed to. And with the Pleistocene, too, would come the Ice Ages and the uncountable changes they wrought – both in the patterns and families of life and in the very shape of our planet Earth.   WAB

Quaternary

From time to time during its history, the Earth has been subject to ice ages during which ice sheets have expanded to cover large areas of its surface. We know of several such episodes during the late *Precambrian, an ice age during late *Paleozoic time, and lastly an ice age during the Quaternary, a period which commenced about 2.5 million years ago and which includes the present.

The base of the Quaternary is normally defined on the basis of a worldwide cooling, and during the period large ice sheets have expanded (glacial periods) and contracted (interglacial periods) on many occasions. The onset of the Quaternary ice age was probably initiated by the drift of the Antarctic continent into a polar position, thus allowing a large ice sheet to build up. The evidence of fossil plants suggests that before the *Miocene the Antarctic enjoyed a temperate climate, but during the late Miocene we find, in deep-sea cores from the southern oceans, evidence of a great increase in iceberg-dropped detritus, suggesting that ice sheets on Antarctica were releasing large icebergs carrying glacially eroded material. By the beginning of the Quaternary, this cooling appears to have become global.

The existence of large ice sheets on Earth exerts a fundamental influence on climatic and oceanic circulation, the distribution and nature of terrestrial environments and the distribution and evolution of animals and plants. The fluctuating nature of these ice masses can be established in a variety of ways, but perhaps one of the most informative is the study of oxygen-isotope ratios in the tests of benthonic *foraminifera found in deep-ocean cores (see *geochemistry). During the onset of a glacial period, water is progressively lost from the oceans and stored in growing ice sheets. At the present day, the isotopes O_{18} and O_{16} occur in seawater in a roughly constant ratio; but when water is lost to growing ice sheets, less of the heavier isotope, O_{18}, is transferred and thus the O_{18}/O_{16} ratio in the oceans increases. This increase depends upon the amount of water lost to the ice sheets, and thus, if we determine the progressive change in the O_{18}/O_{16} ratio in the tests of deep-water foraminifera which have slowly accumulated on the deep ocean floor, we can deduce changes in the volume of ice on Earth. From such studies it has been shown that the last glacial period ended about 10,000 years ago, and that the warmest part of the present interglacial occurred about 4–5000 years ago. Since then our climate has been generally cooling.

The Frozen World. How do environmental conditions on Earth change from periods of maximum ice-sheet exten-

sion to periods of minimum extension? A wide variety of geological and biological techniques applied to sediments in the sea and on land enables us to reconstruct environmental conditions during that period of the last glaciation when ice sheets were at their most extensive.

Most of North America was covered by an ice sheet, as was much of northwest Europe and northern Russia. Most of the world's great highland areas bore ice caps, and extensive areas of floating pack ice blocked the Atlantic to north and south. As a result of the loss of oceanic water to the ice sheets, worldwide sea levels were lowered by about 90m (300ft), leading to the exposure of large areas of shallow continental shelves and thus to an enlargement of many land areas.

The polar fronts migrated in an equatorwards direction, so that latitudinal temperature gradients in middle latitudes were increased, thus increasing oceanic and atmospheric circulation intensity: this led to increased storminess in middle latitudes. The changed pattern of climatic circulation also led to desiccation in the equatorial zones of Africa.

In addition, the compression of climatic zones between expanded polar fronts led to a contraction of floral zones on land. Extensive areas of tundra, underlain by permafrost, lay to the south of the northern European ice sheet as far as the northern shore of the Mediterranean. To the south of this lay the forest zones, first birch and pine and then broad-leafed forest, all displaced some 2000km (1250mi) to the south of their typical interglacial positions.

The enormous changes in climatic and vegetational environments during a glacial period produced concurrent great changes in the distribution of animals. The expansion of tundra to the south enormously extended the range of the great tundra vertebrates, such as reindeer, mammoth, woolly rhinoceros and musk ox, many of which are now extinct – their bones may be found in glacial river gravels throughout much of the modern temperate zone, fossil evidence of earlier glacial conditions.

Chronology. The passage of Quaternary

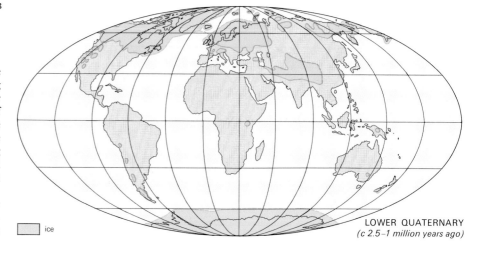

□ ice

LOWER QUATERNARY
(c 2.5–1 million years ago)

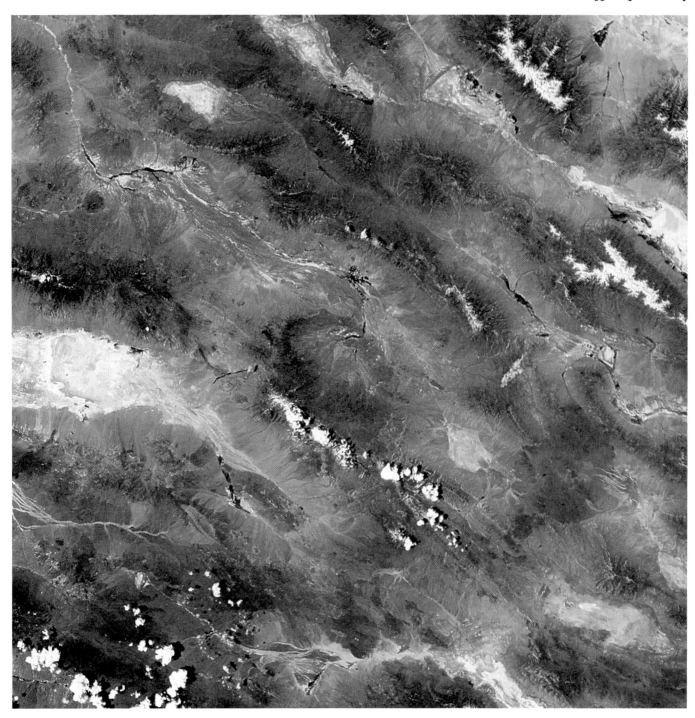

A satellite picture of the Sierra Nevada. These mountains are the result of uplift and faulting during late Pliocene and Pleistocene times.

time is measured by a variety of means. Absolute dates can be obtained from many materials: fossilized organic matter can be measured by carbon-14 dating provided that it is not older than 50–60,000 years; igneous rocks can be dated by the potassium-argon method; and some sediments can be dated by magnetic means (see *age of the Earth). When fine magnetic particles fall out of suspension in still water to be deposited as sediment, they tend to adopt orientations that are dictated by the direction of the Earth's magnetic field. Recent sediment studies have shown that over the last 12,000 years the Earth's magnetic pole has moved from an extreme easterly to extreme westerly position about every 2000 years. From dated lava flows it has also been shown that from time to time the direction of the Earth's magnetic field has reversed. The time-scale of these changes has now been used to date sedimentation in the deep oceans and in other areas of quiet sedimentation (see *geophysics).

Much of the history of the Quaternary can be seen as one of enormous fluctuation between the extremes of environment we have described. Because of these strongly marked environmental changes, the basis of stratigraphic subdivision is different from that of other geological systems and eras: relatively cold glacial stages are interspersed with relatively warm interglacial stages – although a glacial stage may not be without warmer periods, and an interglacial may have cold periods within it.

Plant remains comprise the most important fossil assemblages used for relative dating and correlation of these stages. Because of the relatively short duration of the Quaternary, plants show little evolutionary change and thus assemblages of plants rather than successive species are used to characterize particular periods of time: organic beds within Quaternary sequences may be ascribed to the appropriate interglacial by the similarities in the composition and sequence of the whole floral assemblage.

Animal Life. The Quaternary history of

As the ice retreated during Holocene times, the removal of weight from the Earth's crust led to uplift. Features such as shorelines moved upwards, resulting in raised beaches such as this one on the coast of East Greenland.

vertebrates strongly contrasts with that of the plants in that there are very striking evolutionary changes. These are probably the result of intense selection pressures during this period of rapidly changing environment when, in as little as five thousand years, areas which had been adjacent to the northern hemisphere ice sheets could change from treeless arctic tundra to temperate broad-leaved forest. We see evidence of rapid evolution, especially in the large vertebrates which adapted to a very varied series of ecological niches. Among the elephants we see the evolution of the steppe mammoth, and its disappearance about 12,000 years ago – probably as a result of human predation. Among the others we see the giant Irish deer with its antler span of 4m (13ft), and forms such as the woolly rhinoceros and the saber-toothed tiger.

In the gravels associated with the valleys of many modern rivers of the temperate zone of the northern hemisphere, the teeth and bones of mammoth, reindeer, bison and many other large vertebrates are commonly found. From such deposits as this, and from a study of pollen, beetles, snails and countless other organic remains, the environment in which many Quaternary strata have been deposited can be reconstructed in very great detail.

The vertebrates of, perhaps, greatest in-terest to us are the *hominids. The earliest known fossil hominids date from the Pliocene; but the Quaternary can be regarded as the true "Age of Man". In particular, its latter part – the Holocene – has seen the rise of civilization, as Man has learned to alter and, to some extent, to control his environment.

Tectonics. In geological terms the Quaternary period is of very short duration (2.5 million years), hardly enough for tectonic activity such as plate movements to produce substantial changes in the Earth's surface configuration and structure. Some of the most dramatic tectonic changes reflect the rhythmic growth and decay of the great ice sheets of the northern hemisphere and the concomitant changes in sea level. For instance, large areas of northwestern Europe and North America are currently rising at a considerable rate – 9mm (0.35in) per year in the Gulf of Bothnia between Sweden and Finland, and 30mm (1.17in) per year around the Hudson Bay. Both of these areas lay beneath the centers of ice sheets, which finally disappeared only about 7000 years ago. The effect of loading the Earth's crust with these enormous masses of ice was to depress it, probably by as much as 700–900m (2300–2950ft) beneath the center of the 2500m (8200ft) thick Scandinavian ice sheet and 800–1000m (2600–3300ft) beneath the 2900m (9500ft) thick North American ice sheet. The crustal material displaced from beneath the ice sheet was pushed outwards and caused the crust to rise beyond the margin. After the retreat of the ice masses, the Earth's crust rebounded – upwards beneath the ice sheet and downwards beyond it. The current upward movements of Scandinavia and northern North America are a continuation of this process, although the rate has been decelerating for several millennia.

These glacially induced crustal disturbances have also influenced igneous activity. In Iceland, the frequency of volcanic activity increases strikingly during glacial periods, the volcanoes rising up through the ice sheet. It is probable that magma is squeezed upwards through the crust, because of the additional load of the ice cap.

Pleistocene

The Pleistocene is a subdivision embracing the period from the beginning of the Quaternary until 10,000 years ago. The latter boundary marks a period of rapidly rising temperature – the end of the last glacial period – and the beginning of the present interglacial, called the Holocene.

Holocene

The Holocene is that period of Quaternary time from 10,000 years ago until the present day. It begins with the final decay of the great ice sheets in Scandinavia and North America, and is generally regarded as an interglacial period, on the assumption that it will be followed by yet another glacial period. The warmest part of the Holocene was some 5000 years ago and, since then, substantial changes in climate have occurred, such as the post-medieval cooling – known as the "Little Ice Age" – which ended at the end of the last century.   GSB

The History of Life on Earth

The Origin of Life

Despite its diversity, living matter is characterized by an astonishing biochemical uniformity. All known forms of life have two features in common: they have a cellular structure, in which the cells are bounded by membranes; and they are composed of two kinds of macro-molecule, nucleic acids and proteins. The combination of all these features is a certain indication of life, quite apart from the consideration of func-

tion. It is thus possible to define "life" without having to define the difference between "living" and "dead".

Most theories for the origin of life presuppose that life has evolved from simpler organic entities. The investigation of the origin of life seeks to differentiate true life from some precursory, prebiological entity which may be called "protolife". The most direct evidence for the origin of life would be the discovery of protolife among the early fossils.

Early *Precambrian fossils are found in two situations. Macroscopic fossils are found in the form of stromatolites, formed in dolomite or limestone, and interpreted as blue-green algal reefs. Microscopic fossils of two kinds, filaments and spheres, of

The results of some experiments support the idea that life could have originated under conditions similar to those found today in volcanic environments. Here, gases are brought together at high temperatures in thermal springs of New Zealand.

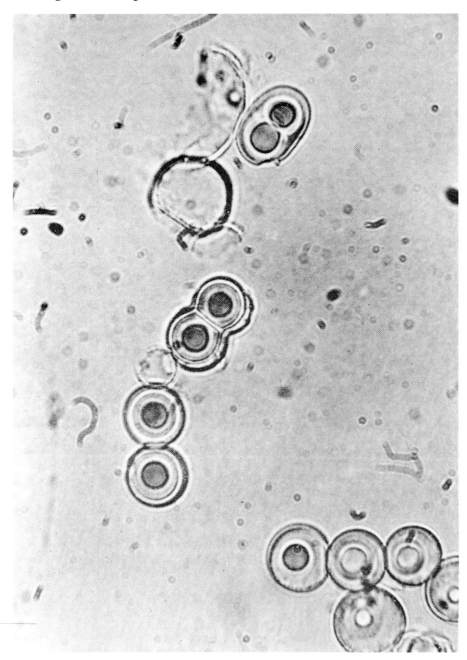

Cell-like structures called "proteinoid microspheres" have been produced by evaporating organic chemicals on hot lava beds. These resemble what were probably the first living organisms.

hydrogen, methane and ammonia are heated in the presence of powdered refractory catalysts, the organic products include most of those found in carbonaceous *meteorites, which are therefore postulated by Anders as having formed in this way. Notable among such products are amino acids, which have been found as racemic mixtures in several carbonaceous chondrites and which, if polymerized, could give rise to proteins. Biologically produced amino acids are never racemic when fresh, so these must be indigenous to the meteorite. Carbonaceous meteorites also contain organic spheres, and mineral grains coated with organic sheaths, that have been likened to "protocells".

Experiments designed to test the hypothesis that organic synthesis can occur in primitive planetary atmospheres have been mounted by several biologists and exobiologists. An electric spark (designed to simulate lightning) is passed through a gas mixture which is kept circulating through hot water. Amino acids and other organic compounds have been identified among the products, and it has since been shown that cell-like bodies are also produced among the insoluble organic products.

Experiments designed to show that polymerization of organic compounds produced in these ways can occur in certain environments have also been mounted, by Sidney Fox of Miami. He has shown that if solutions of amino acids are evaporated on a bed of hot lava, and subsequently heated to the point of fusion before quenching with cold water, cell-like structures are produced which possess a double-layered membrane. Some degree of polymerization has occurred, and they are referred to as "proteinoid microspheres". They can be manipulated in certain ways (e.g., by changing the acidity of the solution) so that they grow and divide.

These experiments suggest that life could have originated in any one of three environments. The first suggestion is that some form of protolife synthesized in the solar nebula infected the atmosphere of the primitive Earth and subsequently invaded a primeval broth, where it evolved into life.

The second alternative is that protolife formed in the atmosphere of the primitive Earth as the result of thunderstorms. This protolife was washed into the sea, which became the primeval broth that gave rise to true life. Objections have been raised that such a broth would be incredibly dilute, but it has been suggested that one form of protolife could have been oil, with the consequent production of an oil-slick which would float on the seas and thus embody its own concentrating device.

The third possible environment would be a volcano in its later eruptive stages. Thunderstorms would occur during the

diameters in the range 10–20μm are found in *cherts.

The oldest stromatolites occur in the Bulawayan Formation and have been dated at 2900 million years. The oldest cherts with microfossils are from the Onverwacht Formation and are over 3300 million years old. The stromatolites are universally regarded as the remains of true life: the earlier microscopic fossils may well also represent the remains of blue-green *algae, but it is perfectly probable that they represent some form of primitive protolife.

The oldest known rocks bear a metamorphic imprint dated at 3760 million years, and are therefore considerably older than the Onverwacht fossils. These rocks include metamorphosed water-laid sediments, some of which have been claimed as derived from biological sources. There are three kinds of such rock: marbles claimed as metamorphosed stromatolites; graphitic

schists claimed as metamorphosed oil-shales; and quartz magnetites claimed as banded ironstones, which are in turn postulated as oxygen receptors for photosynthetic organisms. If these claims are admitted, life cannot have originated after about 4000 million years ago. As the *age of the Earth is supposed to be only about 4600 million years, the origin of life must have been a very early event in the Earth's history.

Experimental evidence has shown that some kinds of protolife could form even before the planets themselves in the solar nebula. It could also form within planetary atmospheres, and on the land-surface of planets in a volcanic environment.

Experiments designed to test the hypothesis that organic synthesis could occur in the solar nebula have been mounted by E. Anders and his collaborators from Chicago. He has shown that, when gases containing eruptions, which would provide gases rich in carbons. Amino acids and other simple organic compounds would be washed down onto hot lava, and subjected to the processes invoked by Fox. Later the combined effects of heat and water would produce sugars and polynucleotides, which could invade the proteinoid microspheres and produce the first living organisms.

Convincing evidence in support of these theories is not easy to come by. Some Russian observers claim to have recognized racemic mixtures of amino acids among the products of volcanoes, but these observations remain to be confirmed.

The most speculative hypothesis of all is the celebrated "bootstraps" theory, that a visitor from elsewhere arrived on Earth, walked around for a while, and then left. However, microorganisms that were brushed off his boots contaminated the surface of the planet, and it was these that evolved toward modern terrestrial life. Of course, this theory begs the question, "What did the spaceman evolve from?", and is unprovable. Nonetheless, some scientists bear it in mind as a remote possibility.

Early stages in the production of protolife could still occur on Earth, although later stages are not likely to survive biological attack from contemporary life. Hydrothermal environments have been explored with such a possibility in view, and it has been claimed that oil and bitumen found associated with hydrothermal mineral veins are non-biological products. It has also been claimed that hydrocarbon globules found in fluid inclusions within vein-quartz may similarly be non-biological in origin, and these may contain associated proteinoid microspheres. Others believe that all these products are the result of contamination by biological products.

PCSB

Evolution

The living world comprises several million plant and animal species, all distinct, and all characterized by the ability to produce fertile offspring. Like gives rise to like, yet at the same time there are always subtle differences that distinguish individuals within any given species; but, despite this variety, the species itself appears immutable.

From this it was until quite recently concluded that all animals and plants were the product of a single act of creation, and destined to last for all time. Early doubts were occasioned by the discovery during the 18th century of the remains of giant animals (e.g., mastodon) that had once existed but did so no longer.

Development of the Theory. Perhaps the most significant contribution to 18th-century evolutionary ideas was made by Linnaeus (1707–78). In 1735 he published his first classification, the 7-page *Systema Naturae*: by 1758, in its 10th edition, this had expanded to 823 pages. The fact that one could erect a hierarchical scheme of all living things suggested certain groups owed their similarities to descent from common ancestors. Many 18th-century scientists assembled a vast amount of data which they claimed demonstrated the divine wisdom of the Creator; but as the evidence accumulated it became apparent that the observations could be better explained by the evolution or transformation of species through time.

*Darwin's great contribution was to assemble the evidence in such a way that it led inevitably to the conclusion that evolution was the only possible rational explanation for the origin of living animals and plants. Darwin served as naturalist on board *H.M.S. Beagle* in its 5-year voyage round the world. The myriad observations that he made could only be understood on the grounds that evolution must have occurred: the succession of life recorded in the rocks, where ancient forms of life became extinct and were replaced by the appearance of more modern; the curious distribution of animals and plants; embryology, which emphasized the basic similarity of fish, reptiles, mammals and Man; the modifications of the vertebrate limb which all possess an underlying basic plan; the existence of vestigial organs such as the pelvic girdle in snakes and whales – and the fact that a hierarchical classification could be erected.

Darwin was well aware of the work of animal and plant breeders in developing new strains by artificial selection. The problem was to discover a mechanism of natural selection. Malthus' essay on population provided the clue – offspring are produced in far greater numbers than their parents yet populations remain relatively stable; i.e., a large proportion of the young cannot survive to maturity. Since all individuals vary from one to another, it seemed evident to Darwin that those better suited to the conditions in which they found themselves stood a better chance of survival than those less suited. Hence the popular notion of Darwinism as "the survival of the fittest".

Natural selection was a process that was easy to grasp and it established an overwhelming case for evolution. The theory of evolution, moreover, coincided with the prevailing philosophy of the new industrial society. Darwin himself was the first to acknowledge that evolution was only a theory, albeit the most reasonable one to account for the natural world. He conceded that the proof could only be provided by the *fossil record and that at the time he was writing such proof was not yet forthcoming.

The key objection to Darwin's theory was that selection could only be on variations that already existed. There was no explanation as to how these variations arose in the first instance, and certainly no suggestion as to the origin of the new characteristics they possessed. Darwin came to believe that characteristics acquired during the lifetime of an animal could somehow be transmitted to the succeeding generations: this belief in the inheritance of acquired characteristics is popularly associated with *Lamarck. While Lamarckism is today generally discredited, it is not unreasonable to imagine that the response of an organism to its environment is in some way built into the hereditary system.

Genetic Basis. Paleontologists record the succession of phenotypes, the structures that are developed during an organism's lifetime, often as a response to the environment. To an extent this must be an expression of the genotype; that is, the hereditary make-up. The hereditary potential may or may not find its full expression in the phenotype. For example, if food is scarce individuals' growth may be stunted, but with the return of optimum conditions they will achieve their full stature: a gradual increase in size may be an evolutionary change or it may be merely a reflection of an improvement of the food supply.

The consensus among biologists is that random changes in the genetic material, "mutations", occur and that these are then subject to natural selection. Normally mutations are disadvantageous and, being recessive, do not find expression in the phenotype unless they occur in both male and female gametes of the zygote. Should circumstances change and the mutation become advantageous, in time it will be favored and become dominant, so that even if present in only one of the contributing gametes it will find expression in the phenotype.

It was recognized that the genetic material was situated in the chromosomes of the cell nucleus and that these could be mapped so that the position of the genes for particular features could be located at a particular position on the chromosomes. But the nature of the gene eluded biologists and no evidence was forthcoming with regard to what constituted mutations.

The nature of the gene and mutations came to be understood once Watson and Crick had elucidated the structure of deoxyribonucleic acid, DNA. This molecule comprises a double helix formed of two complementary strands of purine and pyrimidine bases strung along a phosphate-sugar backbone, each strand acting as the template for its opposite number, thus providing the basic copying mechanism required of any genetic material. Subsequent researches established that, although there were only four bases, their sequences along the DNA molecule formed the genetic code which contained the instructions for protein synthesis. The nature of a protein is a consequence of the sequence of the amino acids. There are only twenty naturally occurring amino acids and these are coded for by triplets of bases. Since there are 64 possible combinations, several triplets will code for the same amino acid: the triplets are the "words" and there are codes for "capital letters" and "full stops".

A mutation consists of a change in the

base on the DNA molecule, changing the genetic code but not necessarily the resulting amino acid for which that triplet coded. Even if a different amino acid is substituted, it may not affect the structure and functioning of the final protein. Each protein has its own shape or configuration – its tertiary structure, which is a consequence of its amino acid sequence. Within certain limits there can be amino acid substitutions at particular places that still allow the protein to function adequately. The sequencing of amino acids in particular proteins such as hemoglobin, the respiratory pigment in the blood of vertebrates and some invertebrates, allows direct comparisons to be made between the same protein occurring

in different animals. For example, the beta chain of hemoglobin of the gorilla differs from that of Man at only one amino acid locus out of about 150, but differs in 17 sites from that of the pig. This indicates that Man is more closely related to the gorilla than to the pig – a conclusion that can be supported on other grounds! By such studies one can draw up family trees of modern animals to show their relationships and hence possible evolution.

At the cellular level, it is not yet feasible to deduce how a simple animal or plant cell first evolved. The origin of the variety of complex organelles, each with their own specialized functions, still awaits an explanation, but it is already clear that this

history is likely to be more complex than previously imagined. Recent evidence suggests that one of the key organelles of all animals and plant cells, the mitochondrion, which is concerned with energy transfer, was originally an independent organism: it still retains its own DNA and the capacity to synthesize protein, and reproduces independently of the rest of the cell. Mitochondria seem to be related to the modern bacterium *Paracoccus*, and it has been postulated that initially mitochondria were parasites, or were perhaps ingested as food by the host organism: now they have become endosymbionts, with the processes of both host and guest so integrated that it is hard to imagine one existing independently

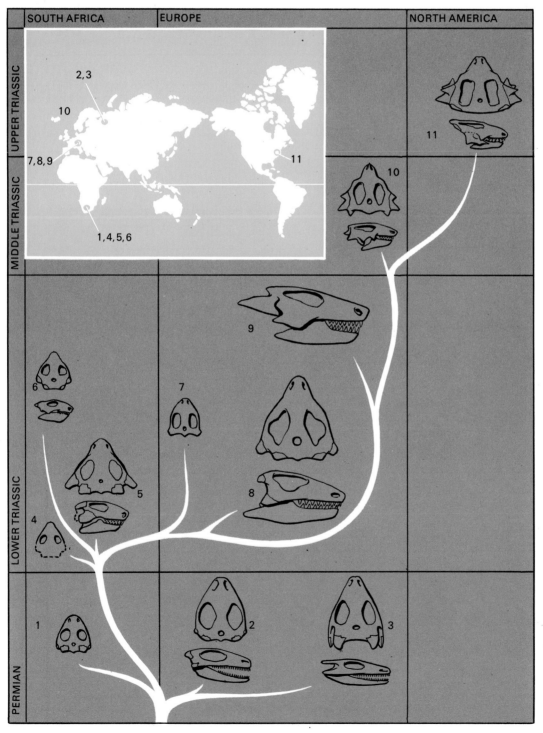

Fossil skulls of small reptiles called procolophonids arranged according to the age of the rocks in which they occur and the continents on which they have been discovered. Paleontologists are able, by comparing the anatomy of related animals or plants, to construct family trees in which the lines linking each type represent directions in which evolution is thought to have taken place.

of the other. A surviving example of this is with the giant freshwater amoeba *Pelomyxa*, which has no mitochondria but instead houses symbiotic bacteria.

There have been few evolutionary studies undertaken at the tissue level, and they are almost exclusively concerned with hard tissues. One such study has documented the gradual evolution of one type of hard tissue into another. The first vertebrates possessed a bone-like tissue, aspidin: this can be traced in the fossil record over a period of 150 million years. In early examples there were no cell inclusions, but eventually cells became incorporated and organized in the same way as in bone. The evolution of the organization of the organic matrix of collagen can be similarly traced. In the early types of aspidin the collagen fibers were laid down as parallel mats of criss-crossing fibers, as in dentine, the material that makes up the bulk of human teeth; but as time progressed they were organized into alternate layers with the fibers orientated as in modern bone. This sequence of changes illustrates a possible way that bone could have originated from a dentine-like tissue.

Evolutionary Histories. By far the best documented accounts of evolution are those of the vertebrates, from fish to Man, taking in such groups as the *dinosaurs. The evolution of the vertebrate skull, teeth and limbs are described in numerous texts on the evolution of the vertebrates, and in all cases it is possible to relate such structural changes to changes in behavior, such as modes of locomotion and feeding habits. From the discovery of fossils and their detailed analysis, evolution by natural selection has become established as fact.

Some of the most remarkable and seemingly mysterious evolutionary events, such as the change from *reptile to *mammal, have recently been elucidated. For example, the bones of the reptilian jaw joint became incorporated into the middle ear of the mammal, and this involved the reduction of the bones of the joint and their migration into the ear apparatus. The reduction of the bones of the jaw joint was the result of the split of the old jaw-closing muscle into two major parts, one with a forward component and the other with a backward, so that there was no force exerted at the joint. The bones, which were now taking no strain, simply shrank in size over the generations. The eardrum was situated close to the joint, and the spike of the articular bone of the lower jaw would have come into contact with the eardrum, automatically amplifying sounds impinging on the membrane. The three-bone system would have given the animal improved hearing – an obvious advantage in life. This remarkable change occurred several times in different evolutionary lines of true mammals.

An equally surprising development is recorded in two quite separate evolutionary lineages – the *dinosaurs and the mammals. Mammals evolved muscular cheeks, which are important in chewing, and a bony

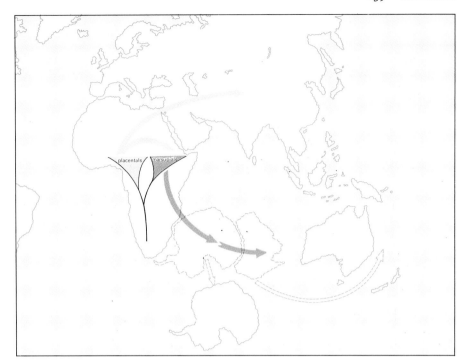

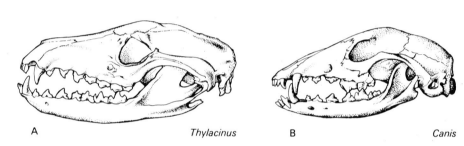

A   *Thylacinus*   B   *Canis*

Plate tectonics has had a profound effect on the evolution and distribution of organisms. Marsupial and placental mammals evolved in Eurasia or possibly Africa. Because marsupials diverged first, they could spread across Antarctica to Australia. By the time the placentals had evolved, continental drift had occurred and Antarctica and Australia had separated from Africa, leaving a marine barrier that could not be crossed. As a result, marsupials (A) and placentals (B) evolved similar forms in regions that are widely separated today.

secondary palate to separate the air and food passages and so allow breathing at the same time as food is retained in the mouth and chewed. In parallel, the herbivorous ornithischian dinosaurs became adapted to deal with the new, tougher plant materials that evolved during the latter part of the *Mesozoic era: unique among all the reptiles, they developed muscular cheeks, a chewing dentition and bony secondary palates.

There are numerous examples of animals evolving similar features which allow them to fill comparable ecological niches. A familiar example is the reptilian marine ichthyosaur that fed on fish and cephalopods, and which had the overall proportions, even to the extent of the triangular dorsal fin, of the modern mammalian dolphin. Both groups display the habit of accumulating the chitinous hooks from cephalopod tentacles in their stomachs and then regurgitating them to dispose of them.

The isolation of South America during most of the *Tertiary resulted in a number of mammalian types which paralleled forms

evolving elsewhere. There was a *marsupial sabertooth cat; and a small horse-like animal, the proterothere *Thoatherium*, in which the reduction of the digits exactly paralleled that in the true horses.

Although evolution can be shown to have taken place at all levels of organization, from the molecular to the whole animal, there is a further way in which it can be seen to have occurred. The entire community of living things can be shown to have undergone significant evolutionary changes. The food chain or web in the different environments has itself evolved from earlier patterns. The first reptiles that ventured onto dry land during the early *Permian fed exclusively on animal protein, the smallest feeding on *invertebrates: there were no vertebrates able to feed directly on plants. The gradual evolution of herbivores eventually allowed the land to support a large population of herbivores preyed upon by a smaller number of carnivores. When the dinosaurs replaced the dominant paramammals or *mammal-like reptiles during the Triassic, the modern-type food chain

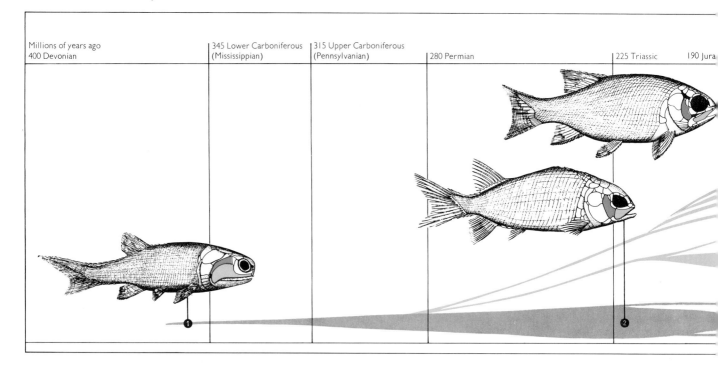

Millions of years ago 400 Devonian	345 Lower Carboniferous (Mississippian)	315 Upper Carboniferous (Pennsylvanian)	280 Permian	225 Triassic	190 Jura

An example of evolutionary trends in a group of animals is seen here in the evolution of actinopterygian fishes. The first well preserved fossil *Moythomasia* (1) occurred 360 million years ago. Two important evolutionary trends are seen in the fossil record. The body lobe of the tail (yellow) became progressively shorter, and the mouth-parts changed considerably. In *Moythomasia*, a bone called the preopercular (red) was firmly attached to the upper jaw or maxilla (blue); the jaws functioned like a pair of scissors. In later chondrosteans such as *Pseudobeaconia* (2) the preopercular slanted forward and jaws became shorter. In the holosteans, for example *Parasemionotus* (3), the maxilla was no longer attached to the preopercular, enabling the mouthparts to function with greater flexibility. This trend culminated in the teleosts, for example *Elops* (4), in which the maxilla was firmly attached to the skull only at its front end. This enabled the whole jaw apparatus to be thrust forward in order to engulf prey.

established by the paramammals was destroyed and initially the land was dominated by exclusively carnivorous dinosaurs. In time there evolved plant-eating forms, so that once again a familiar type of food chain was established.

Radiations and Extinctions. One of the most striking features recorded in the fossil record is the pattern of radiation followed by extinction. From humble beginnings there is a rapid efflorescence and then, more often than not, a sudden decline until the group vanishes from the Earth: time and time again, large groups of highly successful and seemingly well adapted animals suddenly go into decline and disappear.

At the beginning of each major radiation are a number of simple primitive forms: for example, if one considers the vast radiation of the archosaur reptiles which produced the gigantic dinosaurs, one finds that heralding the age of dinosaurs were primitive *crocodiles. But today the only living archosaurs are those selfsame crocodiles; and this is the paradox of evolution. Natural selection appears to favor those animals that become well adapted to their environment – the better this adaptation the more successful the group. The trend is towards greater and greater specialization, greater efficiency and hence greater success. But, by the same token, with environmental change comes disaster. The primitive forms that have not become highly specialized, but have retained a degree of adaptability, are the ones

that survive. In the final analysis the fittest, the survivors, are those forms that have remained variable. Many modern species are polymorphic, the shapes, sizes and colors varying so that, although circumstances may alter to the disadvantage of one form or other, the chances are that at least some will survive.

Evolutionary Centers. The process of evolution is a fundamental feature of the living world. However, one of the curious aspects of the fossil record is that, wherever a paleontologist finds himself, it is obvious that evolution has always taken place somewhere else. Geologists use fossils to distinguish strata from one another, fossil faunas being usually quite distinct from one stratigraphic level to another. But the intermediates are somehow always missing. Most of the time the fossil record documents a succession of migrations.

If all the fossils of any particular group are examined from all over the world, it does become possible to discover where evolution was actually taking place – usually in a single faunal province situated in tropical or subtropical latitudes. From here there were periodic waves of migration, the emigrants then establishing themselves in new areas and continuing with little or no modification until the next wave of immigrants invaded the region.

At the present day the greatest genetic variety of animal and plant life is to be found in tropical latitudes, whereas in more

temperate latitudes the variety is less although the numbers of the forms that are present are greatly increased. Away from the tropics, natural selection undoubtedly operates, but this is essentially a negative process – it eliminates.

The most dramatic evolutionary events seem to have occurred when natural selection was in abeyance. The radiation of the *amphibians during the *Carboniferous took place in a swampy environment that was highly favorable, so much so that virtually *any* new variety was viable. The radiation of the mammal-like reptiles during the Permian and Triassic and the evolution of the Jurassic and Cretaceous dinosaurs also represent times when a new group of animals found itself in a position to exploit a new environment. Once the new environment was fully occupied, and all the available niches had been taken, then and only then would natural selection come into play.

Conclusion. Though the fossil record provides overwhelming proof that evolution has occurred, it does not give any indication as to the mechanism. This is the domain of the geneticist, but any account that he produces of the process of evolution will only be viable if it can explain what the fossil record establishes has actually happened – and when, and where.   LBH

The Fossil Record

The term "fossil" has subtly changed in meaning over the centuries. Until the 18th century it was used to describe any rock or mineral dug up from the Earth, but since then its use has become more and more confined. Today, any relic or trace of a formerly living organism may be called a fossil, from the perfectly preserved insect in amber or Woolly mammoth in ice to the

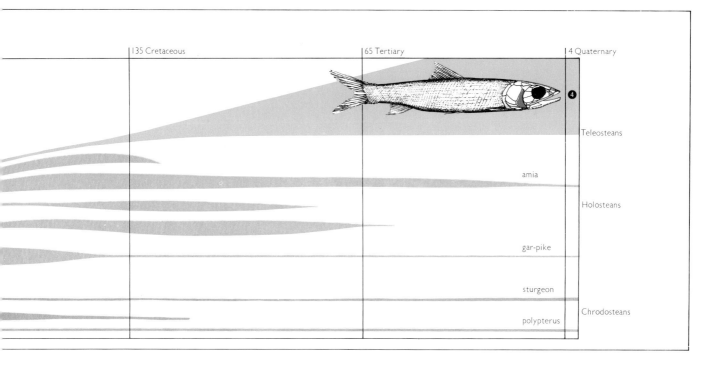

135 Cretaceous | 65 Tertiary | 4 Quaternary

4

Teleosteans

amia

Holosteans

gar-pike

sturgeon

polypterus

Chrodosteans

crawling trails of snails and the excrement of crabs. We should also include the faintest traces of all, chemical imprints of organic molecules left in the rocks.

Whether or not an organism is preserved as a fossil depends primarily on the nature of the materials of which it is made up. Most tissues comprise complex organic molecules that are either consumed after death by other organisms or readily broken down by chemical action. Soft tissues, therefore, leave little for the fossil record.

However, these customary processes may be bypassed in some way by the embalming resins of amber, the refrigerating action of ice or the antibiotic conditions of some stagnant lakes and ocean basins. Occasionally, the soft tissues of plants or animals may be petrified (i.e., their places may be taken, after death, by minerals that were initially dissolved in water which has permeated the organism). More usually, however, only the most resistant and stable organic materials can survive long after death, as with the lignified tissues of fossil land plants and the scleroprotein skeleton of the extinct fossil *graptolites.

Fortunately, a great many animals and plants have developed the ability to mineralize their own tissues. Familiar examples of such mineralization are the silica (SiO_2) and calcium carbonate ($CaCO_3$) spicules of sponges, the calcium carbonate shells of mollusks and skeletons of corals, and the calcium phosphate ($Ca_3(PO_4)_2$) of vertebrate bones and teeth. Not only are these mineral skeletons relatively resistant to decay in comparison with the soft tissues, they may also encapsule a complete life-history of the organism from birth to death, allowing us all fashion of valuable insights.

Despite the preservation potential of skeletons, only a very small proportion are ever fossilized. Fossilization depends on the creature either living in, or being transported to, an area of active sedimentation, circumstances that are rather unevenly distributed about our planet. Most favored for a continuous fossil record are those regions of the seabed below the zone of wave disturbance (the more one approaches the strand line, the more frequently are the sediments reworked or eroded and so the less complete will be the fossil record). Terrestrial environments experience such a dominance of the processes of decay and erosion that our knowledge of terrestrial life at any one time may rest on no more than a few relics washed into rivers or the sea or lodged in the deposits of cave *tufas and fissure fillings.

Changes following death and burial can lead to considerable transformations in the appearance of plants or animals. Not only may bacterial decay and groundwater solutions leach away organic and mineral parts, but the cavities so formed can later fill with new and uncharacteristic minerals. Alternatively, these minerals may replace the original skeletal parts. More usual, though, is a recrystallization of the original materials into more stable forms of the same or similar minerals, with preservation of the fossil's original microstructure if conditions are suitable. In addition, a considerable burden of sediment on top of the fossils will cause compression, and they may be further distorted by folding of the rocks during Earth movements.

Worse can happen. The high temperatures and pressures of metamorphism invariably cause recrystallization of the fossil materials and obliteration of their diagnostic characteristics. For these and many other reasons, the fossil record is neither complete nor easy to interpret.

It should not surprise us, then, that until the 17th century most naturalists were perplexed by the stony nature and curious shapes of many fossils, embedded – as they often are – in hard rocks from outlandish places. Common opinion classed them with meteorites and gemstones, strange manifestations of cosmic forces. Without knowing either the meaning of shape in living organisms or the nature of the geological column, a correct interpretation remained impossible; and although *Steno and *Hooke had by 1667 argued for a natural organic origin of fossils, it was almost 200 years before an acceptable correlation of the fossil record and its meaning was arrived at.

A widely held misconception that persisted for some time took fossiliferous rocks to be antediluvian relics washed up by the Flood. It is to *Cuvier that we owe the demonstration (1801) that more than one "flood" was involved. By 1808 he had sufficient data to outline the basic biological and geological changes involved in the fossil record and he went on to suggest its value to stratigraphy.

However, it was a humble civil engineer called William *Smith who put fossils to the test. In 1815 he published a geological map of England, compiled from many years of study, which traced strata of similar age right across the country by recognition of their similar fossil content. Many years of similar intensive and thorough work followed, so that by 1914 all the geological periods and many of their subdivisions were firmly grounded on a paleontological basis.

WWI saw an increased demand for oil, and the development of subsurface drilling techniques became a priority – to the paleontologist's profit. As the recovered rock chips from such boreholes are rather small, a new field of enquiry, that of micropaleontology (the study of microscopic fossils), was initiated on a grand scale and still forms a considerable proportion of present-day research: the abundance, variety and rapid evolution of tiny fossils such as pollen grains, *ostracods and

How animal remains become incorporated into the geological record. A dinosaur is drowned in a flooded river, and its body comes to rest on the river bed where it is rapidly covered with sand and mud. These sediments form the layers of rocks called sedimentary rocks. In this example the rock strata were tilted before erosion took place.

*foraminifera render them ideal time-markers in rocks of Cambrian age and younger.

As well as being restricted to rocks of a certain age, fossil species resembled their living descendants in being adapted to a restricted range of environmental conditions (see *paleoecology). Information about past environments is not just interesting for its own sake: it provides a useful guide for the economic geologist in search of mineral reserves like oil which tend to accumulate only in rocks formed in certain environments.

Paleontological research has contributed in this way to knowledge of ancient shorelines, ocean currents, climates and latitudes. It has also helped to reveal more remarkable phenomena, including *plate tectonics and the gradual reduction through time of the number of days in the year. But perhaps most significant is the support lent by the fossil record to the hypothesis of *evolution. Family trees have been worked out for a great many organisms, both living and extinct. The rates of evolutionary change for these different lineages have been assessed, and related where possible to the major controlling factors. We are now in a position to reveal the outline of the history of life on this planet, deduced almost entirely from the fossil record.

Although we have evidence that oceans existed at least 3750 million years ago, the first organic remains are much younger. These are unicellular bacteria and blue-green algae an estimated 3100 million years old (though there are earlier microfossils – see *origin of life). At this time the Earth's atmosphere was probably almost devoid of oxygen and the surface itself was bombarded by harmful ultraviolet rays from the Sun and outer space. Possibly because of these adverse conditions, primitive plant cells were the dominant form of life on Earth for at least another 2400 million years.

Nonetheless, considerable advances were

Once discovered, fossils must be protected from damage. Here a specimen is covered with layers of cloth soaked in plaster prior to its removal to the laboratory.

made in the organization of the cell itself. The nucleus appeared about 1600 million years ago and served as an information store – it enabled the cell to function more efficiently. First steps in sexual reproduction were taken between that time and about 900 million years ago: sexual reproduction allows useful information for survival of the organism to be passed on and distributed in time as widely as possible.

This lengthy algae-dominated period was terminated by a series of world-wide glaciations about 700 to 600 million years ago. It is possible that these glaciations, together with the advances already made, may have been responsible for the revolution that followed.

Fossil animals are first known from about 700 million years ago. They are found as impressions left by soft-bodied creatures in shallow-water sediments. Jellyfish and wormlike creatures seem to have predominated, though there is evidence of a planktonic and benthonic flora. After a period of about 100 million years there is clear evidence of a dramatic rise in sea level, possibly the result of melting polar ice caps, so that continents were flooded to an extent which has barely been matched since.

This event heralds the beginnings of the

*Paleozoic. The rapid expansion of sea-floor area was accompanied by the first occurrence of animals with mineral skeletons; and both of these factors contribute to the absolutely staggering increase from this time in the number of known plant and animal groups. Thus the number of invertebrate phyla rises from two in the late Precambrian (Coelenterata and Annelida) to at least twelve in the early *Cambrian.

From that time onward the fortunes of the various plant and animal groups appear to have been linked to changes in geography. Humid land areas became widespread in the *Devonian, providing a habitat for early vascular plants and land arthropods. Fish were also evolving rapidly in lakes and coastal waters, seasonal desiccation of which advanced the development of fish with lungs. By *Carboniferous times, certain of these had all the characteristics of *amphibians and were thriving in the coal-swamp forests of tree ferns, giant clubmosses and horsetails.

The ensuing *Permian was extremely hot and dry in many parts of the world so that those groups which were more or less independent of standing water found themselves at an advantage. Most notable of these were the *reptiles and the seed-bearing plants. The withdrawal of the seas in the Permian is attributed to continental collisions and a major glaciation. Many marine organisms characteristic of the Paleozoic became extinct, especially the more

specialized forms. The ecological niches left vacant by this extinction were gradually filled in the *Triassic by a less varied and less specialized fauna that had managed to hang on. As the seas began to flood back onto the much eroded continents, marine life of a new order began to diversify again – the *Mesozoic had arrived in no uncertain terms.

Cold climates were virtually unknown in the *Mesozoic era, so the radiation of warmth-loving reptile groups went on more or less unimpeded. Mammals are not very significant in the Mesozoic fossil record, possibly because they inhabited cooler, upland areas; their early remains are known from Triassic fissure fillings. Birds are likewise scarce, but they did not appear until late in the era, in the *Jurassic, evolving from either *dinosaur or *crocodile ancestors.

Flowering plants are known from the early *Cretaceous and seem to have dominated land floras by the late Cretaceous. At the same time, both plant and animal plankton were undergoing a revolution that led to the almost world-wide deposition of chalk sediments.

Associated with all these changes is the gradual dwindling of characteristic Mesozoic elements, such as dinosaurs and ammonites. Their final replacement by the new Cenozoic plants and animals was hastened further by global climatic and geographic changes, in turn brought about by

Earth movements. Nevertheless, this late Mesozoic extinction was not as drastic as that at the end of the Paleozoic; and it was followed by a much more rapid and spectacular radiation of plants and animals. The radiation can be explained both by the great variety of habitats that resulted from geographic isolation of the continents and by the development of distinct climatic belts. Mammals, birds, gastropod *mollusks and foraminiferid *protozoa are among those groups which evolved rapidly and spread widely in the early Cenozoic.

Climatic differentiation led gradually to climatic deterioration. About one million years ago, the ice caps at the poles began to expand considerably, and since that time rapid (in geological terms) oscillations between glacial and temperate climates have led to the extinction of a great many plants and animals.

The extremely adaptable *hominids were presumably in a position to take advantage of this. Unfortunately, hominid behavior was such that there is a scarcity of human bones in the fossil record, and so our knowledge of this most interesting period is very incomplete. Nevertheless, the rather slight differences between recent Man and the fossil remains of about 2.6 million years ago seem to suggest that Man's evolution has been, for at least that time, psychological and social rather than biological.

Obviously the study of ancient life from fossils preserved in the rocks has great bearing on our understanding of Earth history, as well as of the evolution of life itself: it follows that this study has more to contribute to our knowledge of Man's true place on the planet than almost any other branch of learning. There cannot, for example, be any support today for the old line of thought which saw Man as the final and glorious culmination of the evolutionary process, atop the tree of life. The fossil record is full of biologically comparable "success" stories, and some of those were very short-lived.

The present-day disturbances of the delicate ecological balance brought about by the activities of Man are also not entirely novel. Geological and biological revolutions of this kind have taken place a great many times. Some geologists have even claimed that these revolutions occur rhythmically through time, being controlled by fluctuations in solar or cosmic energy. We shall be able to confirm this only if similar patterns are observable in the fossil records of other planets. Even the dead planets could well possess a fossil record almost as revealing to us as the discovery of living cosmic neighbors. MDB

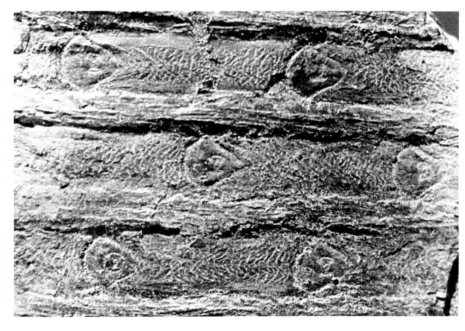

Fossils preserved in a variety of ways. *Above*, a Jurassic nautiloid seen in section clearly shows how mineral replacement has led to the preservation of internal structures. *Center*, a mold of a Jurassic gastropod in which sediment has filled the space once occupied by the mollusk. *Below*, an impression of a fern from the Carboniferous in which only surface features are preserved.

Paleoecology

Ecology is the study of the relation between living organisms and their environment, where they live and how they interact with other organisms and with the environment. Organisms and their environment together make ecosystems. Paleoecology is concerned with the ecosystems of the past, and with their evolution.

The problem is immense, for while we can closely study living organisms and determine modern environmental parameters, the evidence for the past is in the unwieldy form of rocks and fossils. Firstly, the environment must be reconstructed from the evidence remaining in the rocks; and secondly the fossils revitalized by making comparisons or analogies with living forms, or by constructing models. The geologist has only one advantage over the ecologist, that he may have evidence of evolutionary and ecological change with time as represented in a succession of sediments.

Problems. The *fossil record is very incomplete and, clearly, it is absolutely impossible to fossilize completely a community of living things: the nearest that can be achieved is a Pompeian-type catastrophe, but even there many people as well as birds, insects and other animals managed to escape. Little remains of the contemporary vegetation and much is lost of physiological and behavioral details. Nevertheless, the degree to which fossil environments can be reconstructed is often surprisingly good. The paleoecologist has to be paleobiologist, sedimentologist and biologist. In addition to trying to understand the biology of fossil species it is important to appreciate how death occurred, and the events that followed, leading to decomposition of the soft parts, disarticulation and breakup of the skeleton and eventual burial and potential fossilization.

As in ecology, the concern is first with the individual or individual species (autecology), and then investigation proceeds to the assemblage as a whole (synecology). Often, much information can be obtained before the fossil is freed from the sediment, and a determination of whether or not the organism was fossilized in life position is particularly important. The orientation of fossils in the sediment not only reflects changes of orientation that may have taken place after death in response to currents and waves but also orientations adopted by the living organisms themselves in respect to feeding currents, gravity and light. Epizoans, such as parasites, encrusting on other organisms may provide physiological information by indicating, for instance, the region on a brachiopod valve where nutrient-bearing water entered.

As with an ecological study, the way in which sampling is tackled can be critical to results. The problem is complicated in that invariably one is sampling organisms which lived at different times. There are obvious difficulties in determining the diversity,

AGE	DEPOSITS	ENVIRONMENT AND FAUNA
UPPER TRIASSIC	Sandstones	DESERTS
MIDDLE TRIASSIC	Sandstones and shales	*Scaphonyx*
LOWER TRIASSIC		RIVERS AND DRY LAND *Euparkeria* *Dicellopyge*
PERMIAN	Shales, siltstones and fine sandstones	LAKES AND SWAMPS *Muchocephalus*
UPPER CARBONIFEROUS	Glacial deposits	GLACIERS NO FOSSILS

The Karroo deposits of southern Africa provide one of the best examples of a sequence of rocks that indicate a gradual change of climate. Evidence from the sediments is confirmed by the fossils preserved in them. Examples of the fauna are *Muchocephalus*, an aquatic amphibian; *Dicellopyge*, a fresh-water actinopterygian fish; *Euparkeria*, a thecodont that lived on dry land; and *Scaphonyx*, a ground-dwelling mammal-like reptile.

abundance and nature of the coexisting fauna, though some indication of the soft-bodied fauna can be obtained from the form, frequency and abundance of their burrows and other *trace fossils.

The recognition of distinct associations of species which were probably part of an original community is an obvious goal. A number of fossil communities have been described but few if any of these would approach the ecological definition of a community – this may be defined as an assemblage in which the sum of interactions between elements within the assemblage is greater than the sum of those between the assemblage and its surroundings. Fossil communities can be reconstructed with a fair degree of certainty from fossil coral reefs and similar units where a high proportion of the fauna is in life position: fairly good estimations of density and diversity can be made in cases like these. Communities have also been reconstructed, though with less certainty, from shell beds where the organisms have clearly been physically aggregated.

Where fossils are found essentially in life position it is possible to determine substrate preference and also trophic (feeding) position by comparison with living representatives. But extinct groups of organisms pose problems as in some cases we're not certain to which phylum – or even kingdom – they belong. In the Mesozoic the *ammonites probably occupied a number of ecological niches, but unfortunately so little is known of their arm form and jaws that only generalizations can be made. On the other hand, individuals have been found of their contemporaries, *Ichthyosaurus*,

whose digestive systems contain large numbers of fossil undigested hooks from squid arms. This establishes a major food-chain link. The determination of food chains and food webs in fossil marine environments is difficult because of the open nature of the marine system: food webs can be more readily deduced for terrestrial environments. The densities of the fossil remains of any one trophic stage reflect, of course, preservation potential and not necessarily original abundances.

As with living organisms, every fossil once lived under the constraints of various physical, chemical and biological factors, any one of which may have been limiting. In reconstructing the environments of the past we must attempt to determine these factors. Most have not changed appreciably with geological time but other factors may have: for example the length of the day (which may be recorded in the daily growth bands of some fossil corals and bivalves) has been increasing and it is extremely likely that the nature and amount of the Sun's radiation has changed through geological time. Other factors such as temperature, salinity, oxygen and carbon dioxide content of air and water, and trace element concentration all vary locally and may vary periodically. It is not possible to determine any of these chemical and physical factors with anything approaching the precision we might want. For instance, the heavier oxygen isotope O^{18} tends to concentrate in the sea since the generally more common O^{16} can escape more readily. The proportion of O^{18} in calcium carbonate shells tends to increase with rising temperature so that knowing – or at least having an estimate of – the

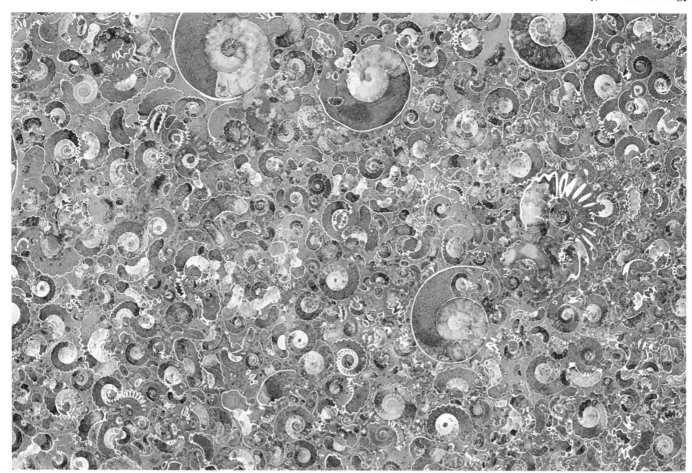

A section through a limestone composed almost entirely of mollusk shells. Such rocks indicate shallow water or shore conditions. Banks of shells are commonly seen on beaches of the present day.

composition of the seawater in which a particular creature lived, the temperature of the water can be estimated.

Depth under the surface of the sea is not in itself a limiting factor although it does, of course, affect pressure and light penetration. However, in reconstructing the geographies of the past it is very useful to be able to draw profiles and show depths with a reasonable degree of certainty. Thus paleoecologists are somewhat mesmerized by depth and, unfortunately, there are few satisfactory criteria that can be used for its determination. Since they require light, marine algae are important: fossil green algae in life position almost certainly indicate depths less than 70m (230ft). Certain bed forms such as underwater sand ripples and dunes may be useful, since the depth of water under which they lay does affect some of the superficial patterns of marine beds. Depth (pressure) has an effect on the solubility of calcium carbonate – but this is

A group of actinopterygian fish preserved in sandstone. Their abundance indicates a sudden change of conditions that led to the simultaneous death of large numbers of individuals, and their exquisite state of preservation is evidence that they were not transported prior to being covered with sand. If they had been transported by currents after their demise, fragile structures such as fin-rays would certainly have been damaged.

also dependent on the amount of dissolved carbon dioxide, so that the relationship is complex. Large eyes or blindness in fossil creatures can be correlated with depth by observations of similar characteristics in creatures alive today.

A major factor affecting nearly all fossils is provided by the changes that may take place after the organism's death and burial by later sediment. Apart from the decay of soft-bodied elements, shells too may undergo change, particularly in porous sediments. More soluble minerals, such as ★aragonite, tend to be lost first; in contrast, ★calcite is relatively stable. Post-Paleozoic corals which have aragonitic skeletons are generally not nearly so well preserved as ★Paleozoic corals which had a calcitic skeleton. All is not gloom: aragonite, including the pearly layer of shells, may well be preserved even from the Paleozoic in muddy and relatively impermeable sediments.

Goals. Most of the world's most famous fossiliferous deposits contain a very biased fauna because of the special conditions of fossilization. It is important to understand how such deposits arise, not only to discover how they themselves were formed, but also to predict where similar deposits might occur elsewhere.

The revolution that has taken place over the last few years in our understanding of the changing distribution of continents and oceans – the concept of ★plate tectonics – has provided paleontologists with a basic

key to the distribution of organisms about the Earth during the geological past. Since different groups of organisms react differently to barriers inhibiting gene-flow, communities also change. Faunal and floral provinces similar to those existing today can be recognized in the past. In the geological record, geographical *connections* that led to important faunal change – such as took place when the Americas were linked up by the Panama isthmus – are mostly easier to detect and appreciate than are the effects of *separation*, though there are exceptions; for example the isolation of the Australian ★marsupial assemblage.

Other barriers leading to speciation may be climatic (the creation of a desert), ecological (the growth of a forest) or physiological; and the effects of these in the past are difficult to determine. Moreover, the relative importance of different types of barriers in geological evolution is not known.

Conclusions. ★Evolution has not proceeded at an even pace, and "crises of evolution" must ultimately have had an ecological cause. For instance, the appearance of calcareous skeletons early in the Cambrian may well be associated with the level of atmospheric oxygen (and hence the level of oxygen dissolved in the seas) having reached about 10% of the present atmospheric level: below this level of concentration, secreted calcium carbonate tends to be easily redissolved. The widespread extinctions at the end of the Paleo-

Dinosaur footprints preserved in mudstone. The presence of infilled cracks indicate a lake bed in the late stages of drying out.

zoic may be largely due to the continentalization and reduction in area of shallow marine shelves, and the extinctions at the end of the Mesozoic to the general climatic deterioration that took place towards the end of the *Cretaceous. In contrast, the widespread decay among the *stromatoporoids and *corals and the temporary termination of reef-building activities towards the end of the *Devonian have not been satisfactorily explained.

As we can never hope to learn the complete story of evolution we can never recover in full its ecological accompaniment, but any information that helps fossils "live again" is worth pursuing. Determining the likely porosity and geographical boundaries of a fossil reef, and hence its *petroleum prospects, is an ecological exercise. Paleoecology is a growing subject, and its significance has extended far beyond the bounds of academia into areas that affect, in terms of simple economics, all of us today. It would appear that this is a trend that will both continue and accelerate in the future.

RG

Plants and Animals

The vast array of species in the *fossil record is the result of the process of *evolution. A classification of plants and animals that reflects the evolutionary process consists of a hierarchy of groups or taxa. Small taxa such as families or genera contain closely related species, that is, species that are thought to have evolved from a common ancestor. Closely related families are classified in larger taxa such as orders or classes and so on. The largest taxa, except for the Plant and Animal Kingdoms, are called phyla.

The entries that follow correspond in most cases to phyla, but for groups that are better known or are of more general interest, entries are based on smaller taxa, usually orders. Each is illustrated with a drawing of a member of the phylum or order concerned. The date of this representative is shown by a red spot on the scale,

and the range of the taxon as a whole by a colored strip. In the example above, the taxon has a range of 500–100 million years ago, and its representative occurred 300 million years ago.

AI

Plants

Algae

The algae are an exceedingly diverse group of plants, ranging in size from single cells measuring just a few thousandths of a millimetre to giant seaweeds 50m (160ft) long. They are usually found in aquatic habitats, both marine and freshwater, but also occur on the surface of soils and on tree trunks. Although they show such variation in external form, internally they all have a simple, rather uniform structure, usually consisting of only soft tissue.

Algae are subdivided into at least six major groups on the basis of the pigments they contain (often reflected in differences in their color; red, green, etc.) together with the nature of the stored food and of the cell wall and, on motile forms, the structure of the hairs (flagella).

The presence of single-celled plants, filaments, plates or complex branching forms in almost every group of the algae makes identification of fossils, which lack their original color, a hazardous process – indeed, the soft nature of most algal tissues mitigates against preservation. Thus, for example, the brown algae (Phaeophyta), which show the most highly differentiated internal vegetative structure of living algae and are the very common and familiar seaweeds of temperate regions, have little or no fossil record. There are notable exceptions: certain Precambrian *cherts contain exquisitely preserved single cells and filaments, while some algae have silica or lime skeletons and were important "rock-builders" in the past. As these latter are overrepresented in geological history, it is impossible to assess the time of maximum diversity of the group as a whole.

Blue-green algae (Cyanophyta).
These, the most ancient of all plants, show the simplest forms of construction: single cells, clumps of cells, or chains (filaments) of cells. Like the bacteria, they have no well defined nucleus within the cell, a feature that contrasts with all other plants and animals. Living blue-greens have the widest habitat range of the algae – indeed, of all plants – surviving such extremes as snowfields and hot springs. Simple spherical cells, attributed to the blue-greens, are recorded from rocks over 3000 million years old (see *origin of life). The younger Bitter Springs Chert flora (800 million years old) contains both single-celled and filamentous types closely resembling living forms, excellent illustrations of the extreme evolutionary conservativeness of the group. Of some stratigraphic importance in the Precambrian are stromatolites, columnar to dome-like structures built around sheets of blue-green algae, which still occur in intertidal regions in some parts of the world.

Green algae (Chlorophyta). These are of enormous importance as they gave rise to all other green plants. Living examples include single cells, filaments, sheets and complex branching forms, sometimes composed of giant cells with a plurality of nuclei. Their fossil history is sporadic except for those which had lime or silica skeletons; for example, *Ordovician representatives added lime to tropical reefs just as *Halimeda* does today. Another calcareous group, with an intricate whorled arrangement of branches (Dasycladiaceae), can be traced from the *Cambrian to the present and shows an increase in complexity and regularity of branching, so that by the end of the Paleozoic the modern verticillate arrangement had evolved. The stonewort *Chara*, which also has whorls of branches and a skeleton of silica, has more limited importance as a rock-builder, but is common in *Chara*-marls from the *Silurian onwards.

Red algae (Rhodophyta). Predominantly deepwater seaweeds, especially in warmer areas, the red algae range in structure from sheets of cells to elaborate branching forms, and have complex life-histories. Their geological history centers on calcified types: *Corallina*-like forms, abundant today, have been reef builders since the late *Carboniferous, while the extant *Lithothamnion* is first recorded from the *Jurassic.

Golden-brown algae (Chrysophyta).
Living golden-browns exhibit a range of forms paralleling that of the green algae, but it is the unicellular types, important constituents of modern planktonic plant life, which are of greatest geological significance. The most familiar are the *diatoms. Another single-celled kind had minute plates of calcium carbonate incorporated into the cell wall: microscopic studies have revealed that the vast *Cretaceous deposits of chalk were built up of countless numbers of these plates (coccoliths – see

*coccolithophores). Acritarchs, small hollow bodies (10–50mm) often brightly ornamented with spines, found in marine sediments from the Precambrian to the present day, are thought to be the reproductive stages (cysts) of planktonic algae. Because of their abundance and wide distribution, they are gaining importance in biostratigraphy. Some of the more recent examples are known to be related to the golden-brown *dinoflagellates and are called hystrichospheres.   DE

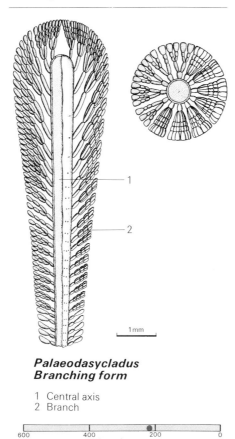

**Palaeodasycladus
Branching form**

1 Central axis
2 Branch

600 ─ 400 ─ 200 ─ 0
millions of years ago

Diatoms

Diatoms are a group of microscopic, single-celled golden-brown *algae. The group are known from the early *Cretaceous to the present, with dubious records from the *Jurassic. They reached their maximum diversity in the *Miocene. Most modern diatoms live in marine or fresh waters, although some live in soil. Again, most are solitary, but some occur associated in filaments or adjoined in colonies. Many are planktonic, but others live attached to other plants.

The cell wall is siliceous and consists of two pieces, the one fitting within the other like a pill-box and lid. Form and sculpture are diverse and variable, and diatoms are among the most beautiful microscopic objects.

They are important contributors to sediments, and diatomaceous oozes are an important category of deep-sea deposit, although shallow-marine and even lake diatom-rich sediments, or *diatomites*, are known.

The group are of considerable stratigraphic value, especially in deep-water late *Mesozoic and *Tertiary successions. They are widely used in the correlation of deep-sea cores.   WJK

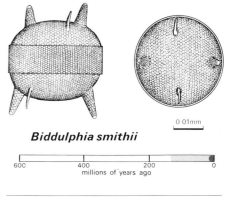

Biddulphia smithii

600 ─ 400 ─ 200 ─ 0
millions of years ago

Dinoflagellates

The dinoflagellates are microscopic single-celled plants, members of the phylum Pyrrhophyta. One of the chief characteristics of the group is the two flagella, long thin appendages, one band-like, extending

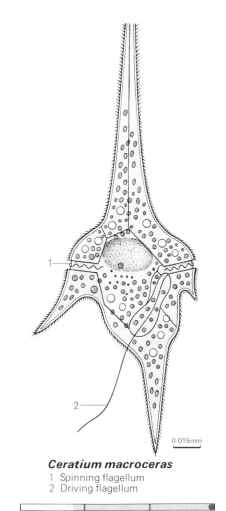

Ceratium macroceras
1 Spinning flagellum
2 Driving flagellum

600 ─ 400 ─ 200 ─ 0
millions of years ago

around the cell and able to spin it about its axis; the other attached by one end, and serving to drive the organism forward.

Most are marine and planktonic, causing so-called Red Tides when abundant, although a few inhabit fresh waters. Others, called zooxanthellae, are important symbionts, living in the tissues of various protozoans, sponges, worms and mollusks.

The earliest record of the group is a cyst (many dinoflagellates produce resistant resting cysts) from the *Silurian. They reached their maximum diversity in the late *Cretaceous, with minor peaks in the *Eocene and *Pliocene. They have great stratigraphic value in the *Mesozoic and *Tertiary.

Morphology is highly variable, with needle-like, globular and top-shaped forms, the theca (outer case) being made up of many plates fused and separated by prominent walls. The surface of the theca is generally ornamented by a network of thickened ridges.

Many marine sediments, from the late *Precambrian onward, yield hollow, organic-walled microfossils generally thought to be cysts of *algae. Almost 3000 species have been described, and referred to an informal group, the *acritarchs*. Many show features in common with dinoflagellates, and may be a related group.   WJK

Coccolithophores

This unique group of microscopic single-celled golden-brown *algae have an excellent fossil record from the late *Triassic right up to the Holocene, with possible records as far back as the *Carboniferous. Recent coccolithophores are mostly marine, although at least two genera are reported from fresh waters. They are typically planktonic at some stage of their life cycle, and constitute nearly half of ocean plankton in temperate waters.   WJK

Psilopsids

The psilopsids, or psilophytes, often considered as a completely extinct group, were the earliest vascular plants: a vascular plant is one possessing water-conducting tissue consisting of dead tubular cells (xylem), as well as a food transporting system (phloem): present-day examples include ferns, clubmosses, horsetails and flowering plants. Xylem is an important structural adaptation for life on land, and indeed *Silurian psilopsids were among the pioneer colonizers of land surfaces.

The psilopsids were strikingly simple plants, just tufts of branching stems, some erect and above ground and the rest horizontal and buried. They lacked leaves and roots, although some of the aerial parts were covered with spines. Each stem had a central rod of xylem tubes (tracheids). Two living rootless genera, *Psilotum* and *Tmesipteris*, were once classified with the psilopsids, but are perhaps best considered as living plants which in overall appearance closely resemble the earliest vascular plants.

Psilopsids are subdivided into three major groups, depending on the complexity of the stem-branching pattern and on the position and number of sporangia (globose or elongate sacs containing spores). The earliest and simplest forms had sporangia on the tips of smooth forking stems (Rhyniophytales) and include the earliest vascular plant, the Silurian *Cooksonia*. In the second group (Zosterophyllales), which appeared at the beginning of the *Devonian, the sporangia were attached to the sides of either spiny or smooth stems – sometimes, as in *Zosterophyllum* itself, forming compact terminal spikes or cones. It was from plants such as these that the *clubmosses are thought to have evolved. The third and youngest group (Trimerophytales) had repeatedly branching stems ending in grapelike trusses of sporangia. Such plants are considered ancestors of the *pterophytes and *horsetails.   DE

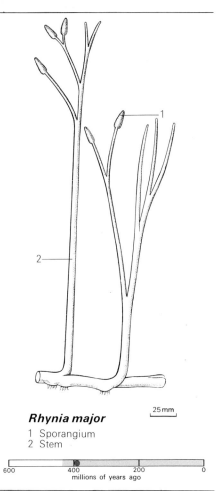

Rhynia major
1 Sporangium
2 Stem

25 mm

600   400   200   0
millions of years ago

Clubmosses

The clubmosses (lycopsids), together with the *horsetails, are truly plants of the past. Today they are rather inconspicuous herbaceous plants, but at their time of maximum diversity, in the *Carboniferous, clubmosses were dominant trees of the swamp forests and important ground-cover plants. Lycopods are characterized by numerous, usually small, moss-like leaves crowded on above-ground stems and sporangia (globose or elongate sacs containing spores) aggregated into cones, each sporangium attached to the base of a special leaf.

There are three types of living lycopods: the clubmosses themselves (Lycopodiales), the spikemosses (Selaginellales), and the rushlike aquatic quillworts (Isoetales). The most important extinct group is the Lepidodendrales. All, except the clubmosses proper and some early representatives from the *Devonian, have two kinds of sporangia, one containing a few large spores and the other numerous small ones. The earliest lycopods, recorded from the early Devonian, were herbaceous, but soon afterwards the forerunners of the Carboniferous trees appeared and evolved alongside the herbaceous forms.

Lepidodendron itself was a tree some 30m (100ft) tall, with a straight columnar trunk up to a metre (3.25ft) wide topped by a dense crown of forking branches, some ending in small cones. The terminal branches were covered with small elongate leaves, but the trunk and larger branches bore characteristic diamond-shaped scars evidencing the loss of leaves through abscission. The base of the trunk was divided into four main branches which subdivided repeatedly and formed a ramifying root system.

Herbaceous lycopods were important ground-cover plants in the Carboniferous, perhaps forming extensive heathland vegetation, and persisted throughout the *Jurassic and *Cretaceous in similar habitats. *Lepidodendron*, however, did not survive the drier conditions at the end of the

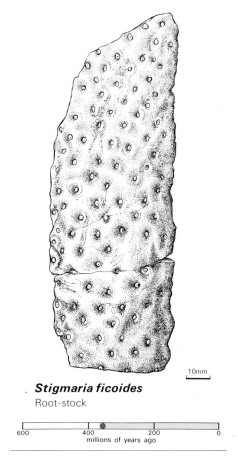

Stigmaria ficoides
Root-stock

10mm

600   400   200   0
millions of years ago

Paleozoic. Some authors consider *Isoetes*, a plant less than a metre tall with squat fleshy root-stocks bearing crowns of quill-like leaves, the living descendants of these Carboniferous giants. DE

Horsetails

The horsetails (sphenopsids), very conspicuous members of the Carboniferous swamp forests, are represented today by a single surviving genus, *Equisetum*. Living horsetails rarely exceed 1.5m (5ft) in height. Deep subterranean rhizomes produce upright, usually ridged, green stems, which are jointed and bear whorls of small scale-like leaves. Distinct cones are present in which sporangia (sacs containing spores), are borne on the lower surfaces of umbrella-like structures (sporangiophores). The jointed nature of the stem and the whorled arrangement of both leaves and sporangiophores are characteristic of all horsetails.

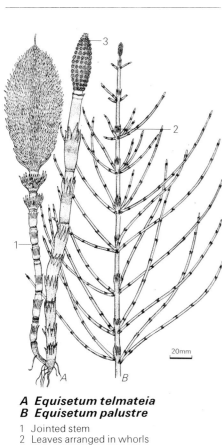

A Equisetum telmateia
B Equisetum palustre

1 Jointed stem
2 Leaves arranged in whorls
3 Cone

600   400   200   0
millions of years ago

Plants very similar to *Equisetum* are recorded from *Carboniferous rocks, but far more spectacular were its close relatives, giant *Equisetum*-like trees (*Calamites*) up to 18m (60ft) tall with trunks sometimes 40cm (1.3ft) wide and whorls of prominent, elongate leaves. There are internal similarities, too: both have hollow stems with a cylinder of strands of dead tubular cells (xylem), although in *Calamites* a further layer of xylem was present.

Not all Carboniferous sphenopsids were trees. *Sphenophyllum* was a scrambling herbaceous plant, an important constituent of the undergrowth. Its slender stems, with whorls of wedge-shaped leaves, were probably supported by surrounding vegetation.

Early sphenopsid history remains uncertain. Possible middle *Devonian members are included in the Hyeniales. *Hyenia* itself was a small herbaceous plant in which a horizontal rhizome gave off slender upright branches, covered with forking appendages some of which were fertile. *Calamites* and *Sphenophyllum* disappeared when the climate became drier during the *Permian, but *Equisetum* persisted and formed extensive stands at the edges of lakes and rivers in the *Jurassic and *Cretaceous. DE

Pterophytes

The pterophytes are the true ferns, in some classifications placed with the *clubmosses and *horsetails in the Pteridophyta, or vascular cryptogams (i.e., seedless vascular plants). Today they are found in a wide variety of localities and climates, though particularly abundant in the tropics. The majority are herbaceous perennial plants, but a few are aquatic and some tree ferns may reach 6m (20ft) in height. Most possess a stem that bears roots and large, often much-divided leaves, or fronds.

Mariopteris nervosa

Frond

600   400   200   0
millions of years ago

Reproduction is by spores which form inside tiny spore-capsules borne on the edges or undersides of the fronds. Each spore grows into a minute, green free-living plant (prothallus), quite unlike the parent fern and bearing the sex organs. External moisture is essential for the male gamete to swim to the immobile egg cell to effect fertilization. After fertilization, a new fern develops.

The leaves of the earliest pterophytes,

Cladoxylales (Devonian) and Coenopteridales (Devonian to Carboniferous), were three-dimensional structures but, by the middle of the *Carboniferous, pterophytes with flattened fronds had appeared (Marattiales). Specimens of fern-like foliage are a conspicuous feature of Carboniferous plant-bearing rocks, and the period has often been referred to as "the Age of Ferns"; but the attribution of these specimens to a particular group of plants is often impossible. Some certainly belonged to ferns, but it is now generally acknowledged that most represent the similar-looking "seed-ferns" (see *cycadophytes).

Living ferns do not inhabit dry arid areas where lack of moisture would prevent completion of the life-cycle. Fossil ferns may, therefore, if their affinities with living forms can be demonstrated, provide clues to paleoclimatic conditions. AW

Cycadophytes

The cycadophytes are the oldest group of gymnosperms, only the cycads (Cycadales) surviving today. The earliest, Pteridospermales or "seed-ferns", combined characteristics of true ferns (*pterophytes) and seed plants, their seeds and pollen-bearing organs being produced on the margins of large fern-like leaves.

Most cycads resemble palmtrees (though not related), having a stout unbranched trunk with a crown of once-divided, leathery evergreen leaves. Their seeds and pollen-bearing organs are borne in separate cones, those of the female often being exceptionally large, as are the seeds. Coexistent with cycads during the Mesozoic, but much more widespread, were the Bennettitales. These cycadophytes were remarkably like cycads in appearance, but bennettite reproductive structures were quite different, being flower-like and usually hermaphrodite. Each "flower" had an axis covered with a large number of naked seeds and intermingled sterile scales, below which was a ring or cup of pollen-bearing organs, the whole being surrounded by a number of protective scales.

Other cycadophytes were the Glossopteridales and the Caytoniales. Glossopteridales were ubiquitous and common in Gondwanaland during the late Paleozoic. Their fossil remains are represented by entire, tongue-shaped leaves to which were attached the reproductive organs. Caytoniales are restricted to the Mesozoic, leaves and reproductive organs alone being found as fossils. Each leaf comprised four leaflets arising from a common point at the end of the leaf-stalk, and the seeds were enclosed in berry-like structures.

Modern cycads are restricted to the subtropics and tropics, which suggests that fossil representatives lived under similar conditions. Glossopteridales occur in beds between glacial tillites. The comparative purity of the flora and paucity of species in the vegetation of which Glossopteridales formed part indicates evolution under severe, but not necessarily glacial, temperature conditions. AW

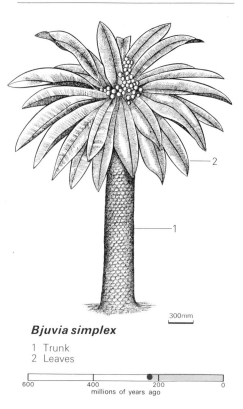

Bjuvia simplex

1 Trunk
2 Leaves

600   400   200   0
millions of years ago

Coniferophytes

The coniferophytes are a group of naked seed plants classified with the *cycadophytes as gymnosperms, though the two groups have had separate past histories. Their only shared feature is the gymnospermous mode of reproduction, which could well have arisen independently in the course of *evolution. Coniferophytes have been dominant plants since the *Carboniferous, and living members still form large forests in north temperate regions. Three orders of living coniferophytes are usually distinguished: Coniferales (conifers), Taxales (yews) and Ginkgoales (maidenhair trees).

Coniferales are cone-bearing trees, much branched and often growing to a height of more than 100m (330ft). The reproductive organs are borne in separate cones, those of the female becoming at maturity either woody (as in pines, firs, western red cedars, etc.) or forming a fleshy berry (junipers). Fossil conifer remains are commonly found in plant-bearing strata from the Permian, and it seems that their evolution was complete by the end of the Mesozoic, little change having occurred since.

Though having the same growth habit and leaf-form as conifers, the yews are distinguished by the lack of female cones, having instead single seeds partially surrounded by an attractive, brightly colored cup, the aril. Fossil yews are recorded from the Triassic onwards.

The living maidenhair tree (*Ginkgo biloba*) is the sole survivor of a once much more widespread and common group. It is doubtful whether it now exists in the wild. The ginkgo is a deciduous tree with leaves

up to 8cm (3in) in breadth and resembling the individual leaflets of the maidenhair fern. The reproductive structures are borne on separate trees, those of the male as catkin-like structures bearing numerous paired pollen-bearing organs, those of the female as a stalk surmounted by a pair of seeds, which after fertilization enlarge to the size of cherries. The ginkgo is remarkable in retaining the primitive feature of motile male gametes.

Cordiatales were coniferophytes which flourished during the Carboniferous. They were tall slender trees with large strap-shaped leaves quite unlike those of the living groups. Their reproductive organs were borne on catkin-like structures.

Living conifers occur in a wide range of latitudes and the usefulness of fossil forms as indicators of climatic conditions is therefore limited.   AW

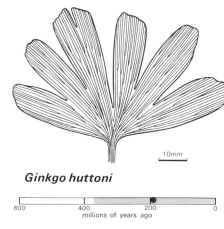

Ginkgo huttoni

600   400   200   0
millions of years ago

Angiosperms

The flowering plants, or angiosperms, are characterized by the production of seeds completely enclosed within the female part of the flower, the ovary. The flower is basically a reproductive shoot bearing several rings of lateral organs. At the base are several protective sepals, frequently green, forming the calyx, above which are the often brightly colored petals of the corolla. Within this are the pollen-producing organs (stamens) of the male part of the flower. The ovary, containing one or more ovules, is central. Pollen grains are carried by insects or wind to a receptive area (stigma) on the surface of the ovary: each produces a pollen tube which grows through the ovary wall and enters an ovule, so that a male gamete is introduced into a female egg cell (ovum) permanently retained with the ovule.

After fertilization the ovule develops into a seed containing, at maturity, the embryo of a new plant. The ovary becomes the seed-containing fruit, and may take one of many forms. Seeds vary in size from the microscopic (as in orchids) to the gigantic double coconut weighing up to 20kg (44lb).

Flowering plants largely replaced the gymnosperms as the dominant group of

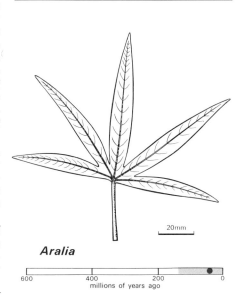

Aralia

600   400   200   0
millions of years ago

seed plants towards the end of the *Cretaceous, and this position they still occupy. They occur in every type of habitat and range in size from gigantic trees to minute plants. Some are climbers, others succulents, and a number have reverted to an aquatic habit. Most are green plants and are able to synthesize food from simple substances in the environment, but a few are partial or complete parasites.

The success of seed plants as land organisms clearly derives from the evolution of a pattern of reproduction by which fertilization can take place independently of external moisture. This has been effected by the elimination of the free-living sexual stage (prothallus) of the pteridophytic plants (e.g., *pterophytes), with retention of the egg cells within the ovule and the evolution of the pollen tube for transfer of the male gamete to the egg. Even those living gymnosperms which still retain the archaic feature of motile male gametes (cycads and *Ginkgo*) no longer require external moisture for fertilization, breakdown of certain tissues of the ovule providing liquid in which the male gametes swim.

Two divisions of angiosperms are recognized, the dicotyledons with two seed leaves and the monocotyledons with only one, other differences being that dicotyledons usually have net-veined leaves and the floral parts in fours or fives, whereas the monocotyledons usually have parallel veins and the floral parts in threes.

Being essentially ephemeral structures, flowers are much less likely to enter the *fossil record than are the more resistant parts of the plant. Knowledge of fossil angiosperms is therefore based mainly on leaves, fruits and seeds, and much can be learned from them about how modern vegetation arose. As an example may be quoted the fine and almost continuous record of fossil angiosperms in the Tertiary of Oregon. This shows a progressive change from a subtropical rain forest during the Eocene, to a temperate hardwood-conifer

An angiosperm leaf preserved in chalk. Such fossils are sometimes so well preserved that they show evidence of leaf-eating insects and plant disease.

forest during the Oligocene-Miocene, to cool temperate forests in the Pliocene, and finally to a semi-arid steppe.

The origin of the angiosperms is a continuing mystery. Perhaps their development and rapid rise was connected with the dramatic increase during the *Cretaceous in the number of insects. Equally it could have been due to the emergence of new habitats to which other plants were unable to adapt.   AW

Invertebrate Animals

Foraminifera

The Foraminiferida, commonly known as foraminifera or forams, are an order of unicellular animals of the class Rhizopoda, subphylum Sarcodina. They are related to the *radiolaria and the familiar *Amoeba*. The earliest undisputed forams are late *Cambrian, and the group survives today. Their maximum diversity has been from the late *Tertiary until modern times.

Individuals are often microscopic, but size ranges from fractions of a millimetre to ten centimetres (4in). All forams are aquatic and most marine, although a few occur in brackish or fresh waters. The soft tissues consist of a mass of cytoplasm, differentiated into an outer ectoplasm and an inner endoplasm, the latter containing one or more nuclei.

Stratigraphic Importance. Most forams live on and in sediments; others are encrusting or cemented or may live on vegetation; and a planktonic mode of life has been adopted by the important superfamily Globigerinacea, which has a good fossil record from the middle *Jurassic onward.

Forams are a major contributor to Holocene sediments, the tests of planktonic species in particular forming a major category of deep sea sediment: *Globigerina* ooze. At various times in the history of the group, large benthonic forms, often several centimetres in length, evolved. These include the fusulinids (*Carboniferous to *Permian), alveolinids (late *Cretaceous and *Tertiary), orbitolines (chiefly Cretaceous) and the nummulitids (chiefly Tertiary). Nummulitid *limestone was widely used as a *building stone in the ancient world, and Herodotus, writing in the 5th century BC, noted their occurrence in the limestone of the Egyptian pyramids. Subsequently, Strabo took weathered-out specimens to be petrified lentils, dropped accidentally by the original pyramid builders! Foraminiferida are one of the most important groups of stratigraphic indicators. Being small in size, and readily extracted from well-cores and chippings, they are of prime importance in the oil industry. Larger forms with complex internal structures provide the basis for detailed subdivisions of the Carboniferous-

Permian and Cretaceous-Tertiary of many parts of the world. The planktonic globigerinids, through their widespread geographic distribution and rapid evolution, are extensively used in late Cretaceous and Tertiary stratigraphic correlation.

The geographic distribution of the Foraminiferida is influenced by such factors as salinity, turbulence, water temperature, bottom conditions and sediment type, and so they can be excellent environmental indicators. Agglutinated forms are often indicative of nearshore or brackish conditions, while faunal diversity generally increases in an offshore direction.   WJK

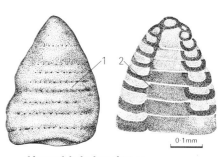

0·1mm

Howchinia bradyana

1 Test
2 Test in section

600   400   200   0
millions of years ago

Radiolaria

Radiolarians are entirely marine, commonly planktonic, microscopic protozoans of the class Rhizopoda, subphylum Sarcodina: they are close relatives of the *foraminifera. The earliest undisputed radiolarians are *Cambrian, although there are possible records from the late *Precambrian. The group are particularly diverse and abundant in the *Cretaceous and *Holocene.

They secrete a delicate skeleton, generally of silica, though members of one group (the suborder Acantharina) secrete *strontium sulfate. The skeleton is partially enclosed in soft tissue, and commonly consists of latticed spheres with radial struts, although there is an enormous variation of shape. The cytoplasm of living radiolarians is exuded outward into radiating, hair-like pseudopodia, and encloses symbiotic unicellular *dinoflagellates. Many radiolarians are dependent on the photosynthetic products of these as a source of nutriment.

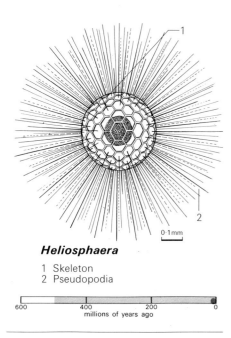

Heliosphaera

1 Skeleton
2 Pseudopodia

600 400 200 0
millions of years ago

Radiolarians inhabit waters of all depths and temperatures, although the requirements of their symbiotic zooxanthellae restrict many to the sunlit surface layers of the oceans. Their remains are an important contributor to deep-sea sediments. Ancient radiolarian-rich sediments are known as *radiolarites*, and many of these are believed to be deep-water deposits. The radiolarian silica is commonly redistributed in *cherts in these deposits.

The group are of great value as stratigraphic indicators, notably in Mesozoic and Tertiary successions, and are widely used in the correlation of oceanic cores.   WJK

Sponges

"Sponge" is the common name for members of the phylum Porifera, a group of invertebrates, predominantly marine (although members of one family are found in freshwater lakes and rivers), which first appeared in the early *Cambrian and are still extant today. They are among the simplest multicellular animals – in fact, until the 18th century they were regarded as plants. Their sizes range from less than a centimetre to several metres across. Most live attached to hard surfaces or have root-like anchorages in soft sediment, though members of the family Clionidae bore into shells and other calcareous substrates.

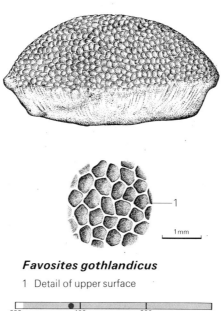

Favosites gothlandicus

1 Detail of upper surface

600 400 200 0
millions of years ago

Sponges are filter-feeding organisms; many have symbiotic single-celled plants in their tissues, and may gain nourishment from them. The body is supported by slender, pointed projections called spicules, which may lie loose in soft tissue or be fused into a rigid framework. These spicules are composed of *calcite or *aragonite, of *silica, or of the protein spongin. They are commonly found as *fossils.

The main groups of sponges are: the Calcisponges (*Devonian to *Holocene), with skeletons of calcareous spicules; the Hexactinellids (Cambrian to Holocene), sometimes also known as Hyalosponges, a wholly siliceous group producing delicate, glass-like skeletons; and the Demosponges (Cambrian to Holocene), with skeletons of spongin (as in bath sponges), or of silica with or without spongin. Some organisms closely resembling *corals (*Ordovician to Holocene) as well as *stromataporoids (Cambrian to Holocene), both formerly regarded as colonial coelenterates, are now believed also to belong to the phylum Porifera.

Sponges have limited stratigraphic value, but can be important rock-forming organisms, and some are important framework organisms, building reef-like structures. Siliceous sponges are numerous in many *Paleozoic and Mesozoic sequences, their spicules being regarded as an important source of silica in *chert and *flint formation.   WJK

Corals

The word "coral" has been applied to a variety of organisms that produce calcareous skeletons, but normally it is restricted to coelenterates of the class Anthozoa which have a calcareous supporting skeleton. Many anthozoans (e.g., sea anemones) have no hard parts, and different groups of corals appear to have been derived from these soft-bodied ancestors at different times.

All anthozoans are marine. They have a nervous system, often with stinging cells, but their body consists of a single cavity with only one opening (there is no separate anus). The corals are unlike many other shallow marine fossils in that their skeleton is primarily for support, not for protection: perhaps the ability to sting was sufficient protection.

The earliest anthozoans, from the late *Precambrian, are unbranched individuals anchored by a calcareous stalk, or possibly (in the case of the late Precambrian *Charnia* and *Rangea*) by a circular calcareous disc. These early anthozoans are included in the subclass Octocorallia along with a group of corals with eight tentacles and eight fleshy partitions which extend radially into the central cavity, which appeared in the *Cretaceous. Modern octocorals are mostly shallow-water dwellers, but some have been recovered from very great depths.

The remaining three groups of anthozoan corals are of the subclass Zoantharia: they are the Tabulata and Rugosa, both extinct, and the Scleractinia.

The tabulate corals are all colonial forms in which the individual polyps (fleshy bodies) had their tubes partitioned by horizontal plates (tabulae): the vertical radiating plates (septa) seen in most other corals are absent. This group is confined to the Paleozoic, being most abundant in some *Silurian, *Devonian and *Carboniferous limestones, where they can form a large proportion of some coral reefs, though more frequently occurring as isolated colonies spread thinly over the floor of shallow seas. Some (possibly most) of this extinct group may, in fact, be not coelenterates but *sponges.

The rugose corals were the dominant corals in the Paleozoic, and are especially common from the Silurian (when the earliest coral reefs were developed) to the *Permian. Like many modern corals, they come in different shapes according to the environment: simple conical corals can occur in quite deep water, whereas the colonial corals are more typical of shallower environments. The colonial corals with loose cylindrical branches occurred either in deeper water below the areas affected by severe storms, or in quiet lagoons where they were protected by a reef or other barrier from the open sea. Many rugose corals had polygonal individuals crowded together to form a massive colony; these were capable of forming reefs in exposed sites, but more commonly occurred as isolated colonies on a flat sea floor.

The Scleractinia first appear in the Mesozoic (after the extinction of the Rugosa), and probably developed separately from some soft-bodied sea anemone. They are solitary or colonial corals with radial septa in successive cycles (starting with six in the first cycle). Abundant small plates or rods are present between the septa. Many scleractinians, which tolerate only warm subtropical or tropical waters, are dependent for their existence on the presence of a large number of single-celled *algae in their tissues; these corals have a maximum depth range of about 90m (300ft), and include most of the modern reef-builders.   WSM

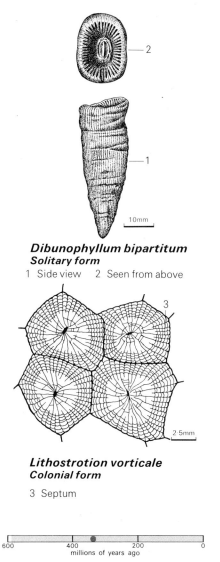

Dibunophyllum bipartitum
Solitary form
1 Side view   2 Seen from above

Lithostrotion vorticale
Colonial form

3 Septum

Stromatoporoidea
The stromatoporoids are, fairly certainly, an extinct group of *sponges. They are among the most conspicuous rock-forming organisms from the middle Silurian to the upper Devonian, representing a timespan

of perhaps 75 million years; and first appeared in the early Cambrian, becoming extinct in the early Eocene. Some ten families and seventy genera have been recognized.   RG

Archaeocyathids
The archaeocyathids are a group of sponge-like organisms that inhabited the shelf seas for, geologically speaking, a very short time, from the early Cambrian until the start of the middle Cambrian. They contributed extensively to carbonate sediments and in places formed extensive banks, though not reefs.

The biological affinities of the group are far from clear, but it is generally considered as an extinct phylum – possibly the only phylum ever to have become extinct.   RG

Bryozoans
Members of two closely related phyla of minute colonial animals are commonly termed bryozoans or polyzoans: the Entoprocta are entirely freshwater, and have no fossil record; while members of the phylum Ectoprocta, which commonly produce a calcareous skeleton and have an excellent fossil record from the Ordovician to the present, are predominantly marine. The marine ectoprocts inhabit all depths and latitudes, with a peak in diversity during the *Ordovician, though the acme of the group is during the late *Cretaceous to present. The status of supposed *Cambrian bryozoans is equivocal: they may be ectoprocts, or entoprocts, or neither.

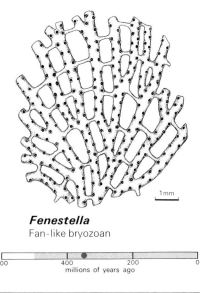

Fenestella
Fan-like bryozoan

Individual animals are of millimetre size, producing a double-walled membranous or calcareous sac which contains the soft parts, including a U-shaped alimentary canal and the reproductive organs. The mouth is surrounded by a hollow, circular or horseshoe-shaped structure that bears hair-like tentacles with which the animals capture food particles and microorganisms.

Hundreds to thousands of individuals produce colonies from less than a centimetre to more than a metre across. These may be like miniature plants, fan-like, or

encrusting on plants, rocks and the like. A few lie loose on the sediment surface, while some produce characteristic borings in shells.

Bryozoans are common fossils and may occur in rock-forming proportions, as they do in some of the Cretaceous and early Tertiary limestones of Scandinavia. They are an important contributor to carbonate sands in temperate latitudes at the present day.   WJK

Brachiopods
The members of the phylum Brachiopoda are marine animals with a shell consisting of two valves, one normally much larger than the other. The name brachiopod is derived from two Greek words (*brachia*, arms; *pod*, foot), and refers to the fact that the animals have internal arms in a similar position to the feet seen in *bivalves. In most brachiopods a pedicle emerges *via* an aperture (foramen) in the larger valve near the hinge and attaches the animal to the sea floor, but some forms have lost this pedicle and rest on the sea floor directly. All brachiopods generate a current in the space between the valves by means of a fleshy arm (the lophophore); this current supplies oxygen to the gills and also carries organic matter to the mouth. Brachiopods have a free-swimming larval stage (of 1–20 days), which allows them to spread about over the sea floor.

Today, brachiopods are not very common compared with the *bivalves and other

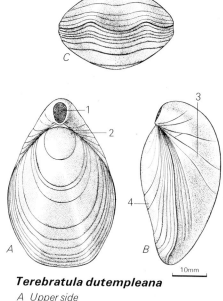

Terebratula dutempleana

A *Upper side*
1 *Aperture for pedicle*
2 *Hinge*
B *Side view*
3 *Lower valve*
4 *Upper valve*
C *End view*

*mollusks. In the *Cambrian, brachiopod fossils are subordinate in numbers to the dominant *trilobites, though they can be locally abundant in some shallow marine environments. From the Cambrian onward, two classes are present: the Inarticulata, with no teeth and usually a shell composed of chitin and phosphate; and the Articulata, with two teeth in the large (pedicle) valve which fit into sockets in the smaller (brachial) valve and hold the valves in place. All articulates have a *calcite shell.

WSM

Echinoderms

The Echinodermata are an entirely marine phylum of invertebrate animals which first appeared in the early *Cambrian and range to the present day. They are important elements of many shallow-water marine Paleozoic, Mesozoic and Tertiary faunas, and include the *echinoids (sea urchins), particularly important in Mesozoic and Tertiary rocks; crinoids (sea lilies), with their acme in the *Carboniferous; asteroids (starfish), ophiuroids (brittle-stars) and holothuroids (sea cucumbers), all of which have a poor fossil record; and a number of now-extinct groups.

These diverse animals are linked by three common features: a basic five-fold (pentameral) symmetry unique in the animal kingdom; hard parts covered and permeated by soft tissues and composed of a series of plates of *calcite, each a single crystallographic unit; and a system of thin, fluid-filled tubes produced into finger-like extensions (tube feet), which generally extend through to the outside of the body and aid in locomotion, digging, gathering food, respiration and chemical sensing.

The group are of some stratigraphic value in the correlation of some Paleozoic sequences and more especially in Mesozoic rocks (in the absence of *ammonites) and in some Tertiary successions. WJK

Blastoids

The Blastoidea are a wholly extinct class of stemmed, sessile *echinoderms, rarely exceeding 30cm (12in) in total height. They range from the Silurian to the Permian, reaching their peak in the lower *Carboniferous. The group seems to have preferred shallow, agitated marine environments. They are rather rare fossils. The body consisted of a stem of thin, disc-shaped plates attached to the sea floor by a branching-system of rootlets. A cup-like calyx at the summit of the stem bore numerous threadlike brachioles ("arms") in five pairs of rows along the sides of petal-shaped ambulacra. Perforations along the sides of the ambulacra led to a complex internal pouch, made up of thin-walled folds of tissue, which served for respiration.

About 80 genera of blastoids have been described, encompassing some 380 species. The evolutionary origin of the group is unknown. WJK

A Triassic crinoid displays two features characteristic of echinoderms: five-fold symmetry, and an outer surface protected by calcite plates.

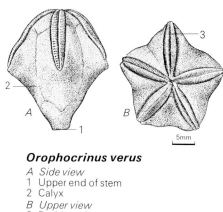

Orophocrinus verus

A Side view
1 Upper end of stem
2 Calyx
B Upper view
3 Position of arms

600   400   200   0
millions of years ago

Cystoids

The Cystoidea are a wholly extinct class of *echinoderms ranging from the lower *Ordovician to the upper *Devonian, with their peak in the Ordovician. The group are usually rare fossils, although some occur in rock-forming proportions. They appear to have favored clear, quiet-water conditions.

Most cystoids were anchored to the bottom by a short stem of disc-like plates, the summit of which bore a many-plated shell which presumably enclosed the bulk of the

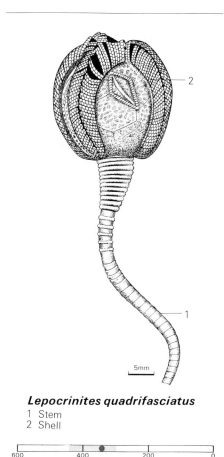

Lepocrinites quadrifasciatus

1 Stem
2 Shell

600   400   200   0
millions of years ago

soft tissues. The plates of the shell had pores, some nearly or completely closed by a thin coating of calcium carbonate, a feature which has led to debate on their precise function, although they were undoubtedly part of the respiratory system.

The cystoid shell also bore brachioles ("arms") to gather food – presumably floating organic debris or microorganisms. The food was transferred, probably via ciliated tracts, to the shell and from there to the mouth. WJK

Crinoids

The cystoid shell also bore brachioles ("arms") to gather food – presumably floating organic debris or microorganisms. The food was transferred, probably via ciliated phylum Crinozoa, and are important as the key to our understanding of the ecology and functional morphology of such extinct groups as the *cystoids and *blastoids.

Crinoids are the most diverse of the stemmed echinoderms. Typical members of the group have a basal series of root-like processes which anchor the animal to the sea bed or to vegetation, coral heads, shell debris and the like. From this arises a stem of articulated plates of diverse morphology, some disc-like, others resembling beads, and yet others taking the form of five-rayed stars. In some crinoids, there are side-branches on the stem. Adults have stems ranging up to 18m (nearly 60ft) in length.

At the top of the stem, the calyx, a cup-like structure comprising a series of cycles of plates, encloses the body cavity, gut, reproductive organs and so forth. From articulatory facets at the top of the cup arise arms, in turn built of articulating plates – these arms may be built simply of rows of plates, or may branch into hundreds of branchlets, giving a crown up to 1.5m (5ft) in length. The arms bear the water vascular system and function as a feeding device: microorganisms and food particles are trapped and transferred down grooves to the mouth, which is borne on the upper surface of the calyx.

All early crinoids and many present-day forms are sessile. During the Mesozoic, however, there arose secondarily free-swimming forms in which the arms are modified to power the creature through the water. Some of these have widespread distribution and have been regarded as planktonic; others, among them the comatulid crinoids which survive to the present, have a circlet of flexible appendages called cirri at the base of the calyx, with which they can temporarily secure themselves to the bottom.

Crinoids were at their height during the *Carboniferous, and throughout the *Paleozoic were successful inhabitants of many marine environments.

They are excellent stratigraphic indicators of marine environments. They have been utilized as indicators in several areas: their plates can be used as index fossils in some of the thick Paleozoic shale sequences of the US mid-continent area, as can the microcrinoids, whose adult calyx was only a few millimetres across. WJK

Woodocrinus macrodactylus
1 Stem
2 Calyx
3 Arms

600   400   200   0
millions of years ago

Echinoids

The sea urchins, or echinoids, are a class of benthonic *echinoderms ranging from the *Ordovician to the present day, being most abundant and diverse from middle *Jurassic times onward.

The exoskeleton (or test) is generally globular, and is built of twenty rows of interlocking calcareous plates. These are generally arranged in two sets of five pairs of plates, extending from the apex to the base of the test. The plates are either perforate, allowing access of the extensions of the internal water vascular system, the tube feet, to the exterior; non-perforate; or bear tubercles which form articulation points for spines usually modified for walking, burrowing, protection and defense.

In regular echinoids, there is a basic pentameral symmetry, with a large opening (the peristome) at the center of the lower surface of the test. This is covered in life by a leathery membrane in the center of which is the strong-jawed mouth. Regular echinoids are all epifaunal grazers, and the powerful jaws rasp at encrusting organisms such as bryozoans, algae and the like, or are used to cut vegetation. At the opposite end of the test, the anus opens in the middle of a further leathery membrane, which covers a second large opening in the test. Around it are specialized plates from which the gonads discharge. All early echinoids show this very regular symmetry, and all appear to have lived epifaunally, generally in quiet-water situations.

During the early Jurassic, however, a major radiation occurred, and forms evolved in which the anus migrated to a position away from the summit of the test. This was accompanied by a change in life habits from epifaunal grazing to infaunal deposit feeding, and specialized spines and tubercles were evolved for digging and burrowing, while the perforated plates and tube feet were modified for respiration, burrowing, crawling and feeding. The greatest specialization is shown by the heart urchins: burrowing is accompanied by assumption of a deposit-feeding habit and disappearance of a functional jaw apparatus in adults.

Echinoids are common fossils in many shallow-water Jurassic, Cretaceous and Tertiary sediments in spite of their fragile tests, and are excellent indicators of sea-

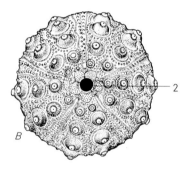

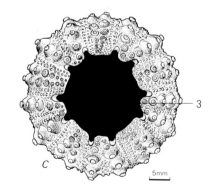

Hemicidaris intermedia
A Side view
1 Tubercle
B Upper side
2 Anus
C Lower side
3 Peristome

600   400   200   0
millions of years ago

floor conditions. In the absence of *ammonites, they have been widely used as stratigraphic indicators.   WJK

Mollusks

The mollusks, the phylum Mollusca, are a major invertebrate group, in numbers second only to the *arthropods. They appear to be of almost unparalleled antiquity: *trace fossils attributed to them occur in the late *Precambrian marine sediments.

Included in the phylum are the chitons, or coat-of-mail shells (class Amphineura); elephant's-tusk shells (class Scaphoda); oysters, mussels and other shells with two movable halves (*bivalves, of the class Bivalvia); snails, slugs and limpets (class Gastropoda); squids, cuttles, octopuses, belemnites, *nautiloids and *ammonoids (class Cephalopoda); the class Monoplacophora of superficially limpet-like organisms; and a number of minor, extinct groups.

Because of this great diversity of form, it is rather hard to describe a "typical mollusk"; but the basic morphology is fundamentally similar. There is little or no segmentation; and they have bilateral symmetry – i.e., one half mirrors the other. The internal body organs (viscera) are enclosed by a body wall, the lower part of which, the foot, is modified to provide locomotion, while the upper part, the mantle, hangs down as a fold so that there is a free space between it and the viscera. This mantle cavity houses the gills and aids also in feeding and locomotion. Sensory organs are concentrated in the head region (except in bivalves).

Some secrete a shell, generally external, though in certain cases it is secondarily enclosed in tissue. When external, the shell more or less totally covers the soft tissues. In some mollusks the shell has been totally lost.

Mollusks inhabit marine, brackish, freshwater and sometimes even terrestrial environments. They are exceptionally important since many produce shells of calcium carbonate: this has often happened in sufficient concentrations to form limestones.   WJK

Ammonoids

The ammonoids are a wholly extinct class of cephalopod *mollusks known from the early *Devonian to late *Cretaceous, their acme being during the *Triassic. Superficially similar to *nautiloids, they had an external aragonitic shell subdivided internally into chambers and so serving for both flotation and protection. The two groups can generally be distinguished in that ammonoids had a globular or barrel-shaped rather than saucer-shaped larval shell, and a ventral rather than sub-central siphuncle (a tissue-filled tube extending back to the larval shell). In addition, the sutures (the trace where the septa, or internal partitions, were attached to the inside of the shell) are complexly crenulated rather than broadly curving; while the shell exterior is often strongly ornamented rather than smooth.

Three broad groups are recognized, and

Jurassic specimens of the ammonite *Metophioceras* preserved on the lower surface of a limestone bed. Ammonites are sometimes over a metre in diameter and include some of the largest mollusks.

most readily distinguished on the basis of the suture lines. In the goniatites, an essentially *Paleozoic group, the lobes and saddles are simple and entire, angular or rounded. In the ceratites, essentially *Triassic, the saddles are entire, and the lobes serrated; while in the ammonites (late Triassic, *Jurassic and *Cretaceous) both lobes and saddles are subdivided.

Ammonoids were a diverse, variable and successful group of planktonic or free-swimming organisms, entirely restricted to marine environments. They were probably rather poor swimmers, being better adapted for vertical movements through the water by varying the amount of fluid within their chambers. Many possessed a well-developed jaw apparatus: some may have been carnivores and scavengers; others were probably plankton feeders or herbivores. The creatures were sexual and the sexes quite distinct: in many species the females (termed macroconchs) were often several times larger than the males (microconchs).

These later, Cretaceous ammonite faunas are notable for the diversity of straight, helical and loosely coiled or heteromorph groups, some of which in turn give rise to normally coiled descendants. These heteromorphs are definitely not "degenerate" forms or evolutionary "dead ends" as is so commonly stated, for they

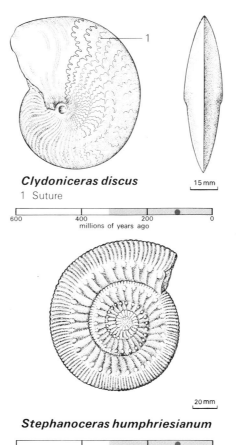

Clydoniceras discus

1 Suture

15 mm

```
600        400        200         0
       millions of years ago
```

Stephanoceras humphriesianum

20 mm

```
600        400        200         0
       millions of years ago
```

include some of the most diverse, long and widely ranging species, genera and families.

Although of limited value as environmental indicators, ammonoids are the most important group of larger invertebrates for the correlation of Devonian to Cretaceous marine sediments, forming the basis of global zonal schemes during this interval.   WJK

Nautiloids

The term nautiloid is commonly applied to *mollusks of three subclasses (Nautiloidea, Endoceratoidea, Actinoceratoidea) of the class Cephalopoda. The earliest nautiloids appeared in the late *Cambrian, and the group reached their acme in the *Ordovician. There is a single extant genus, *Nautilus*.

All nautiloids had external shells of *aragonite, consisting of a hollow cone, the regions toward the apex being divided into chambers (or camerae) by simple, transverse concave, saucer-like partitions (septa). This body chamber housed the bulk of the tissues. The foot was modified into tentacles which surrounded the head and were used in feeding, and there were well developed eyes and jaws. The mantle cavity, housing the gills, acted as an organ of jet propulsion: water was expelled through the restricted aperture to boost the creature forward.

Early nautiloids were straight or slightly curved, and maintained an equilibrium position by partial flooding of chambers or the deposition of aragonitic "ballast" in the chambers and siphuncle (a tissue-filled tube extending through the septae back to the larval shell). Coiled forms evolved independently on several occasions during the Paleozoic. Modern *Nautilus* reaches a maximum diameter of 25cm (10in), though some straight Paleozoic forms reached lengths in excess of 9m (30ft), and are among the largest mollusks.

Nautiloids are good indicators of fully marine conditions.   WJK

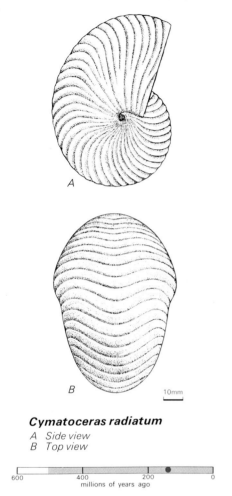

Cymatoceras radiatum
A Side view
B Top view

Belemnoids

The belemnoids are a group of wholly extinct, *Carboniferous to *Oligocene cephalopods referred to the subclass Coleioida, order Belemnitida. The best known representatives are the belemnites of the *Jurassic and *Cretaceous, at which time the group reached the acme of its 300-million-year history. They evolved from the Bactritoida, a group of straight, *nautiloid-like cephalopods with external shells, by a process of total envelopment of the hard parts by soft tissue. They were squid-like in external appearance, with lateral fins, well-developed eyes, and tentacles. Internally there was a well-developed muscular mantle cavity, expulsion of water from which enabled rapid jet propulsion. Some possessed ink sacs.

The hard part most commonly found fossil is the guard, a tapering, bullet-shaped object composed of radiating fibers of *calcite (there may have been some

*aragonite present as well). The pointed end lay at the posterior of the body, while a conical cavity at the anterior end of the guard housed the phragmocone, a reduced equivalent of the conical, chambered, ancestral bactritid shell. The phragmocone probably acted as a buoyancy apparatus, while the guard may have served as a counterweight. In front of the phragmocone lay the proostracum, a quill-like extension which served to protect the viscera.

Belemnites have been used widely in the correlation of Cretaceous marine sediments, especially the European chalk sequences, and to a lesser degree in the Jurassic. They are excellent marine indicators.   WJK

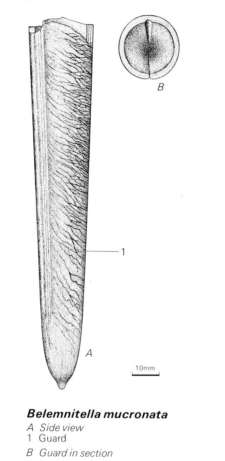

Belemnitella mucronata
A Side view
1 Guard
B Guard in section

Gastropods

The gastropods are a class of marine, freshwater and terrestrial *mollusks typified by a single, conical, variously coiled external calcareous shell made of *aragonite or aragonite and *calcite, over whose open end there is often a calcareous or organic "door", the operculum. Some have lost the shell altogether (nudibranchs), and in others (slugs) it is internal: a good example of a gastropod with an external shell is provided by the common snail.

Gastropods show distinct anatomical

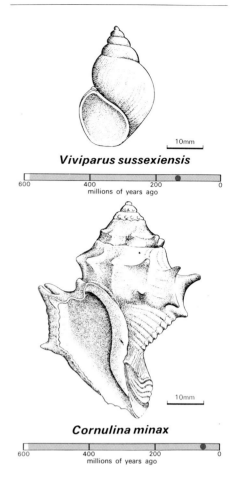

Viviparus sussexiensis

Cornulina minax

differences from other mollusks. The viscera are concentrated together, and there is a head with well-developed sensory organs, in which is a mouth containing a specialized radula (see below). The foot is flattened and muscular and is used in locomotion, be it crawling, burrowing or swimming. The mantle cavity contains the gills, and in some forms (the pulmonates) is modified into an airbreathing "lung".

The basic subdivision of the gastropods is made on features of the heart, gills and other soft tissues, and in many cases this makes classification of fossil material difficult. Three subclasses are generally recognized, the Prosobranchia, Opisthobranchia and Mesogastropoda.

Gastropods first appeared in the early *Cambrian, underwent a major radiation in the late *Cretaceous and have reached their peak today: the Holocene will probably be known to future paleontologists as the age of gastropods!

The group occupy both aquatic and terrestrial habitats, the first land snail (*Palaeopupa*) appearing in the *Carboniferous. Some are pelagic and most have pelagic larvae. Most bottom-living forms creep on hard or soft substrates, and their range is from the intertidal regions to the abyssal depths. Pelagic forms swim using a highly modified foot; some are epiplanktonic on *algae, while the truly planktonic *Janthina* supports itself in the water by a float of air bubbles which it produces.

A few float, others bore, are cemented, or are parasitic. The majority graze on vegetation or on encrusting animals such as *bryozoans and *sponges; but some have a long, out-thrust proboscis and become active predators on other mollusks, while others have a specialized horny, toothed band called a radula and can drill into shells. Still others have a poisonous bristle called a style, and feed on fish; while some are parasitic on bivalves or are symbionts living on the feces of *echinoderms and other mollusks: a number of fossil examples of such symbiosis is known. A few gastropods occupy a deposit-feeding niche (*Aphorrhais*), whilst *Turrilella* are ciliary feeders, living part-buried in sediment and drawing in detritus or mucus strands, or filtering such material out of seawater.

In spite of their varied habitats, little use has been made of gastropods as environmental indicators, and they have no wide application as stratigraphic markers.

WJK

Bivalves

The Bivalvia (also known as Pelecypoda and Lamellibranchiata) are an aquatic class of *mollusks ranging from the middle *Cambrian to the present. They are an important element in many shallow-water late *Paleozoic and *Mesozoic sequences, reaching their acme during the *Tertiary.

Anatomically, they have diverged considerably from the basic molluscan pattern. The mantle has expanded markedly, and is modified into voluminous folds which hang down on either side of the visceral mass. The body has become compressed laterally and the head reduced to the point of disappearance; and they have adopted a passive mode of life. The shell has two valves, or halves, joined dorsally along the hinge by an elastic ligament, and is built of layers of *aragonite, or aragonite and *calcite in separate layers.

Bivalves are superficially similar to *brachiopods. Typical members may be distinguished by, in bivalves, the lack of an opening for the pedicle and of internal calcareous supports; and by the plane of bilateral symmetry, which generally runs through the plane of contact of the valves in bivalves, but bisects the valves in a front-to-back direction in brachiopods.

Bivalves are essentially suspension feeders, the gills being expanded into a filtration device. The foot is commonly modified into a hatchet-shaped burrowing organ. Most are infaunal burrowers, living buried at various depths in sediment, although maintaining contact with the surface *via* tubular extensions of the mantle called siphons. Others are cemented; lie on the sediment surface; attach by a cluster of organic fibers secreted by a gland near the foot; or bore down into the sediment. A few can swim for brief periods, usually to escape predators.

Externally, bivalves are ornamented by diverse ribs, spines, tubercles and growth lines, which function as aids in burrowing, stabilization and attachment. Internally,

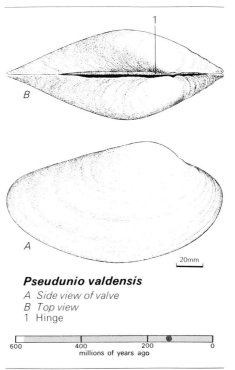

Pseudunio valdensis
A Side view of valve
B Top view
1 Hinge

600   400   200   0
millions of years ago

there are along the dorsal margin tooth- and socket-like structures associated with the articulation of the valves, and generally large scars indicating the site of attachment of the adductor muscles which work, in opposition to the elastic ligament, to close the shell.

Bivalves are excellent environmental indicators, their morphological features giving information of substrate conditions; while knowledge of modern forms allows recognition of ancient freshwater, brackish and fully marine environments. Overall geographic distribution and diversity patterns of bivalves have been used to determine climatic zonation during parts of the Mesozoic, while they have been widely used as paleoclimatic indicators for the Tertiary and Quaternary. The group are also of great stratigraphic value from the

*Carboniferous onwards. Freshwater mussels are one of the keys to the correlation of the coal-bearing strata, whilst inoceramids and the giant, abberant rudistids are widely used in the correlation of the *Cretaceous. In the absence of larger *foraminifera or planktonic forms, bivalves form the basis for much Tertiary correlation.

WJK

Arthropods

The Arthropoda are Man's only serious competitor on this planet. As a group, they are biologically more successful than the vertebrates in being able to inhabit more extreme environments; but they are relatively small animals compared with vertebrates and mollusks, their size being limited by their characteristic exoskeleton. The largest size achieved by an arthropod was in the *Devonian, a giant *chelicerate almost 1.8m (5.9ft) in length, though the giant Japanese crabs today can, with legs outstretched, extend to 5m (16ft). About 75% of all known living animals are arthropods, most being insects: however, because of the relative paucity of fossil insects, arthropods are less dominant in the fossil record.

The fossil history of the phylum goes back at least to the early *Cambrian, and there can be no doubt that it had a long Precambrian history. The fossil remains do not demonstrate the origin of the phylum, and biologists favor the suggestion that it derived from more than one ancestral stock. The common features of arthropods (the exoskeleton – and therefore molting –, the articulated limbs and the compound eyes) are thus likely due to convergent evolution.

Nine superclasses of Arthropoda are recognized though two of these are unknown as fossils. The *trilobites are probably close

The arthropods, of which this Jurassic lobster from the Solnhofen limestone of Bavaria is a good example, are characterized by jointed limbs and an exoskeleton. Being hard, the exoskeleton is sometimes preserved in such a way as to reveal considerable anatomical detail.

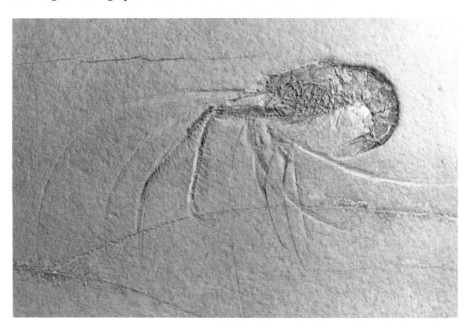

to the *chelicerates; the *crustaceans (crabs, lobsters) appear to be a quite distinct group; and the *insects can be grouped with several other, small classes into the Hexapoda (six legs). However, there are a number of fossils which do not fit into any of these groups. RG

Trilobites

This fascinating group of aquatic *arthropods appeared early in the *Cambrian, but must have been in existence long before. They reached their acme in the *Ordovician, thereafter gradually decreasing in diversity and in numbers of ecological niches occupied. Only one order survived into the *Carboniferous and the class became extinct late in the *Permian.

They are probably the best known fossil invertebrates: their relative abundance and distinctiveness, especially in lower *Paleozoic sediments, means that they are among the easiest fossils to find.

The name "trilobite" derives from the longitudinal division of the animal into three parts seen on the upper surface of the body. Apart from a narrow border zone and a large mouth-plate, the lower surface of the creature was not mineralized, and so is only rarely preserved in fossil form.

Each segment carried paired appendages, all similar except for the first two pairs behind the mouth, which appear to have carried a denticulated jaw. Each appendage is in two parts; a jointed section, which performed a walking and digging function, and a bristly section, probably used mainly for food collecting. The bristly section has often been considered as a respiratory organ, but it seems likely that respiration took place through the extensive thin integument on the underside. The organization of the appendage suggests a relationship between the trilobites and the *chelicerates.

Like other arthropods, trilobites had to molt to grow. They did not assimilate the exoskeleton as *crustaceans do when molting, and trilobite molts probably form the bulk of their fossilized remains. Molts have provided much information on the trilobite life-cycle. Shedding occurred at various places on the head-section, along the back (like shrimps) or, more frequently, along a line just behind the eye. Trilobites are relatively rarely found complete as, obviously, the highly segmented skeleton is easily dispersed.

Trilobites were essentially filter feeders, but some were probably able to deal with small prey. They show great diversity in shape and size and different forms were adapted for burrowing, swimming and many other ecological niches. Deeper-water forms often show a considerable reduction of the eye surface. The largest trilobite reached a length of 70cm (27in).

Most trilobite eyes, like those of other arthropods, have several thousand lenses covered by a single cornea. The lenses are of *calcite with the central axis normal to the visual surface, thus largely eliminating double refraction.

Apart from the more primitive forms most trilobites protected themselves like the modern woodlouse – by rolling up into a ball – and various devices were evolved to ensure a complete envelope.

The activities of the bottom-dwelling trilobites have been preserved as *trace fossils in shallow marine sandy environments (where trilobites were most common).

The reasons for their extinction are not understood. There is no obvious competitor for their ecological niches, though the crustaceans began to take their place later, in the Mesozoic. RG

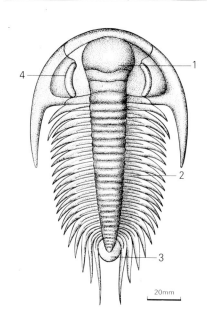

Paradoxides davidis

1 Head
2 Thorax
3 Pygidium
4 Eye

600   400   200   0
millions of years ago

Crustaceans

The crustaceans are the major group of marine arthropods today, and rival the *insects in abundance, if not in diversity. They are distinguished from other arthropods by having two pairs of antennae on the head. The exoskeleton is generally strengthened by patches of calcium carbonate, but to a variable extent. The crustaceans were and are mostly marine, but the superclass as a whole shows an astonishing range of ecological tolerance and adaptation.

Found from the *Cambrian onward, they comprise over 900 genera in 9 classes. The more important fossil classes (apart from the *ostracods) are:

The *Branchiopoda* (which includes the brine shrimps), a relatively primitive filter-feeding group with numerous longitudinal segments, and often partially protected by a bivalved carapace. They first appeared in the early *Devonian.

The *Copepoda* are the marine equivalent

of insects. Although extremely abundant in marine waters today, the only fossils known are from *Miocene lake deposits.

Complete *cirrepedes* (barnacles) are rather uncommon as fossils, because the skeletal plates separate quickly after death. Stalked goose-barnacles are known from the late Silurian, where complete specimens were fossilized still attached to *Eurypterus* (an aquatic arachnid). Acorn barnacles are known from the *Cretaceous.

The *Malacostraca*, which includes the crabs, lobsters and shrimps, is a group known from the early *Cambrian. They show considerable diversity, though good fossils are relatively uncommon, partly due to the incompleteness of calcification. Many forms burrow into sand or mud, sometimes plugging the burrow wall with muddy pellets to produce characteristic *trace fossils. RG

Meyeria magna

[10mm]

600   400   200   0
millions of years ago

Ostracods

The Ostracoda form an extant subclass of *crustaceans. They are common fossils from the lower *Cambrian onward; and, because of their small size, often rapid evolution and the frequent environmental restriction of some groups, are important stratigraphical and environmental indicators. Of the over 900 genera which have been recognized about a quarter still survive.

Ostracods have an egg-shaped bivalved shell of calcium carbonate generally less than 1mm (0.04in) in length. This gives little indication of the form of the animal's soft parts except for the adductor muscles (used for closing the shell), eyes and certain structures which are regarded as brood chambers. Nevertheless there is considerable variation of the shell, with lobes, ridges, pits, spines and frills. Freshwater ostracods are typically smooth-shelled, whereas those of the shallow marine shelf are often deeply sculptured. Most are bottom-dwellers but a small group is planktonic.

The abundance of their shells in the fossil record is undoubtedly influenced by the fact that, like other arthropods, growth is accomplished by molting, this occurring six to nine times during the lifespan, so that each individual may be represented by several shells. RG

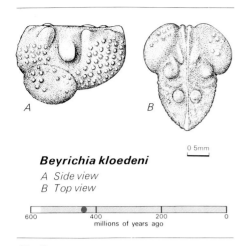

Beyrichia kloedeni

A Side view
B Top view

600   400   200   0
millions of years ago

Chelicerates

The group of *arthropods with a pair of chelicerae, pincers in front of the mouth, is made up of the arachnids, which include spiders and scorpions, and the merostomates, which include the much less formidable horseshoe crabs and the extinct eurypterids. While the merostomes have always inhabited inland or coastal waters the arachnids are typically air breathers. They are relatively uncommon fossils.

The best-known fossil arachnids are spiders fossilized in *Oligocene amber from the Baltic coast; but the first mites are known from the lower *Devonian. Scorpions date from the *Silurian, but were then relatively large animals – sometimes attaining a length of 0.9m (about 3ft). Probably such forms were aquatic, breathing through gills. Around 60,000 species of arachnids are known.

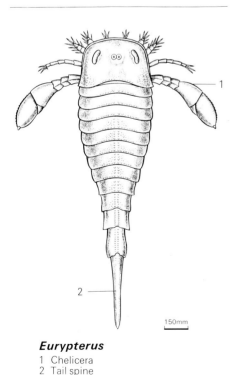

Eurypterus

1 Chelicera
2 Tail spine

600   400   200   0
millions of years ago

The horseshoe crabs show only minor evolutionary changes since the late *Paleozoic: fossils date from the lower *Cambrian. The broad, low-vaulted three-piece shield at the front is followed by a number of segments and a prominent spine toward the rear. Locomotory appendages are borne on this shield; the segments that follow bear the gills.

The eurypterids (represented by 30 genera) range in size from small to very large. They have an elongate body which may be as long as 1.8m (about 6ft), making them the largest known arthropods of all time. The body terminates with either a spine or a spine modified to form a paddle-like tail. The last pair of locomotory appendages is often modified for swimming, and the chelicerae may be very large and obviously predatory.   RG

Insects

Perhaps the most beautiful of fossils is an insect trapped in golden *amber; and many insects are found thus in *Oligocene amber from the Baltic shores. However, in spite of their numerical abundance among living animals, insects have a rather poor fossil record. The reason is obvious: the adults are air breathers and, on death, only rarely will the intact insect sink to the substrate of a fluvial pool or lagoon.

The earliest insect known is a collembolid (springtail) from the lower *Devonian. Winged insects are known from the upper *Carboniferous, associated with *coal beds, though they had probably evolved by late Devonian times (when plants had reached tree size). The largest known insect, found in the Carboniferous, had a wing span of almost 600mm (23in) and was rather like a giant dragonfly. Insects with folded wings and fore-wings modified to become protective elytra (as in the beetles) are also known from the upper Carboniferous. They probably occupied a similar scavenging role to that of today. Butterflies, moths and ancestral dipterans (flies) are known from the lower *Permian, and thus arrived on the scene prior to the evolution of the *angiosperms (flowering plants).   RG

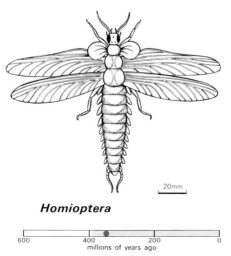

Homioptera

600   400   200   0
millions of years ago

Conodonts

The conodont animal is one of the most elusive fossils. While the small toothed or platey structures of calcium phosphate, known also as conodonts, can be readily found in most marine sediments (especially *limestones) from the *Ordovician to the upper *Triassic, there is still no positive evidence of what their owners were like.

With their rapid evolution, the Conodontophorida are important for stratigraphic zonation, particularly during the upper *Devonian and lower *Carboniferous. Individual conodonts represent an internal secretion built up plate by plate, with a "pulp" cavity at the base. The crown varies from being fang-like to serrated blades, bars and serrated plates.

Six types of natural associations are known, though these do not particularly aid interpretation of the biological relationships between them. The six assemblages are assigned to six genera.

In the absence of the conodont animal, the relationships of the group are in doubt, but they were possibly supporting structures for teeth or, less probably, respiratory organs, of a fish-like or perhaps hagfish-like animal. There are a number of other interpretations.   RG

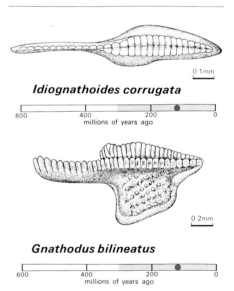

Idiognathoides corrugata

600   400   200   0
millions of years ago

Gnathodus bilineatus

600   400   200   0
millions of years ago

Graptolites

The name graptolite is commonly applied to members of the wholly extinct class Graptolithinia. These somewhat enigmatic fossils first appear in rocks of middle and late *Cambrian age, and finally disappear in the late Carboniferous. Graptolites (the name means literally "rock writing") generally occur as silvery carbonaceous or pyritic films on bedding planes in fine-grained black *shales. The group were originally regarded as inorganic or vegetable in origin, and have subsequently been referred to the Cephalopoda, Coelenterata and Bryozoa. We now know from detailed studies of their microarchitecture that they should be referred to the subphylum Hemichordata, phylum Chordata.

Graptolites were colonial organisms: individuals secreted minute, millimetre-sized cups built up of half-rings of protein or protein-like material. Through complex budding and branching, colonies of from one to many branches were produced. In some graptoloids the contributory cups are of different shapes, suggesting several distinct types of individual were present.

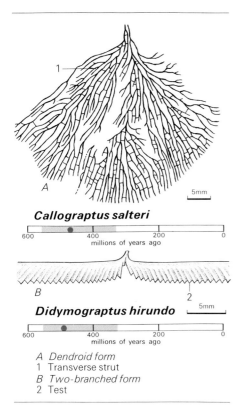

Callograptus salteri

600   400   200   0
millions of years ago

B   2

Didymograptus hirundo   5mm

600   400   200   0
millions of years ago

A Dendroid form
1 Transverse strut
B Two-branched form
2 Test

The earliest graptoloids are referred to a group known as the dendroids, which exhibit cups of differing shapes and developed into shrubby, lace-like colonies of moderate size (usually less than 15cm (6in)), in which the branches were commonly joined by transverse struts. Perhaps the best known is the many-branched *Dictyonema*, which is believed to have been either planktonic or epiplanktonic in its life habits: other dendroids seem to have been sessile, living attached to the sea floor.

From the dendroids arose the graptolites proper, again a planktonic or epiplanktonic group, characterized by colonial structures with generally only a few branches, and only a single type of cup. The earliest graptolites were many-branched, but parallel evolution led to progressive reduction in the number of branches. In addition, evolutionary studies have demonstrated progressive modification of the position of the branches and changes in details of the shape of the cups.

The graptolites are key stratigraphic indicators from the upper Cambrian through to the lowest Devonian: their distribution was very wide, and they are often common in offshore black *shales and siltstones, although generally scarce (through ecologi-

cal exclusion or non-preservation) in nearer-shore sediments such as coarse sandstones and limestones. Because nothing is known of their soft tissues, little is known of their precise life habits. They are believed to have fed on organic debris or planktonic microorganisms, by analogy with their nearest relatives, and float-like devices may have been developed. In some cases several colonies have been found associated in star formations.

Three smaller groups, the Tuboidea, Camaroidea, and Crustoidea have also been described, mostly from Ordovician rocks of the European area.   WJK

Trace Fossils

Trace fossils are structures left in sediment or hard substrates by living organisms. In a ghostly way, they indicate the presence, at one time, of an organism whose bodily remains have long disappeared. Trace fossils therefore include tracks, trails, burrows, borings, fecal pellets (either the fossil excrement itself or castings), resting impressions and other structures actively produced by animals (and, more rarely, plants).

Trace fossils thus represent fossilized behavior – they give an indication of what the animal *did*, despite the fact that only rarely can a trace be attributed to a particular genus or species.

Organisms are only rarely preserved in association with the traces they produce, and much of our inferences about the affinities of trace fossils depends on observation of modern organisms. Since trace fossils are commonly produced by soft-bodied organisms, they frequently give an indication of the presence of elements of fauna that would rarely, if ever, be otherwise preserved.

Trace fossils first appear in rocks approximately 1000 million years old, and provide the earliest evidence for the existence of metazoan (many-celled) animals; also, they give us our only evidence of the early history of several invertebrate phyla during the late Precambrian. Trace fossils are found from all ages in terrestrial, marine and freshwater sediments. Tracks and trails attributed to groups as diverse as polychaetes and dinosaurs, trilobites and fish, have been recognized, and in the case of wholly extinct groups such as the trilobites provide our only concrete evidence of the behavior of the living organisms.

Most trace fossils cannot be attributed to particular organisms, or even to groups of organisms. They can, however, be divided into various behavior categories. These include: *Domichnia*, simple dwelling burrows, more or less permanent habitations of the animals producing them; *Fodinichnia*, tunnel systems produced by semi-sessile sediment-mining organisms; *Pascichnia*, complex meandering trails which systematically cover surfaces without crossing, and which record grazing and feeding activities; *Cubichnia*, generally shallow resting traces which correspond approximately to the form of the producer; and *Repichnia*, which are simple locomotion traces.

In addition to this behavioral grouping of trace fossils, recurrent associations of traces, often dominated by particular behavioral types, have been recognized. The composition of these associations (or trace-fossil facies) was controlled by environmental conditions. They are: terrestrial (non-marine) *Scoyenia* facies; *Skolithos* facies, which represent high energy littoral conditions of rapid sedimentation and transport; *Glossifungites* facies, associated with shallow-water submarine erosion surfaces; *Cruziana* facies, or shallow-water, but sub-littoral conditions; *Zoophycos* facies, from deeper (or quieter) waters transitional to the abyssal zones; and *Nereites* facies, which represent abyssal or deep basinal conditions, often associated with turbidites.

Trace fossils are of only limited stratigraphic value; and individual traces are often (in time) long-ranging, since they were produced by several members of a group or even by quite unrelated organisms carrying out the same type of activity. In a few cases, however, notably *trilobite tracks, there is close correlation between hard-part morphology and trace fossils.

Although at first sight an individual trace fossil tells us little, we can find from them much more about the history of life on Earth than from some of the perfectly preserved fossil animals of (comparatively) recent time.   WJK

Vertebrate Animals

Vertebrates are animals with backbones. They comprise five groups; fish, amphibians, reptiles, birds and mammals. Entries dealing with these groups are in evolutionary sequence; that is, the order in which they first appear in the *fossil record and in which they formed dominant elements of the faunas of the past – an order that is so impressive that the history of the Earth is commonly divided into the Age of Fishes, the Age of Amphibians, and so on.

Where more detailed entries, on the varied fish or reptilian types, are included, their sequence is more arbitrary but nevertheless corresponds roughly to the order of their appearance.   AI

Jawless Fishes

Small plates of bone-like material found in middle Ordovician rocks of the USA represent the earliest evidence of vertebrate life. Little is known about the animals that possessed these plates except that they were members of the class Agnatha, the jawless fishes. In addition to their lack of jaws, another primitive feature that they show is the absence of true paired fins, although in some of them fleshy flaps were developed. Both of these features must have imposed limitations on feeding and movement: it is unlikely that they could have chased and captured active, struggling prey.

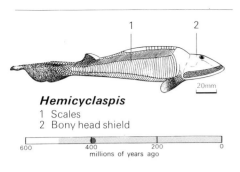

Hemicyclaspis
1 Scales
2 Bony head shield

600   400   200   0
millions of years ago

Evidence of agnathans is most abundant in *Silurian and *Devonian rocks, where they are represented by a great variety of forms commonly known as ostracoderms (shell-skinned). Ostracoderms were small creatures, rarely exceeding 45cm (18in) in length. *Cephalaspis*, a Devonian form, is a well known example. There was a small mouth on the underside of the head and this, together with the general shape of the body, suggests that ostracoderms lived on the sea bottom and fed on organic debris that accumulated on the surface of the mud. The probable method of feeding was to suck in or scoop up the surface mud and the organic debris together. The nutritional disadvantages are obvious.

The earliest ostracoderms lived in shallow marine waters, and this was to remain their chief environment until the end of the Silurian, when several groups adopted a life in the lower reaches of rivers and streams. By the middle Devonian they had added lakes to their range of habitats. Ostracoderms became extinct by the end of the Devonian, and today the lamprey and the hagfish are the only survivors of the ancient agnathan lineage. Both have lost the bony covering seen in their forerunners.

The lamprey is specialized, having adopted a semi-parasitic lifestyle: the mouth is surrounded by a suction disk which is used for attachment to other fish, and there is a rasping tongue to scrape away the host's flesh. Such a mode of feeding may well have developed from the habit that one or two ostracoderms seem to have had of scraping algae off rock. A larval lamprey, however, feeds as the majority of ostracoderms must have done, by sucking in mud.

Studies of ostracoderms have suggested that the original environment of the vertebrates was the sea, and that bone, such a typical vertebrate feature, was first formed in the skin.   PF

Jawed Fishes

The appearance of jaws and associated teeth was one of the most important steps in vertebrate history. Jaws allowed the vertebrates, at that time confined to the water, to capture active prey and so do away with the necessity of sucking in mud, which has generally only a low food-content.

Evidence from the studies of embryology and comparative anatomy suggests that jaws were developed from the front members of the gill-arch series, which lies beneath and behind the braincase. The gill-arches are skeletal structures which support the gills, and when seen from the side each is shaped like a "V" lying on its side, with the apex of the "V" directed backwards.

During the evolution of jaws the front gill-arch became enlarged and the two limbs of the "V" became joined at the apex. This enlarged gill-arch was attached to the braincase by ligaments and moveable joints. In most groups of fish the gill-arch next in line also became modified to provide further support for the jaws.

Together with the development of jaws came the development of two pairs of fins corresponding in position to our arms and legs. These paired fins, the pectorals just behind the head and the pelvics further along the body, helped to control and steer the fish through the water.

Many jawed fishes had thick, heavy scales—seen clearly in this specimen of *Dapedius* from the lower Jurassic beds of Dorset, England.

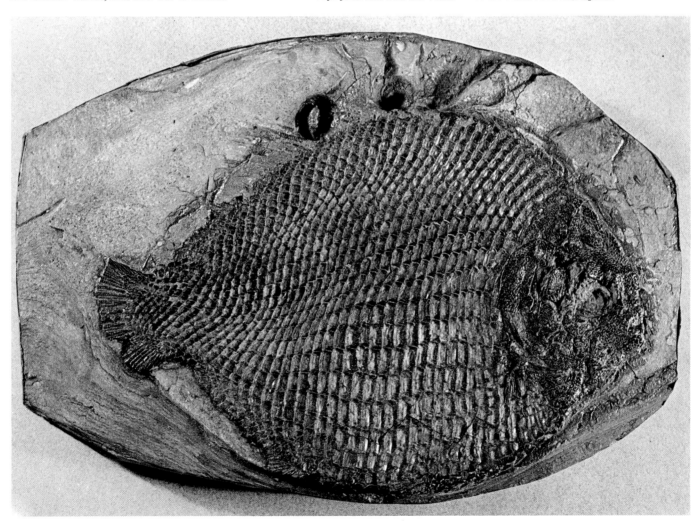

Placoderms

The placoderms or "armored fish" were an exclusively Devonian group of *jawed fishes. A thick armor made of tightly interlocking bony plates covered the head, and this shield was articulated by way of a neck joint with a similarly constructed shield that covered the front part of the trunk. The rest of the trunk and the tail were usually devoid of any bony covering.

There were several different kinds of placoderm. Arthrodires (meaning literally "jointed-neck") were of streamlined shape; and instead of true teeth had large bony shearing blades anchored to the jaws. These features suggest that the arthrodires were active predators – and some, up to 8m (26ft) long, must have been fearsome inhabitants of the late *Devonian seas. The majority were, however, of moderate size, rarely exceeding 1m (3.25ft) in length.

A somewhat curious order of placoderms were the antiarchs. As well as showing the typical placoderm armor, they had a pectoral fin that took the form of a bony appendage rather like a crab's leg. These "limbs" may have been used to raise the front part of the body off the muddy bottom.

The antiarchs were exclusively freshwater dwellers and their remains are found in upper Devonian deposits from all over the world: this distribution is an important addition to the circumstantial evidence suggesting that the land masses of the present world were once conjoined into larger units.

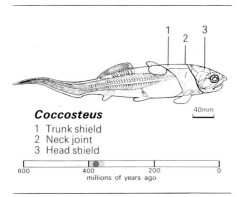

Coccosteus
1 Trunk shield
2 Neck joint
3 Head shield

600   400   200   0
millions of years ago

Cartilaginous Fishes

The cartilaginous fishes, the sharks, rays and the highly specialized ratfish, are first recorded from the middle *Devonian. The name of the group is derived from the fact that the internal skeleton is composed of cartilage, bone being totally absent. The surface of the body is covered with many tiny scales (denticles) instead of large bony plates.

From the beginning the cartilaginous fish were swift, agile predators. The jaws were equipped with many rows of teeth which if lost could be rapidly replaced so that, at any one time, there was always a complete set of razor-sharp teeth.

Most cartilaginous fish of the Paleozoic and early Mesozoic were superficially similar to modern sharks – except for one obvious feature: the base of the pectoral fin

of a Paleozoic shark was very wide, in contrast to the narrow-based fin of later types: the broad-based fin is relatively immobile while the narrow-based fin is more flexible and so allows more rapid turning and upward and downward movement.

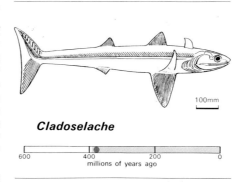

Cladoselache

600   400   200   0
millions of years ago

The first rays are found in the middle *Jurassic, and looked very similar to modern rays. It is probable that they evolved from a shark-like ancestor sometime in the early Mesozoic. The pectoral fins, which were used as the chief propulsive organ, are enormously expanded and merge with the body outline. The tail is reduced to a whip which, in forms such as the stingray, may bear a poisonous spine. Rays feed on hard-shelled invertebrates and for this purpose the jaws are equipped with low-crowned teeth used for crushing.

The cartilaginous fishes were and still are a predominantly marine group. Details of their fossil history are not well known since little of their skeleton is capable of being preserved, so that it is only in occasional circumstances that anything more than teeth and scales is found.

Spiny Sharks

During the Silurian a small group of fishes, the acanthodians or "spiny sharks", made their appearance. Most less than 30cm (12in) in length, they were streamlined and had a body covered with tiny scales which superficially resembled those of sharks: however, the microscopic structure of these scales is completely different, being more like that of *bony-fish scales. The fins comprised a leading spine followed by a thin web of skin. Some early spiny sharks had several small intermediate fins between the pectoral and pelvic fins. Sight must have been important in the life of a spiny shark since the eyes were large.

Spiny sharks lived in medium-depth and surface waters and, although they had large jaws, they frequently did not have any teeth. From this it seems possible that they were plankton feeders, straining food from the surface waters.

Bony Fishes

By far the most numerous and most diverse group of fishes are the bony fishes, the class Osteichthyes; familiar examples include the eel, herring, guppy, perch and cod. Less familiar are the lungfish, coelacanth and the extinct rhipidistians. These last are very

Bony fishes during and after the Cretaceous period underwent a phase of rapid evolution which produced a considerable diversity of body form. *Acanthonemus* has large dorsal and anal fins composed of a small number of fin rays, contrasting with the larger numbers found in more primitive bony fishes.

important as they were probably very similar to the ancestors of the *amphibians, the first vertebrates to conquer the land.

The name of the group is derived from the fact that in the adult the internal and external skeleton is made of bone. The external covering in early bony fishes consisted of thick, tightly fitting, rhomboid scales on the body, and bony plates over the head. Unlike the bony shield of the *placoderms, the head plates could move relative to one another, providing protection with the minimum sacrifice of flexibility. Internally, the early bony fishes had sac-like outgrowths of the esophagus which represent the first stages in the development of lungs.

By middle *Devonian times three major groups of bony fish had become differentiated; the *actinopterygians, the *dipnoans and the *crossopterygians.

Dipnoans

The dipnoans or lungfish are a small group of specialized *bony fish first recorded from the middle *Devonian: today the few surviving species are found in freshwater environments in South America, Africa and Australia. Devonian lungfish show several differences from their actinopterygian contemporaries. The paired fins are limb-like, with the internal skeleton and the muscles extending outside the body. The upper jaw is fused with the braincase, and instead of small teeth there are a few large tooth plates

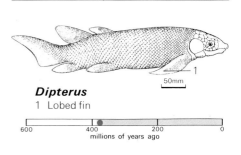

Dipterus
1 Lobed fin

600   400   200   0
millions of years ago

used to crush aquatic plants and hard-shelled invertebrates.

Modern forms have retained the original function of the lung and can breathe air when the pools they inhabit become stagnant. The South American and African lungfish breathe air exclusively for months when encased in cocoons following the seasonal drying-up of their native pools. Since the Devonian, lungfish have changed very little, the chief difference being that in the later forms much of the internal skeleton fails to ossify in the adult and is instead represented by cartilage.

At the turn of the century it was held that lungfish were involved with the ancestry of the *amphibians. This was founded on

superficial resemblances of lungfish fins to the amphibian limb and the common habit of air-breathing. However, more thorough studies have shown that the lungfish are not amphibian relatives but represent a highly specialized lineage of bony fish.

Crossopterygians

The crossopterygians or tassel-fin fishes include the coelacanths and the extinct rhipidistians. They have paired fins outwardly like those of *dipnoans, but lack the latters' skull specializations.

The coelacanths, like the lungfish, are a conservative group, the Devonian forms being little different from later types. The body is plump and the tail has equally developed upper and lower lobes with, between them, a small central lobe. Most of the Paleozoic coelacanths inhabited fresh water but during the Mesozoic many adopted a life in the sea.

It was long believed that the coelacanths had died out in the upper Cretaceous: however, in 1938, a fisherman off the coast of South Africa hauled in a strange-looking fish quickly recognized on reaching port as a coelacanth. Many more specimens of *Latimeria* have now been recovered. Great interest centers around "old fourlegs" since it is the nearest living fish relative of the land-dwelling vertebrates and the extinct rhipidistians.

The earliest rhipidistians are found in lower *Devonian rocks and the latest from the lower *Permian. Most lived in fresh water. Their most interesting feature is that they very probably share a common ancestor with the amphibians.

An example is *Eusthenopteron*, one of the best known fossil fishes. The plump body was about 0.5m (1.6ft) long, with the dorsal and anal fins far back near the tail. The head was large and the mouth particularly so. The overall picture suggests a lifestyle similar to that of the modern pike, lying quietly among the vegetation and darting out to snap at passing prey. Inside the mouth were batteries of small teeth as well as a few large tusks. If one of these tusks is cut across and examined under a microscope, it can be seen that the enamel coat is infolded into the dentine core in a labyrinth pattern. This is significantly similar to the kind of tooth found in labyrinthodont *amphibians.

Another characteristic seen in both rhipidistians and land-dwelling vertebrates is the internal nostril, a feature associated with air-breathing; so it is generally assumed that rhipidistians used lungs much as do the modern amphibians and lungfish (although lungfish do not have true internal nostrils). The characteristic red color of many Devonian rocks is often taken to indicate that arid conditions were prevailing at that time, so that the presumed ability of rhipidistians to breathe air would have had definite survival value.

Behind the head, the spine was composed of bony vertebrae which were very similar in construction to those of the early amphibians. In the paired fins were bones perfectly comparable with the bones of the upper and lower segments of the amphibian fore and hind limb: however, fingers and toes were absent. Significantly, the shape of the limb bones indicates that the limb muscles of *Eusthenopteron* were capable of rotating the limb to a considerable degree (a feature actually observed in a modern coelacanth). This rotation is a prerequisite for walking on land. Indeed *Eusthenopteron* and its allies appear to have been capable of taking short excursions onto the land, and may well have done so to escape the stagnant pools in which they were living.

By the later Devonian the amphibian had appeared, and so the history of the land-dwelling vertebrates had begun. The history of the fishes saw a gradual replacement of primitive Devonian types by fishes superior in swimming and feeding abilities. The present fish fauna is dominated by the *actinopterygians and the *cartilaginous fishes.

Actinopterygians

The ray-finned fishes or actinopterygians are distinguished from other *bony fish by the structure of their paired fins: the muscles which move the fins are inside the body, so that the only parts of the fin visible from the outside are the slender rays which support the web of skin.

*Devonian actinopterygians were streamlined in shape with large eyes, indicating that sight rather than smell was important in the life of these predatory fishes. The jaws were long and had many pointed teeth. These early forms are called *chondrosteans* and were predominantly a Paleozoic group, though a few, such as the sturgeon, survive today. Though the sturgeon is a highly specialized form, it still has the long, upturned tail seen in extinct chondrosteans.

During the Mesozoic the actinopterygian lineage was represented by the holosteans and teleosts. Most holosteans still had the early type of scale forming a tough body covering, but other features indicate improvements in swimming and feeding ability. The lungs became modified to form a swim-bladder, lying above the center of gravity of the fish and functioning as a float chamber, so that the body of the fish approached neutral buoyancy, an energy-saving development since it eliminated the

need for constant swimming in order to stay at one level in the water. Concomitant with this development, the tail became shorter and less upturned and produced a more nearly horizontal thrust than did the chondrostean tail which tended to drive the head of the animal downwards, an obvious disadvantage.

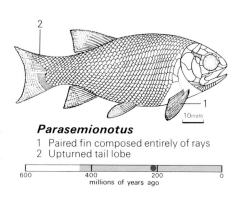

Parasemionotus
1 Paired fin composed entirely of rays
2 Upturned tail lobe

600   400   200   0
millions of years ago

Holosteans were a varied group. Some forms such as the *Jurassic *Dapedius* were deep-bodied with teeth adapted for crushing, while *Aspidorhynchus*, also a Jurassic type, was elongate with a long, bill-like nose. Most became extinct by the end of the Mesozoic, though two forms, the bowfin and the garpike, survive today in freshwater environments in North America.

By the end of the *Triassic early teleosts had appeared. The teleosts are the largest and most diverse group of actinopterygians, about 20,000 species being extant today.

The evolutionary trends seen in the teleost fishes parallel those in the holosteans, but went much further. The jaws became shorter and more mobile. The tail became outwardly symmetrical resulting in a perfectly horizontal thrust, and this, plus the fact that there was a well developed swimbladder, obviated the need for rigid pectoral fins to act as hydrofoils: indeed, in the great majority, the pectorals moved to lie high up on the side of the body so that they could be used as brakes for rapid stopping and to aid in quick turning. The scales became thin, circular and overlapping, protection thus being sacrificed for flexibility, in contrast to the holosteans and chondrosteans. Most of these trends resulted in unsurpassed ability to maneuver in the water.

Although the teleosts are known from the early Mesozoic, it is not until the lower *Cretaceous that their remains are abundant. By this time several surviving groups had arisen: the tarpons, the herrings, the osteoglossomorphs, and extinct relatives of the salmon and trout. Later in the Cretaceous there is evidence of primitive members of two important lineages. One, the spiny-rayed teleosts, comprising the largest and most diversified group of teleosts, is today exemplified by the perch, mackerel, plaice and angelfish: as the name suggests, these fishes have spines in the dorsal and

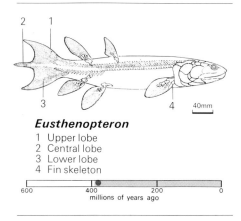

Eusthenopteron
1 Upper lobe
2 Central lobe
3 Lower lobe
4 Fin skeleton

600   400   200   0
millions of years ago

anal fins. The other includes the cod, whiting and anglerfish. The Cenozoic saw the appearance of a very large group of freshwater fishes which includes the minnows, carps and catfish. Recent research suggests they had marine ancestors in the Cretaceous.   PF

Amphibians

It is in many ways unfortunate that the word "amphibian" is used to describe the early, Paleozoic vertebrates which first colonized the land. The word certainly describes their ability to live both on land and in the water. Like most modern amphibians, they also laid and fertilized their eggs in the water, and these eggs developed into aquatic larvae with gills, which only later emerged onto the land. But "amphibian" makes most people think of the little living forms, such as frogs, newts and salamanders, which are very different from their remote Paleozoic ancestors. The main difference lies in the fact that living amphibians respire not only through the lungs but also through the skin: the skin must therefore be kept moist, and amphibians consequently continuously lose water through it. This limits both their size and the range of habitats they can colonize; they are unable to survive for long in dry environments or in the sea.

Paleozoic amphibians cannot yet have developed this peculiar specialization, for many had a covering of scales or, more rarely, a dry leathery skin. They were therefore able to grow to a much greater size, and a few took to the seas.

Origin of the Amphibia. Amphibians evolved from the rhipidistian lobe-finned fishes (see *crossopterygians). These fish had not only lungs but also, as their name implies, rather stout, muscular fins which could have supported them on land. They probably first crawled ashore as juveniles, trying to escape the attacks of larger predatory fish. Able both to breathe and to move about on land, they would have found also a new and unexploited source of food in the insects and other invertebrates then beginning their own invasion of the land.

This situation provided the opportunity for the first step in the vertebrates' conquest of the land. The amphibian step was, however, a very hesitant one. It appears that the overwhelming majority of the Paleozoic amphibians spent all, or nearly all, of their lives in the water. This may be partly illusion: the majority of the deposits from which we know *Carboniferous amphibians were laid down in the coalswamps of the northern hemisphere, whose inhabitants would of necessity have been aquatic. Possibly other, more terrestrial amphibians lived on higher, drier ground. But *reptiles evolved from amphibians in the middle to late Carboniferous; so the reason that we know very few land-dwelling amphibians is probably that competition from the reptiles excluded them from a

terrestrial way of life. This is also the reason for the gradual disappearance of nearly all the amphibians during the late Permian and the Triassic.

Paleozoic amphibians can be subdivided into two groups according to a difference in the structure of the vertebral column. In the first group, the lepospondyls, that part of each vertebra that lies below the spinal cord is made up of a single bone. In the other group, the apsidospondyls, this region is made up instead of several blocks of bone: they are often called labyrinthodonts because the enamel tooth surface is folded into the center of the tooth in labyrinthine fashion.

Labyrinthodonts. The earliest known amphibian is *Ichthyostega*, from the latest Devonian of Greenland. Its skull had no trace of the "operculum" which covers and protects the gills of fish, so it presumably did not have gills (at least during adulthood), and must have breathed through its lungs alone – its body was covered with fish-like scales so it cannot have respired through the skin. The fins of its fish ancestors had become stout limbs, the upper parts projecting sideways from the body, which was thus suspended between them rather than supported above them: this posture is found in most amphibians and reptiles.

Later labyrinthodonts are divided into two groups, the temnospondyls and the anthracosaurs, which differ in the details of construction of the vertebrae. The temnospondyls include the greater number and variety. Most were aquatic, with a rounded body and slightly elongated skull with many teeth: they probably fed on fish. The rather small size of their limbs shows they would have found it difficult to move on land.

By the early Permian had appeared such semiaquatic crocodile-like types as *Eryops*, whose broad skulls suggest that they may have preyed on other amphibians or even lurked in the water for reptiles that came down to drink. Some little amphibians with gills, once placed in a separate group known as the branchiosaurs, were in fact the aquatic larvae of such forms as *Eryops*, showing that the life history of these Paleozoic forms was indeed the same as that of living amphibians. A few totally terrestrial types, the dissorophids such as little *Cacops*, are also known from the early Permian.

Other temnospondyls developed an extremely elongate, narrow many-toothed snout, like that of the modern fish-eating crocodile, the gavial. Some of these, the Permian archegosaurs, lived in fresh waters, but the Triassic trematosaurs took to the seas – the only marine amphibians known.

The Triassic temnospondyls were all semiaquatic, with wide, flattened skulls. In some, the capitosaurs and metoposaurs, the skulls were quite long; while in others, the plagiosaurs, it was very short: types such as *Gerrothorax* probably lay open-mouthed on the bottom, attracting fish by means of a

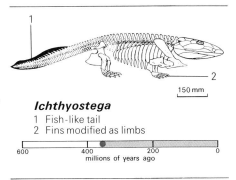

Ichthyostega
1 Fish-like tail
2 Fins modified as limbs

600   400   200   0
millions of years ago

colored fleshy lure.

The other major group of labyrinthodonts, the anthracosaurs, contains both aquatic and terrestrial forms: aquatic types are called embolomeres. Some were fish-eaters, such as *Eogyrinus*, an elongate Carboniferous amphibian about 4m (13ft) long. Its relative, *Archeria*, of the early Permian, about 2m (6.5ft) long and with a fish-like fin on its long tail, had a number of small, peg-like teeth and may have fed on tiny animals which it strained out of the water.

The terrestrial anthracosaurs, known as seymouriamorphs, include the only large terrestrial amphibians known. *Seymouria* itself, from the early Permian of Texas, was 1.5m (5ft) long and seemed so completely adapted to a terrestrial way of life that it was classified as a reptile until it was discovered that its young lived in the water, showing that it was really an amphibian. Its relative *Diadectes* was even larger, some 3m (10ft) long: its bulky body and flattened grinding teeth suggest that it was a herbivore – if so, it is the only known herbivorous amphibian. Though reptiles are probably descended from the seymouriamorph anthracosaurs, they evolved from that group much earlier, during the Carboniferous.

Lepospondyls. Unlike the labyrinthodonts, the lepospondyls were all quite small – most were only 10–15cm (4–6in) long. The majority were completely aquatic and lived during the Carboniferous. Some, such as *Ophiderpeton* and *Phlegethontia*, were limbless and elongate, swimming by side-to-side undulations of the body, like water snakes today. Others, such as *Microbrachis*, retained their limbs and were more newt-like. Only a few early Permian relatives of *Microbrachis* had sturdier limbs and were, at least as adults, apparently wholly terrestrial.

Modern Amphibians. Modern forms are in three groups: the tailless frogs and toads; newts and salamanders; and the limbless, burrowing apodans. These groups represent the only ways of life to which amphibians have been able to cling in face of competition from the reptiles.

Though quite different from one another in appearance, they share a number of specializations, such as peculiar, hinged teeth and use of the skin for respiration, suggesting that they probably evolved from a single Paleozoic group by a single original line of descent, diverging later; but the identity of this group is still uncertain. Both

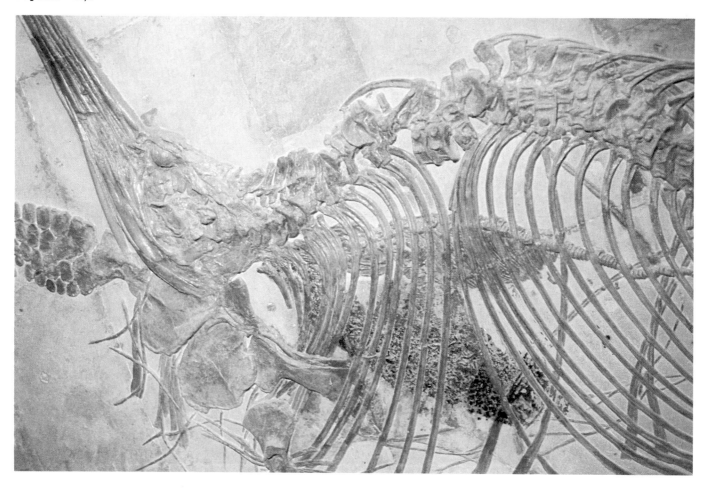

The vast majority of living reptiles lay eggs. Here we see evidence that marine reptiles of the Mesozoic, the ichthyosaurs, were able to bear live young. An immature *Ichthyosaurus* can be seen within the body cavity of its parent. Below this unborn young it is possible to see the remains of the adult's stomach contents.

the dissorophid temnospondyls and the *Cardiocephalus*-like lepospondyls have features that suggest that they may have been ancestors of the modern amphibians, but intermediate forms are unknown.

The earliest known frog is the Triassic *Triadobatrachus*, already showing the start of the great shortening of the body and enlargement of the hind limbs associated with the jumping habits of this group. Newts and salamanders did not appear until the late Cretaceous; and the apodans not until the Paleocene, but it is not surprising that few remains of these little, frailboned creatures have been preserved. CBC

Reptiles

The hard or leathery egg-shell, protecting its contents from desiccation and small predators, is the characteristic feature of reptiles. They are classified primarily by the skull structure. In the most primitive, the *cotylosaurs, and their specialized descendants the *turtles, the skull roof is solid except for openings for the eyes and nostrils: this type of skull is known as *anapsid*.

In all other reptiles, openings appeared

behind the eyes to reduce the weight of the skull. The *synapsid* type is found in the *mammal-like reptiles, the *parapsid* type in the aquatic *placodonts, *plesiosaurs and nothosaurs.

The commonest skull type is *diapsid*, present not only in *lizards and *snakes but also in the great group of archosaurian reptiles – the *dinosaurs and their ancestors the *thecodonts, their relatives the *crocodiles and *pterosaurs, and their descendants, the *birds.

Reptiles evolved in the early *Carboniferous, and were the dominant land vertebrates from the early *Permian until the end of the *Mesozoic, when the mammals became dominant.

Cotylosaurs

This group of anapsid *reptiles includes three different types. The captorhinomorphs include the earliest known reptiles, from the early *Carboniferous. Similar to modern lizards in size and appearance, and probably also in diet, they survived until the late *Permian.

The second group were the pareiasaurs, which were up to 3m (10ft) long. The many teeth and clumsy, lumbering body suggest that they were herbivores: if so, they were the first successful reptilian herbivores; but their success was short-lived, for they are found only from the late *Permian. Their extinction may have been due to competition from the dicynodont *mammal-like reptiles.

The third type are the procolophonids,

half a metre (20in) long, with skulls often ornamented with protective spikes of bone. Probably omnivorous, procolophonids appeared in the late *Permian and survived until the end of the *Triassic. It is possible that they were the ancestors of the *turtles.

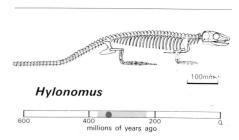

Hylonomus

100mm

600   400   200   o
millions of years ago

Turtles and Tortoises

Their unique bony shell, fused to the ribs and the outer surfaces of the vertebrae, is the most obvious characteristic of these animals, the order Chelonia. They are the only surviving anapsid *reptiles, and probably owe their survival to the protective shell. The first chelonian, *Proganochelys*, appeared in the late *Triassic. It was already completely tortoise-like except that it still had teeth. Many of the *Jurassic and some *Cretaceous chelonians, in which the specializations of the neck vertebrae found in later forms had not yet appeared, are placed in a suborder, Amphichelydia.

These specializations involve two different methods, both of which appeared in the early Cretaceous, of withdrawing the head into the shell. The neck is flexed sideways in the suborder Pleurodira, which is mainly confined to the southern hemisphere and includes only a few living forms; while in the suborder Cryptodira the neck is flexed vertically. This latter suborder includes most living chelonians, including terrestrial, freshwater and marine forms.

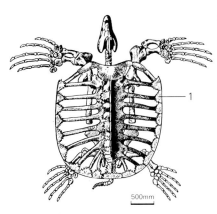

Archelon
1 Armor plates attached to ribs

Ichthyosaurs

Just as seals and *whales evolved from land mammals, so some *reptiles found new opportunities in returning to the sea. One of these groups, the ichthyosaurs (the others were the *plesiosaurs and nothosaurs, and the *placodonts), developed a very porpoise-like appearance. They had fish-like propulsive tails, soft dorsal fins and streamlined bodies, which were leathery and had no scales. The limbs have become small, inflexible paddles which can have been used only for steering and certainly not for support on land, so it is not surprising to find some specimens containing the skeletons of young individuals, indicating that ichthyosaurs almost certainly bore living young, as do sea-snakes today.

Most ichthyosaurs were about 3m (10ft) long, but specimens 12m (40ft) long are known. They had a long, narrow snout with many short teeth, and must have fed on fish and ammonoids.

The first ichthyosaurs, known as mixosaurs, evolved in the middle *Triassic. At that time the tail was still low and elongate. An unusual Triassic group, the omphalosaurs, had a short, strongly-built skull and stout, blunt teeth, which they probably used for crushing mollusk shells.

Ichthyosaurs became common and diverse in the *Jurassic, and are found in marine deposits in most parts of the world. They are divided into "longipinnate" forms, in which the paddles contained only five elongated fingers, and "latipinnate" forms, in which the paddles were shorter but broader, containing up to four extra fingers. The latipinnate forms were the more successful: the longipinnates did not survive beyond the early Cretaceous.

Ichthyosaurs continued into the *Cretaceous, but were less common. Though they became extinct somewhat before the end of the Cretaceous, their disappearance is part of a general pattern of extinction at that time of so many creatures on land and in the sea and air.

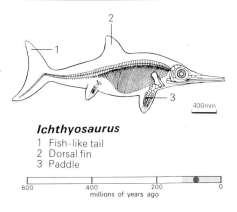

Ichthyosaurus
1 Fish-like tail
2 Dorsal fin
3 Paddle

Plesiosaurs and Nothosaurs

Plesiosaurs and nothosaurs had the characteristic parapsid skull and a long series of pointed teeth. Like the *ichthyosaurs and *placodonts, they were reptiles which returned to the sea; but, unlike the ichthyosaurs, they retained their limbs as a method of propulsion instead of developing a fish-like tail. They had a rather inflexible spindle-shaped body, and their limbs became enlarged into powerful paddles. In life, the paddles were probably used rather

like those of turtles today, with a graceful up and down stroke rather like a bird's wing in slow motion. Most must have fed on fish.

The nothosaurs were the first to appear, in the middle *Triassic. Up to a metre (3.25ft) long, they seem to have lived mainly in the coastal waters of Tethys, the sea that covered parts of Europe and the Middle East at that time. But they have been found also in shore deposits, and so had not yet become as thoroughly adapted to marine life as did the plesiosaurs. This is also apparent from their limbs, for their hands and feet were merely webbed and slightly elongated, and did not develop into large paddles. Nothosaurs ranged from 30cm (1ft) to 6m (20ft) in length, and became extinct before the end of the Triassic, probably due to competition from the plesiosaurs.

These appeared in the late Triassic, probably from a nothosaur ancestor. They can be divided into two groups whose most obvious difference is in the length of neck. The Plesiosauroidea had a long neck, with up to 44 vertebrae, and a rather small head. They may have used the flexible neck in catching their maneuverable food, fish. The creatures themselves were from 3 to 13m (10–42ft) long. The Pliosauroidea, in contrast, had a large head on a shorter neck. They became common in the middle Jurassic. Some, like *Stretosaurus* and *Kronosaurus*, were 12m (40ft) long with a massive head. Just as sperm whales today feed on squid, these pliosaurs fed mainly on a type of cephalopod, the *ammonoids, whose remains have been found inside their skeletons.

All the plesiosaurs became extinct at the end of the *Cretaceous.

Placodonts

Like the *ichthyosaurs and *plesiosaurs, placodonts were reptiles which returned to the sea to live. But, unlike most other marine reptiles, they were specialized for a diet of *mollusks. To pull the mollusks from rocks, or to dig them up, the front end of the jaws bore forwardly directed teeth or horny beaks. The shells were then cracked between enormous flattened teeth, set into strongly built jaws.

Placodonts lived in the Tethys, and propelled themselves by their long tails and their flipper-like paddles, the body being short and inflexible. Later forms, such as

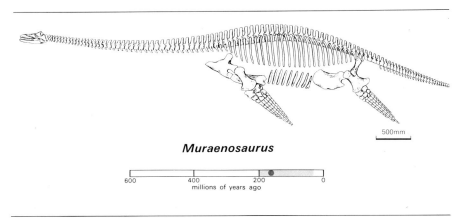

Muraenosaurus

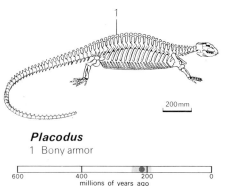

Placodus
1 Bony armor

An almost perfectly preserved specimen of the Triassic nothosaur *Ceresiosaurus*. Unlike the related plesiosaurs, nothosaurs did not have paddles; their limbs were relatively unmodified although these reptiles were undoubtedly aquatic.

Placodus and *Henodus*, had a very turtle-like armor of bone, but differing from that of *turtles in being composed of a large number of bony plates, and in not being attached to the ribs and vertebrae. The ventral armor was also flexible, to allow the animal to breathe by expanding and contracting the volume of the body. *Henodus* also resembled turtles in having strong horny plates in place of teeth.

Placodonts ranged in size from 1 to 2m (3.25–6.5ft). They were a short-lived group, becoming extinct by the end of the Triassic.

Snakes

These reptiles, which originally had a diapsid skull, are descended from *lizards, in which the amount of bone in the skull is reduced. The process is taken even further in snakes, both pairs of holes in the skull losing their lower borders so that the skull becomes very flexible. Together with hinges in the lower jaw, whose right and left halves can also swing apart at the front end, these features make it possible for snakes to swallow in one piece prey that is larger than their own body diameter.

Apart from a short-lived middle *Cretaceous group, the earliest known snake is *Dinilysia* from the late Cretaceous. This appears to be related to snakes such as boas which kill their prey by constricting and so suffocating them. The smaller type of snake, known as colubroids, which includes 90% of living snakes, are not known until the *Eocene. However, their remains are more difficult to find than those of larger snakes, and the group almost certainly originated much earlier, probably in the Cretaceous.

Several types of colubroid snake developed poison, which can be injected into their prey by means of enlarged fangs. The elapids, in which the fangs are fixed, and the viperids, with fangs which can be rotated back up into the roof of the mouth when not in use, are both first known from the *Miocene but probably evolved much earlier.

Lizards

These evolved from the later captorhinomorph *cotylosaurs. Their earliest known ancestors are the paliguanids from the late *Permian to early *Triassic: the skulls of these had already evolved from diapsid to distinctly lizard-like. An unchanged diapsid skull is found only in the rhynchocephalian line, which led to the peculiar herbivorous rhynchosaurs of the Triassic, and to the modern tuatara, *Sphenodon*.

The hallmark of true lizards was the rapid lightening of the skull by the loss of bone. Despite its lightness, the skull retains strength by becoming very flexible. The earliest known true lizards are the very specialized *Kuehneosaurus* and *Icarosaurus* from the late Triassic. Our knowledge of *Jurassic and *Cretaceous lizards is still very imperfect. At least six of the eleven or twelve modern superfamilies had appeared by the late Jurassic, and they must have increased in variety and abundance in the early Jurassic to achieve an importance similar to that which they show today. They also gave rise to a group of marine lizards,

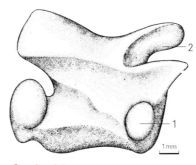

Coniophis

Snake vertebra
1 Articulation facet for rib
2 Articulation facet for adjoining vertebra

600   400   200   0
millions of years ago

the early Cretaceous aigialosaurs, whose late Cretaceous descendants, the mosasaurs, were up to 6m (20ft) long and swam by means of a powerful tail. It was also from the lizards that *snakes evolved.

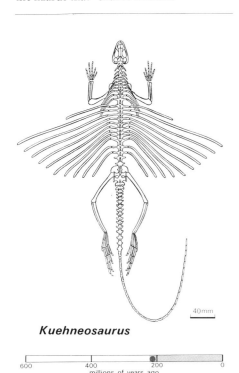

Kuehneosaurus

600   400   200   0
millions of years ago

Thecodonts

The thecodonts were the ancestral group of all the archosaurian reptiles, and therefore directly or indirectly the ancestors not only of the *dinosaurs and *pterosaurs, but also of the *crocodiles and *birds. They had strongly-built diapsid skulls, and diversified into a wide variety of Triassic reptiles whose interrelationships are still only imperfectly understood. However, three main types can be recognized: amphibious forms, bipedal terrestrial forms, and quadrupedal terrestrial forms. Though such a division may not accurately reflect their evolutionary relationships, the thecodonts as a whole are best understood by considering them in this way.

Surprisingly, the earliest known thecodonts, *Proterosuchus* (*Chasmatosaurus*) and its relatives, were already amphibious forms. One to two metres (3.25–6.5ft) long, these animals were crocodile-like in appearance, with powerful elongate jaws, though the front end of the upper jaw was turned down in a strange fashion. They are found only in the early *Triassic; but may have given rise to the two other semiaquatic types, the poorly-known proterochampsids of the middle and late Triassic, and the late-Triassic phytosaurs. The latter, too, looked very like crocodiles, but have an obvious distinguishing feature: their nostrils are positioned just in front of their eyes rather than at the end of the elongate snout. Phytosaurs are little known from the southern hemisphere.

Though it probably normally walked on all fours, *Euparkeria* of the early Triassic could move on two legs when running fast. It may have been the ancestor of the later large bipedal carnivorous thecodonts, such as *Ornithosuchus*. Three metres (10ft) long, this late Triassic thecodont is barely distinguishable from a small dinosaur.

The quadrupedal thecodonts, too, appeared quite early – in the form of *Erythrosuchus*. Stockily-built and 3.5m (11.5ft) long, this early Triassic animal had a massive head and must have been a formidable carnivore, preying on *mammal-like reptiles. It was probably ancestral to the variety of quadrupedal thecodonts found later in the Triassic. Most of these were carnivorous, but they include also two late Triassic herbivorous groups – the aetosaurs, pig-like root-grubbers with bands of bony armor, and the stagonolepids.

Though the dinosaurs, crocodiles and probably pterosaurs descended from the thecodonts, the thecodont group from which each evolved is still uncertain. But the disappearance of the thecodonts at the end of the Triassic must have been due to competition from their descendants.

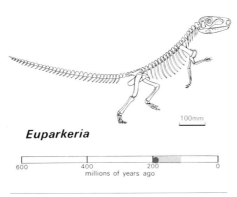

Euparkeria

600   400   200   0
millions of years ago

Pterosaurs

Though these flying reptiles, relatives of the dinosaurs, are often called pterodactyls, that name should really only be applied to the more advanced of the two suborders of pterosaur, the more primitive being known as the rhamphorhynchoids.

Though they had a leathery flight membrane like a bat's, that of pterosaurs was borne behind the elongated wrist and fourth finger, not spread across several fingers. The first three fingers are of normal size, and must have been used for grasping and for support when at rest. The flight membrane continued alongside the body to the hind limbs, which were small and weak.

Pterosaurs are usually found in marine deposits, and it seems likely that they were fish-eaters, flying out to sea in search of their prey. Their wings were large, and they must have mostly glided. They probably lived on sea cliffs, since on flat terrain they would have been vulnerable to predators and would have found it difficult to become airborne. They had diapsid skulls, and probably evolved from the *Triassic *thecodonts, but nothing is known of their ancestry.

The primitive rhamphorhynchoids appeared first at the end of the *Triassic and survived throughout the *Jurassic. They had long tails ending in a vertical rudder membrane, a relatively short wrist, and a wing-span of up to a metre (3.25ft).

They were replaced fairly suddenly in the late Jurassic by their descendants the pterodactyloids, in which the tail was reduced to a stump and the wrist bones were more elongate. In many pterodactyloids the teeth were reduced or lost and replaced by a horny beak.

They ranged from the sparrow-sized *Pterodactylus* to *Pteranodon*, with a wing-span of nearly 7m (23ft) and a large crest projecting from the back of its head. An unusual recent find was the partial skeleton of an enormous pterodactyloid, *Quetzalcoatlus*, whose wing-span may have been 12m (40ft): apart from its great size, it was exceptional in being found far inland.

The sparrow-sized pterodactyloids are found only in the late Jurassic, but the pteranodontids survived until the end of the *Cretaceous. Though the rise of the *birds from the late Jurassic onward may have contributed to their extinction, it is probably significant that their final disappearance at the end of the Cretaceous coincided with the extinction of so many other animals both on land and in the sea.

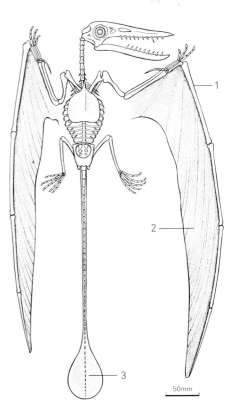

Rhamphorhynchus
1 Fourth finger
2 Flight membrane
3 Rudder membrane

600   400   200   0
millions of years ago

Some of the finest dinosaur fossils have been recovered from Dinosaur National Monument, Utah, USA. These bones are the remains of individuals drowned in a river and deposited on a sandbank where accumulation of sediment was extremely rapid.

Dinosaurs

No other time has such a romantic appeal as the period between 200 million and 65 million years ago, when the world was dominated by dinosaurs. Not only because of the great varieties of type, including some of immense size and others with a bizarre armor of plates and spikes of bone, but also because of their still unexplained sudden disappearance, dinosaurs fascinate our imaginations.

Dinosaurs were reptiles, and therefore laid eggs, some of which have been found in the sands of the Gobi Desert and elsewhere. But not all large reptiles are dinosaurs, and dinosaurs themselves can be divided into two great groups which may have evolved independently from ancestral *thecodonts. These two groups, the Saurischia and the Ornithischia, can be easily distinguished from one another by their pelvic girdles.

Both groups appeared first during the second half of the *Triassic and had evolved into a variety of different types by the end of that period. Because the Earth then contained only a single supercontinent, Pangaea, with few great mountain ranges, and because the climate was mild and uniform, these early dinosaurs soon spread throughout the world.

Saurischian Dinosaurs. These were the first dinosaurs to appear, in the middle Triassic. Though several different saurischians are known from the late Triassic, most of these were not ancestors of the later, Jurassic, dinosaurs, but were instead members of a separate group known as the prosauropods. They had a long neck and tail, and a fairly bulky body. Though some may have been flesh-eaters, most had smaller, serrated teeth better suited to a diet of plants. Like most herbivores, they had begun to increase in size. Some, such as *Thecodontosaurus* were only 2–3m (6.5–10ft) long and, though they normally walked on all four legs, probably reared onto their hind legs to feed on higher foliage or to run fast, their long tails helping to balance the rest of the body. Larger forms, such as *Messospondylous* and *Plateosaurus*, 6m (20ft) long, may have been able to rear up on two legs when feeding, but would not have been able to run in this fashion.

The prosauropods became extinct at the end of the Triassic: their successors probably evolved from early prosauropods. These successors, the sauropods, are the great quadrupedal herbivores which dominate the museums of the world today as they must have dominated Jurassic and Cretaceous landscapes. From the 28m (91ft) long *Diplodocus* to the 12m (39ft) high, 100 tonne *Brachiosaurus*, these animals hold all the size and weight records for terrestrial animals.

With massive legs and long necks and tails, sauropods have often been portrayed as amphibious animals, living in lakes and streams. This has been based mainly on the belief that their limbs would have been unable to support the weight of their bodies. But it is difficult to see how these great animals would have been able to find enough plant food in such an environment. Also, their comparatively small feet would have sunk into the mud, fatally trapping them. It has similarly been suggested that the long neck of the sauropod evolved to enable it to move in deep water and still keep its head above water. But the body would then have been so far below the surface that water-pressure would have made it impossible for it to expand its chest to breathe air into its lungs. For these reasons, many paleontologists now believe that sauropods lived on land, where plant food is much more abundant, and used their long necks in giraffe-like fashion to feed on the foliage of tall trees.

Alongside these quadrupedal herbivores lived a variety of both large and small carnivorous saurischians, all bipedal. The smaller ones, called coelurosaurs, first appeared in the late Triassic. Such little, slenderly built forms as *Coelophysis* had long flexible necks and a long balancing tail. Even smaller coelurosaurs existed in the Jurassic, *Compsognathus* being only the size of a chicken.

The coelurosaurs had long hind legs and must have been fast-moving. They survived until the end of the Cretaceous, latterly giving rise to the ornithomimids, which looked rather like an ostrich with long arms. Nearly 3m (10ft) long, these very slender, long-legged dinosaurs clearly did not prey on other animals of similar size, for they had no claws or teeth, their rather weak jaws bearing a horny beak.

The existence of such massive herbivores provided the opportunity for the evolution of carnivores large enough to take advantage of this source of food. Like their great prey, the carnosaurs appeared at the start of the Jurassic, when such forms as *Megalosaurus* were already up to 9m (30ft) long. Carnosaurs had massive heads and hind limbs and short, heavy necks, but the forelimbs were comparatively small and bore three powerful clawed fingers.

Towards the end of the Cretaceous, the largest megalosaurids were replaced by their descendants, the tyrannosaurids, the largest and most formidable carnivores the world has ever known. This type of carnosaur is distinguished by having very tiny forelimbs, ending in two clawed fingers: the function of these little limbs is uncertain. The best known tyrannosaur is *Tyrannosaurus* itself, which was 15m (50ft) long and 5–6m (16–20ft) high.

Ornithischian Dinosaurs. The first of the ornithischians appeared in the late Triassic, with such little metre-long bipedal forms as *Fabrosaurus*. These not only had the characteristic pelvis of ornithischians, but also, like all members of that group, had a horny beak in place of front teeth. This

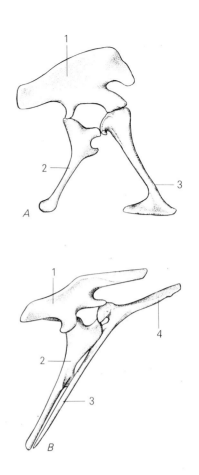

The two types of pelvic girdle found in dinosaurs. (A) The saurischian type and (B) the ornithischian type. Ilium (1), ischium (2), pubis (3) and front process of the pubis (4).

may have been one of the reasons why all the ornithischians seem to have been herbivores or omnivores. Their success as herbivores may also have been due to the fact that, like mammals but unlike most reptiles, they apparently had muscular cheeks which helped retain the food and force it into the battery of many leaf-like teeth. Most ornithischians probably stayed on all fours when at rest or moving slowly, but could rise up onto their hind legs for fast movement, the body being balanced by a rather stiff, inflexible tail.

Few early Jurassic forms are known. Larger ornithischians, known as ornithopods, had appeared in the middle Jurassic, but the best-known is even later, *Hypsilophodon* from the early Cretaceous. Very like *Fabrosaurus* in general proportions, it was about 1.5m (5ft) long. Though at one time it was thought that it was semi-arboreal, it has now been recognized as a normal terrestrial dinosaur.

Larger ornithopods evolved from the hypsilophodonts during the Jurassic and Cretaceous. The most conservative were the iguanodonts, which were little more than enlarged versions of *Hypsilophodon*. *Iguanodon* itself was about 5m (16ft) high

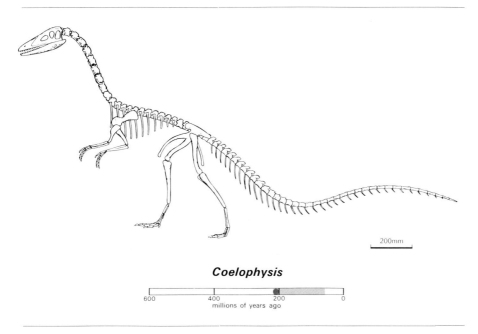

Coelophysis

600   400   200   0
millions of years ago

and 11m (36ft) long, and weighed about 4.5 tonnes. Its forelimbs were quite large, and the thumb had become modified into a spike-like structure whose function is still uncertain.

From the iguanodonts in turn came the bizarre hadrosaurs, which first appeared in the early Cretaceous. 9m (30ft) long and weighing about 3 tonnes, they are sometimes called the duck-billed dinosaurs because of the flattened shape of their horny beaks, behind which was an even more extensive battery of teeth than that found in other ornithischians. But the most remarkable feature of most of them was the extension of the bones of the forehead into a variety of weird crests and horns, through which ran the nasal passages. The most likely explanation of these structures is that they served a social function in these herbivores, which probably lived in herds, much as antlers affect the social structure of modern herds of deer. Moreover, the crest's hollow cavities would have magnified the bellows and cries of the animals. Some scientists have suggested that hadrosaurs

were semi-aquatic, but the facts that their fingers and toes bore hooves, and that the fossilized stomach contents of one hadrosaur contains the remains of land plants and conifer needles, indicate otherwise.

A rather similar function has been suggested for the greatly thickened skull roofs that give the name "dome-heads" to another group of ornithopods, the pachycephalosaurs. Otherwise rather like *Hypsilophodon*, these animals had skull roofs 5cm (2in) thick, and it is thought that these were used in mating battles, the males charging each other head on, as rams do today, until the lighter male submitted to his more powerful adversary.

The remaining ornithischians were all completely quadrupedal descendants of the ornithopods, having developed defensive plates and spikes of bone. The earliest, the 4m (13ft) long stegosaurs of the middle and late Jurassic, had a line of projecting triangular plates or spikes protecting their backs. This group was replaced in the early Cretaceous by the ankylosaurs, whose backs were covered by bony plates, knobs

or spikes, forming a close protective carapace. Up to 5m (16ft) in length and 5 tonnes in weight, they must have been cumbersome and slow-moving.

The ceratopians, which appeared in the late Cretaceous, were better equipped to defend themselves, for the head of most was armed with rhinoceros-like horns. They originally owed their success to their very large heads, with powerful jaws whose muscles were attached to a bony frill which projected backwards over the creature's neck. The earliest, 2m (6.5ft) long *Protoceratops*, was hornless, but later forms had a nasal horn and usually also a pair of horns above the eyes. The largest, *Triceratops*, was 11m (36ft) long, and must have weighed about 8.5 tonnes.

The Distribution of Dinosaurs in Space and Time. Most of the major types of dinosaur seem to have appeared in the late Triassic or early Jurassic. Apart from the extinction of the prosauropods at the end of the Triassic, and the replacement of the Jurassic stegosaurs by the Cretaceous ankylosaurs, all the dinosaurs survived until the late Cretaceous. At that time, the world's climate and vegetation underwent profound changes.

The floral change was the replacement of the gymnosperms, such as pines, ginkgos, araucarians and cycads, by the *angiosperms, or flowering plants. This commenced during the early Cretaceous, and became complete during the late Cretaceous. It seems likely that the herbivorous dinosaurs which appeared during this time, such as hadrosaurs and ceratopians, did so as a response to the availability of the new type of food. The carnivorous tyrannosaurids and omnivorous ornithomimids also appeared during the late Cretaceous.

But all these new forms were appearing in a world geographically very different from that in which the first dinosaurs had evolved. *Continental drift, together with the spread of seas across the continents themselves, resulted in the appearance of a number of different land areas in the late Cretaceous. For some reason, all the new types evolved in Asiamerica. Though some may have reached South America at the end of the Cretaceous, none are known to have reached other parts of the world.

The End of the Dinosaur World. The fact that dinosaurs were still actively evolving new types, both herbivores and carnivores, in the late Cretaceous makes it all the more difficult to understand why, only a few million years later, they had all disappeared. Clearly it was not because they could not cope with the vegetational changes of the late Cretaceous, and this would seem to indicate the improbability of any mysterious "racial senescence" that would make them incapable of responding to change in the biological world around them having played a part. Equally, their disappearance cannot be blamed on the little *mammals that were to replace them: the mammals had evolved at the same time as the dinosaurs themselves, in the late Tri-

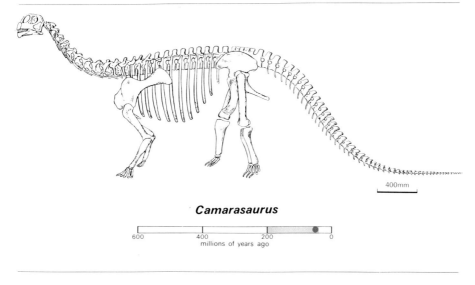

Camarasaurus

600   400   200   0
millions of years ago

assic, and the two groups had been coexisting without any signs of incompatibility for some 135 million years.

It is logical, therefore, to seek the explanation in the physical world. And here a possible explanation of the dinosaurs' extinction can be found in current interpretations of their probable physiology. Dinosaurs were clearly active creatures, like birds and mammals. Like them, they were probably "warm-blooded" – that is to say, could maintain a constant, warm body temperature. But, unlike birds and mammals, they had no warm, insulatory covering of feathers or hair. Instead they probably relied on their great bulk, which contained such an enormous reservoir of warmth that day-to-day heat losses were comparatively insignificant – as long as they were not too great and did not continue for too long.

But there is evidence, from floral changes in the late Cretaceous, that climates became cooler at that time. This may have caused temperatures to fluctuate outside the limits of control of the rather crude temperature-regulation of the dinosaurs, and so caused their sudden disappearance. But this conclusion is based on a tenuous series of hypotheses, and we should be unwise to conclude that the certain key to this 65-million-year-old mystery has been found.

Crocodiles
Most crocodiles are amphibious reptiles, feeding in fresh waters or (more rarely) the sea, swimming by means of the powerful tail, but resting and laying their eggs on land. They are, however, archosaurian diapsids which, like the dinosaurs, evolved from terrestrial *thecodonts. The earliest crocodiles may even have been quite fast-moving land animals, for the bones of their ankles are elongated in a fashion normally found only in animals with this way of life. Like thecodonts, crocodiles have an armor of bone under the skin of the back and sometimes also on the belly.

The earliest known crocodile is the late *Triassic *Protosuchus*, which is very close to its thecodont ancestors in structure. Together with a few other late Triassic relatives, it is placed in a separate crocodilian suborder, the Protosuchia.

It is not until the *Jurassic that typical crocodiles appear, with an elongate, flattened skull and the beginnings of a "secondary palate" – a partition of bone separating the mouth from the air passage that leads from the nostrils to the back of the throat. This secondary palate is still short in the suborder Mesosuchia, first known from the early Jurassic. Several Jurassic and *Cretaceous mesosuchians took up a marine existence, feeding on fish, and developed a narrow, many-toothed snout like that of living gavials. Some marine forms, such as the teleosaurs, looked otherwise like normal crocodiles; but the thalattosuchians of the late Jurassic and early Cretaceous developed a vertical tail-fin like that of an *ichthyosaur.

Though some marine forms survived until the early *Cenozoic, most mesosuchians were replaced during the Cretaceous by their descendants the eusuchians, in which the secondary palate isolates the air passage all the way to the rear end of the skull. This final suborder is known first from the early Cretaceous, and includes the lines leading to the living carnivorous crocodiles, alligators and caimans, which appeared in the late Cretaceous, and also to the fish-eating gavials, which are unknown before the *Pliocene.

During the milder *Mesozoic climates, crocodilians extended much further north than their modern relatives, which are almost unknown outside the tropics. The late-Cretaceous eusuchians also included the giant form *Deinosuchus*, some 15m (50ft) long, which must have preyed on dinosaurs.

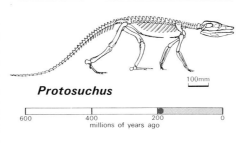

Protosuchus

600   400   200   0
millions of years ago

Mammal-Like Reptiles
Though less glamorous than the *dinosaurs, the mammal-like reptiles were in many ways the more important group of the two. They proliferated from the late Paleozoic, earlier than did the dinosaurs, and they were the first to establish a complex terrestrial ecosystem including herbivores, carnivores, insectivores and omnivores. Furthermore, they were in a way more successful than the dinosaurs: they were the ancestors of the mammals, which fill the modern world, while the dinosaurs came to an abrupt end some 65 million years ago.

Mammal-like reptiles form a completely independent line of reptile evolution which can be traced back to the late *Carboniferous. They are distinguished by their synapsid skull, and the group is known as the subclass Synapsida. Another notable feature is that, while many dinosaurs moved on two legs, this is unknown for any synapsid.

Synapsida is divided into two orders: the Pelycosauria, which include the earlier and more primitive forms, with a sprawling posture; and the Therapsida, their descendants, in which the limbs have become somewhat more erect and more efficient in support and locomotion.

Pelycosaurs. Three different types of pelycosaur are known from the early Permian but, though representatives of all are known also from the late Carboniferous, their interrelationships are still uncertain. Most primitive appear to be the ophiacodonts, little elongate forms 1–2m (3.25–6.5ft) long, whose long, low skulls bore many teeth: they may have been semi-aquatic fish-eaters. Their larger and more strongly-built relatives, the sphenacodonts, were up to 3m (10ft) long. Such sphenacodonts as *Dimetrodon* had powerful teeth and were the dominant carnivores of the early Permian community. Their herbivorous prey included the edaphosaurs, up to 3m (10ft) long, whose mouths contained many blunt chewing teeth, and such giant pelycosaurs as *Cotylorhynchus*, of similar length but very much greater mass, weighing nearly 350kg (775lb).

A remarkable feature of *Dimetrodon* and many edaphosaurs is the extension of the vertebrae into a row of spines interconnected by a sail-like membrane. This was probably for temperature regulation, the animal standing broadside to the Sun when cold, the heat being absorbed and circulated by blood vessels in the "sail": some modern animals employ a similar system.

Therapsids. The pelycosaurs died out at the end of the early *Permian. Intermediates between them and the therapsids are known from the middle Permian: some features suggest that therapsids evolved from the sphenacodontid pelycosaurs. The therapsid radiation otherwise appears suddenly in the late Permian Karroo Beds of southern Africa, rapidly spreading throughout the world.

Like their ancestors, the therapsids included both herbivores and carnivores. The most successful herbivores were the dicynodonts, 10cm–2m (4in–6.5ft) long, with teeth partially or wholly replaced by a turtle-like horny beak. Their only therapsid competitors were some members of the group known as dinocephalians, which had a bulky body, blunt herbivorous teeth and a thickened skull which may have been used for butting with the head during mating combats. Not all dinocephalians were herbivores, for genera such as *Titanophoneus* had a carnivore-type dentition.

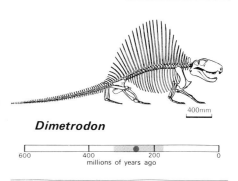

Dimetrodon

600   400   200   0
millions of years ago

The therapsids also included more exclusively carnivorous groups, such as the gorgonopsids. Like saber-toothed tigers, these animals had enlarged, stabbing canine teeth, the teeth further back in the mouth being reduced in size. The gorgonopsids were about 1m (3.25ft) long. Less specialized carnivores included the therocephalians and their descendants the bauriamorphs. These were of similar size but had a normal reptilian dentition.

Apart from some large dicynodonts

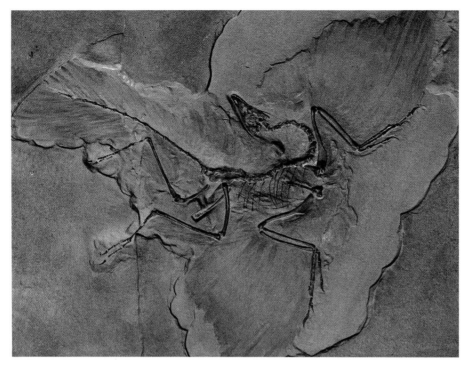

One of the few known specimens of the early bird *Archaeopteryx*. Preservation is so complete that impressions of the feathers of the wings and tail are clearly seen.

which persisted into the late Triassic, and some early-Triassic bauriamorphs, the only synapsids that survived the end of the Permian were members of the only remaining group, the cynodonts. These began in the late Permian as small insectivores and carnivores, about 0.5m (1.6ft) long. Some large Triassic cynodonts developed wide, peg-like cheek teeth and must have been omnivores or even herbivores. Rather rodent-like descendants of these animals, known as tritylodonts, survived until as late as the middle Jurassic.

Yet other Triassic cynodonts became active little insectivores with sharp, many-cusped cheek teeth, and it is from these that, in the late Triassic, the mammals evolved. Like *mammals, they had a bony partition or "secondary palate" separating the mouth cavity from the air passage leading from the nostrils to the throat. Unlike the cold-blooded reptiles, mammals cannot stop breathing for long, and so need a secondary palate in order to continue to breathe while chewing. Its presence in the Triassic cynodonts (and in the bauriamorphs) suggests that these ancient reptiles were becoming mammal-like in physiology as well as in anatomy.   CBC

Birds

Compared with other groups, birds appear rather late in the *fossil record. However, since their first representative, *Archaeopteryx* from the upper *Jurassic, is partly reptilian, partly avian, it would appear that what we have here is indeed the first true ancestor of the birds – or, if not, then at least we can be fairly confident it is a close relative of the as yet unknown ancestor of

birds.

Archaeopteryx is linked by only a few scattered bone fragments to the main avian sequence of fossil remains that extends onward from the middle Cretaceous. Birds suffer paleontologically from having small, fragile and fragmentary remains that can often be overlooked and which are, in any case, difficult to identify; and from the very fact that the kind of strata in which fossils are likely to occur favors the larger water birds at the expense of other forms.

Evolution. From the fossil record we can deduce three stages in bird *evolution – although we lack evidence of any species that might link them together. These three stages are usually treated, taxonomically, as subclasses of Aves, the class of birds. The first subclass, Sauriurae or lizard-like birds, contains only *Archaeopteryx*. This creature, known from six specimens from Bavaria, shows the intermediate stage between reptile and bird – not with each characteristic intermediate but with the overall structure made up of a mosaic of bird and lizard characteristics. The toothed jaws are relatively short and stout, the limbs bird-like, though the digital bones of wings and legs are not fused as in modern birds. The bony tail is long and slender, and there is no evidence of a bird-like keeled breastbone for the attachment of flight muscles. The wings are composed of long flight feathers exactly like those of later birds, and the presence of a row of similar feathers along either side of the tail shows that these had a support function in movement even if, as some people suggest, *Archaeopteryx* was incapable of true flight.

The second subclass, the Odontoholcae or toothed birds, is known from the North American middle to late Cretaceous, al-

though a recently described incomplete humerus from the lower Cretaceous of the English Weald may also be referred to this group. Like *Archaeopteryx* they differ from all other birds in possessing true teeth in the jaws.

All the known forms are seabirds. The remains of the seven species in the genera *Ichthyornis* and *Apatornis* are very like those of other birds, but with a few osteological peculiarities. In the remaining genera, *Hesperornis* and *Coniornis*, are three species of diving birds, with legs set far back on the body and tiny, vestigial wings: evolutionary diversification had clearly already progressed a long way.

All other known species are contained in the third subclass, Ornithurae or typical birds. We have at present no knowledge of the forms that might link them, in terms of evolution, with the toothed birds because the fossil record is so incomplete.

A wide range of fragmentary remains from the lower *Eocene indicates that most modern orders had evolved by then: in view of the incompleteness of the remains and the number of species that are still being discovered, we cannot assume that the apparent absence of any particular group is good evidence that it had not yet evolved. The evidence we have suggests that the divergent evolution that was to produce the diversity we know today occurred mainly during the Cretaceous, so that by the early Tertiary a typical range of avian forms was already in existence.

The Tertiary. The Passeriformes and the mainly arboreal Coraciiformes and Piciformes are very poorly represented in the early Tertiary: their main radiation may have occurred much later, or their absence may merely reflect that most of the bird remains we have are from sea or coastal deposits. In other groups, early Tertiary bones are very similar to those of modern birds of the same orders, but may be a little less specialized in structure or, as in the Pelecaniformes, lack the air-spaces of later bird bones. There is evidence to suggest a rather wider range of forms, resulting in a more even gradation between different kinds of birds, and that loss of many of these forms has produced the "gaps" now apparent between discrete modern orders of birds.

In general, however, the picture is one of early diversification with repeated extinction and adaptive radiation. And there is a tendency for certain types of specialization to recur several times in different groups of birds. For example, after the toothed birds had disappeared there emerged a group with bony tooth-like projections along the jaws, the Odontopterygiformes. These were giant seabirds found from all over the world from the lower Eocene to the upper Miocene or possibly even the Pliocene. It is of interest to note that, among modern birds, ducks of the genus *Mergus* have small tooth-like processes in the bill. These are formed from the bill sheath.

Greater aquatic adaptation, with increase

in leg size and reduction of wings, recurs repeatedly through the evolutionary history. Apparent from early times in penguins, and briefly apparent in Eocene divers and Miocene gannets, it produced the flightless auks of the genus *Mancalla* in the Pliocene – long before the famous, recently extinct, Great Auk.

Another recurrent feature is the appearance of huge, flightless running birds. The ostrich is an example of a group persisting from early Tertiary times to the present, while a jaw of what is possibly a giant running bird is known from the North American upper Cretaceous. The Diatrymidae occur from the Eocene of both Europe and North America. Most are presumed to be grazers, but the Phororuscids of the South and Central American Oligocene and Miocene, with huge heads and eagle-like bills, seem to have been scavengers or raptors (snatching at passing prey) that proliferated in the absence of predatory *mammals. Later grazers were the Holocene elephantbirds of Madagascar, the Aepyornithiformes; and the Moas, Emeidae and Diornithidae from the Pliocene to Holocene of New Zealand.

Today. With this constant adaptive fluctuation it is incautious to suggest a time of maximum diversity, but overall there is a suggestion of increase from the late Cretaceous to a possible maximum in the early Pleistocene, although by then some of the groups we have mentioned had become extinct. Subsequently there is evidence of the disappearance of many larger forms – possibly related to Man's hunting of them. Since many modern species are of late Pleistocene origin and show external differences which are small and may not be reflected in the skeleton (and hence the fossil), strict comparisons between the number of species extant today and the number present in past geological periods are not possible.   CH

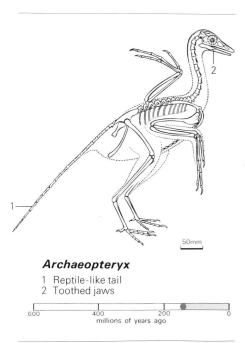

Archaeopteryx
1 Reptile-like tail
2 Toothed jaws

600   400   200   0
millions of years ago

Mammals

The mammals are treated here in more detail than the rest of the Animal Kingdom, and the entries that follow correspond to orders. The reasons for doing so are twofold: mammals comprise most of the best known animal groups living today and, secondly, they are of particular interest because they include in their number Man himself.   AI

Primitive Mammals
In the late *Triassic, 200 million years ago, the first mammals appeared. From Wales, South Africa and China come remains of these small shrew-like beasts. They evolved from small *mammal-like reptiles, and the transition is so gradual that the point at which a reptile became a mammal is often difficult to fix. Most methods of differentiating living reptiles from mammals depend on soft anatomy; but, in fossils, bone and tooth characteristics must be used. Mammals have three small middle ear bones; their lower jaw is a single bone; and their teeth are differentiated into incisors, canines, premolars and molars, and there are usually two sets during the lifespan, a milk dentition followed by the permanent set.

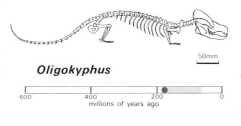

Oligokyphus

600   400   200   0
millions of years ago

During the *Mesozoic, six major mammalian stocks evolved, some "dead ends" and some giving rise to later mammals. The dead ends include triconodonts, with teeth in which three cusps were arranged lengthwise on each tooth; and multituberculates, which grew as large as beavers and were adapted like *rodents to gnawing with chisel-shaped incisors and multicusped cheek teeth. The most important Mesozoic stock were the pantotheres, with a complex tooth pattern not unlike that of a shrew; from these, in the *Cretaceous, evolved the three groups of living mammals, the *monotremes, the *marsupials and the placentals, or "modern" mammals.   RJGS

Monotremes
Monotremes, the egg-laying mammals, are represented by the Spiny anteaters and the Duck-billed platypus of Australia. They are considered to be mammals because they possess hair and the females secrete milk with which to feed the young. Many of their anatomical features are, however, reptilian in character. The earliest fossil monotreme is from Miocene deposits, but it is likely the group has a longer history which extends back to, perhaps, the late Mesozoic. AI

Marsupials
Marsupials, the pouched mammals, include

opossums, koalas, wombats and kangaroos. They are an ancient group extending back to the *Cretaceous when they evolved from pantothere ancestors (see *primitive mammals): they live today in Australia and the Americas. Their young are born very prematurely and continue development in the mother's pouch, and this distinguishes them from the placental *mammals.

There are differences in the skeleton and dentition (teeth) which enable us to differentiate fossilized marsupials and placentals. Marsupials do not have a set of milk teeth later replaced by a set of permanent teeth: there is only one set, though sometimes a single tooth may be replaced. In general, also, marsupials have more incisor and molar teeth than do placental mammals.

In *Tertiary times, marsupials radiated to fill a wide range of ecological niches in South America and Australia, and insectivorous, carnivorous and herbivorous stocks developed. They also lived in North America and Western Europe, but no remains have been found in either Asia or Africa.

The South American fossil record displays a vast array of species, some of which range back as far as the Cretaceous. Early marsupials were insectivore-like and it was only later that from these all the more specialized kinds evolved.

During much of Tertiary time the marsupials evolved species that paralleled placental types on other continents. Among these was *Borhyaena*, which was hyena-like; *Thylacosmilus*, with a great sabertooth, paralleling the saber-tooth cats of North America and Europe; and *Lycopsis*, a marsupial version of the dog. Among the largely extinct suborder Caenolestoidea were many rodent-like creatures, including bipedal species rather like the modern jerboa.

In Australia, the fossil marsupial record is rich in the late Pliocene and *Pleistocene, but poor in earlier times: the earliest known are a few *Oligocene specimens, including a primitive kangaroo. Most fossils fit into or close to, living Australian families, so we are still in ignorance of the earlier differentiation of marsupials here.

There are eight families of Australian marsupials of which two are extinct, namely the thylacoleonids and the diprotodontids, fossils of both dating from the Pliocene and Pleistocene. *Thylacoleo* was a large lion-like beast, with very large tusk-like incisor teeth and enormous shearing cheek teeth; it is thought to have been a carnivore. *Diprotodon* was bigger than a rhinoceros and again had tusked incisors, but its cheek teeth were suited for grinding vegetation. Other fossils include gigantic kangaroos 3m (10ft) high.

Of great interest to zoogeographers is the peculiar distribution in time and space of marsupials. Probably marsupials originated in South America in Cretaceous times and soon radiated into North America. During the *Eocene, some migrated from there *via* a land bridge across the Arctic circle into western Europe. By the end of the *Miocene they were extinct in both North America and Europe. In South America, where they faced less competition from placentals, they were more successful; and from here they migrated *via* Antarctica to Australia, possibly in *Paleocene or Eocene times. RJGS

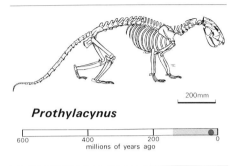

Prothylacynus

600   400   200   0
millions of years ago

Insectivores

Insectivores are the most primitive living placental *mammals: they include hedge-hogs, moles and shrews. They have an ancient origin and a good fossil record from the *Cretaceous onward. They are all small, and sometimes have a poison gland. They feed on a wide variety of items, including insects as well as other small invertebrates, small vertebrates and even fruit. Their classification is exceptionally complex and they are more a ragbag collection of small insectivorous mammals than a true natural order of mammals. We can group them into three divisions: Deltatheridioids, extinct shrew-like mammals known mainly from North America and Asia from the Creta-ceous to the *Oligocene, and including the ancestors of carnivores and ungulates, and hence most other modern placental mam-mals; Proteutheres, seven families of ex-tinct insectivores, ranging from the Creta-ceous to the Oligocene; and the Lip-otyphlans, comprising about seven families of living insectivores with a fossil record dating back to the *Eocene. RJGS

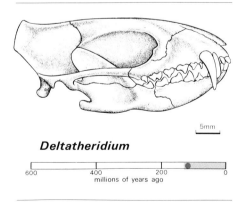

Deltatheridium

600   400   200   0
millions of years ago

Bats

Bats are essentially flying *insectivores with front legs modified for flight by elongation of four of the fingers, between which is the leathery membrane. They have a poor fossil record, though occasionally almost com-plete specimens are preserved, like *Icar-onycteris* from the *Eocene of the USA. We know nothing of the evolution of flight among the insectivores before this.

There are two groups: microchirop-terans ("small bats"), including a wide variety of carnivorous, bloodsucking and insectivorous bats; and the larger mega-chiropterans ("big bats"), mainly fruit eaters, which evolved from them. RJGS

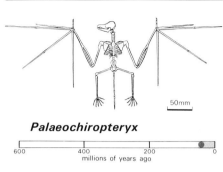

Palaeochiropteryx

600   400   200   0
millions of years ago

Carnivores

The lion is the king of beasts, but 25 million years ago there was a beast that would have made a lion look like a toy dog – *Megisto-therium*, a gigantic hyena-like mammal, the largest land carnivore that has ever lived.

Carnivores, the meat-eating placental *mammals, classified in the order Carn-ivora, come in many sizes and shapes; there are running, climbing, burrowing and swimming kinds. Some are exclusively flesheaters while others are more catholic in their tastes: a few are secondarily veg-etarian. The feet usually retain four toes which carry prominent claws; the eye or "canine" tooth is often a large stabbing tooth; and in the cheek area some teeth are usually modified for slicing meat. The earl-iest occur in *Paleocene sediments, and they have been fairly abundant ever since in North America, Eurasia and Africa. Until relatively recently their place was taken in South America and Australia by carni-vorous *marsupials.

There are two groups of living carn-ivores, fissipeds and pinnipeds, and one extinct group, the creodonts.

Fissipeds. In this group are placed dogs, bears, raccoons, weasels, mongooses, cats and hyenas. Most of the families can be traced back to the *Miocene and some, like dogs and mongooses, to the *Eocene: bears and raccoons evolved from dog-like re-latives, while cats and hyenas evolved from mongoose relatives. Among the extinct kinds the most spectacular are the saber-tooth cats which prowled America, Eurasia and Africa from the Miocene to *Pleistocene.

Bears tend to be large, heavy, slow and omnivorous beasts. Raccoons and pandas are mainly arboreal and often vegetarian. Mongooses and civets are small and feed on small lizards, snakes, birds and mammals. The weasel family is very varied, with swimming otters, arboreal martens, bur-rowing badgers, skunks and ratels. Hyenas have specialized in bone crushing; and cats can be regarded as the ultimate in carnivore evolution.

Pinnipeds. These aquatic carnivores – the seals, sea-lions and walruses – are not a natural group, since seals evolved from otter-like ancestors and sea-lions evolved separately from dog-like ancestors, with walruses as a sideline. Seals and sea-lions have numerous simple peg-like teeth, feed mainly on fish, and swim with flippers which are modified hind feet. Walruses break off mollusks from the sea floor with their huge tusks and crush them with their cheek teeth. All are relatively rare fossils though their ancestry can be traced back to the Miocene.

Creodonts. These prowled around the world early in the *Tertiary, and in Mioc-ene times almost all had become extinct, having been replaced everywhere by the more intelligent fissipeds. Creodonts had a separate ancestry from other carnivores, and indeed they evolved animals which paralleled dogs, weasels, bears, cats and hyenas – including the gigantic *Megistoth-erium*. They are distinguished from fiss-ipeds in having small brains, a different type of slicing dentition. RJGS

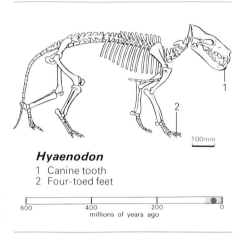

Hyaenodon
1 Canine tooth
2 Four-toed feet

600   400   200   0
millions of years ago

Subungulates

Grouped together as subungulates are no less than 16 orders, of which 11 are extinct: the survivors are aardvarks, pangolins, hy-races, seacows and elephants, of which elephants are the most important. All are basically hooved animals feeding on a var-iety of plants and adapted to many ecologies in different parts of the world.

In the late *Cretaceous animals with preferences for plant food and having broad flattish teeth evolved from insectivorous stock – they also developed a digestive system to cope with the cellulose in their diet. These, the condylarths, looked much like small piglets. By *Paleocene times they had reached all continents except Australia and Antarctica. They evolved into all the later hooved or ungulate animals.

Condylarths reached South America

The skull and part of the body of a mammoth preserved in Russia on the banks of the Berezovka river and excavated during a thaw in 1902.

before it was cut off from North America in the *Eocene and evolved there in isolation from most other placental *mammals for much of the *Tertiary. They developed a vast array of bizarre and now extinct stocks. Most successful were the notoungulates: some, such as *Toxodon*, were large rhinoceros-like animals, and others, such as *Typotherium*, small rodent-like kinds. One stock migrated into North America and across into China.

Hyraces and elephants had their early evolutionary centers in Africa. By the *Oligocene hyraces were the dominant herbivores, but never achieved this position again, giving way first to giraffes and rhinoceroses and then to bovids.

The story of elephant evolution is well-preserved in their fossil bones and teeth. It is one of increase in size – the bigger they became the less able were predators to attack them. The earliest elephant was *Moeritherium* from the Eocene, about the size of a pig and weighing 200kg (4cwt). As elephants grew bigger, so the legs lengthened or the animals would have been immobilized. At the same time they were developing tusks, imposing a considerable extra weight on the skull, which had thus to be held close to the body since the strain of carrying 90kg (200lb) of tusk at the end of a long neck would have been intolerable. With long legs and a short neck the animals

would not normally be able to drink without kneeling, thus putting themselves in mortal danger from carnivores; and so a trunk developed for both drinking and food gathering. Increased size made more food necessary, and the teeth adapted to this by becoming much bigger and emerging cyclically: only one tooth is present in each half of the jaw at any one time but, as it wears out, another moves up from behind to replace it.

Until the *Miocene Africa was the center of elephant evolution, and then they migrated into Europe and Asia and, later, North America. By the *Pliocene they had developed teeth that could cope with tough siliceous grasses, and all later kinds were primarily grazers. *Pleistocene mammoths were a very widespread and successful stock. In the Arctic, some fell down crev-

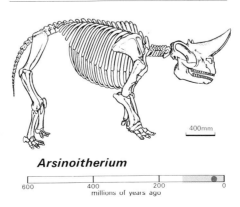

Arsinoitherium

600   400   200   0
millions of years ago

asses and were "deep frozen", remaining intact for thousands of years. So well preserved are some that even their last meal is still present in the stomach.

Among the minor subungulate stocks are the seacows (manatees and dugongs); aardvarks, pig-sized burrowing anteaters of Africa; scaly pangolins; and the extinct embrithopods.   RJGS

Perissodactyls

Horses, rhinoceroses and tapirs, together with the extinct brontotheres and chalicotheres, make up this large order of herbivorous *mammals. They are odd-toed (as opposed to the even-toed *artiodactyls): the toes on either side of the hoof are reduced, and weight is taken mainly on the third or middle toe, sometimes with support from the second and fourth toe. They arose from condylarth ancestors (see *subungulates) in the *Paleocene and reached their peak in the *Eocene; since then they have dwindled in diversity, except for the horse family, which has continued to expand.

Tapirs are today restricted to two species in central America and southeast Asia; they are the most primitive members of the order and have never been spectacularly successful. Rhinoceroses, on the other hand, roamed America, Eurasia and Africa and adopted many forms; some with horns, some without, some with tusks, some without, some large, some small, some semiaquatic, some running types. *Baluchitherium*, an extinct hornless rhinoceros from the Oligocene of central Asia, was the largest land mammal that has ever lived; it stood about 5.5m (18ft) high at the shoulder and must have weighed about 16 tonnes.

The extinct brontotheres or titanotheres of the Oligocene deposits were also huge rhinoceros-like animals reaching 2.4m (8ft) at the shoulder: they had a pair of bony horns.

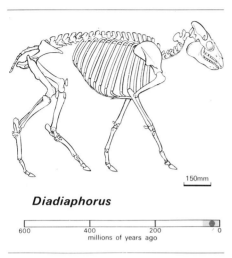

Diadiaphorus

600	400	200	0

millions of years ago

Chalicotheres are rare but curious fossils. Related most closely to horses, they have big, wide feet with claws rather than hooves. They occur occasionally in Eurasia, while in Africa the last record is in the Pleistocene. Some reports appear to suggest the chalicotheres may not be extinct but are still living in the African forests.

The horse family is certainly the most important perissodactyl stock. Horses are known in detail from their origin in the Eocene through to today. The main center of evolution was North America, from where they migrated at various times: Australia and Antarctica are the only continents they did not inhabit during the course of the Cenozoic.

The story of the horse is like that of an Olympic runner; faster and faster, each time improving technique and performance. Early Eocene horses were about the size of a terrier. Longer legs developed for faster running: as these were used solely in running and were not required for grasping, the ability to twist was lost, so reducing the chances of sprained ankles – this was achieved by fusing the two bones in the lower limb. The smaller an animal's feet the better it can run: the horse lost the toes at the side of the feet, leaving the third or middle toe to bear all the weight; they had achieved a fully functional one-toed condition by the late Miocene. Then again, an animal should not grow too heavy or, like elephants, it will be able only to amble. The optimum size/speed ratio is achieved in the modern racehorse.

Another important evolutionary aspect is feeding adaptation. Early horses were browsers on shrub vegetation, but as they grew bigger they needed more food, and so the teeth had to change to grind more efficiently the tough fibers. In the early Miocene grasses appeared and horses quick-

ly took to feeding on these, and so living on plains where they could make maximum use of their speed.

From *Eohippus* 60 million years ago to *Equus* today the horses have become one of the most successful and most intelligent of mammals. RJGS

Artiodactyls

The artiodactyls, the even-toed or cleft-hoof ungulates, include pigs and cattle. They have four or more usually two toes on each foot, each toe carrying a horny hoof. They first appeared in the *Eocene and rapidly expanded at the expense of the *perissodactyls, which declined in numbers: they had an explosive radiation in the *Miocene when grasses appeared and the bovids adapted to grazing them. Sizes ranged from tiny hare-like antelopes to gigantic pigs, giraffes and cattle.

The stomach has usually several compartments so that the maximum value is extracted from the food by ruminating – food that has already been chewed once is passed back to the mouth from the stomach, *via* a compartment called the rumen, to be rechewed before digestion. On the skull often develop outgrowths of bone (giraffe), antler (deer) or horn (antelope).

There are over 500 genera of which about 400 are extinct. The 25 families can be considered in 3 groups.

Suina. Pigs, peccaries and hippopotamuses are the living families in this group, anthracotheres and entelodonts the extinct ones. All are essentially stocky, short-legged animals, usually with four toes on each foot; some are semiaquatic. They are ground browsers and rooters, with low-crowned teeth.

Suids, or pigs, have a fairly good *fossil record and are stratigraphically valuable. Entelodonts were gigantic pig-like animals found mainly in the *Oligocene of North America and Eurasia. Anthracotheres, also pig-like, are known from Eocene to *Pleistocene times. Hippos may have originated in the African Miocene, and are found fossilized in the Pleistocene from England to Jawa.

Tylopods. Camels and two related extinct families make up this group. Members have either four or two toes on each foot, the canine tooth is vestigial, and the stomach has three compartments. Cheek teeth have crescentic cusp patterns and the animals usually browse on shrub vegetation. Wild camels occur today only in central Asia. In South America live their cousins, llamas and vicunas.

Camels have a good fossil record, and their many kinds include *Stenomylus*, a slender gazelle-like animal, and *Oxydactylus*, a long-legged, long-necked species. The extinct cainotheres of the European Oligocene and Miocene were much like hares; and the extinct Oligocene oreodonts were abundant browsing antelope-like animals.

Pecora. This, the largest, most advanced group, includes bovids, chevrotains, giraffes, deer and antelopes. All are two-

toed with a fully ruminating four-compartment stomach. Skull outgrowths are common.

The chevrotains are small, primitive, tropical gazelle-like animals: some North American fossil cousins had bizarre bony nasal outgrowths. The Giant Irish Deer, *Megaloceros*, of the Pleistocene is the largest of the deer family, with branching antlers which often spanned 4m (13ft); borne only by the males, these were shed annually. Some fossil giraffes also had great bony outgrowths from the skull, but these were present on both sexes and not shed.

The bovids are the largest family of pecorans, containing muskox, bison, cattle, sheep, goats, antelopes and gazelles. RJGS

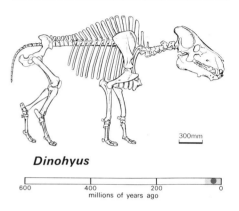

Dinohyus

600	400	200	0

millions of years ago

Edentates

The name Edentate implies "toothless" mammals, but this is true only of some varieties of this exotic order of herbivorous and insectivorous American mammals. Included in the order today are giant anteaters, armadillos and tree sloths. The extinct members include ground sloths and glyptodons, known from the *Eocene to the *Pleistocene.

Megatherium, a giant ground sloth from the *Pliocene and Pleistocene, reached over 7m (23ft) in length, bigger than an elephant. It had large claws on front and hind feet, probably used for digging. *Glyptodon*, a mammalian version of the turtle, was equipped with a truncheon-like tail to fight off predators. RJGS

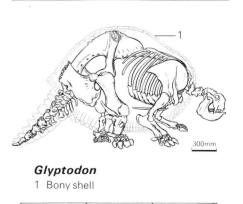

Glyptodon
1 Bony shell

600	400	200	0

millions of years ago

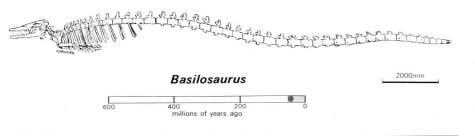

Basilosaurus

2000mm

600 400 200 0
millions of years ago

Whales

The Blue whale is the largest animal that has ever lived; it reaches lengths of up to 30m (100ft) and weighs around 100 tonnes. There are, however, many small whales, such as dolphins and porpoises. Most live in the oceans, but some small dolphins inhabit rivers. All are totally aquatic, never coming ashore – they give birth in the water. Next to primates they are the most intelligent *mammals.

The earliest fossils, the archaeocetes, occur in *Eocene and *Oligocene sediments. Some were over 15m (50ft) long and their jaws bore saw-edged teeth. They are so like the modern whale that the ancestry of the group is difficult to establish: possibly they evolved from a group of *carnivores during the Paleocene.

From archaeocetes evolved the two living groups, odontocetes and mysticetes. Odontocetes or toothed whales also appear in the Eocene and are characterized by having many peg-like teeth in the jaws. The mysticetes or baleen whales appeared in the Oligocene and grew to enormous size (they include the Blue whale). Rather than teeth they have plates of chitinous baleen, or "whalebone", which trap the minute plankton on which they feed. Since baleen does not fossilize, fossil jaws of these whales look toothless. RJGS

Rodents

Of the 5000 living species of mammals, half are rodents; and there are over 400 extinct rodent genera. They are the most successful *mammals, adapting to the widest range of climates, altitudes and ecological niches. Their striking characteristic is their adaptation to gnaw, using a single pair of incisors in each jaw to chisel off nuts, bark and bone. Enamel is restricted to the front side of the teeth and, as the softer dentine behind wears more rapidly, a chisel edge is produced. The incisors continually grow replacement tissue to counter the rapid wear. Most rodents, but far from all, are small scampering animals.

The harvest mice are among the smallest of all mammals with a head and body only 50mm (2in) long: the largest living rodent is the capybara from South America, about 1.3m (4ft) long. Some of its fossil cousins were twice as large.

The 45 families can be considered in 4 groups.

Sciuromorphs. These include the most primitive rodents, the extinct Paleocene paramyids, which were already distinctly rodent-like, and the squirrels.

Myomorphs. This, the largest group, contains rats, mice, voles, lemmings, jerboas, mole rats and many other small animals. They are most useful in stratigraphy from the *Eocene onward. Their geographic spread has been so rapid that they give a finer indication of the age of a sediment than do any other mammalian fossils.

Caviomorphs. This group includes chinchillas, coypus, agoutis, guinea pigs and the capybara. Fossil members include gigantic forms like *Artigasia* and *Eumegamys* with skulls over 50cm (20in) long. Members of this group evolved in South America from the Oligocene onward.

Other families. Grouped together under this ragbag heading are fossil and living families such as the Eocene and Oligocene theridomyids; the Oligocene phiomyids; beavers, known from the Oligocene onward; and porcupines. RJGS

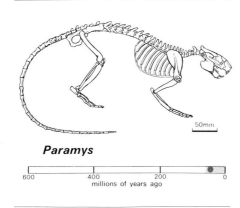

Paramys

50mm

600 400 200 0
millions of years ago

Lagomorphs

Pikas, hares and rabbits are members of the small order Lagomorpha. While superficially like rodents, they are not closely related: lagomorphs have two pairs of incisor teeth in each jaw and the enamel forms a continuous band around the tooth, rather than the single pair, with enamel only at the front, characteristic of *rodents.

The earliest known lagomorphs are from the *Paleocene: from this stock evolved two families, the leporids (hares and rabbits) and the ochotonids (pikas). In *Eocene times they spread into North America, in *Oligocene times into Europe, in *Miocene times into Africa and in *Pleistocene times into South America. Early kinds were small scampering animals from which later evolved running and leaping varieties. Though lagomorphs are never plentiful in the fossil record, this may not truly reflect their

natural abundance at any time in the past. The present widespread distribution of the rabbit is due to Man rather than to its celebrated breeding habits. RJGS

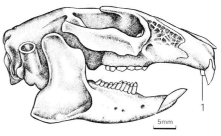

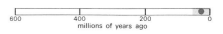

Palaeolagus

1 Second pair of incisor teeth

5mm

600 400 200 0
millions of years ago

Primates

Man, apes, lemurs and monkeys are all members of the order Primates, one of the oldest mammalian stocks, going back some 70 million years to a time when *dinosaurs were still plodding the Earth.

The characteristics distinguishing primates from other mammals are mostly primitive unspecialized features – if you like, their lack of specialization. In many ways they resemble primitive Eocene mammals in their retention of digits and a fairly complete and unspecialized dentition. Features which mark them out are the opposable thumb, the flattened nails, stereoscopic vision and the large brain.

There are two major groups of primates, prosimians and anthropoids. Prosimians are the more primitive, and living kinds include lemurs, lorises and the tarsier. The anthropoids comprise monkeys, apes and Man.

Early Evolution. Primates probably originated directly from an insectivore stock or indirectly *via* primitive condylarths, a now extinct group transitional between *insectivores and herbivorous *mammals. The earliest known primate is *Purgatorius* from the late *Cretaceous, known only from some isolated teeth that show a few primate characteristics. In the early *Paleocene are four families of primates, three of which appear to be dead ends: carpolestids were very peculiar primates with some resemblances to multituberculates; picrodontids were a fruit-eating stock; and plesiadapids were essentially squirrel-like with claws and laterally placed eyes. The paromomyids constitute the fourth family, and includes *Purgatorius*: they are probably close to the stem of the later primates. All were small generalized feeders and nothing is known of their skeletal anatomy.

In the *Eocene, primates faced increasing competition from *rodents and the two stocks diverged – primates tending to become more arboreal while rodents remained ground-dwelling. There is a persistence of some Paleocene stocks – paromomyids and plesiadapids – and the appearance of two new groups, lemuroids and tarsioids. The lemuroids survive today only on Madagascar and we have nothing in the fossil record to link them with the Eocene forms. Eocene lemuroids were all lemur-like animals with opposed first digit, stereoscopic vision and fairly large brain. From the Isle of Wight come remains of a small tarsier-like primate, *Microchoerus*, but again there is a gap in our knowledge of the group from the Eocene until the present.

The Oligocene. After the Eocene, the new- and old-world primates evolved separately: the opening of the north Atlantic made migrations impossible between Canada and western Europe. While all our knowledge to the end of the Eocene is derived from North America, Europe or Asia, in the succeeding *Oligocene virtually all our information comes from Africa, a continent from this time on of great importance in primate evolution.

At Fayum Oasis, 100km (60mi) southwest of Cairo, is a succession of lake and river sediments that has yielded all we know of primate life in the Oligocene. Dense tropical forests existed around the Fayum in Oligocene times, 30 million years ago. In the forests lived a vast array of mammals, including many kinds of primate.

Most of the primate fossils belong to two families, each represented by two genera. In the parapithecids there are *Apidium* and *Parapithecus*, and in the pongids are *Aegyptopithecus* and *Propliopithecus*.

Parapithecids have three premolar teeth, and molar teeth that are constricted across the middle – characteristics unique among old-world anthropoids. In *Apidium* the face was short, smell poorly developed, and the eyes placed close together giving stereoscopic vision. It had a leaping gait like living lemurs.

Among the pongids, *Aegyptopithecus* is known from a nearly complete skull: the snout is long, the eyes placed well forward, the canine teeth large and the sense of smell weak; the molar teeth are larger toward the back of the mouth. It was probably rather like living lorises in appearance. The other genus, *Propliopithecus*, is rather different: the face is short and the canine teeth small; in addition the molar teeth do not become larger towards the back of the mouth. *Propliopithecus* could well be an ancestor of the *hominids.

Among the other primates is *Oligopithecus*, which may be close to the ancestry of cercopithecids or old-world monkeys.

The Miocene. The next phase of primate evolution of which we have evidence is from the early *Miocene of East Africa. Sediments around Lake Victoria have yielded a rich fauna, due to the untiring efforts of *Leakey and his colleagues. These fossils are about ten million years younger than those of the Fayum, but we know nothing of intermediate primates.

Proconsul, an animal about the size of a small baboon, is known from a good skull, jaws and some limb bones: the teeth are in some ways similar to those of chimpanzees. It may well stand in line of ancestry to the great apes, and was probably mainly a ground-dweller, like baboons, living on forested volcanic slopes. Also from Kenya comes *Limnopithecus*, a gibbon-like animal and possibly an ancestor of the gibbons, the smallest living apes.

During the Miocene in both Europe and Asia there were other baboon-like primates referred to collectively as *Dryopithecus*. They were essentially similar to *Proconsul*. One of the Indian types may be ancestral to the orang-utan.

Later Evolution. The last five million years have seen the gradual evolution of present-day species and some extinct sidelines: there are ceboid monkeys from Columbia, cercopithecoid monkeys from Kenya, a giant lemur, *Megaladapis*, from Madagascar, and from China a gigantic gorilla, *Gigantopithecus*.

So, in the fossil record, we can trace the history of the four living types of ape back 20 million years, and recognize even from 30 million years ago animals that could be hominid ancestors.   RJGS

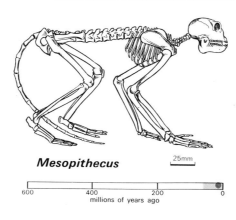

Mesopithecus   25mm

600   400   200   0
millions of years ago

Hominids

Hominid fossils discovered near Olduvai Gorge in Tanzania during 1975 by Dr Mary Leakey have been dated at around 3.5 million years old, and thus are certainly the earliest reliably dated hominid fossils known. But what do we mean by hominid? The hominids are a family containing Man and his close relatives, distinct from the family of apes, the pongids.

Man and the Apes. Man shares many characteristics with the other *primates, and some which clearly distinguish him from all others. Only bones and teeth fossilize, so difference in soft tissues cannot be checked out in extinct animals and need not concern us here (see *fossil record). The major skeletal differences between Man and the apes relate to locomotion and to feeding. Man stands upright and walks on two legs: all apes walk on four legs, though they can if need be walk short distances on their hind legs. Man's bipedalism has influenced the shape of his hip bones, thigh bones and foot bones, and given him a characteristic S-curved backbone. Further, the skull is carried on top of the backbone rather than being hooked on the end as in apes.

In Man the skull roof is domed over the very large brain, while in apes, which have smaller brains, there is often along the skull roof a prominent ridge, to which are attached strong chewing muscles. The short human face contrasts with the projecting face and jaws of apes, and these differences mirror dental characteristics. Human teeth are arranged in a continuous arcade, all being on the same level, and the canine is not enlarged. Apes have two parallel rows of cheek teeth, terminating in large projecting canines, with a gap before the transverse incisors – this gap is necessary to accommodate the tips of the lower canine teeth when the jaws are closed.

Early Hominids. These differences enable us to easily distinguish the skeletons of present-day Man and present-day apes. But both shared a common ancestry, perhaps 30 million years ago, differences being only gradually acquired. So when we examine any monkey-like fossil we are looking for evidence of specialized hominid characteristics.

One member of the large fauna of anthropoids from the *Oligocene of the Fayum, *Propliopithecus*, might be considered a hominid ancestor; but we must beware of reading too much into a few broken jaw bones from 30 million years ago. For the next evidence of hominids we have to travel some 15 million years or more to the middle *Miocene of Kenya, where *Kenyapithecus* was found. From sediments of similar age in northern India comes another animal named *Ramapithecus*, so like *Kenyapithecus* that they probably belong to the same genus. They have flat, broad, squarish cheek teeth, molars that are all of the same pattern, and small canines; and the dentition appears to have been arranged in an arcade: these are all hominid characteristics not seen in apes. They lived in forests but we cannot tell whether they were arboreal or ground dwellers. It was, however, a time when the pongids were well established and these fossils are quite different from them, so there are good grounds for placing them in the family of Man.

Following the mid-Miocene fossils, there is again a long gap in the record. There are a couple of isolated teeth from the Lake Baringo area of Kenya which can be aged at 10 and 6 million years and which show hominid features, but isolated teeth are not enough to enable us to say that there were hominids there.

The Pliocene and Pleistocene. From cave sites in the Transvaal have come hundreds of hominid fossils of the genus *Australopithecus*. At least two kinds existed, a gracile form named *A. africanus* and a robust form named *A. robustus*. The cave *breccias are impossible to age precisely

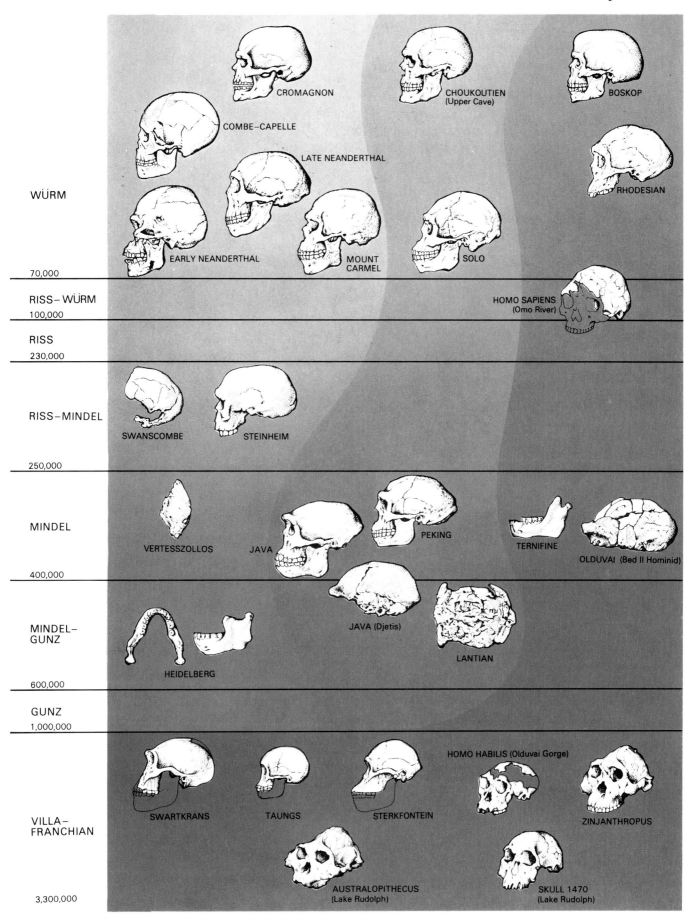

The chronological position of the later hominids in relation to the glacial and interglacial periods of Europe. Colored bands show, from left to right, Europe, Asia and Africa.

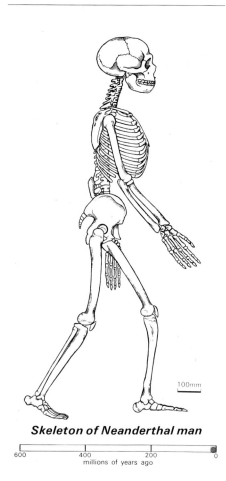

Skeleton of Neanderthal man

600　　　　400　　　　200　　　　0
millions of years ago

with radiometric methods, but on other grounds we believe the fauna to be between 1 and 2 million years old.

With such large quantities of material – one site alone has yielded over two hundred individuals – we can learn a great deal about these primates. They were short, averaging 1.4m (4.5ft) in height and weighing around 32kg (70lbs). Their brains range in size from 280 to 750cm^3 (16.6–44.5in^3) with an average of 480cm^3 (28.5in^3). For comparison, the gorilla's brain is around 500cm^3 (29.7in^3) and modern Man has a range of 1000 to 2000cm^3 (59.3–118.6in^3) with an average of 1350cm^3 (80.1in^3). So the largest *Australopithecus* brain was only a little smaller than the smallest human brain.

The robust species had heavy crests along the skull roof similar to those found in apes and indicating strong chewing muscles. The teeth were similar to those of modern Man, but much bigger, suggesting a heavy grinding dentition. They probably had a life expectancy of only 18 years.

In addition to these is a jaw which represents another line, closer to modern Man. It has been named *Telanthropus* and may be comparable with some specimens from East Africa.

By far the richest discoveries of early Man have been made by members of the *Leakey family at Olduvai Gorge in Tanzania and in the East Rudolf area of northern Kenya. The Olduvai Gorge sites give

an almost continuous succession over the past two million years. Well dated by radiometric methods, they have yielded a prodigious number of implements, a wealth of information on the mammals that lived around the early hominids, and a dazzling array of hominids themselves.

From the early beds (between 1 and 2 million years old) are recognized two kinds of hominids. One is *A. boisei* (sometimes called *Zinjanthropus*), which has strong similarities with *A. robustus* from South Africa: its brain capacity of 530cm^3 (31.4in^3) is within the range of the Transvaal specimens. The second kind is *Homo habilis*, a smaller, much more human-like species, only about 1.23m (4ft) high but with a brain capacity of around 650cm^3 (38.6in^3): *Telanthropus* could also belong to this stock.

Associated with these are vast quantities of tools belonging to the Oldowan culture. They are the most primitive known tools, stones flaked to give a cutting edge sharp enough to cut through a mammalian hide.

Further north along the shores of Lake Rudolf (now Lake Turkana) have been found rich collections of hominid fossils. There appear to be three kinds: a robust australopithecine like the Olduvai *A. boisei* and the South African *A. robustus*; a gracile australopithecine like *A. africanus* from South Africa; and a third kind with resemblances to *H. habilis* from Olduvai. These fossils are similar to those from the early beds at Olduvai, though some may be rather older.

Other but less extensive australopithecine discoveries have come from west of Lake Rudolf; from the Omo valley north of Lake Rudolf; and from the Lake Chad area. Some hominid fossils from northern Israel may be australopithecine and if so they are the only known occurrence of the stock outside Africa.

Recent emphasis has been placed on earlier sediments, and there have been two notable successes: from the Afar region of Ethiopia has been reported a fairly complete skeleton of a hominid which is provisionally dated as over 3 million years old; and from Laetolil near Olduvai in Tanzania Dr Mary Leakey has reported finding partial dentitions of eleven individuals confidently dated at around 3.5 million years. It is too early yet to pronounce judgment on the relationships of these specimens to other hominids, but it can be said that they carry the history of Man's ancestry back a further million years.

Outside Africa there is almost no evidence of hominids until middle Pleistocene times, with then the rich faunas of *H. erectus* from Jawa and China. During the 1930s, scientists in Jawa discovered skulls often referred to as Jawa apeman, *Pithecanthropus erectus* or *H. erectus*: the beds in which they were found are between 0.5 and 0.75 million years old. Similar hominids have been found in the Choukoutien caves south of Pekin in China, where there are tools and evidence of the use of fire. Fossils of the

same species have been identified in the higher levels of Olduvai Gorge, and elsewhere. *H. erectus* was much more advanced than *Australopithecus*: he had a brain capacity of 750–1000cm^3 (44.5–59.3in^3), the top of which range is close to that of modern Man; but he may represent a dead end rather than an ancestral link with ourselves.

Homo Sapiens. The final phase of human evolution took place in the late Pleistocene. Many fossils are known from Europe, Asia and Africa, but as yet none until rather later in the Americas or Australasia. The earliest claims to the name *Homo sapiens* are the partial skull from the Thames estuary at Swanscombe, England, and a skull from Steinheim in West Germany. Both are almost indistinguishable from modern Man and are probably about 250,000 years old.

One of the most numerous kinds was Neanderthal Man – his remains have been found all round the Mediterranean, and specimens from as far afield as Jawa and Rhodesia may also belong to this stock. Neanderthal Man is sometimes regarded as a separate species of *Homo* and at other times as only a subspecies. His average height was only around 1.5m (5ft) but he had a brain capacity of around 1600cm^3 (95in^3). His heavy eyebrow ridges and sloping forehead give him a rather ape-like appearance. Although the average brain capacity was higher than in modern Man the brain itself seems to have been less complex and so he was probably not so intelligent.

Another paleolithic culture was that of Cro-Magnon Man, physically indistinguishable from modern Man, and responsible for the famous cave paintings of northern Spain and southern France. There are many African late Paleolithic sites which have yielded remains of *H. sapiens* – for example, the youngest beds at Olduvai.

The lineages leading to modern Man are not clear. It is not known whether Neanderthal Man was a direct ancestor of some or all modern men. At Mount Carmel in Israel, in deposits about 40,000 years old, there is a population with characteristics intermediate between Neanderthal and Cro-Magnon Man, and these could be interpreted as representing a transitional link between the two kinds or the product of interbreeding between them. It seems possible that the Rhodesian neanderthaloid man might be an ancestor of the bushman and perhaps also of the negroid types. However, the differences between present-day human "races" are scarcely more than skin-deep; using only bones and teeth and allowing for the wide range of variation in populations, there is so much overlap that racial differences are virtually undetectable.

We can now trace the major developments in human evolution over the course of millions of years. By the middle Miocene we see evidence of a stock differentiated from the apes in being less fully vegetarian; that possibly lived on the edge of the forests,

View across the Serengeti Plain, Tanzania. Olduvai Gorge is in the middle of the plain.

kept mainly to the ground and hunted small mammals to supplement its diet.

By about 2 or 3 million years ago we see three kinds of hominid. Each has a number of human characteristics – upright gait, teeth very like those of modern Man, and brain capacities greater than those of the apes but still, generally, less than that of modern Man. The presence of tools in association with these fossils clearly defines a cultural level distinctly human.

In the later stages we see a complex of types, some of which could have contributed to present-day stocks and others of which may have been dead ends. In spite of exciting finds over the past decades, there is still much we have to discover about our ancestry. RJGS

The Making of Geology

Man and the Earth

From earliest times, Man has tried to understand, and by understanding to control, the planet of his destiny, the environment which he inhabits, the ground which he treads: the Earth. And investigation of the Earth has figured large in the development of many basic sciences, sciences of the universe in general, such as cosmology, cosmogony and general natural philosophy; sciences of elements, such as alchemy and chemistry; sciences of structures, such as mineralogy, gemmology and cystallography; sciences of the environment, such as meteorology, physical geography, topography and oceanography; and the sciences of the living creatures and plants which inhabit the Earth, natural history, biology and ecology.

But most of these sciences treat the Earth as only a fraction of their subject-matter – or they focus upon only a few of its features and materials. By contrast, distinct and special investigation of the Earth itself, as the object of an autonomous science – geology – is a quite recent development. The same is true for the subdivisions of geology, such as geomorphology, paleontology and petrology. The currency of the very word "geology" (literally "Earth-knowledge") does not date back more than two hundred years. For practical and philosophical reasons, both the study of the planets – and indeed of the universe in general – and the minute analysis of individual fossils and rare minerals advanced earlier, faster and with greater confidence than the science of planet Earth. Not until quite recently have men wanted, or been able, to create a science which takes pre-

A 17th-century illustration of the geocentric universe, the sphere of the fixed stars being outermost, then Saturn, Jupiter, Mars, Sun, Venus, Mercury and Moon.

Thales' view of the flat Earth floating on water (air and fire above complete the four elements).

cisely the Earth as its object, and which poses the questions of geology familiar to us today.

From Early Times to the Renaissance. Like so much of European culture, the modern sciences of the Earth are indebted to traditions of thought which crystallized in the Near and Middle East and the Mediterranean societies in the first two millennia BC. These traditions differed between civilizations, but embodied common outlooks which were suggested by common patterns of life. All civilizations, from the Assyrian through to Medieval Latin Christendom, were fundamentally dependent on the land, whether agrarian or pastoral. Man was close to the soil, in daily contact with the face of the Earth, aware of being part of an economy of Nature which nourished his body – a body which, from dust, in due course would to dust return.

On the one hand, this situation bred an insistent sense of the precariousness of the human condition in the face of the Earth's features and forces. Men recognized the curse of ceaseless toil, the need to bow to the course of the seasons and to the unarguable brute facts of volcanoes and earthquakes, droughts and floods, barren deserts and mountains. Yet, at the same time, men of the Mesopotamian twin rivers, of the Nile Valley and of the Mediterranean coastline could also experience Nature at her most benign, an Earth which provided a bountiful and comfortable habitat, a landscape long-settled by Man, redolent with human associations, tamed and assimilated by generations of agriculture.

Above all, these civilizations experienced the Earth only on a minute – and homely – scale. Beyond adjacent parts of the Eurasian and North African landmass, the globe was *terra incognita*. And, in the absence of concrete experience, mythological alternative worlds were conjured up – of burning tropics, lost continents and exotic monsters. There was little perspective on the Earth's past, nor solid knowledge of the depths of the crust. The known Earth was the here-and-now Earth. The starry firmament was a far more visible training-ground for the infant natural sciences.

Not surprisingly, such civilizations realized the Earth in their sciences very differently from subsequent interpreters. Above all, up to the Renaissance practically all schools of philosophy took a unified and highly integrated view of the whole cosmos. The Earth could be understood only in terms of its place and function within all the bodies which made up the universe. In fact, the Earth was most commonly seen as being at the *center* of them: the science of the Earth was part of cosmology and cosmogony, because the cosmos was thought to be geocentric. The Earth was influenced by planets and stars – for example, through astrological powers, and through the control which supralunary bodies exercised

over the growth of minerals and plants on Earth, and over the life of Man. Metals, for instance, had their controlling planets: iron was the metal of Mars, lead of Saturn, gold of the Sun. Gems reflected cosmic light.

Similarly, the Earth was also seen not just as a planet but also as an element which pervaded all matter within the corruptible regions of the cosmos. In turn, the planet Earth was composed not solely of its own element, but also of three other elements – water, air and fire. Ancient science stressed the unity and relatedness of all parts of the environment: the atmosphere, crust, oceans and "bowels" of the Earth. Thus Lucretius saw clouds as the caverns of the skies, the lairs of the winds; and *Aristotle thought earthquakes and volcanoes the terrestrial forms of thunder and lightning: significantly, the great Greek scientist set down most of his ideas about the Earth in a work entitled *Meteorology*.

Early science thus saw the Earth as a central, integrated part of the entire cosmos. Similarly, it also regarded the nature, operations and destiny of the Earth as inseparable from those of Man. The cosmos was more than geocentric, it was

anthropocentric. The Earth had clearly been designed as a habitat for Man, the rational and superior creature who tilled its soil, dug its mineral resources, explored and conquered it, mapped it and enjoyed it. Within the Christian account of the Creation, for example, God had created the Earth – indeed, Paradise – specifically for Man's needs, immediately before creating Man himself: once God had created Man, creation ceased. God willed that Man should go forth and multiply, and have dominion "over all the Earth and over every creeping thing that creepeth upon the Earth".

Furthermore, in their operations, Man and the Earth had the same nature, and reflected and responded to each other. For example, within Classical Greek medicine, Man's humors, his disease symptoms, his temper and disposition were seen as the equivalents in the microcosmic human body of the combinations of dryness and wetness, cold and heat, of the macrocosmic elements of earth, water, air and fire which made up the body of the Earth. The Hippocratic medical tradition saw many diseases as occasioned by climatic and en-

Li. VII Trac. I

An early 16th-century view of Man, the microcosm, in relation to the universe, the macrocosm.

be at the center of the Earth.

Thus, the Earth was as nothing compared with the Deity who had created it, and who would terminate its existence in the fulness of time. Nevertheless, it bore evident marks of divine intervention and purposes. Practically all cultures saw volcanoes and earthquakes, floods, droughts and storms either as gods themselves or as the work of gods. Special mountains, islands, wells and springs were set aside as holy places. The Earth was enchanted and populated by good and evil spirits. In a Christian formulation, the Earth was the Book of God's Works, which the intelligent could read and find full of emblems and tokens of God's purposes.

Such was the general framework for understanding the Earth which arose from the geographically confined and technically quite simple societies which sprang up in the Mediterranean and Middle Eastern regions a few thousand years ago; an outlook which carried over into European civilization. Within this general philosophy, certain more specific investigations were carried out and observations made on particular terrestrial phenomena. Practical mining experience, and the notion that the Earth was a living organism, posed the question whether metals and minerals were generated and continually grew – and if so, how? Seeing all things beneath the Moon as compounded of the four elements created interest in the relations of the Earth's core, surface and atmosphere. For example, volcanoes suggested that the core was full of fire, trapped, trying to escape: on the other hand, the fact that rivers originated from springs which welled out of the Earth suggested to others that there must be great internal reservoirs connecting the ocean beds to the sources of rivers in a continuously flowing system. The Roman philosopher Seneca, among many others, was fascinated by the parallels between atmospheric explosions – thunder – and earthquakes and volcanoes.

Similarly, viewing the Earth as a theater of ceaseless elemental imbalance focused interest upon change: the continual ebbings and flowings of the tides; the changing courses of rivers, estuaries and deltas, such as the Nile; the creation and destruction of land. Pythagorean and Stoical philosophy, as expressed by naturalists like Strabo and Pliny, and by poets such as Ovid, suggested that continents and oceans, mountains and valleys, were in constant revolution. Such a viewpoint looked favorably upon interpreting the appearance of bones and shells embedded in solid rock as genuine organic remains which had been petrified in a former state of the world, perhaps when such strata had been the seabed. Debates as to whether fossils were organic remains or were, like minerals and crystals, forms produced in her own womb by great creating Nature, continued throughout the Mid-

vironmental imbalances – effluvia given off by the Earth, dampness caused by rivers or marshy areas, or poisonous airs around mines. Catastrophes in the natural world – such as comets, storms, and earthquakes – presaged human disasters.

In turn, the Earth was explained as an organic, living replica of the human body, and of other living things. Its round shape suggested an egg, with a hard, thin shell and – it was supposed – successive layers of different fluids beneath the shell. Chains of mountains were frequently called its bones, lowlands its flesh, rivers its veins and arteries, the tides the coursing of the blood caused by the pulse of its "heart" somewhere at the "core" of the Earth, and vegetation its hair. Hills were sometimes seen as warts on the face of the Earth.

Belief in the harmonious integration of Man, Earth and cosmos could offer a pastoral vision. Classical and Christian culture both had their rosy myths of Man in harmony with the Earth: Paradise, the Garden of Eden, the state of nature, the Golden Age, where Nature spontaneously yielded her bounty for Man, who knew no shortage, no toil, no disease, no old age. But

not wholly so. For Nature was also thought, like Man, to have cycles of old age and decay, when the Earth became barren and enfeebled, grew cold or feverish or decomposed. Natural disasters could be interpreted as the wrath of the gods directed against wayward Man. Within the Christian myth, Man had been expelled from Paradise at the Fall and condemned to work a refractory Earth by the sweat of his brow.

Thus early science saw the Earth wholly in relation to the cosmos and to Man. But it also saw it in a larger, transcendental vision, a divine framework. Before and after, above and beyond the Earth and the cosmos, was some overriding Deity – or deities. By contrast with things divine, which were eternal, changeless, spiritual, perfect and heavenly, the Earth was a theater of change, decay, transience, imperfection. Plato saw the Earth as "unreal", a mere reflection of Ideal truth. St Augustine contrasted the mundane vale of tears of the *civitas terrena* (the Earthly city) with the heavenly bliss of the *civitas Dei* (the city of God). Early Christians urged contempt for the world. Within much popular Medieval Catholicism, the location of Hell was believed to

The dominant medieval schools of thought accepted that space was finite. Here a man looks through the boundary to God's realm beyond.

dle Ages and the Renaissance in the work of naturalists such as *Leonardo and *Gesner.

Ancient and Medieval science did not produce distinguished investigation of the relief of the Earth, its deep structure, its processes or its historical alterations over time comparable in quantity and quality to contemporary work in astronomy or medicine (for example). Too little of the Earth's surface was known, too little of the Earth's depths was accessible. Understanding advanced on two other levels.

Firstly, the period down to and including the Renaissance achieved a steadily accumulating body of knowledge on particular products of the Earth – minerals, gems, fossils, metals, crystals, useful chemicals and medicaments. This was originally set out in encyclopedic natural histories such as those of Pliny (23–79) and Isidore of Seville (560–636). It was later embellished, reorganized and sometimes corrupted by later Arabic commentators such as Avicenna (980–1037) and Averroës (1126–98), and by Medieval scholastics such as *Albertus Magnus (1193–1280) and Thomas Aquinas (1225–74). Natural-history accounts of particular objects described and defined them, investigated their chemical and medical properties, and listed their locations, as well as being fascinated with their etymologies and with marvellous anecdotes associated with them.

Secondly, comprehensive philosophies of the Earth were being established. The one which was to have most influence upon later Western science was the Christian revelation of Creation as set out in the first chapter of Genesis; and the Christian view of the subsequent development, purposes and destiny of the Earth which could be digested from many parts of the Bible, perhaps above all from the Psalms and the Book of Revelation.

It saw the Earth – and indeed, the cosmos – as the product of a single, purposeful act of Creation, answering to God's ends for Man. Whether or not the Biblical chronology of the "days" of Creation, and the continuous descent from the first man, Adam, through to contemporary times was taken precisely literally, this Christian view saw the Earth as recent – perhaps no more than 6000 years old – in contradistinction to the apparent belief of Aristotle and other pagan philosophers that the Earth was eternal. It thought of the Earth and Man as practically coterminous. It envisaged time as directional, not cyclical. God had made the Earth perfect but, in response to sin, had sent Noah's Flood to punish and reform Man by setting him in a newly harsh environment, one of decay, struggle and the niggardliness of Nature. Such decay would continue until God had completed His purposes with Man, when God would

either destroy or transform the Earth, probably by fire, in the consummation of all things.

From the Renaissance to the End of the 18th Century. From the time of the Italian Renaissance of the 15th and 16th centuries onward, scientific study of the Earth gradually but steadily developed a new impulse, and interpretation of the Earth began to be transformed. This was, of course, part of a very general intellectual and scientific revolution stimulated by the recovery of writings from classical Greece and Rome, the invention of printing, the vast increase in education, literacy and communication, and growing faith in the goodness, dignity and progress of Man. But it was also quite specifically a product of European society's changing relations with,

and experience of, the globe.

For, within Europe, the surface of the Earth was being increasingly altered by agricultural advance and urban development. The depths of the crust were ransacked for useful metals and coal. Travel and exploration blossomed. As human technical powers over Nature grew, attitudes towards the Earth itself changed in their wake. The environment was tamed, civilized, brought within human experience. Mountains, once thought desolate, threatening and ugly, were transformed by a mental revolution into objects of awe and beauty – and sport for the climber.

At the same time, Europe spread out by voyages and conquest to become by the 18th century the first civilization bounded only by the limits of the globe. The de-

Travelers brought back tall tales of their encounters: these dog-headed and long-eared individuals were said to have been seen in India.

John Woodward's theory of the internal structure of the Earth, with regular strata throughout. Water from depth rises to the surface to provide the sources for rivers.

Below, Kircher's view of the Earth: a central fire powers the subterranean circulation of water. Where the fire approaches the surface, volcanism results.

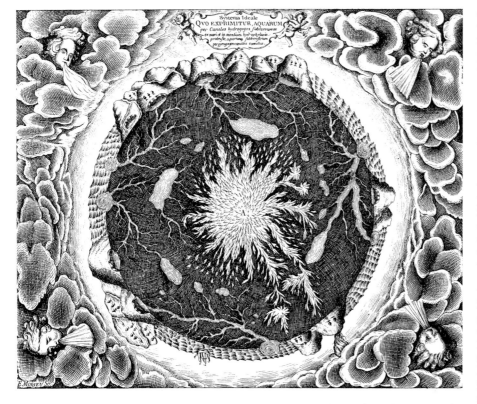

traditional macrocosm-microcosm analogies and the animistic belief that the Earth was "alive". The four Aristotelian elements were abandoned. All these increasingly led to the separation of the scientific study of the Earth from that of the cosmos at large. Thus, when late in the 17th century scientists like ★Steno, ★Hooke and ★Woodward developed "theories of the Earth", they focused attention on the issues of the origin and destiny of the Earth, independently of the larger canvas of the universe in general. The Earth was increasingly conceived as having its own internal laws and economy. The geocosm gave way to the planet, Earth.

Furthermore, between the 15th-century Renaissance and the 18th-century Enlightenment, Christians gradually came to take up more liberal and rational stances on the relations between Scripture and scientific truth. Many European scientists abandoned Christianity for less dogmatically specific forms of theism and deism. In such a climate, the broad Christian pattern of Earth history, moving from original divine Creation through to final destruction, was usually maintained. Nevertheless, taboos against believing that the Earth was extremely old, and that it might have an indefinitely long future, began to be broken down in the work of scientists such as Hooke, ★Buffon, ★Hutton and ★Lamarck.

In the 18th century it became generally accepted for the first time that the Earth had long predated the creation of Man. Furthermore, the Scriptural vision of a personal God of love and wrath yielded to the Supreme Architect Deity of the Enlightenment, a God of order, reason, and benevolence. By reflection, this meant that the Earth itself could be envisaged as possessing order and permanence in its fundamental laws and economy; as being a stable and functional system of processes deploying materials in good mechanical order. Divine interventions receded. Anthropocentrism was watered down.

Impressed with the power and success of the great 17th-century synthesizers – Galileo, Descartes, Newton and Leibniz – to give comprehensive rational order to the celestial world, scientists now looked for a Newton of the terrestrial globe. Hence the ambition was developed to grasp the Earth's economy in terms of a minimum of powerful natural laws applicable to all periods of the Earth's history, uniting past and present, and also to all parts of the globe, without the need for extraordinary or miraculous events beyond the normal course of nature. Originating in the 17th-century theories of the Earth, this movement culminated in the work of Buffon, Hutton and Lamarck.

Empirical examination of the Earth was now being pursued with ever greater energy and thoroughness. In part, this was founded upon the analysis of specimens in pursuit of the traditional disciplines of natural history, and spurred on by ever grander collections and museums. This in turn provoked the urgent need to introduce

signedly close fit between inhabitant and habitat, which had characterized the anthropocentric Christian view, was thrown off balance by new accounts of greater extremes of relief and climate than previously experienced. Areas of the globe such as much of North America were found scarcely inhabited – and sometimes almost uninhabitable. Entirely new fauna and flora were discovered. The globe began to appear a much more complicated planet. The stable world of Mediterranean Europe was dwarfed. Furthermore, old myths were dissolved, such as that the tropics were an impassable region of fire, or that there could

be no Antipodes. In the wake of circumnavigations, the globe had increasingly to be seen as an integrated Earth system; and new problems, such as that of the appearance and disappearance of whole continents, loomed larger.

In this situation, the science of the Earth was reconceived. The development of Copernican astronomy denied that the Earth was at the center of the planetary system. Belief grew – aided by the telescope – that the universe was indefinitely – or even infinitely – large, and that the Earth and the Solar System occupied no special place in it. The new mechanical philosophy rejected comprehensive systems of classification into the diversity of minerals, rocks and fossils. Once agreement was reached – by the late 17th century – that fossils were indeed organic remains, meticulous examination of fossils by naturalists such as *Hooke and *Steno established that many were the remains of creatures and plants lacking extant analogues, and hence there emerged Hooke's notion of a succession of former faunas and floras now extinct.

A series of great 18th-century mineralogists, including *Lehmann and *Werner, established the principle of a correlation between particular "mineral" and lithological characters and location of that specimen within the horizontal sequence of the strata. Furthermore, the vertical order of the strata themselves was increasingly acknowledged to be an index of temporal succession. For rock-forms were now being categorized according to their composition, origin and location into an ordering principle – such as Werner's Primary, Transitional, Secondary, Recent and Volcanic.

Primary rocks were broadly identified as being massive, frequently crystalline, non-fossiliferous, non-stratified and found at the base of rock-formations, often at steep angles to the horizon. They had supposedly been deposited out of a universal ocean by chemical means. They commonly made up rugged uplands and mountainous areas. Secondary rocks were believed mechanical in their deposition; were stratified and were thought to have been consolidated on former seabeds out of the detritus of earlier Primary formations. They were commonly fossiliferous, and frequently bedded more nearly parallel to the horizon. Transitional rocks shared some of the characteristics of both. Recent rock formations were less consolidated, and were to be found adjacent to the surface. For Werner, few other rocks than true lavas were volcanic or igneous. These were essentially extraneous to the above-mentioned succession, localized, and exceptional.

Of course, there were many rival classifications of rocks, and all were highly controversial. In particular, battle raged in the late 18th century over the nature of *basalt, and by extension over a range of other massive rocks which revealed vitreous and/or crystalline structure. Were they of aqueous or igneous origin? Not only was the classification of a number of rocks at stake, but so also were comprehensive rival conceptions of Earth history. One, the Wernerian, saw the Earth's crust uniquely and successively precipitated out of aqueous solution. The other, culminating in the work of *Hutton, asserted the continual formation of all rock types proceeding from the power of the Earth's supposed central heat over an indefinite timespan.

At the same time as collection, analysis and classification of specimens were bounding in popularity, meticulous examination of the Earth *in situ* by scientific traveler-naturalists also enormously expanded. It led to the detailed investigation for the first

A mid-19th-century illustration of a coral reef. Charles Darwin's explanation of their origin was a major contribution to physical geology.

A mid-18th-century diagram showing the formation of surface features through the buckling of rock strata.

time of European landscape and crustal features; but it also stimulated exploration of global phenomena not encountered in Europe, such as the volcanic islands and the coral atolls of the Pacific. In the 18th century, the work in particular of Lehmann, Pallas, Arduino, Sir William Hamilton, de Saussure, de Luc, Dolomieu and Hutton built up a tradition of *fieldwork*. Their attention shifted away from planet Earth in its entirety, focusing specifically on its crust.

Such fieldwork could take many forms. In some cases it simply recorded the terrestrial phenomena visible along the length of travels and traverses. In a more systematic way, fieldworkers began exhaustively

to comb the topography of entire regions. They thus began to trace the distribution and continuities of strata, fossils, landscape features and structural faults across a terrain. Furthermore, fieldworkers increasingly saw the importance of probing perpendicularly down through the Earth's crust, making use of exposures at cliffs, mountainsides, mines, canal diggings, quarries and the like. When incorporated into a regional survey, such sections added a grasp of the composition of the crust. This in turn was ever more successfully visualized through the development of maps, sections, panoramas and block diagrams.

Above all, the growth of fieldwork focused attention for the first time on a hitherto

The Flood, by Francis Danby (1793–1861). The Flood was long regarded as an established event in Earth history.

little-investigated feature of the crust: the strata themselves. From the mid-18th century the strata became perhaps the chief focus of Earth science. Some geologists like *Lehmann studied them mainly to establish their general vertical order of succession; others organized their fieldwork around pinpointing the distribution of their outcrops throughout a region.

Furthermore, investigation of strata in the field increasingly provided an interpretative framework for grasping other crustal features. This was especially so because it came to be recognized that the vertical succession of the strata piled on top of each other represented, in general, the temporal sequence of their formation or deposition. Hence strata provided an index of the history of the Earth. Thus, by the end of the 18th century, consensus had been achieved among geologists that there was a correlation between the incidence of particular fossils and particular strata (or, in William *Smith's terms, that strata could be identified by organized fossils). Similarly, strata, landforms, and processes at work on the face of and beneath the surface of the globe were recognized as being interconnected, and their relations were explored. For, as *Hutton and his followers *Playfair and *Hall insisted, forces of denudation were destroying all existing rock formations. Yet, out of this detritus of denuded continents, the strata of future landmasses would be consolidated.

The Golden Age. Thus, by the turn of the 19th century, powerful traditions of examining the Earth's crust, both in the field and as specimens, had grown up. Attention was being focused on the strata, for strata gave meaning to other features of the Earth, and were the index of the history of the globe. By now, it was agreed the Earth had had an extremely long history, characterized by ceaseless change and profound revolutions. In short, an organized, integrated science of the Earth had developed, which contemporaries were, for the first time, beginning to call "geology". From the early 19th century there is a reasonably smooth development of Earth science up to the present.

The 19th century was the golden age of geology. It was taught at major institutions such as *Werner's Mining Academy at Freiberg, Saxony, and at the School of Mines in Paris. In England, the Geological Society of London (founded 1807) orchestrated pursuit of the science. In all major nations, state-financed Geological Surveys were set up. These were led by, and employed, first-rate geologists such as *De la Beche, *Murchison and *Geikie in Britain; *Élie de Beaumont in France; *Logan in Canada; and *Hall, *Powell and *Gilbert in the USA. Such surveys conducted the task of mapping the stratigraphy of their appointed regions. But they also in turn pioneered conceptual advance, as, for example, *Lapworth's work on the Silurian and Ordovician systems. Publicly-funded

oceanic voyages also enabled scientists like *Dana, *Darwin (Charles), *Huxley and *Murray to make substantial contributions to understanding oceans, islands and marine life, and the general balance of Nature.

Moreover, geology became a science deeply attractive to the amateur. For several generations, gentlemen amateur geologists such as *Scrope, *Lyell, *Murchison, von *Buch and von *Humboldt held the field, giving way only gradually to the professional and the academic. And, of course, throughout the century geology was to the fore in public controversy, especially in the religious debates which raged over the historicity of the Biblical Deluge, the contesting realities of miracles and the uniformity of Nature, the theory of *evolution by natural selection, and the question of the antiquity of Man. Industrial and economic applications of geology were also increasingly capitalized upon, at first for the discovery of coalfields to feed the hungry steam engines of the Industrial Revolution and then, later, ever more in the search for *petroleum and metals, particularly in imperial territories overseas.

Geology advanced in many important fields in the first half of the 19th century. The detailed order of the strata was explored for particular regions. Georges *Cuvier and Alexandre Brongniart undertook classic work unravelling the Cretaceous succession of the Paris Basin. In England, William *Smith's pioneering labors established the sequence of outcrops

A minor eruption at Krakatoa about six months before the disastrous explosion of August 1883.

of Secondary strata from the coal measures (i.e., ★Carboniferous) up to the Chalk (i.e., ★Cretaceous). ★Sedgwick and ★Murchison went to work on the complex formations of the western parts of Britain, hitherto confusingly known as Transition Grauwacke, and proposed – with much conflict – a division into ★Cambrian, ★Silurian and ★Devonian systems. ★Lyell established in all essentials the modern concept, and divisions, of the ★Tertiary: in Lyell's classifications, ★Smith's principle of identifying strata by their type fossils was used – but extended. For Lyell, the divisions of the Tertiary hinged upon their relative balance of extinct and currently existing fossils. The principle of (relatively) dating strata by type fossils rather than by lithology was triumphantly employed by Albert Oppel in his classic comparative studies of the Jurassic in France, Switzerland, Germany and Britain.

Perhaps more trail-blazing was the growing success in using fossil evidence to reconstruct a history of life. 18th-century naturalists had gained a general awareness that the Earth had formerly been populated by creatures now extinct, ranging from enormous vertebrates of elephant-like form (mastodons) down to huge shells, such as ammonites. But this was given new precision and perspective, above all by the work of ★Cuvier. Using the rich collections of fossil bones available to him in the Paris Museum of Natural History, he applied the principle of the correlation of function and structure within the method of comparative anatomy to reconstruct hitherto unimagined extinct vertebrates – saurians, birds and mammals.

Furthermore, Cuvier was concerned to integrate zoology and geology by reconstructing the *total* physical environment of each epoch of such extinct creatures, viewed as particular stages in the development of life on Earth; each with its own distinct climate, terrestrial conditions, fauna and flora. In Cuvier's work, and that of those who followed in his footsteps, entire populations of extinct crustaceans and saurians, birds and mammals, were reconstructed. As the paleontological record was pieced together, rising upward through the strata, it suggested that such populations had existed in a temporal succession which led from Azoic times to invertebrates, and then on to vertebrates; from fish to reptiles to mammals; from simple to complex; from extinct to living. Early generalizations, such as that warm-blooded creatures had not existed in the Secondary epoch, or that pre-Silurian rocks were azoic (without life), were shaken, of course, by subsequent research: nevertheless, through the first half of the century the ★fossil record seemed to yield ever more solid evidence of a progressive succession of forms of life, culminating in Man, each population apparently separated by periods

of "revolution".

Hotly debated, however, were questions as to the precise method of geological inquiry. These in turn were deeply connected with the fundamental problem of constructing a theoretical interpretation of the Earth. The challenging difficulty of geological inquiry was to reconstruct past stages of the Earth which had never been directly observed by Man. As a guide, geologists had relics of this past, in the form of rocks, fossils, and relief features such as rivers, valleys, and erratic boulders, from which deductions could be made. They also had the analogy of the present processes and economy of the Earth.

One school of geological thought, the "Uniformitarian" school, developed by ★Hutton and, later, ★Playfair and ★Hall – and taken up by ★Lyell – declared that the only properly scientific method of inquiry was to explain the Earth's past in terms of causes of the same kind and intensity as those currently active: to go beyond was to enter the realm of mere speculation and

religious miracles. "The present was the key to the past."

Moreover, claimed the Uniformitarians, *given enough time* the Earth's past could actually be explained in terms of the cumulative effects of gradual causes. In time, rivers would excavate their own valleys; in time, the sequence of earthquakes would build whole continents out of the sea, or raise mountain-ranges; in time, ecological pressures would gradually cause the extinction or migration of entire biological populations.

Other geologists, including ★Cuvier, ★Buckland and ★Sedgwick, denied that it was unscientific to postulate that sudden, catastrophic events had occurred far back in geological times, events for which we had no current analogy. In fact, they claimed, evidence *demanded* such interpretations. The magnitude of the disruptions of strata throughout mountainous areas of the globe testified to sudden, catastrophic dislocations. Dry valleys equally seemed to prove that the theory of gradual fluvial

*erosion was false. The sheer extent of alluvial materials and erratic blocks scattered across northwest Europe could not be explained by any known cause.

This debate, however, was not confined to issues of method in geology, for it also embodied rival conceptions of the very patterns and direction of Earth history. *Hutton and *Lyell were postulating a steadystate Earth – ceaseless piecemeal local change was occurring. But the consequence of such change was to maintain from the indefinite past to the indefinite future an overall, constant equilibrium in the terrestrial economy. Lyell, for example, believed that even though particular species had come into being or become extinct, there was no general direction or development in the history of life – Man alone excepted.

Geologists such as Cuvier, Élie de Beaumont, Buckland, Sedgwick, De la Beche and Murchison, however, were putting forward a completely different, "directional" picture of Earth history. In this, time was not a cycle but an arrow: they postulated a gradually cooling Earth, a solidifying crust, an irreversible and unrepeatable succession of strata, a diminution of energy located within the globe. They believed that the Earth had supported successive, discrete populations of flora and fauna, usually regarded as specially created by God.

In fact, both points of view made their mark upon the approaches of most geologists. The justice of Lyell's claim that most "catastrophist" explanations reflected ignorance and explained nothing was increasingly admitted. On the other hand, paleontology seemed to provide mounting evidence for a profound historical sequence of lifeforms. Very few geologists were prepared to accept Lyell's claim that this was a gigantic illusion created by the random and misleading accident of the survival of fossil evidence: once the "catastrophist" case had been freed from the taint that it was "unscientific" in introducing miracles and in seeking to confirm Scripture, most geologists were prepared to admit a genuine succession of the stages of life. A typical compromise was that of *Scrope, who demanded that geology proceed by actual causes, but conceded that these had diminished in intensity in the course of Earth history.

The geology of the second half of the 19th century was to solve many of these dilemmas by putting them in new perspective. The principle of evolutionary organic transformation was almost universally adopted in the years following the publication of *Darwin's *Origin of Species* (1859). This helped to confirm the claim of the "directionalists" that life had proceeded through a significant succession of levels, while bearing out Lyell's emphasis on the gradualness of change. The

band of great palaeontologists – *Huxley, *Marsh, *Cope and *Osborn – employed fossil evidence to chart the branching directions and the pulse of evolutionary progress, now that enough "missing links" had come to light to trace a solidly connected progression of forms.

Secondly, the great surge of interest in geophysics in the second half of the 19th century imposed powerful pressure from outside upon geologists searching for the correct pattern of Earth history. Above all, Thomson (later, Lord *Kelvin) applied the Second Law of Thermodynamics and the developing concept of entropy to the cooling of the Earth. Such work not only gave strong support to a "directionalist" view of the Earth but pointed to an Earth markedly less old than the Uniformitarians, or than Darwinian evolutionists, demanded. For Kelvin, thermodynamics applied within geophysics showed – contrary to the Uniformitarians – that the Earth had had a beginning, and would certainly, as a habitable planet, have an end. Hence, it gave some renewed credit to "catastrophic"

Earth and life histories.

Fierce debate ensued over the applicability of Kelvin's methods of physics and mathematics to "geological" problems. Many geologists claimed that the Earth had its own laws which were not reducible to the laws of general physics. And the discovery of radioactivity early in the 20th century reassuringly indicated that the Earth was cooling – if at all – at a far slower rate than Kelvin had believed, and so restored to the geologists their vast timescale. But the intervention of Kelvin was important in breaking down the dogmatically stubborn Uniformitarian preference for time-over-violence in Earth history, and in bringing narrowly geological speculations (such as Darwin's suggestion of 300 million years for the denudation of the English Weald – i.e., for Tertiary time alone) before the bar of other sciences.

In the second half of the 19th century, in a different area of geology, another theory, which transcended the polarity between the earlier "steady state" and "directional" theories of the Earth, gained importance.

This was the elaboration of glacial theory. In its original form, most famously advanced by *Agassiz, the theory postulated a former, more extensive distribution of ice across Northern Europe to account for effects such as erratic boulders, striated mountainsides and alluvium. In its later, more thoroughly developed shape, glacial theory began to suggest a whole succession of Ice Ages, in which a substantial portion of the European continent had been covered with glaciers. (See *glaciation.)

Many hypothetical causes were offered. Some were purely terrestrial, hinging upon the suggestion that the changing distribution of land and sea could have a major effect in transforming climate. Some, like *Penck, suggested external causes such as fluctuations in the Sun's heat.

The importance of glacial theory was that it demanded a more sophisticated conception of the pattern of Earth history than either "Uniformitarians" or "Directionalists" had originally offered. For Ice Ages were in a sense catastrophes – yet they were, at the same time, evidence that the path of Earth history was subject to major oscillations rather than being uniformly progressive.

Indeed, the most important and exciting development in theoretical geology in the period leading up to WWI was precisely the growing awareness that the real complexity of the present crust of the Earth and its history could not be explained by a rigid insistence upon uniformity of causation and an overall "steady state"; nor by a simple view of progress, with its neatly-tailored succession of stages leading up to the present. The great interest newly taken in geophysics was highly influential in this respect, for it demanded that geologists should envisage the forms and materials of the crust in terms which went beyond merely identifying strata and projecting surface causes back into the past. Thus the question of the deeper physical energies governing the rhythm and balance of continents and oceans became increasingly raised – i.e., the question of "isostasy". Kelvin's arguments as to the *age of the Earth revived physical controversy over the original nature of the Earth. Had the Earth once been incandescent? or had it taken its origin from the adherence of cool particles? Was the great mass of the Earth beneath the solid crust to be inferred as being solid? or as a fluid, magmatic core?

Similarly, in the work of *Dana and *Murray, theories were constructed of large-scale, counterbalancing raising and subsidence of the ocean bed in the Pacific. *Davis advanced comprehensive physical theories of continental denudation, and so attempted to put geomorphology upon a basis of general laws of landscape. Scientific sedimentology was systematized by the work of *Murray, *Grabau and *Barrell. The physical bases of lithogeny and orogeny were investigated.

Furthermore, stratigraphical observation in the field was growingly confirming this perception of the extremely complex rhythm of geological causation. Sophisticated investigation of type fossils, as, for example, by *Lapworth in his work on graptolites in the Southern Uplands of Scotland, was revealing evidence of hitherto unsuspected unconformities, reversed faults, thrust planes, and lateral displacements of formations on a scale previously hardly contemplated. The work of *Argand and *Haug revealed the nappe structure of the Alps. Peach and Horne showed the gigantic thrust planes and faults of the northwestern Highlands of Scotland, the products of forces contemporaneous with and subsequent to the laying-down of the strata. Such new perspectives indicated how incomplete was the former stratigraphical stress upon simply tabulating the *order* of the strata: more attention needed to be paid to the *deformations* of the rocks. In other words, the modern emphasis on tectonics was emerging.

The new focus on tectonics was closely associated with a growing analysis, in the field and in the laboratory, on metamorphic rocks; and, in its turn, such work on internal dynamics hinged upon the emergence in the second half of the century of petrology, the science of the physical and chemical understanding of rock materials. *Cloos initiated the study of flow textures of solidifying magma. And the rise of petrology was aided by the development of microscopic analysis of thin slices as pioneered by *Sorby.

The 20th Century. Thus, in some ways, geology marched into the 20th century having achieved a solid foundation of basic conceptions and empirical knowledge. Important progress had been made in charting the order of the strata and the succession of fossils. Such work was to continue through the present century. National surveys were mapping the face of the Earth. Related and subordinate sciences, like seismology and physical geography, were expanding. The study of human prehistory developed. Yet in other respects, and above all when faced with the problem of grasping the patterns and causes of the physical forces which governed the globe's dynamic tectonics, geologists were still groping in the dark.

Geology was to be transformed after WWI by the formulation of the theory of *continental drift, above all by *Wegener. This theory was, however, hardly commonly accepted until its incorporation within a comprehensive vision of *plate tectonics after WWII. In many ways, however, the rise of petrology, and the development of new interest in geophysics in the late 19th century had provided the basis for such a geological revolution. RP

Great Geologists

Here follow, in alphabetical order, biographical notices of those scientists and philosophers who have made considerable contributions to our modern understanding of the Earth sciences.

Agassiz, Jean Louis Rodolphe (1807–1873)

Agassiz received his scientific training at Zurich, Heidelberg, Munich and Paris, where he was deeply influenced by *Cuvier. He became professor at Harvard in 1847, where he founded the Museum of Comparative Zoology. His fundamental *Researches on Fossil Fish* (1833–44) used Cuvierian comparative anatomy to describe and classify over 1700 species.

In his *Studies on Glaciers* (1840) he pioneered the concept of the Ice Age, which he saw as an agent of extinction separating past from present flora and fauna.

The later years of his life were spent on *Contributions to the Natural History of the United States* (1857–62), an exhaustive study of the American natural environment. He also became one of the most powerful opponents of *Darwin's theory of *evolution on religious, philosophical and paleontological grounds. RP

Albertus Magnus (c. 1200–1280)

Born of good family in Bavaria, Albertus was educated in the liberal arts at Padua, and joined the Dominican Order. His life was divided between intellectual studies and administration within the Church. His importance to science lay in commenting on and popularizing the philosophy of *Aristotle. He accepted the general Aristotelian doctrine of the four elements; and seems to have believed that Democritus' doctrine of atoms might be incorporated. He was interested in alchemical experiments and in the possible transmutation of metals.

Aristotle wrote no work of mineralogy, but Albertus' *Book of Minerals* (c. 1261) took over many Aristotelian ideas. He accepted, for example, that there were subterranean exhalations: a dry one which produced earths and stones, and a moist one which produced metals. He also explored the relations between the Earth and the heavens, and was interested in climate. Most important, he listed about 100 minerals, describing them, citing classical authorities on them, and sometimes adding his own observations. Almost all subsequent mineralogies and lapidaries stemmed from Albertus. RP

Argand, Émile (1879–1940)

Born in Geneva, Argand devoted himself to understanding the tectonics of Alpine regions, whose structure is complex through intense folding and metamorphism and difficult to unravel because of stratigraphical ambiguity and the relative absence of fossil markers. He carried his researches to the interior of the Alps; and developed techniques of geometrical projection to explain missing structures. He opened up as a major tool of structural analysis the dynamic direction and axes of tectonic organization and movement. Among his highest achievements lay his pioneering, theoretically sophisticated, block diagrams.

In later years, he developed a greater concern with the time-dimension of tectonics, especially with orogeny, and made

use of *Wegener's continental-drift hypothesis as a framework for his ideas of Eurasian structural development.　RP

Aristotle (384–322BC)

Born in Chalcidice, Aristotle studied in Athens under Plato and became tutor to Alexander the Great. He founded the Lyceum as a center of teaching and research. His influence on the study of the Earth lasted until the 18th century, his influence on cosmology until Copernicus.

He thought the Earth, composed of four elements (earth, water, air and fire), was a sphere at the center of the universe. He was particularly interested in comparing earthquake and volcanic phenomena to climatic events like storms, explaining them as the produce of conmingling of dampness from rain with underground winds.

He believed that earthly objects were also due to exhalations from the Earth, fossils and minerals being produced by a dry exhalation, metals by a moist one. He seems to have attributed organic fossil forms to some kind of plastic force in the Earth, which imitated the workings of nature on the surface.

Aristotle was concerned also with the origins of springs and rivers, seeming to have accepted that rainwater was not the sole source of springs, and holding that the Earth itself intrinsically produced water as a result of its coldness.　RP

Arrhenius, Svante August (1859–1927)

Arrhenius' importance to geology rests on three works. His 1896 paper, "On the influence of carbonic acid in the air upon the temperature of the ground", postulated that the variable capacity of carbon dioxide to absorb infrared radiation from the Earth's surface might explain fundamental climatic changes in the Earth's history (e.g., Ice Ages). The paper was largely ignored.

His "Towards a physics of volcanism" (1901) sought to explain volcanic eruptions: at very high temperatures, he claimed, water forced its way into the magma, causing massive expansion and penetration into volcanic fissures. As this magma cooled, water was liberated, causing explosions.

In his *Textbook of Cosmic Physics* (1903) he turned to the problem of the *origins of life on Earth, developing the theory that seeds had been transported here from other cosmic systems by light-pressure.　RP

Bailey, Sir Edward Battersby (1881–1965)

Bailey served in the Geological Survey of Great Britain until 1929, when he became professor at Glasgow, only to return to the Survey as its Director in 1937.

His chief contributions were in the fields of tectonics, metamorphism and igneous geology. His most important metamorphic studies were of northwest Scotland, where

he produced major reinterpretations of late Precambrian schists. In igneous geology, he helped to formulate the theory of the "cauldron subsidence" of Glen Coe, Scotland, and investigated the intrusion tectonics of Arran granite and the volcanic complex of Rhum.

He was a distinguished writer on the history of his science, publishing biographies of *Hutton and *Lyell and an account of the Geological Survey of Great Britain.　RP

Barrell, Joseph (1869–1919)

Born at New Providence, New Jersey, Barrell had a thorough general education in the natural sciences before specializing in geology. This may explain the two most marked features of his approach: he was always primarily concerned with the broad problem of the effects of the totality of physical agents upon the evolution of the Earth; and he was a philosophical geologist, more interested in a conceptual grasp of the problems of the Earth than in pioneering new fieldwork. Much of his work was concerned with the relations between volcanic phenomena, magma and metamorphism. His main interpretation of these forces lies in *Geology of the Maryville Mining District, Montana* (1907).

He was influential as a theorist of sedimentology. The traditional assumption was that almost all sedimentary rocks were of marine origin: Barrell put forward the view that a substantial proportion are in fact continental, fluvial or aeolian in origin.　RP

Barrois, Charles Eugène (1851–1939)

Born in Lille, France, Barrois spent his working life attached to the Lille Faculty of Sciences. He excelled in all aspects of field geology and paleontology.

His chief love became Primary geology (see *Werner). In the late 1870s he undertook minute research on the Primary formations of northern Spain, and later the Sierra Nevada. And, throughout his career, he occupied himself with mapping and interpreting the formations of Brittany, a task made especially difficult by metamorphism and the absence of fossils.

Though a distinguished descriptive paleontologist, he made little use of fossils to unravel the history of life, avoiding entanglement in evolutionary controversy because of religious commitments.　RP

Bowen, Norman Levi (1887–1956)

Norman Bowen was born and educated in Kingston, Ontario. As a graduate student, he worked at the Geophysical Laboratory in Washington, then recently founded, and in 1912 published his first results on the experimental melting of silicates and their crystallization behavior. He remained at the Geophysical Laboratory for most of his life, and over a period of more than forty years reported on and interpreted results from innumerable more extensive experiments.

During this period Bowen and his associates, notably O. F. Tuttle and J. F.

William Buckland outfitted to explore a glacier. His contributions to glacial theory arose from his pre-occupations with catastrophism.

Schairer, established the physico-chemical principles relevant to the fractional crystallization of magmas and the formation of magmas by partial melting. They transformed igneous petrology from the descriptive science created by Zirkel and Rosenbusch into its current state, in which the origins of rock types and the evolution of magma are of prime interest. KGC

Buch, Christian Leopold von (1774-1853)

Born at Stolpe, near Berlin, Buch passed his life as one of Europe's eminent amateur traveler-geologists. He studied mineralogy and chemistry at Berlin, and then under the great *Werner at Freiberg: though modifying and extending Werner's work in many respects, Buch remained a lifelong admirer. His great strength lay in his ability as an observer: his findings were published in a number of volumes of travels (1802, 1808, 1810, 1825).

In 1802 he developed his theory of elevation craters to explain volcano-like phenomena, maintaining that Auvergne *basalts had once been lava. He continued, however, to accept Werner's teaching that Saxon basalts were of aqueous origin, finding no volcanic phenomena associated with them. RP

Buckland, William (1784-1896)

Born at Axminster, England, Buckland attended Oxford University. Elected Reader in Mineralogy in 1813, he became Reader in Geology in 1818, and subsequently Dean of Westminster in 1845. His geological work has three main foci.

Firstly, he made important contributions to the descriptive stratigraphy of the British Isles, inferring from the strata a stage-by-stage history of the surface of the globe. Secondly, he became a distinguished paleontologist. He used the methods of Cuvierian comparative anatomy to reconstruct *Megalosaurus* and the history of the hyena cave den, Kirkdale Cavern, in his *Relics of the Deluge* (1823).

But his greatest concern lay in exploring evidence for catastrophic transformations of the Earth's surface in the geologically recent past, as indicated by features of relief, fossil bones, erratics and gravel displacement. In his *Geology Vindicated* (1819) he confidently attributed these to the Flood – an assertion he later withdrew. His concern for such phenomena helped him to become, however, a leading British exponent of glacial theory. RP

Buffon, Georges Louis LeClerc, Comte de (1707-1788)

Though educated for the law, Buffon turned towards a scientific career, becoming a member of the Royal Academy of Sciences, Paris, in 1734. He contributed to many areas of science, but his major work was the *Natural History,* in 36 volumes, published from 1749. In the first volume he set out a theory of the Earth, stressing the gradual and ceaseless change of the Earth's crust produced by natural causes; that new land was continually forming, and continents being destroyed by the sea. In his *Epochs of Nature* (1779) he somewhat modified this theory, now emphasizing that the Earth had been a fragment thrown off by the Sun, and had cooled gradually in seven stages, so accounting for the existence of primitive vitreous rocks, subsequent volcanic action, and more recent aqueous formations. The cooling Earth also explained the successive appearance of different forms of life, beginning with gigantic forms, now extinct, and ending with Man.

He was one of the first to put an age to the Earth on the basis of experiments conducted with cooling iron balls, suggesting 75,000 years in public but as much as 3 million years in his private manuscripts. RP

Chamberlin, Thomas Chrowder (1843-1928)

Chamberlin's most important contribution to geological thinking lay in his attack on Lord *Kelvin. Kelvin had postulated that the Earth was rather young (less than 100 million years), basing his views on the

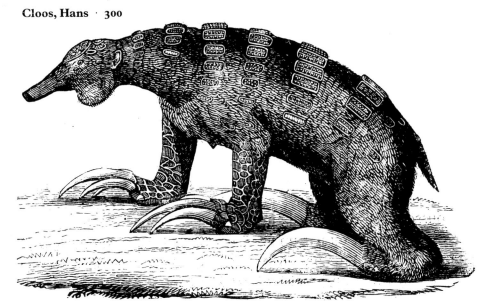

A reconstruction according to Cuvier of *Megatherium*, the largest of the extinct group of edentate mammals known as ground sloths.

assumption, derived from the nebular hypothesis, that the Earth had steadily cooled from a molten mass.

Chamberlin rebuked Kelvin for the dogmatic confidence he placed in extrapolations from a single hypothesis, and stressed that geological reasoning must follow from a plurality of working hypotheses. He also believed geological evidence anyway suggested the Earth to be older than Kelvin had estimated, backing his case against the nebular hypothesis by developing with the celestial physicist, F. R. Moulton, the planetesimal hypothesis. This postulated a gradual origin, by accretion of particles, for the Earth and other planetary bodies, an origin which was therefore cool and solid.   RP

Cloos, Hans (1885–1951)
Cloos made tectonics his lifetime's study. His pioneering work on the tectonics of Silesian granite revealed that, far from being structureless as commonly thought, it bore marks of features acquired soon after its intrusion: he opened up the study of the flow textures of solidifying magma. He later extended his tectonic studies to the jointing and cleavage features of the deformation of solid rocks. He was also one of the pioneers of the reproduction of tectonic processes in the laboratory.   RP

Cope, Edward Drinker (1840–1897)
Born in Philadelphia, Cope, as editor of *The American Naturalist*, was to the fore in airing and popularizing evolutionary views in the USA.

His chief work was as a student of the fauna – and above all, the fossil fauna – of the little-explored American Midwest, particularly in regard to fishes, amphibia, reptiles and mammals, pursuing the techniques of comparative anatomy, classification and systematics as laid down by *Cuvier, *Owen and *Huxley. He made pioneering studies of the dinosaurs of the *Cretaceous of New Jersey and the plesiosaurs of the Kansas Cretaceous.

A convinced evolutionist, he championed Lamarckism; and argued for the creative role of consciousness in organic development, believing that favored variations were shaped by will. He wrote widely in his mature years on religion, psychology and general philosophy, interpreting the significance of evolution.   RP

Cuvier, Georges (1769–1832)
Born at Montbêliard in the principality of Württemberg, Cuvier received his training in natural history at Stuttgart, spent six years as a private tutor in Normandy, and came to Paris in 1795.

His great contribution was to systematize the principles of comparative anatomy to apply them to fossil vertebrates, thus effectively founding modern paleontology. In his *Researches on the Fossil Bones of Quadrupeds* (1812) and his *Animal Kingdom* (1817) he reconstructed extinct fossil quadrupeds such as the mastodon and the paleotherium.

His stratigraphical work in the Paris Basin demonstrated that fossil flora and fauna were specific to particular strata. Cuvier interpreted this as implying that the history of the Earth involved a series of revolutions ("catastrophes") which periodically swept away whole living populations, their place being taken either by migration or by the creation of new species. This theory, set out in his *Preliminary Discourse* (1812), expressly countered the evolutionary views of *Lamarck and the paleontologist Geoffroy Saint-Hilaire (1772–1844).   RP

Daly, Reginald Aldworth (1871–1957)
Born at Napanee, Ontario, and educated at Harvard, Daly became a widely experienced field geologist and one of the great synthesizers of geological fact in the early part of the 20th century.

He once remarked that geology was "drowning in facts" and urged more concern about the theoretical framework behind Earth processes. His principal field of study was *igneous rocks and he developed many of the ideas which are still current on such subjects as the mechanics of magmatic intrusion; he added greatly to our knowledge of basalts and the ocean basins with his studies of oceanic islands such as St. Helena. Daly was among the first to look upon continental ice sheets as a means of testing the reaction of the crust to loading, and hence deducing information about the physical properties of the crust and upper mantle. In addition, it was Daly who suggested that submarine canyons might have been eroded by the action of turbidity currents. He also developed a hypothesis of the glacial control of sea level.   KGC

Dana, James Dwight (1813–1895)
The formative period in Dana's career came with his service with the Wilkes expedition of 1838–42, a natural history survey of Polynesia financed by the US government. He spent much of the 1840s writing reports of this expedition (*Geology*, 1849; *Crustacea*, 1852).

This experience led to a lifelong interest in coral and volcanic phenomena – his identification and classifications of Pacific corals have survived remarkably well. He was also preoccupied with the problem of the diversity of coral formations: why were there atolls, barrier reefs and fringe reefs? He became a firm supporter of *Darwin's theory that atolls were the result of subsidence, and confirmed it independently. From this he deduced the likely enormous extent of downward movements of the crust in the tropical Pacific.   RP

Dart, Raymond Arthur (1893–   )
Born in Toowong, Brisbane, Dart studied medicine at the University of Queensland, held teaching posts in anatomy in England just after WWI, and became Professor of Anatomy at Witwatersrand from 1923 to 1958.

His major work lay in the investigation of African fossil *hominids. In 1924 he discovered at Taung, South Africa, the skull of a man-ape child which he recognized represented an extinct ape group, called by him *Australopithecus africanus*. Subsequently more specimens were uncovered.

Later work by *Leakey and others has confirmed Dart's identification of the "missing link" and that the African continent seems to have housed the earliest specimens of Man.   RP

Darwin, Charles (1809–1882)
Up to the publication of his *Origin of Species* (1859), Darwin was best known as a geologist. Born in Shrewsbury, he was educated at Edinburgh and Cambridge Universities, before serving from 1831 on the famous five-year voyage of the *Beagle*. At Cambridge, under *Sedgwick, he had developed an interest in geology. His own extensive observations on the voyage, however, won him over to Uniformitarianism, as expressed in *Lyell's *Principles of Geology*. In particular, his South American experiences of the interconnectedness of volcanoes, igneous rocks, raised beaches and land uplift convinced him of both the paramountcy and the adequacy of the slow action of present-day causes.

This led to Darwin's major contribution to geology, his theory of coral reefs. Darwin saw that the perfect condition for the formation of coral atolls was the slow subsidence of former islands; whereas fringing reefs were the product of uplift of similar land areas. He showed how the Pacific could be divided between zones of fringing reefs and active volcanoes on the one hand and areas of atolls with extinct volcanoes on the other.

Darwin's theory of *evolution by natural selection was utterly dependent upon the indefinitely long and gradual Earth history postulated by Uniformitarian geology. Darwin himself believed that the *fossil record was not sufficiently full to adjudicate between rival theories of the succession of forms of life.   RP

Darwin, George Howard (1845–1912)

The fifth child of Charles *Darwin, George Darwin was essentially a mathematical cosmogonist, concerned with the application of detailed quantified analysis to cosmological and geological problems. His first major contribution was his 1876 paper "On the influence of geological changes in the Earth's axis of rotation". He then carried out celebrated work on the origins of the Moon, believing it to be the product of the fission of the parent Earth as a result of instability produced by solar tides. He dated the event as at least 50 million years old, seeing this as compatible with his friend *Kelvin's relatively low estimate of the antiquity of the Earth. However, by the early 20th century, Darwin was one of the first to recognize how the discovery of radioactivity overthrew Kelvin's estimates.   RP

Davis, William Morris (1850–1934)

In his day, Davis was the leading American scientist of the physical environment. He made contributions to three fields of science, meteorology, geology and above all geomorphology. He developed the organizing conception of the regular cycle of *erosion, a concept that was to influence the discipline for fifty years. He assumed a standard life-cycle for a *river valley, marked by youth (steep-sided valleys), maturity (flood-plain floors), and old age as the river valley was worn lower and lower into a "peneplain". Such cycles could be interrupted by uplift, which would rejuvenate the river and impose fresh cycles.

The Davisian cycle presupposed a strongly uniformitarian view of Earth history in which the present was key to the past, and gradual causes were paramount.   RP

De Geer, Gerard Jacob, Baron (1858–1943)

De Geer is remembered in science as the founder of geochronology, the dating of past events in the geological record.

In 1882 he conceived the idea that the finely banded deposits known as varved clays might hold within them a seasonal record of glacial melting during the retreat of the great Scandinavian ice sheets at the end of the glacial period. According to this hypothesis, each sandy layer represents·

rapid sedimentation in a water body at the front of the glacier, while the clay layers represent the slow settling of fine particles after the sediment supply has stopped during winter. By painstaking measurement and comparison from one small exposure to another, De Geer and his colleagues were over a number of years able to build up a chronology of the glacial period which extended back for some 15,000 years. By 1920 he had extended his studies to North America and was able to correlate deposits there with those in Europe.   KGC

De la Beche, Henry Thomas (1796–1855)

Born in London, De la Beche became a gentleman amateur geologist, traveling extensively during the 1820s through Great Britain and Europe and publishing widely in descriptive stratigraphy. He became a scrupulous fieldworker, stressing the primacy of facts and distrusting theories, as can be seen from his *Sections and Views Illustrative of Geological Phenomena* (1830) and *How to Observe* (1835).

In the 1830s he conceived the idea of government-sponsored geological investigations of areas of Britain. He personally undertook a survey of Devon, for which he was paid £500, and then persuaded the government to formalize this arrangement. In 1835 the Geological Survey was founded, with De la Beche as director.

The Survey flourished and expanded. De la Beche's career reached its peak with the establishment of a Mines Record Office and then the opening in 1851, under the aegis of the Geological Survey, of the Museum of Practical Geology and the School of Mines in London.   RP

du Toit, Alexander Logie (1878–1948)

du Toit was the greatest of South African geologists. Born near Cape Town, he studied at Cape Town, Glasgow and the Royal College of Science, London. He spent the 17 most creative years of his career from 1903 mapping for the Geological Commission of the Cape of Good Hope.

At the height of his career, and following a visit to South America, he began to take seriously *Wegener's theory of continental drift. In his *A Geological Comparison of South America with South Africa* (1927) he systematically set out the numerous similarities in the geologies of the two continents, suggesting an original contiguity. These ideas were most fully and popularly stated in his important book, *Our Wandering Continents* (1937), in which he suggested that the southern continents had at one time formed the supercontinent of Gondwanaland, which was distinct from the northern supercontinent of Laurasia. This notion, though originally discounted, has grown in acceptance since.   RP

Élie de Beaumont, Leonce (1798–1874)

Élie de Beaumont studied mathematics and physics at the École Polytechnique in Paris before entering the School of Mines in 1819. His greatest work for the School was to take charge of drawing up the eastern division of the official geological map of

MONKEYANA.

Punch's contribution to the debate over Darwin's *Origin of Species*.

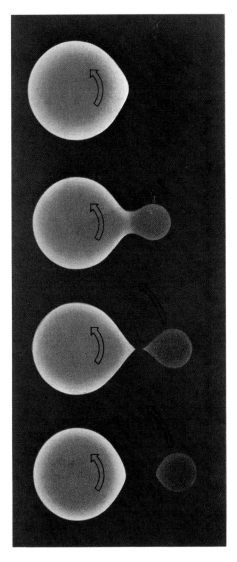

George Darwin's theory of the origin of the Moon. The length of the month compared with that of the day was explained by conservation of angular momentum.

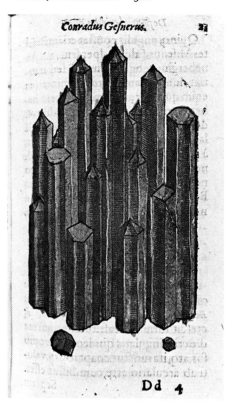

Conradus Gesnerus.

Dd 4

One of the illustrations from Conrad Gesner's *On Fossil Objects, Chiefly Stones, and Gems, Their Shapes and Appearances* (1565).

France, completed in 1831 on six sheets to a scale of 1:500,000.

Élie de Beaumont's main contribution to geological thought was set out in his *Researches on Some of the Revolutions of the Globe* (1829–30). He tried to show how the mountain ranges of the globe were of different epochs and sudden elevation. The shock of elevation had had global consequences which were probably the cause of the revolutions of flora and fauna (including extinctions) which *Cuvier's paleontology demonstrated. The theory attracted contemporary support, since it plausibly linked the history of the globe with that of life. RP

Eskola, Pentti Eelis (1883–1964)

Born in Lellainen, Finland, Eskola was educated as a chemist at Helsinki. He was a life-long student of metamorphic rocks with particular reference to the *Precambrian of Finland. He was one of the first to apply physico-chemical principles extensively to the study of metamorphism, and laid the foundations of most subsequent studies in metamorphic petrology. The main feature of his work was that it began to define the varying pressure and temperature regimes under which *metamorphic rocks are formed, and so enabled rocks of widely differing compositions to be compared in terms of their pressure and temperature of origin.　KGC

Geikie, Archibald (1835–1924)

Geikie first became interested in geology as a hobby. In 1855 he managed to secure a post on the Geological Survey, and by 1867 had become head of its Scottish branch. In 1881 he was promoted to become its Director General, and in 1908 he became the first

– and so far only – geologist ever to be President of the Royal Society.

He contributed to many areas of geology. His fundamental work was on ancient volcanic phenomena in Great Britain, set out most fully in *The Ancient Volcanoes of Great Britain* (1897). Not least among his achievements were many works on the history of geology, including his popular *Founders of Geology* (1905), and a series of distinguished textbooks.　RP

Gesner, Conrad (1516–1565)

Born in Zurich, Gesner was perhaps the greatest of the encyclopedic naturalists of the 16th century.

His interest in the Earth issued in *On Fossil Objects, Chiefly Stones, and Gems, Their Shapes and Appearances* (1565): by "fossils", Gesner meant any objects of special interest dug up from the Earth, not merely organic remains, though he did compare many of his fossils to living specimens. He was less interested in questions of the origins of fossils than in describing them – and their powers (e.g., medicinal virtues). His work was a milestone in the enumeration, identification and description of mineral objects, not least because of the many accurate woodcuts which accompanied his text.　RP

Gilbert, Grove Karl (1843–1918)

Born in Rochester, New York, Gilbert joined the Geological Survey of Ohio and subsequently worked with *Powell's Rocky Mountain expedition, the beginning of a long and fruitful association with Powell. He rose to become the chief geologist of the US Geological Survey.

His main work was on subaerial *erosion, especially on the problem of fluvial erosion in valleys, where his work to some extent paralleled, and to some extent influenced, that of *Davis. His theories on river valleys are most fully set out in his *Report on the Geology of the Henry Mountains* (1877). In the early 20th century he pioneered laboratory experiments replicating and simulating erosive situations.　RP

Goldschmidt, Victor Moritz (1888–1947)

Born at Zurich, Switzerland, Goldschmidt is regarded as the father of *geochemistry.

Working in Oslo from 1920 to 1928, and in Göttingen from 1929 to 1935, Goldschmidt and his colleagues set about a massive analytical program directed at all types of geological materials. As a result of their labors, the distribution of most elements in most rocks and minerals became well known; and Goldschmidt proposed rules governing the distribution of trace elements in minerals which did much to set geochemistry on a firm theoretical basis.　KGC

Grabau, Amadeus William (1870–1946)

Born at Cedarburg, Wisconsin, Grabau studied at MIT and Harvard. He became one of the great pioneers of *paleoecology. He stressed the importance of the environment of deposition in determining the characters of rocks and their fossil assemblages. He also devoted much work to a petrologi-

cal classification of *sedimentary rocks, placing particular emphasis on the need to arrange them according to origin as well as to composition and texture.

Grabau was concerned to explain the development of the distribution of land and sea throughout the globe, and devised an early theory of *continental drift. In his *Rhythm of the Ages* (1940) he developed a "pulsation theory" which saw the distribution of land masses as the product of major rhythmic marine advances and recessions. To a large extent his theory has been superseded by modern ideas.　RP

Guettard, Jean-Étienne (1715–1786)

Born at Étampes, Guettard showed deep interest in botany and medicine, but his chief scientific pursuits were geological. He was the first to recognize the volcanic nature of the Auvergne district of Central France, and his discovery led to the enormous geological interest subsequently shown in the area.

Secondly, he played an important part in originating the debate on the origin of *basalt, at first asserting that columnar basalt was not volcanic in origin: however, after visits to Italy in the 1770s, he began to doubt his earlier views.

Thirdly, he pioneered geological cartography. In 1746 he presented his first mineralogical map of France to the Academy; and in 1766 he and Lavoisier were commissioned to prepare a geological survey of France, of which they completed about an eighth.　RP

Gutenberg, Beno (1889–1960)

Together with Charles F. Richter (with whom he collaborated), Gutenberg stands as the leading seismologist of the 20th century. Much of his work centered on the problem of making inferences about the various layers of the Earth's mantle and core from the study of seismic waves, their patterns and velocities. He deduced the existence of a low-velocity core at a depth of some 2900km (1800mi), computed from the travel-time of waves. This provided an important hinge for theories of crustal movements.　RP

Hall, Sir James (1761–1832)

The son and heir of Sir John Hall of Dunglass, Berwickshire, Hall spent much of the 1780s traveling in Europe, becoming a convert to the new chemistry of Lavoisier and undertaking lengthy geological observation in the Alps and in Italy and Sicily.

He was a friend of *Hutton and an early defender of his theories. Hall's geological work centered upon finding experimental proofs for Hutton's speculations. In a series of furnace experiments, he showed with some success that Hutton was correct to claim that *igneous rocks could assume crystalline structures if cooled very slowly; that there was a degree of interconvertibility between basaltine and granitic rocks; and that limestone subjected to enormous heat would not decompose if maintained under sufficient pressure.　RP

Hall, James (1811–1898)

Born in Hingham, Massachusetts, Hall was

employed from 1836 on the newly founded Geological Survey of New York State. Five years' work for this led to a stratigraphical outline and an unrivalled collection of invertebrate fossils. The fruit was Hall's massive 13-volume *Paleontology of New York (1874–94)*.

In 1843 he was appointed State Paleontologist, holding the post until his death. Partly under his prompting a state museum was established, of which he was made curator in 1864 and first director in 1871. Hall was also active in the establishment of other state geological surveys.

In 1857 he outlined the important idea of crustal downfolds at the edges of continents. He believed that these filled with sediments and in course of time could become mountain ranges (such as the Appalachians). *Dana was one who took up this notion. RP

Haug, Gustave Emile (1861–1927)

Born at Drusenheim, Alsace, Haug in 1887 migrated to Paris, holding a succession of posts at the Sorbonne, and becoming full professor in 1911. His research integrated many fields of geology, and the monument of his vision was his massive *Treatise of Geology* (1907–11). He made specific contributions to many fields. Among his lasting contributions to geological thought is his work on *geosynclines, including the formulation of Haug's rule, that subsidence in a geosyncline results in the regression of the sea over the adjacent epicontinental areas, and that folding in a geosyncline gives rise to marine encroachment. RP

Haüy, René-Just (1743–1822)

Born at St Just-en-Chaussée, Oise, Haüy entered the priesthood before becoming interested in mineralogy and crystallography. His two major works are the *Treatise of Mineralogy* (1801) and the *Treatise of Crystallography* (1822).

The regular forms of *crystals had drawn throughout the 17th and 18th centuries a plethora of causal explanations, in terms of shaping forces – some chemical, some physical, some atomistic – but none satisfactory. Haüy's approach was not to explain the causes of the regularly varied forms of crystals, but to try to classify those forms in terms of geometry – and above all, he hoped, through the geometry of simple relationships between integers.

He saw crystals as structured assemblages of secondary bodies (integrant molecules) that grouped themselves according to regular geometric laws. He proposed six types of primary forms: parallelepiped, rhombic dodecahedron, hexagonal dipyramid, right hexagonal prism, octahedron, and tetrahedron, and spent much of his career elaborating on this typology of forms. RP

Hess, Harry Hammond (1906–1969)

Harry Hess was a mineralogist, geologist, geophysicist and oceanographer. Born in New York, he received his geological education at Yale and Princeton after first studying electrical engineering. Hess was one of those rare scientists who achieved distinc-

tion in several fields.

He first began to study the oceans and geophysics in 1931, when he accompanied F. A. Vening Meinesz in an expedition by submarine to take soundings and measure gravity in the West Indies. During WWII he continued his oceanographic investigations while captain of the assault transport U.S.S. *Cape Johnson*; and it was at this time that he discovered the flat-topped sea mounts known as guyots. After the war he was an active sponsor of the Mohole project, in which an attempt was made to drill through the Earth's crust to reach the upper mantle.

His long interest in the oceans led in 1960 to the idea that convection within the Earth might lead to the generation of new ocean floor at mid-ocean ridges, the hypothesis known as sea-floor spreading. The interpretation of the magnetic anomalies of the sea floor by F. J. Vine and D. H. Matthews in 1963 provided the confirmation needed, and by 1968 *plate tectonics had become a reality.

The credit for this, the so-called "revolution in the earth sciences", representing the largest single advance in the earth sciences in the last sixty years, belongs largely to Hess. KGC

Holmes, Arthur (1890–1965)

Born at Hebburn, Newcastle-upon-Tyne, Holmes received his scientific education at Imperial College, London, first in physics and mathematics, then in geology.

He was distinguished in many branches of geology but is best remembered for his studies of geochronology, giving the first modern estimates of the *age of the Earth. Radioactivity had been recognized as promising a method by which the actual ages of minerals could be determined. In 1911 the first tentative mineral ages had been determined from uranium-lead ratios by Boltwood; and R. J. Strutt (later Lord Rayleigh) had discovered that the abundance of radioactive minerals in the Earth's crust was sufficient to provide a significant heat source, and thus that the assumptions used by *Kelvin in his estimate of the *age of the Earth were ill-founded.

Holmes began work on the uranium-lead method and in 1913 published *The Age of the Earth*, in which existing data were summarized and the first proper time scale for the Phanerozoic was presented. He continued work on the Phanerozoic time scale until 1959 but his original estimates did not require any fundamental changes.

Holmes' other important contributions included a notable text-book, *Principles of Physical Geology* (1st edn., 1944). KGC

Hooke, Robert (1635–1703)

Educated at Westminster and Christ Church, Oxford, Hooke showed genius in a wide range of sciences, falling short of Newton only in respect of his mathematical deficiencies. In geology his importance is twofold.

He was among the leading champions in his age of the belief that fossils were genuine organic remains. He was well aware that

many fossils were probably the remains of species which had become extinct.

Secondly, he advanced a theory of the Earth which stressed the power of natural causes to create, destroy and hold the Earth in equilibrium. He accepted the reality of denudation, while believing that volcanoes and earthquakes were creative processes in the reconstruction of the Earth. RP

Horne, John (1848–1928)

Horne entered the Geological Survey at the age of 19. He received his geological training from his superior, *Peach, and this initiated a lifetime's distinguished collaboration. Peach was the more imaginative and speculative member of the partnership, Horne the more judicious geologist, and the more articulate and lucid in print.

Their great work was on the northwest Highlands of Scotland. The orthodox view, established by *Murchison, was that the underlying fossiliferous Durness limestone passed up conformably into the eastern metamorphic schists, though voices had been raised against this. It took the deep researches of Horne and Peach to demonstrate conclusively the profound structural break separating the two orders of rock. RP

Humboldt, Friedrich Heinrich Alexander, Baron von (1769–1859)

Humboldt has exercised an enormous influence on the study of the globe, as an explorer, as a theorist, and as one who set out to chart the history of Man's interrelations with his planet Earth. In 1799 he set out on a pioneering and immensely productive expedition across Latin America, studying physical geography above all, but also collecting great quantities of geological, botanical and zoological material. The next 20 years were occupied in writing up his results in his *Narrative of Travels* (1818–19).

His ambition was to construct a new science, a "physics of the globe" which would demonstrate the deep interconnectedness of all terrestrial phenomena. He sought ultimately to understand relief in terms of Earth history, and geological phenomena in terms of more basic physical causes (e.g., the Earth's magnetism or rotation). He patiently collected evidence of similar geological phenomena from each continent.

His most popular work, *Cosmos*, begun 1845, is a profound and moving statement of Man's relationship with the Earth. RP

Hutton, James (1726–1797)

Son of an Edinburgh merchant, Hutton studied at Edinburgh University, Paris and Leiden, taking his doctorate in medicine in 1749. He spent the next two decades traveling and in farming. During this time he developed his taste in geology. About 1768 he moved back to Edinburgh and became a leading member of the scientific and literary establishment, playing a large role in the early history of the Royal Society of Edinburgh.

Hutton wrote widely on many fields of natural science, but he is best known for his

Baron von Humboldt, whose monumental *Narrative of Travels* (1818–19) contributed enormously to studies of botany, zoology and geology.

geology, set out in his *Theory of the Earth*, of which a short version appeared in 1788, followed by the definitive statement in 1795.

Hutton attempted to demonstrate a steady-state Earth, in which causes had always been of the same kind as at present, acting with the same intensity (Uniformitarianism). In the Earth's economy there was no vestige of a beginning, no prospect of an end. All continents were being gradually eroded by rivers and weather. Denuded debris accumulated on the sea bed, to be consolidated into strata and subsequently thrust upwards to form new continents by the Earth's central heat. Non-stratified rocks such as granite were of igneous origin. All the Earth's processes were very gradual. The Earth was incalculably old. The catch-phrase was: "The past is the key to the present."

Much attacked in its own day, Hutton's theory found more favor when popularized by *Playfair and *Lyell, and still forms the basis for much geological reasoning. RP

Huxley, Thomas Henry (1825–1895)

Huxley trained for medicine and obtained a succession of teaching and honorific posts within the London scientific establishment. His most important role was as *Darwin's bulldog in defence of *Origin of Species* (1859). In his early vindications, Huxley emphasized in an orthodox Darwinian way that the *fossil record could not present positive evidence for evolutionary modifications of organic forms. By the 1870s, however, Huxley believed that the newly discovered extensive series of North American extinct horses, dating back to the upper Eocene, could now offer fossil proof that the gradual modifications demanded by Darwin's theory had actually occurred. RP

Kelvin, Lord (1824–1907)

Lord Kelvin, born William Thomson, was the son of a successful professor of mathematics and showed early mathematical aptitude. Throughout a long career he made enormous contributions to the development of modern physics. His one brief but notable foray into the Earth sciences had

profound consequences.

In the mid-19th century the doctrine of Uniformitarianism (see *Hutton) was widespread among geologists, one of the accepted tenets being that the range of past time available for the explanation of geological phenomena was unlimited. Kelvin questioned this, and maintained that the Earth had a finite age. This he calculated as between 20 and 40 million years by considering the time required for a body like the Earth to cool from a molten state.

Kelvin underestimated the *age of the Earth by a factor of more than 100, because he was unaware that an important extra heat source existed within the Earth – that is to say, its content of radioactive elements, notably uranium, thorium and potassium. Even in error Lord Kelvin performed a notable service for the developing Earth sciences, for he focused attention onto one of its most fundamental problems.   KGC

Lamarck, Jean Baptiste Pierre Antoine De Monet De (1744–1829)

Of impoverished noble stock from Picardy, Lamarck saw himself as a general natural philosopher, aiming to grasp the deep emergent unity of nature. He pursued studies in botany, zoology, paleontology, meteorology and chemistry, as well as developing the evolutionary theory for which he is most remembered, that acquired characteristics could be inherited.

He set out his geological views in his *Hydrogeology* (1802). The Earth was indefinitely old, and was being continually changed within a balance of gradual organic, climatic, and marine forces. Tidal forces were gradually eroding old continents and building up new ones. The sea and land masses were slowly progressing around the Earth. The sedimentary matter out of which rocks were built was essentially organic detritus (e.g., limestone from shells).

Lamarck seems to have believed that this geological theory obviated the need for extinction (the reality of which he fiercely denied) but produced those environmental pressures which explained complex diversities in the direction of organic *evolution.

RP

Lapworth, Charles (1842–1920)

Lapworth's early work was on the Southern Uplands of Scotland. The shale-band outcrops had been thought to represent successive horizons in a very thick ascending series of strata. But, employing exact mapping and using the *graptolites as an index, Lapworth was able to show how in reality the outcrops were repetitions of comparatively few bands, constituting a series of overfolds which had become parallel as a result of compaction. By examining graptolite evidence from many parts of the world, he showed his method of identification to be generally applicable.

We owe the conception of the *Ordovician, filling the gray area between the Cambrian and the Silurian so much disputed by *Murchison and *Sedgwick, to Lapworth's suggestion.   RP

Leakey, Louis Seymour Bazett (1903–1972)

Son of a Kenya missionary, Leakey studied at St John's College, Cambridge. Working with his wife on the Miocene deposits of western Kenya, he discovered the skull of *Proconsul africanus*. His archaeological investigations lighted upon the Acheulian site of Olduvai, in the Rift Valley, where the skull of *Australopithecus boisei* and the first remains of *Homo habilis*, a *hominid dated at some 1.7 million years, were found.

Other skulls of the founders of the Acheulian culture at Olduvai, dubbed *Homo erectus*, have since been discovered. Leakey's son Richard (1944–) now appears to have discovered hominid remains dating back perhaps 4 million years. RP

Le Gros Clark, Sir Wilfrid Edward (1895–1971)

Le Gros Clark excelled in two main fields of science, the comparative anatomy and physiology of the *primates, and questions of early Man and his antecedents. He was initially skeptical of the claims of *Dart and others for the hominid affinities of *Australopithecus*. Having visited South African sites for himself, however, he confirmed Dart's conviction that *Australopithecus* was close to the main line of *hominid evolution and separate from that of anthropoid apes. He also played a large part in exposing the Piltdown Man forgery. RP

Lehmann, Johann Gottlob (1719–1767)

Lehmann has a strong claim to be called one of the founding fathers of stratigraphy. In his *Attempt at a History of Sedimentary Rock-Forms* (1756), he set out his view that there were fundamental distinctions between Ganggebürgen (masses formed of veined rock) and Flötz-Gebürgen (masses formed of stratified rock). These distinctions represented different modes and times of origin, strata being found in historical sequence. The older strata were chemically precipitated out of water, the more recent strata mechanically deposited by the separation of sediment.

Lehmann's work in this direction laid the foundations for *Werner's more refined stratigraphy. RP

Leonardo da Vinci (1452–1519)

Leonardo's all-round genius brought him face-to-face with problems of understanding the Earth. He saw the Earth undergoing perpetual change, largely occasioned by the forces of weather and water (both marine *erosion of coasts and river erosion of hills). Solid land was perpetually decaying into alluvial plains. The creation by rivers of their own valleys, which they then silted up, fascinated him. Land loss was being compensated for by a steady rise of the continents from the sea (as erosion made them lighter, they were able to rise).

His awareness of the denudatory power of water enabled him to recognize fossils as organic remains buried in strata debris, and he pointed to the similarities between fossil and living specimens. He denied that fossils were due to the Flood and privately speculated on the high antiquity of the Earth. RP

Logan, Sir William Edmond (1798–1875)

Born at Montreal of Scottish descent, Logan was educated at Edinburgh University.

In 1842 the Canadian government set up its own Geological Survey, with Logan as its director. His lifework thereafter was on Canadian geology, up to 1869 for the Survey, and privately after his retirement: he spent much of his personal fortune in promoting the Survey.

Logan's main qualities as a geologist were the accuracy and detail of his descriptive fieldwork. One of his main interpretative achievements was to explain the structural anomaly by which the older Quebec rocks overlie the younger Ordovician beds of the St Lawrence lowlands. He showed this was due to a massive thrust fault, which he traced from Alabama to the Canadian border. RP

Lyell, Sir Charles (1797–1875)

Born on his father's estate at Kinnordy, Angus, Lyell was educated at Oxford, where his growing appetite for geology was fed by *Buckland. He trained for a career in law, but used his weak eyesight as a convenient pretext to pursue his favorite subject, geology, full-time.

Much of the 1820s Lyell spent geologizing, culminating in his expedition in 1828 with *Murchison through the Auvergne and on his own through Italy and Sicily. These experiences led him to a geological theory bearing strong resemblances to *Hutton's, which he set out in his *Principles of Geology* (3 vols, 1830–33), perhaps the most influential book in the history of geology. He argued for a steady-state, Uniformitarian view of the Earth and of the history of life (Man excluded). Present Earth processes – denudation, land-building, volcanic action – were supposed, given time, to be adequate to explain past changes. He argued that the "catastrophes" then so popular in *Cuvier's directional biostratigraphy were unscientific – and unnecessary.

Lyell argued that living populations became extinct gradually. His examples from recent geological history established the concept of the *Tertiary and its subdivisions.

In writing the *Principles*, Lyell was a powerful enemy of *Lamarck's evolutionary theories. In later years, however, he was won over, somewhat grudgingly, to accept his friend *Darwin's new theory of *evolution. His *Antiquity of man* (1863) assembled the evidence for a high antiquity for Man, and cautiously hinted at the possibility of human evolution from the higher primates. RP

Mantell, Gideon Algernon (1790–1852)

A man of enormous energies and enthusiasms, Mantell became a successful surgeon, but his hobby of geology steadily grew until it monopolized his life. He was a prolific writer, not least of popular texts. His contemporary fame rested on having identified for the first time fossils of a land

saurian (i.e., dinosaur) – the enormous Mesozoic saurians reconstructed earlier by *Cuvier and others had been aquatic. Among his other work on extinct giant vertebrates was his discovery of armored dinosaurs (*Hylaeosaurus*). RP

Marsh, Othniel Charles (1831–1899)

Son of a farmer of modest means, Marsh was an important public figure in US science. He acted as vertebrate paleontologist to the US Geological Survey (1882–92), and built up the vast fossil collections of the Peabody Museum at Yale (established by his uncle). His paleontological interpretations depended on the completeness and magnitude of phylogenetic series of fossil specimens housed in his huge collections.

His early work hinged upon four great expeditions through the Western territories between 1870 and 1873. His greatest achievement lay in his remarkable series of fossil horses from the Eocene (*Orohippus*) to the Pleistocene. He showed how European horses were an offshoot of the main American line of evolution. *Huxley used this succession as the best fossil evidence available in his day for Darwinian evolutionary change. RP

Murchison, Roderick Impey (1792–1871)

A gentleman amateur, Murchison became a "professional" in 1855, succeeding *De la Beche as Director of the Geological Survey. He spent a generation in the field surveying strata according to clear interpretative principles. He believed in a near universal order of deposition, indicated by fossils rather than purely by lithological features. Fossils themselves would show a clear progression in complexity from Azoic (pre-life) times to invertebrates, and only then up to vertebrate forms, Man being created last: this progression was allied to the Earth's cooling.

The great triumph of these principles was to unravel the *Silurian system (i.e., those strata beneath the Old Red Sandstone). For Murchison, the Silurian contained remains of the earliest life (though no fossils of vertebrates or land plants were to be expected). Controversy with De la Beche over the younger end of the Silurian led to Murchison introducing, with *Sedgwick, the fruitful concept of the *Devonian, which incorporated the Old Red Sandstone.

But Murchison also quarreled with Sedgwick over the lower limits of the Silurian, seemingly seeking to incorporate Sedgwick's independent *Cambrian: on this he proved in error. He stubbornly denied there were fossiliferous systems underlying the Silurian.

Late in life, Murchison became highly dogmatic. His growing campaign against *Lyell's Uniformitarianism turned into a rigid denial of Darwinian evolution. RP

Murray, John (1841–1914)

Born at Cobourg, Ontario, Murray was educated in medicine at Edinburgh. However, after a voyage on an Arctic whaler in

1868 the sea became his great love. Between 1872 and 1876 he was one of C. Wyville-Thomson's assistants on the epoch-making voyage of the *Challenger*. After Thomson's death, Murray edited the 50-volume *Report* of the voyage (published 1880–95), so influential in the establishment of oceanography as an organized science.

Murray's chief scientific work lay in investigating the sedimentary deposits on the ocean floor. He identified the main contributors to organic sediment, and showed they were chiefly surface dwellers. He also traced the extent of inorganic mud, which he believed originated from volcanic dust. His work on the *ocean floor emphasized the slowness of deposition.   RP

D'Orbigny, Alcide Charles Victor des Sallines (1802–1857)

Born at Couëron, near the mouth of the Loire, d'Orbigny obtained his scientific education in Paris.

His career falls into two parts. In 1826 he was commissioned to explore South America. He was away for eight years, making arguably the most thorough natural-history investigation of that continent yet undertaken, publishing his findings in the 10 volumes of his *Travels in South America* (1834–47).

His other chief concern was in paleontology, undertaking two major ventures. From 1840 he worked on the *Paléontologie française*, which was to be a complete account of all known forms of mollusks, echinoderms, brachiopods and bryozoans found in French Jurassic and Cretaceous deposits. An even grander project was his *Introduction to Universal Palaeontology* (1850–52), which sought to divide up all known fossils into 27 successive, and essentially distinct, extinct faunas. Such a view was based upon the Cuvierian notion of successive destructions and creations in Earth history, and so was rather undercut by the subsequent acceptance of *evolution.   RP

Osborn, Henry Fairfield (1857–1935)

Osborn's main work was in vertebrate paleontology. He eagerly accepted evolutionary theory and was concerned to fill out its main trends and details. He continued *Cope's work on the evolution of mammalian molar teeth and wrote an influential textbook, *The Age of Mammals* (1910).

His evolutionary studies focused on the problem of the adaptive diversification of life. He was particularly concerned with the parallel but independent evolution of related lines of descent, and with the explanation of the gradual appearance of new structural units of adaptive value. He always stressed the way evolution results from pressures from four major directions: external environment, internal environment, heredity, selection.

A convinced Christian, he was deeply involved with the interpretation of evolution in religious and moral terms.   RP

Owen, Sir Richard (1804–1892)

Owen's early career was marked by a phenomenal quantity of zoological identification and classification in the style of *Cuvier. He then became progressively more interested in paleontology, his work on the reconstruction of the New Zealand moa and on *Archaeopteryx* being classics of their kind. He published an important *History of British Fossil Reptiles* (1849–84) and a popular textbook, *Palaeontology* (1860).

Owen came into collision with the Darwinians on two important issues. He made Man the single example of a special subclass of Mammalia: *Huxley, in reply, showed plausibly that the anatomical grounds for this classification were illusory. Owen also fiercely attacked Darwin's natural-selection mechanism for evolution.   RP

Peach, Benjamin Neeve (1842–1926)

Much of Peach's childhood was spent in the north of Scotland, near where he was to accomplish his most important work, with *Horne on the question of the succession of strata in the northwest of Scotland. *Murchison, as Director of the Geological Survey, had interpreted the Moine Schists as part of the *Silurian, and asserted they succeeded conformably upon the fossiliferous Durness limestone. Peach was at first inclined to accept this interpretation; but, seeing on the shore of Loch Eireboll Cambrian zones repeated over and over again by thrusting, he started a major reinterpretation. He began to see the remarkable tectonic features which signified the fundamental unconformability between the two orders: overfolds, reversed faults and gigantic lateral thrusts. In 1884 *Geikie, the new Director of the Survey, publicly acknowledged Peach's new assessment, which signalled the yielding of traditional stratigraphy to the new discipline of petrological tectonics.   RP

Penck, Albrecht (1858–1945)

Penck was a leading student of the *Quaternary. Almost all aspects of current relief occupied his attention, and he ranged freely into prehistory, anthropogeography and climatology as well as in the mainstream of geomorphology.

His principal work was on the Alps, and is summed up in *The Alps in Times of Ice Ages* (3 vols., 1901–09), in which he established the basic fourfold division of the Pleistocene glaciation of the Alps. Penck never minimized the complexities of this period, seeking rather to study their causes: glacial isostasy, tectonic movements and changing climatic conditions.   RP

Playfair, John (1748–1819)

Playfair was born near Dundee, son of a Scottish minister of religion. Educated at St Andrews and Edinburgh universities, he followed his father into the ministry. From 1785 to 1805 he was Professor of Mathematics at Edinburgh, vacating the chair for that of Natural Philosophy. He is chiefly remembered, however, as a friend, early supporter and popularizer of the ideas of *Hutton, particularly in his *Illustration of the Huttonian Theory of the Earth* (1802). Almost all 19th-century geologists learned

about Hutton through reading Playfair. RP

Powell, John Wesley (1834–1902)

Powell was intended by his Methodist farmer father for the Methodist ministry, but he early developed a love for natural history. In the 1850s he became secretary of the Illinois Society of Natural History.

In 1870 Congress placed him at the head of an official survey of the resources of the Utah, Colorado, and Arizona area, the fruits of which were published in his *The Exploration of the Colorado River* (1875) and *The Geology of the Eastern Portion of the Uinta Mountains* (1876). This work produced lasting insights on fluvial erosion, volcanism, isostasy and orogeny; but, even more significantly, Powell grasped how geological and climatic causes together produce the essential aridity of the region.

For almost 20 years thereafter he campaigned for massive funds for irrigation projects and dams, and for the geological surveys necessary to work out a water policy. He also argued the need for a change in land policy and farming techniques in the "drylands". Suffering political defeats on these issues, he resigned in 1894 from the Geological Survey, of which he had become director in 1881.   RP

Rosenbusch, Karl Harry Ferdinand (1836–1914)

Born at Einbeck, Hannover, Rosenbusch received his scientific education at Freiburg and Heidelberg.

He graduated at a time when the study of rocks in thin section by microscopic means was in its infancy. It was clearly necessary to correlate all the available information relating to the optical properties of minerals as exhibited in thin sections. This was done by Rosenbusch in 1873 in *Microscopic Structure of the Petrographically Important Minerals*, a work which became a standard text in optical mineralogy.

Rosenbusch also devoted much effort to the description and classification of rocks, particularly *igneous rocks, and is well known for his *Microscopic Structure of the Bulky Rocks*, which appeared in various editions between 1877 and 1908.

Rosenbusch and *Zirkel stand together as the founders of petrography, the descriptive aspect of the study of rocks.   KGC

Scrope, George Julius Poullett (1797–1876)

Scrope studied at Cambridge, being influenced by *Sedgwick, and for half a dozen years thoroughly explored the volcanic areas of Italy and Sicily, the Auvergne, Vivarais and Velay areas of France, and the Eifel region of Germany. The fruits of this exploration were his *Geology and Extinct Volcanoes of Central France* (1826) and *Considerations upon Volcanoes* (1828).

His interpretations of volcanism followed *Hutton and anticipated *Lyell. He stressed the vast extent of volcanic phenomena and showed the volcanic origin of *basalt and related rocks. He argued that volcanoes had been continuously active through geological time, thus emphasizing the "actualist" conception that nature always acts by the same order of causation. He saw how volcanoes were instrumental in mountain-building. His work on the Auvergne also convinced him of continual massive fluvial *erosion.   RP

Sedgwick, Adam (1785–1873)

Son of the curate of Dent, in North West Yorkshire, Sedgwick attended Trinity College, Cambridge, where he became fellow in 1810. Though knowing no geology, he was appointed Woodwardian professor in 1818, holding the chair for 55 years. He soon made himself one of the foremost British geologists.

Sedgwick combined a highly Baconian approach to geological investigation, seeing facts founded on fieldwork as the essence of the science, with strongly held Christian sentiments. He never fully approved of *Hutton's Uniformitarianism; was suspicious of glacial theory; and set his face completely against all theories of evolution, not least those of his former pupil and friend, Charles *Darwin.

Sedgwick excelled in two fields. He was a foremost student of paleontology, especially of Paleozoic fossils, and he also contributed greatly to understanding the stratigraphy of the British Isles, using fossils as an index of relative time, and assuming relatively distinct fauna and flora for each period. His major work lay in the geology of Wales, and bringing to birth the concept of a *Cambrian system, over which he eventually quarrelled with *Murchison. With Murchison, Sedgwick was instrumental in formulating the idea of a *Devonian system encompassing the Old Red Sandstone.   RP

Smith, William (1769–1839)

"Strata" Smith was a pioneering figure in British stratigraphy.

He received little formal education. Born at Churchill, Oxfordshire, he became skilled in the arts of land drainage and improvement, canal construction and mine surveying and prospecting. He also gained close familiarity with the terrain in a broad swathe of England from Somersetshire through to the northeast. Throughout the 1790s his geological ideas developed, and by 1799 he was able to set out a fairly comprehensive list of the secondary strata of England and had started to construct geological maps.

Smith was not the first geologist to recognize the *principles of stratigraphy, or the usefulness of type fossils. His achievement was to actually determine the succession of English strata from the Carboniferous up to the Cretaceous in greater detail than previously and to establish their fossil specimens.

Beyond this, perhaps, lay his achievements in mapping. Smith rightly saw the map as the perfect medium for presenting stratigraphical knowledge. In developing a form of map which showed outcrops in block, he set the essential pattern for geological mapping throughout the 19th century.

His relations with the main British geological community remained ambiguous

A contemporary caricature of the 19th-century paleontologist Richard Owen, shown here "riding his hobby".

and even strained up to the 1820s. In 1831, however, his work was belatedly recognized by the Geological Society of London in the award to him of the first Wollaston Medal.

Sorby, Henry Clifton (1826–1908)

Sorby may be called the founder of microscopic petrology in Britain. Before Sorby, Sir David Brewster (1781–1868) had explored the molecular structure of minerals by investigating how light passed through them, but Brewster had been limited to the investigation of well crystallized specimens.

Sorby overcame this problem by adapting the art of thin-slicing of hard minerals, analyzing under the microscope specimens treated thus. His early papers met with much hostility amongst contemporaries, but came to be recognized as opening up a new science, though one to be developed

more in Germany – by *Zirkel, *Rosenbusch and Vogelsang – than in Britain.

Sorby's interests spread over into metallurgy, where he employed his microscopic techniques to investigate the structures of iron and steel under stress.   RP

Steno, Nicolaus (1638–1686)

Steno (or Niels Stensen, as he was known in his native Denmark) studied medicine at Leiden in the Dutch Republic, migrated to Paris and then obtained a post in Florence under Duke Ferdinand. In 1666, he was struck by the similarity between shark's teeth and certain objects, found on or near the surface of the Earth, called glossopetrae, "tonguestones". He concluded that the stones were the petrified remains of sharks' teeth. In his *Sample of the Elements of*

A geological map by William Smith showing the rock types to be found in different parts of Oxfordshire, England.

Mylogy (1667), he demonstrated how they had come to be mineralized.

From there he was led to consider the problem of how such fossils came to be deeply embedded in rocks. He assumed six successive periods of Earth history: first, a period of deposition of non-fossiliferous strata from an ocean; then several peiods of undermining and collapse; then another period of the deposition of strata, this time fossiliferous; followed by further undermining and collapse. This explained why the deepest strata contain no fossils but are overlain by fossiliferous strata, and also why certain strata are found horizontal while others are tilted to the horizon.

Steno's work is important as one of the earliest directional accounts of both Earth and life history, and it was of considerable influence in the 17th and 18th centuries. RP

Suess, Edward (1831–1914)

Born in London, Suess was educated in Vienna and at the University of Prague. As well as his geological interests, he occupied himself with public affairs.

His geological researches took many di-rections. As paleontologist, he studied graptolites, brachiopods, ammonites and the fossil mammals of the Danube Basin. He wrote a pioneering text on economic geology. He carried out important research on the structure of the Alps, the tectonic geology of Italy and seismology. He was concerned with a possible former land-

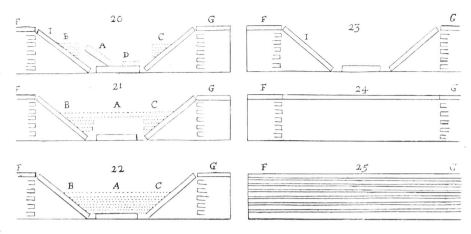

Drawing by Steno showing the effects of folding and faulting (running from 25 back to 20).

bridge between North Africa and southern Europe.

The fruit of these interests was *The Face of the Earth* (1885–1909), a work devoted to investigating the physical agencies which had determined the Earth's geographical evolution. Suess offered a comprehensive view of crustal movement, of the structure and distribution of mountain chains, of foundered continents, of the history of the oceans; and he rewrote on new lines the structural geology of each continent. This essentially held good until the coming of the theory of *plate tectonics in the 20th century.   RP

Ussher, James (1581–1656)

A student of Trinity College Dublin, James Ussher followed a career in the Church of England, and eventually rose to become Archbishop of Dublin. He was an important historical, philological, and Biblical scholar. One of the fields to which he turned was universal chronology, in particular in his posthumously published *Sacred Chronology* (1660).

Contemporary scholars were deeply concerned with the problem of relative and absolute datings of the history of mankind, beginning from Adam. Their methods were to compute backwards, largely using written evidence, and, above all, the various texts of the Bible. Because the Bible pronounced human and earthly history inseparable, and that the Earth had been created only six days before Man, it was widely assumed that the age of the Earth could be calculated from that of Man.

Ussher arrived at a date of creation of the Earth of 4004BC, a figure which (within a few hundred years either way) was widely acceptable in the 17th century; and which has since become famous.

It is often assumed that Ussher's figure was in some way "official" and that not till the 19th century did any Christian dare postulate much higher figures for the Earth's antiquity. This is not true. Several 17th-century British geologists advanced much higher estimates.   RP

Walther, Johannes (1860–1938)

Walther was born in 1860 at Neustadt-an-Orla. The first part of his career was chiefly spent on paleontological research into fossil shells, corals and fossil fish. His interests lay less with the reconstruction of the forms of life than with the problems of the fossilization of specimens within sediments on the sea floor.

Walther also became occupied with lithogeny. He took up the wider question of the origin, nature and formation of deserts, and became possibly the world's greatest authority on the geology of deserts, in a series of books beginning with his *Denudation in Deserts and their Geological Meaning* (1891). He is well known also for his important statement of the nature of geology as a science, *Introduction to Geology as a Historical Science* (1890–93).   RP

Wegener, Alfred (1880–1930)

Wegener seems at this time to be the most influential geologist of the 20th century. Though by no means the first theorist of *continental drift, Wegener's expositions of the theory, especially his *The Origin of Continents and Oceans* (1929), endowed the hypothesis with scientific plausibility.

Born in Berlin in 1880, Wegener studied at Heidelberg, Innsbruck and Berlin. Before WWI he taught at Marburg. From 1924 until his death on his third expedition to Greenland he held a specially created chair in meteorology and geophysics at Graz, Austria.

From 1910 he began to develop his theory of continental drift, or continental displacement. Empirical evidence for this was the close jigsaw fit between coastlines on either side of the Atlantic, and paleontological similarities between Brazil and Africa. But Wegener had strong convictions that geophysical and geodetical considerations would also support a theory of wandering continents – though he himself was rather unclear about the causes of such displacement, believing partly in tidal forces and partly in a "flight from the poles".

Wegener supposed that the Mesozoic had seen the existence of a united supercontinent, Pangaea. This had developed numerous rifts and had drifted apart. During the Cretaceous, South America and Africa had effectively split, but not until the end of the Quaternary had North America and Europe finally separated, or South America from Antarctica. Australia had been severed from Antarctica during the Eocene.

Wegener's hypothesis met general hostility in its own day. Only with the development of a satisfactory mechanism for displacement, i.e., with the rise of the theory of *plate tectonics since WWII, has the (modified) hypothesis won support.   RP

Werner, Abraham Gottlob (1749–1817)

Born in Silesia, Werner studied between 1769 and 1771 at the Mining School at Freiberg, Saxony, and then at Leipzig. In 1775 he was appointed to the Freiberg Akademie, where he continued to teach for the rest of his life. He was the most influential teacher in the history of geology, most of the leading geologists of the next generation having studied under him.

Though wrong on many points, and much attacked (and misunderstood) in the early 19th century, Werner's geology was of fundamental importance in its day for establishing a physically-based stratigraphy. He proposed a general succession of the creation of rocks, beginning with Primary Rocks (precipitated from the water of a universal ocean), then passing through Transition, Flötz (sedimentary), and finally Recent and Volcanic. The oldest rocks were chemically deposited, crystalline and fossilless. Later rocks were mechanically deposited, fossiliferous and superincumbent, formed out of the denuded debris of the first creations.

Werner's approach was particularly important for linking the order of the strata to the history of the Earth, and relating the studies of mineralogy and strata.   RP

Woodward, John (1665–1728)

A vain and jealous man, Woodward crossed swords with many of his contemporaries. In his will he established the Woodwardian Chair of Geology at Cambridge, and left to the university his magnificent fossil and mineral collections.

He was a theorist of the Earth, a pioneer of stratigraphy, and an important collector and systematizer of fossils. He asserted that the strata lay in regular order throughout the globe and that fossils were organic remains (contrary to the general view of the time, which saw them as sports of nature). Both phenomena he attributed to the effects of the Flood, which he explained by the action and suspension of the Newtonian force of gravity.   RP

Zirkel, Ferdinand (1838–1912)

Ferdinand Zirkel was born and educated in Bonn. He trained as a mining engineer but as a young man acquired an interest in the microscopic study of rocks in thin section, a technique which had recently been developed by *Sorby.

His detailed studies of a great variety of rocks culminated in the publication in 1866 of his *Textbook of Petrography*, which later appeared as an expanded second edition in 1893–94. This work, together with those of *Rosenbusch, marked the founding of the science of petrography, the descriptive and classificatory study of rocks.   KGC

Glossary

Ablation, the loss of snow and ice from the surface of a glacier by melting, evaporation or sublimation (the direct passage from solid to vapor state).

Accessory, adjective applied to minerals present in a rock in such small quantities that they are of negligible importance when considering the mineral composition of the rock.

Acid, adjective applied to igneous rocks that contain more than 10% free quartz.

Aeolian, literally, of the winds. The term is applied especially to sediments that have been eroded and transported by the wind.

Aerobic, term applied both to organisms that require oxygen for their survival, especially bacteria, and to an environment in which oxygen is present in significant quantities.

Agglomerate, a rock made up of angular fragments of lava in a matrix of smaller, often ashy, particles.

Alkaline rock, an *igneous rock* in which the feldspar content is comprised primarily of sodium and potassium *silicates*.

Alluvium, gravel, silt, sand and similar material deposited by streams and rivers, mainly near their mouths. As alluvium is good agricultural soil, the earliest civilizations had their origins as farming communities centered on alluvial *flood plains*.

Amorphous, an adjective applied to structures or formations that have no apparent order, whether at an atomic, molecular, crystalline or higher order.

Anaerobic, a term applied both to organisms that do not require the presence of oxygen in order to survive, especially bacteria, and to environments, such as the deep ocean floor, in which free oxygen is not present in any significant quantities.

Aqueous solution, a solution of a substance in water.

Aquifer, an underground rock formation through which *groundwater* can easily percolate. Sandstones, gravel beds and jointed limestones make good aquifers.

Aureole, or metamorphic aureole, the zone around an igneous *intrusion* in which the heat of the intrusion has caused local thermal metamorphism. The size of the aureole depends both on the nature of the intrusion and on the nature of the *country rock*.

Axis of symmetry, in crystallography, a line drawn through a crystal such that the crystal is symmetrical about it. See *symmetry*.

Azoic, that part of the Precambrian in which, it is thought, there was no life on Earth: it came immediately prior to the *Cryptozoic*. The term is increasingly disused.

Banded, a term applied to formations and rocks in which there are recognizable bands of different chemical composition, physical nature, or both. The term is often used synonymously with *laminar*, although the latter term implies that the bands are thin.

Basic rock, an *igneous rock* which contains no (or very little) free quartz and whose feld-spars are dominantly *silicates* of calcium rather than of sodium or potassium.

Batholith, a large subterranean intrusion of *igneous rock*, having an areal extent of upwards of 100km² (40mi²).

Benthonic, adjective applied to organisms which live on, or very near to, the ocean floor.

Biogenic, a term applied to materials (usually sediments or sedimentary rocks) which owe their origin to living organisms. As example, coal is of biogenic origin.

Boreal, a term applied to northerly *faunas* or faunal realms. In some cases boreal is equated with "cold", referring to the temperature of the environment in which the fauna thrived.

Calc-alkaline rock, an *igneous rock* in which the most prevalent feldspar is rich in calcium.

Caldera, an extremely large crater of volcanic origin, resulting from repeated or massive explosion, collapse, or the amalgamation of a number of smaller craters.

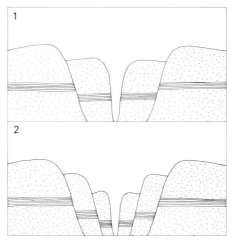

Caldera: formation of a collapse caldera.

Canyon, a steep-sided valley formed where a river has eroded down through horizontal strata of hard rock. The most famous is the Grand Canyon, Ariz., which is about 350km (220mi) long and has a greatest depth of about 1.75km (1.1mi).

Carbonates, minerals which contain the ion CO_3^{2-} – that is, one carbon atom and three oxygen atoms having a net negative charge of two. Among the more important minerals are calcite ($CaCO_3$), aragonite ($CaCO_3$), dolomite ($CaMg(CO_3)_2$), magnesite ($MgCO_3$) and malachite ($CuCo_3.Cu(OH)_2$).

Cirque, a steep sided hollow eroded by a glacier. Where the glacier has retreated, cirques are commonly occupied by lakes: where it is still present, they are generally occupied by névé, compacted snow. Cirques are also known as *corries* or *cwms*.

Clastic rock, a rock composed of weathered or eroded particles of other rocks. In general, these particles have been transported to their present position from the areas in which the erosion or weathering took place.

Cleavage, of a mineral, the tendency to split along a definite plane parallel to an actual or possible crystal face: for example, galena, whose crystals are cubic, cleaves along three mutually perpendicular planes. Such cleavage is useful in identifying minerals: for example, aragonite and calcite have the same chemical composition ($CaCO_3$) but quite different cleavages.

Rock cleavage most commonly takes place between roughly parallel beds whose resistances under deformation to internal shearing differ.

Colloid, or colloidal solution, a system in which two (or more) substances are uniformly mixed such that one is extremely finely dispersed throughout the other. A colloid may be viewed intuitively as a halfway stage between a suspension and a solution, the size of the dispersed particles being larger than simple molecules but still too small to be viewed through an optical microscope. A typical example of a colloid is fog (water in air).

Compaction, the "pressing together" of the individual grains of a sediment, usually resulting from the weight of overlying sediment, to form a *sedimentary rock*.

Consolidation, any process whereby a soft, loose material is transformed into a harder, denser one. A typical process is *compaction*.

Cordillera, an extended mountain system, often comprising a number of parallel ranges, associated with a geosyncline. In many cases, a cordillera appears as a string of islands.

Core, the innermost sphere of the Earth, with a diameter of about 7000km (4350mi). It is thought to be subdivided into an inner core, which is solid, and an outer core, which is predominantly liquid. The main constituents of the core are thought to be iron and nickel.

Country rock, the local rock surrounding a particular rock body, most commonly an igneous *intrusion* or an *ore* deposit.

Cross-bedding, in sedimentary structures, *laminar* records of usually short-lived changes in the velocity of the current in which the sediment was laid down. The angles of such laminae to the main bedding plane give an indication of the direction of current flow.

Crust, the outermost layer of the Earth, having a typical depth of around 35km (22mi).

Cryptozoic, the aeon of geological time in which life first appeared and in which the rocks of the Precambrian were formed. The rocks do not contain any fossils that can be used for dating – hence the name Cryptozoic, which means "hidden life". The aeon of visible life is called, in contrast, the *Phanerozoic*.

Crystals, homogeneous solid objects having naturally formed plane faces. This order in their external appearance reflects the regularity of their internal structure, an internal regularity which is the keynote of the crystalline state. The study of crystals and of the crystalline state is the province of crystallography.

Deflation, the transportation of loose surface debris by the wind. See also *aeolian*.

Degradation, the lowering of the level of the land by any or all of the processes that together comprise *denudation*: erosion, weathering and transportation.

Dendritic, an adjective applied to any form reminiscent of a tree – with branches, subsidiary branches, etc. A typical application is to a drainage system. A *dendrite* is a branched crystal form, common in ice (especially in the form of frost) and certain minerals, and of prime importance in metals, which often consist of dendrites embedded in a matrix of the same or (for alloys) different composition.

Dendrochronology, the dating of past events by the study of annual rings in trees. A hollow tube is inserted into the tree trunk and a core from bark to center removed. The rings are counted, examined and compared with rings from dead trees so that the chronology may be extended further back in time. Through such studies important corrections have been made to the system of radiocarbon dating; though of course, in geological terms, the timespan covered is extremely short. See also *radiometric dating*.

Denudation, those processes that contribute to the *degradation* of the land, namely erosion, weathering and transportation.

Deposition, the end result of *transportation*.

Detrital, adjective describing particles of rocks or more usually minerals derived by *erosion* or *weathering* from pre-existing rocks. Rocks made up of detrital particles are known as *clastic rocks*.

Diagenesis, an envelope term for the processes occurring close to the Earth's surface and affecting a particular sediment. Deeper within the Earth, where temperatures and pressures are higher, metamorphism takes place; and there is no sharp borderline between the two processes. The end-result of diagenesis is the formation of a coherent *sedimentary rock*.

Diastrophism, the large-scale deformation of the crust of the Earth to produce such features as continents, oceans, mountains and rift valleys. Typical processes are faulting, folding and *plate tectonics*.

Differentiation, in a magma, the separation (usually through crystallization at different times) of the various constituents to produce different varieties of *igneous rock*. The term is used analogously for the concentration of the constituents of a rock during metamorphism, resulting in, for example, the banding of a *gneiss*.

Dip, the angle between an inclined plane and the horizontal plane. The dip of a fault is the angle between the fault plane and the horizontal (its complement is the *hade*, the angle between the fault plane and the vertical). The dip of a slope is regarded as perpendicular to the strike, the direction in which a horizontal straight line may be drawn on the slope.

Discontinuity, a layer within the Earth where the speed of transmission of seismic waves changes. Best known, though not most important, is the *Mohorovičić discontinuity*.

Dorsal, term used to describe the back parts of an animal (e.g., the dorsal fins of a fish), or those parts that are generally uppermost. In botany, the dorsal part of, say, a leaf is that side turned away from the stem.

Drift, the material left behind when a glacier retreats. The unstratified material deposited directly onto the land is called *till*. Fluvioglacial drift, material which has been transported by melted waters of the glacier, is in contrast well stratified. Drift may be up to 100m (330ft) deep,

composed of particles that vary from fine sand up to huge boulders.

Ecological niche, the environment most favorable for a particular organism which is better adapted to that environment than any other organism.

Dendritic: typical dendritic drainage pattern.

Emplacement, the development of an igneous rock body surrounded by another rock, most usually through *intrusion*.

Envelope, the rock surrounding an igneous *intrusion*, usually comprising a metamorphic *aureole* surrounded by *country rock*.

Epeirogeny, large-scale upward or downward movement of a landmass; not to be confused with the more dynamic processes that comprise *orogeny*.

Epicontinental, adjective describing, for example, a sea situated within a continental mass or covering the continental shelf.

Erosion, the wearing away of the Earth's surface by natural agents. Running water constitutes the most effective eroding agent, the process being accelerated by the transportation of particles eroded or weathered further upstream. Other important agents of erosion include groundwater, ocean waves and glaciers. Rocks exposed to the atmosphere undergo the closely related process of *weathering*.

Eustatic, adjective describing worldwide, rather than local, sea-level changes.

Evaporites, sedimentary deposits of salts that have fallen out of solution owing to the evaporation (and thus concentration) of a body of water. Evaporite deposits have the least soluble salts at the bottom, followed by progressively more soluble salts.

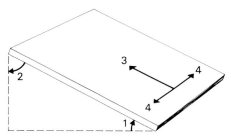

Dip: dip (1) and hade (2), the dip angle being measured in the direction (3) perpendicular to the strike direction (4).

Extrusion, the consolidation of magma on the surface of the Earth to form volcanic igneous rocks. See also *intrusion*.

Facies (plural facies), of a sediment, the total of those of its characteristics that uniquely indicate the conditions of its deposition. *Biofacies* or fossil facies are sediments characterized by the type of fossil organisms which they contain.

Fault, a fracture in the Earth's crust on either side of which there has been relative movement. Faults seldom occur in a single plane: usually a vast number of roughly parallel faults take place in a belt (fault zone) a few hundred metres across. Faults may be large enough to be responsible for such features as rift valleys, or, in contrast, microscopically small.

Fauna, the animals occurring in a particular region and/or, to the paleontologist, period of time. See also *flora*.

Flood plain, a plain bordering usually the lower reaches of a river, initially formed by the downstream migration of meanders, widening the river valley. When the river floods, sediment is carried over its banks and deposited over the flood plain, so that the level of the plain gradually rises, especially near to the channel to form raised banks called *levees*. Further deposition of sediment may raise both the levees and the river channel itself. See also *alluvium*.

Flora, the plants occurring in a particular region and/or, to the paleontologist, period of time. See also *fauna*.

Fold, a buckling in rock strata occurring as a result of horizontal pressures in the Earth's crust. Folds convex upwards are called anticlines; those convex downwards, synclines. Folds may be tiny or up to hundreds of kilometres across.

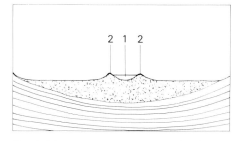

Flood plain: A cross-section through a flood plain, showing the river (1) flowing between levees (2) above the plain of alluvium deposited during flooding.

Fossils, the remains, traces or impressions of living organisms that inhabited the Earth during past ages. There are a number of mechanisms whereby fossil remains may be preserved. *Trace fossils* may take the form of, for example, footprints, burrows or preserved droppings.

Fracture, a breakage of a rock or mineral in a direction other than the *cleavage* direction.

Gangue, the unwanted part of an *ore* deposit, both the minerals that are left unexcavated and those that are separated from the desired material after excavation.

Glacier, a mass of ice that survives for several years. There are three recognized types of glacier: ice sheets and caps; mountain or valley glaciers; and piedmont glaciers. They form wherever conditions are such that annual precipitation of snow, sleet and hail is greater than the amount that can be lost through evaporation or otherwise. Mountain glaciers usually result from the coalescing of bodies of snow that have accumulated in *cirques*; and a piedmont glacier occurs when a mountain glacier spreads out of its valley into a contiguous lowland area. Glaciers account for about 75% of the world's fresh water.

Gneiss, a general term to describe any of a

number of highly regionally metamorphosed, usually coarse-grained rocks with a banded, laminated structure. See also *schist*.

Gondwanaland, the southern-hemisphere continent formed as *Pangaea* split up. Stratigraphic and other evidence suggest it comprised what are now Antarctica, Australia, India, South America and other, smaller, units.

Groundwater, water that accumulates beneath the Earth's surface. It may be *meteoric*, where rainwater has percolated down from above, or *juvenile*, where water has risen from beneath.

Homeomorphy, similarities of morphology between members of different genera within the same *phylum*. It is distinct from *homomorphy*, similarities of morphology (usually only superficial) between members of different phyla.

Homomorphy. See *homeomorphy*.

Hydrothermal, adjective describing processes, and their products, that involve the action of heated or superheated water. Hydrothermal ore deposits occur around igneous *intrusions*.

Igneous rocks, those rocks which form directly from a molten silicate melt, comprising one of the three main types of the rocks of the Earth. They crystallize from the *magma* either at the Earth's surface (*extrusion*) or beneath it (*intrusion*), resulting in the two principal classes, *volcanic rocks* and *plutonic rocks*.

Inclusion, a foreign body enclosed within a rock or mineral: the foreign body may be solid, liquid or gaseous.

Insolation, the amount of radiant heat and light received from the Sun by a particular place at a particular time.

Intensity, Mercalli, a measure of the extent of the effects of an earthquake on a particular area. Clearly the Mercalli intensity depends both on the *magnitude* of the earthquake and on the position (especially distance) of the observer. The scale runs from I, where the shocks can be detected only by seismograph, to XII, "catastrophic".

Intrusion, the forcing of magma into preexisting rocks beneath the surface of the Earth to form a body of *igneous rock*, itself termed an intrusion. See also *extrusion*.

Island arc, a curving chain of islands found associated with an ocean trench, earthquake activity and volcanism: an example is Japan. The arcs and their associated features are manifestations of plate underthrusting (see *plate tectonics*).

Isostasy, the theoretical tendency of the Earth's crust to maintain equilibrium as it floats on the mantle, assumed to result from flows of the dense plastic *sima* in the lower crust in response to local changes in the pressure on it of the lighter *sial* above. Local differences in the proportion of sima to sial thus maintain an equal weight of crust all round the Earth. The mechanism of isostasy is nowadays being increasingly questioned.

Joint, a fracture in a rock distinct from a *fault* in that there has been no lateral displacement between the two sides. Best known are shrinkage joints, which can occur in extrusive igneous rocks to form distinctive columnar structures: as the surface of the lava cools, there are surface contractions toward discrete centers giving rise to a pattern of polygonal, generally hexagonal, cracks which, with further cooling, develop in depth to produce the columns.

Kraton, craton or **shield,** a large mass of igneous and metamorphic rock, generally of Precambrian age and covered with at most a very superficial layer of sediment, which comprises a major crustal unit. Usage of all three terms varies widely.

Laminar, made up of or resembling laminae,

thin sheets or plates, usually parallel.

Land bridge, a short-lived land connection between continents, of primary importance in that it permits migration of fauna between the continents concerned. The status of land bridges in the history of life on Earth – and even their very existence – has recently been questioned.

Laurasia, the northern-hemisphere continent formed as *Pangaea* split up. Stratigraphic and other evidences suggest it comprised what are now Europe, North America and northern Asia.

Leaching, the action of percolating rainwater on rocks and soils. Rainwater dissolves various substances during its descent through the atmosphere (most important of which is carbon dioxide, CO_2): the resulting solution is usually acidic, and can selectively dissolve various substances as it percolates down through the soil or rock, carrying them down to a lower level. See also *weathering*.

Levee. See *flood plain*.

Lithosphere, a term used either for the rocks of the Earth – as contrasted with the atmosphere and hydrosphere (waters of the Earth) – or for the upper part of the Earth's crust.

Littoral, the region of the shore covered at the highest spring tide but uncovered at the lowest spring tide. The term is used also adjectivally to describe organisms, deposits, etc., characteristic of this region.

Load, of a stream, the solid material carried along by the water, either in solution or suspension or dragged along the bed. See also *transportation*.

Loess, wind-deposited silt found in deposits up to 50m (160ft) thick. Extremely porous, it forms a highly fertile topsoil.

Long profile, of a stream, a plot from source to mouth of the vertical height of the bed.

Magma, molten material formed in the upper mantle or crust of the Earth, composed of a mixture of various complex *silicates* in which are dissolved various *volatiles*, including notably water. In suitable circumstances magma crystallizes to form *igneous rocks*, the gaseous materials being lost during the solidifying process. The term magma is also, though rarely, applied to other fluid substances, such as molten salt, in the Earth's crust.

Magnitude, Richter, a measure of the size of an earthquake event. The Richter scale was originally devised by C. F. Richter (1900–   ),

and runs from 0 to 10, about one quake a year registering 8 or over. The magnitude M is calculated from

$$M = \log_{10}\frac{A}{T} + B;$$

where A and T are respectively the amplitude and period of either the P pulse or the portion of the surface-wave packet with greatest amplitude, and B is a correction factor that takes account of the distance of the recording station from the event.

Mantle, the layer of the Earth between the internal *core* and the external *crust*: its internal radius is about 3500km (2200mi) and its outer radius around 6345km (3940mi), the latter corresponding to a depth of about 35km (22mi) beneath the surface of the Earth.

Metamorphic rocks, rocks that have undergone change through the agencies of heat, pressure or chemical action, and comprising one of the three main types of the rocks of the Earth. *Sedimentary rocks* undergo prograde metamorphism, by which they lose *volatiles* such as water and carbon dioxide, under conditions of heat and pressure beneath the Earth's surface. *Igneous rocks*, and rocks that have already been metamorphosed once, undergo retrograde metamorphism, absorbing volatiles, usually from sediments that are being metamorphosed nearby.

Meteorite, the part of a meteoroid that has survived the passage through the atmosphere to reach the Earth's surface. Meteoroids are small particles of interplanetary matter believed to consist of asteroidal and cometary debris. Most burn up due to friction in the Earth's atmosphere, showing a trail of fire, known as a meteor, in the night sky. Parts of the larger meteoroids survive this fiery passage to impact with the surface of the Earth, sometimes producing large craters like that in Arizona. Meteorites are of two types: "stones", whose composition is not unlike that of the Earth's crust; and "irons", which contain about 80–95% iron, 20–5% nickel and traces of other elements. Intermediate types also exist.

Mohorovičić discontinuity, or **Moho,** a layer of the Earth once regarded as marking the boundary between *crust* and *mantle*, evidenced by a change in the velocity of seismic waves. Its physical significance is uncertain. Perhaps more important are the discontinuities between *core* and mantle (Gutenberg or Oldham Discontinuity), with a radius of about 3500km (2200mi); and that between inner and outer core, with a radius of about 1200–1650km (750–1050mi).

Neoteny, in an organism, the retention of larval characteristics at and beyond sexual maturity, especially in those organisms whose larvae never completely metamorphose into the adult form. It has been suggested that neoteny may be a powerful evolutionary agent, sessile adult forms being gradually excluded from the lifecycle; but evidence is lacking.

Old Red Sandstone, the continental *facies* of the British Devonian.

Oolite, a limestone made up of *ooliths*.

Oolith, a more or less spherical particle of rock which has developed by the accretion of material about an initial nucleus. The accretion may be concentric (so that in cross-section circular bands of material may be seen) or radial, and combinations are known. Larger ooliths are termed *pisoliths*. Concentration of ooliths can form, for example, oolitic limestones, often called *oolites*.

Ore, an aggregate of minerals and rocks from which it is commercially worth while to extract

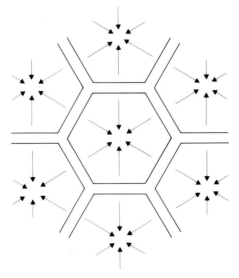

Joint: the development on the surface of cooling lava of hexagonal joints caused by contractions toward discrete centers.

minerals – usually metals. An ore has three parts: the *country rock* in which the deposit is found; the *gangue*, the unwanted rocks and minerals of the deposit; and the desired mineral itself.

Orogeny, any process which results in the formation of mountains; i.e., relative changes of level within a landmass. Overall changes in the level of a landmass constitute *epeirogeny*. A period of mountain-building in a particular area may be described as an orogeny.

Paleo-, a prefix meaning "ancient". Thus paleogeography is the geography of the Earth in past ages; paleoclimatology the study of climates in past ages; and paleontology the study of fossil life. Special cases are: Paleozoic, the era of ancient life; and paleomagnetism, the study of the information within certain rocks relating to the magnetic field of the Earth in past ages. This latter area of study has been of paramount importance in many of the Earth sciences, not least in validating the theory of *plate tectonics*.

Palynology, a branch of paleontology embracing the study of fossil spores, especially of pollen. Its importance as an aid to correlation is rapidly increasing.

Pangaea, the primeval supercontinent which split up to give Laurasia to the north and Gondwanaland to the south. Pangaea formed in the Permian, gradually disintegrating during the succeeding periods.

Pegmatite, a term to describe a coarse-grained *igneous rock*, usually applied to those having a granitic composition.

Pelagic, an adjective applied to marine creatures that are not *benthonic*; i.e., that are free-swimming or *planktonic*.

Permafrost, permanently frozen ground, typical of the treeless tundra of Siberia and common throughout polar regions. The permafrost layer may attain depths of up to 600m (2000ft).

Permeability, a measure of the ease with which water may pass through or into a rock.

Petrogenesis, an envelope term embracing all aspects and features of the formation of rocks.

Petrology, the study of rocks in all aspects.

Phanerozoic, the aeon of visible life, the period of time represented by rock strata in which fossils appear, running from about 600 million years ago through to the present and containing the Paleozoic, Mesozoic and Cenozoic eras. It is contrasted with the *Azoic* ("no life") and *Cryptozoic* ("hidden life").

Phylum (plural phyla), a major taxonomic subdivision of either the plant or the animal kingdom (for plants, the term *division* is commonly used in its place).

Placer, an ore deposit which owes its origin to the action of water. In general the water has both transported the mineral to its present position and removed extraneous matter about it.

Planktonic, adjective describing those organisms that float (usually at the surface) in marine or other waters, generally possessing at best weak locomotive abilities.

Plate tectonics, theory of, fundamental theory of modern geology, arising from studies of continental drift, earthquake and volcano distributions, and sea-floor spreading – which phenomena it largely explains. The Earth's *crust* is viewed as consisting of a number of semi-rigid plates in motion relative to each other. Where plates meet, one edge is subducted beneath the other: in mid-ocean, this results in ocean trenches, deep seismic activity and arcs of volcanic islands; at continental margins, similar subduction of the oceanic plate results also in *orogenies*. Where lighter continental blocks are forced together, neither edge is subducted and more complex orogeny results. Belts of shallow earthquakes define the mid-ocean ridges where new material is emerging.

Pluton, or **plutonic intrusion,** a large-scale mass of igneous rock that has consolidated at great depth beneath the surface of the Earth.

Plutonic rock, an *igneous rock* formed, usually as a large *intrusion*, at depth beneath the surface of the Earth.

Porosity, a measure of the amount of empty space within a given volume of rock, and hence of the amount of water that that rock body can absorb. Since certain porous rocks will not permit the transmission of water, porosity cannot be considered a synonym of *permeability*.

Porphyry, adjective describing a rock in which large isolated crystals are set in a rather fine matrix.

Province (faunal or floral), at a particular time, a distinct ecological region.

Pyroclastic rock, a rock made up of particles thrown into the air by volcanic eruption.

Radiation, an increase (through time) in diversity and abundance of a type of organism.

Radiometric dating, any technique whereby the age of a rock or artifact may be determined by comparison of the amount of a particular radioactive substance (of known half-life) present there, and the amount of one or more of its decay products. Best known is the carbon-14 technique though, as C^{14} has a half-life of only about 5730 years, it is of interest primarily to the archaeologist rather than to the geologist.

Red beds, an assemblage of sedimentary rocks characterized by a red color resulting from formation in a highly oxidizing environment, the redness being due to the presence of iron in a ferric, rather than ferrous, state.

Regolith, a collective term for the unconsolidated material of the Earth's surface lying upon the bedrock. The most important component of the regolith is soil.

Regression, a withdrawal of the sea from the land; the converse of *transgression*.

Rejuvenation, the effect of an uplift of the land or a fall in the sea level on a drainage system: the stream initially reaches the sea by way of frequent waterfalls, which migrate upstream as a result of rapid downcutting.

Replacement, in fossils, any process whereby the original skeletal material of the organism is replaced by another mineral, often such that all the details of the skeleton are preserved.

Schist, a metamorphic rock in which the tendency to split into layers along perfect *cleavage* planes is pronounced.

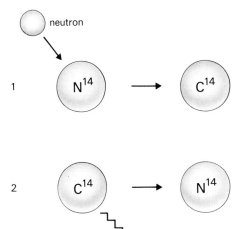

Radiometric dating: (1) the impact of a fast neutron generated by a primary cosmic ray particle converts a nitrogen-14 atom in the upper atmosphere to an atom of carbon-14. Carbon-14 is radioactive, with a half-life of about 5730 years. In (2) an atom of carbon-14 loses an electron, decaying back to the stable nitrogen-14.

Sedimentary rocks, rocks which consist of weathered particles of igneous, metamorphic or even former sedimentary rock transported, usually by water, and deposited in distinct strata. They may also be of organic origin, as is chalk, or of volcanic origin, as are *pyroclastic rocks*. Sedimentary rocks are important as they contain fossils as well as most of the Earth's mineral resources: the order of their strata is clearly of importance to the stratigrapher. They comprise one of the three main types of rocks of the Earth's crust.

Segregation, any process which results in a concentration of a particular mineral (or group of minerals) in a rock of different composition. In igneous and metamorphic rocks, an example of segregation is *differentiation*.

Shield. See *kraton*.

Sial (*si*lica-*al*uminum), a collective term for those rocks, lighter and more rigid than those of the *sima* and composed to a great extent of silica and aluminum, that form the upper portion of the Earth's crust.

Silicates, minerals containing the SiO_4^{4-} ion – that is, one atom of silicon and four of oxygen, having a net negative charge of four – and including also silica (SiO_2) itself. The silicates, together with the aluminosilicates, are the most important class of minerals and comprise about 90% of the rocks of the Earth's crust.

Sima (*si*lica-*m*agnesium), collective term for those rocks, denser and more plastic than the *sial* and composed to a great extent of silica and magnesium, that form the lower portion of the Earth's crust.

Strain, deformation produced by *stress*.

Stress, the internal forces within a body resulting from the external forces acting upon it. The four main types of stress are: shearing; bending; tension and compression; and torsion. The deformations resulting from stress are described as *strain*.

Subduction, the forcing of one plate under another (see *plate tectonics*).

Subjacent, literally, "lying under". The term is used also in geology to mean bottomless: a batholith can be termed subjacent if its base cannot be detected.

Symmetry. If a plane can be drawn through a crystal such that the halves of the crystal on

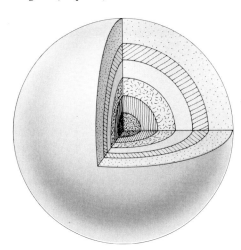

Oolith: idealized view of the concentric bands of accretional material in an oolith.

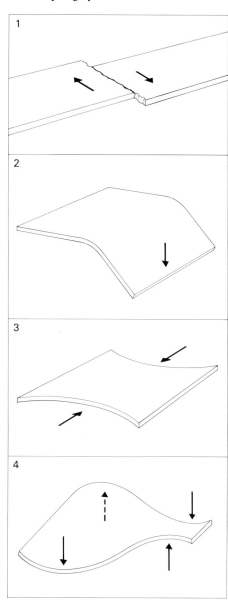

Stress: strains produced by different stresses – shearing (1), bending (2), compression (3) and torsion (4).

either side of it are exact mirror images of each other, the plane is a plane of symmetry, denoted *m*. If a straight line (axis) can be drawn through the crystal such that, when the crystal is rotated through an angle of 60°, 90°, 120°, 180° or 360° about the axis, it fills exactly the same space as it

did originally, then the axis is a (rotation) axis of symmetry, denoted 6, 4, 3, 2 or 1 (called hexad, tetrad, triad, diad or identity axes respectively) depending on how many times the "symmetry operation" must be performed to bring the crystal back to its original orientation (i.e., a rotation of 360°). All crystals have an infinite number of identity axes.

The crystal may also have an inversion axis of symmetry about which, after inversion, the crystal may be rotated through a certain angle to occupy exactly its original space. Inversion axes are denoted $\bar{6}$, $\bar{4}$, $\bar{3}$ and $\bar{2}$ (a $\bar{2}$ axis is equivalent to a plane of symmetry and so the description is rarely used). A $\bar{1}$ axis implies that the crystal has a center of symmetry, which lies at the center of any line through it joining opposite faces.

Crystals are usually classified according to the symmetry they display.

Tectonic, an adjective pertaining to deformation of the Earth's crust and the processes whose effect is such deformation.

Terrigenous, an adjective applied to sediments derived from the land, whether mixed in with marine material or deposited on the land.

Tethys, the sea that lay between the supercontinents of *Laurasia* and *Gondwanaland*. As the continents evolved toward their present form, Tethys was closed off, though the Mediterranean is regarded as a remnant. The sediments of the Tethys geosyncline are to be found in folded mountain chains such as the Himalayas.

Transgression, the advance by the sea over the surface of the land, for *eustatic* or other reasons. Transgression is the converse of *regression*.

Transportation, the conveyance of eroded or other material from one place to another by one of the following agencies (or a combination thereof): running water, the sea, the wind, glaciers or gravity. The end result of transportation is deposition, the depositing of the material in a new locale.

Turbidite, the sedimentary deposit formed by the action of a turbidity current. Such currents, moving mixtures of fine particles and water, are extremely fluid and so can run for long distances down submarine slopes, spreading sediment over a wide area.

Ultrabasic rock, usually plutonic though sometimes volcanic *igneous rock* containing little or no quartz, feldspars or feldspathoids but rich in the *silicates* of iron and magnesium.

Valve, one of the distinct pieces, usually articulated, of which a shell is made up. Many shells have only a single valve, while members of the class Bivalvia and many other creatures have shells with two.

Varve, a layer of sediment deposited in the course of a single year, specifically in a lake

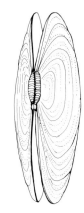

Valve: a bivalve shell.

formed of glacial meltwater. Characteristically, a varve has a silt layer overlying a sand layer. Study of varves is of considerable importance in geological dating.

Vein, a mineral formation that is of far greater extent in two dimensions than in the third. Sheetlike *fissure veins* occur where fissures formed in the rock become filled with minerals. *Ladder veins* form in series of fractures in, for example, dikes. *Saddle veins* are lens-shaped, concave below and convex above. Veins containing economically important *ores* are termed *lodes*.

Ventral, term used to describe the front parts of an animal, or those parts generally turned toward the ground. In botany, the ventral side of, say, a leaf is that side turned toward the stem. See also *dorsal*.

Volatile, adjective used of substances that readily vaporize.

Volatiles, substances present in the magma that would in normal circumstances be gaseous at the temperature of the magma: they are inhibited from vaporizing both because of the high pressure and because of their tendency to be dissolved in the melt. Most common are carbon dioxide and water.

Volcanic rock, an extrusive igneous rock (see *extrusion*), though the term is sometimes taken to include certain associated intrusive rocks (see *intrusion*).

Water table, the uppermost level of *groundwater* saturation.

Weathering, the breaking down of rocks through contact with the atmosphere (e.g., by wind, by rain). There are two types of weathering: chemical, including processes such as *leaching*; and physical, usually related to the temperature changes though other factors may play a part. Unlike *erosion*, weathering does not imply *transportation*.

Index

(Page numbers in *italics* indicate relevant illustrations; those in **bold** indicate major references.)

aa 96
ablation 24, 27
abyssal hills 98, **99–100**
abyssal plains 97, 98, **100**, 102
acanthodians (spiny sharks) 201–2, **266**
achondrites 25, 26
acritarchs 200, 249, 250
actinolite **121**, 122, 123, 133, 138
actinopterygians 218, *240–1*, 245, 246, 266, **268–9**
adamellite 165
adamite *133*
aegirine 164
Aegyptopithecus 284
Agassiz, J. **297**
agate *120*, **121**, 125, 135, 139
agglomerate 160
acritarchs see corals — *(omit)*
achondrites — *(omit)*
Agnatha 201, **264–5**
aigialosaurs 273
alabaster 129
Albertus Magnus 291, **297**, *298*
albite 34, *118*, 132, 134, 141, 176
algae 35, 82, 155, *169*, 172, 190, 191, 192, 200, 201, 203, 206, 219, 223, 243, **248–50**, 255; blue-green 189, 219, 235, 236, 242, 249; brown 248; golden-brown 249, 250; green 247, 249; red 249
Allosaurus 217
alluvial fans 85, 86, 171
almandine 128, 129
Alps 39, 73, 74, 75, 189, 216, 297, 306
alumina 140, *141*, 161
aluminosilicates see clay minerals
aluminum 10, 11, 18, 33, 34, 59, 83, 112, 127, **141**, 144, 160, 166, 168, *173*, 179
alundum 126
alveolinids 253
amazonite 132
amber **121**, 140
amblypods 229
amethyst *121*, **121–2**, 135
ammonites 212, 213, **218–9**, 220, 223, 225, 245, 257, 258, **259**, 295
ammonoids 201, 202, 213, 214, 223, **258–9**
amphibians *245*, 266, 268, **269–70**; Carboniferous 206, 209, 240; Permian 212; Triassic 214
amphiboles 10, 121, **122**, 124, 126, 127, 129, 130, 135, 164, 174, 175, *176*, 177, 178
amphibolites 130, 175, **178**
amygdales 139
analcite 139
andalusite **122**, 131, 137, 176
Andes 43, *74*, 79, 111, 164, 165, 189, 208, 224, 232
andesine 134, 164
andesites 72, 123, 130, 149, 165, 188, 200, 216
andradite 128, 129
Angara (Angaraland) 204, **209**, 211, 212, 215
angiosperms (flowering plants) 219, 222–3, 243, 250, **252–3**, *253*, 263, 276
anhydrite 54, 121, **122**, 129, 142, 173
ankerite *129*

Annelida 243
anorthite 34, 134, 189
anorthosite 17, 188, 189
anthozoans see corals
anthracite 155
anthracosaurs 269
anthracotheres 282
Anthracotherium 229
anthropoids 283
antiarchs 266
anticlines *151*
antidunes 55
antigorite 137
antimony 137, **147**
apatite *116*, 118, 120, 126, 142, *173*, 178
Apidium 284
apsidospondyls see labyrinthodonts
aquamarine 118, **122**, 124
aquifers 108
aragonite 53, 59, 122, 140, 142, 172, 247, 254, 259
Archaean **188–9**, 190
archaeocetes 283
archaeocyathids 190, 192, 193, **255**
Archaeopteryx 217, **278**, 306
archegosaurs 269
arches 94
archosaurs 240, 270, 273, 277
arenaceous rocks *161*, 168, **169–70**; see also sandstones (quartz), arkoses, graywackes
arêtes 75, 86
arfvedsonite 164
Argand, É. **297–8**
argentite 146
argillaceous rocks 168–9, *176*; see also clays, marls, mudstones, shales
argon 20, 26, 35
arid landscapes **83–6**, *84, 85*
Aristotle 289, 297, **298**
arkoses 169, **170**
Armorican geosyncline 203
Arrhenius, S. **298**
arroyos 86
arsenic 122, 136, 144, **145–6**
arsenopyrite **122**, 125, 145
artesian flow 82, 108
arthrodires 201–2, 266
arthropyrites 211
arthropods 223, 243, 258, **261–3**
artiodactyls 229, 231, 281, **282**
asbestos 119, 121, **122–3**, 125, 137, 138; see also crocidolite (blue asbestos)
asparagus stone 122
asphalt 150
asteroid belt 10, 21, 26, *27*
asteroids (fauna) 223, 257
Aswan Dam 90, *106*
atmosphere **34–5**, 47, 72, 76, 160, 181, 189, 191; origin of life in 236
atolls 94, **101–2**, *102*, 204, 300, 301
atomic number 35, 139, 180
augengneisses 177
augite 117, **123**, 127, 130, 135
aureole, metamorphic 177
Australopithecus 232, **284–6**, 300, 305
autecology 245
Ayers Rock 81

azurite **123**, 127, 132, 144

bacteria 59, 152–3, 242; sulfur 134
Bactritoida 260
baddeleyite 146
badlands *221*
Bailey, Sir E. **298**
Baltic Shield 198, 225
Baluchitherium 281–2
barchans 84
barite (barytes) *85*, 117, *123*, **123–4**, 125, 139, 142, 147
barium 34, 124, **147–8**
Barrell, J. **297**
Barrois, C. **298**
barytes see barite
basalt 11, 17, 20, 34, 39, *46*, 72, 80, 95, 98, 102, 121, 123, *161*, 162, 163, 164, **165–6**, *166*, 167, 170, 188, 217; debates over nature of 293, 299, 302; tholeiitic 130
bastnaesite 142
batholiths 69, *70*, 163, 164
bats 280
bauriamorphs 277–8
bauxites 48, 83, 112, 125, 127, 141, 144, **173**
belemnites 258, 260
belemnoids 223, **260**
Bennettitales 251
bentonite 132, 224
bergschrunds 86
beryl *118*, 122, **124**, 127, 140
beryllium 33, **140**
bioherms 225
biotite 47, 106, *120*, **124**, 132, 133, 139, 162, *163*, 164, 176, *177*
birds 217, 221, 223, 225, 243, 244, 270, 273, *278*, **278–9**
birefringence see double refraction
bismuth **149**
bismuthinite 149
bivalves 53, 192, 194, 203, 214, 219, 222, 223, 255, 258, **261**
blastoids **257**
blastomylonites 177
bloodstone 125
blowholes *52*, 94
Blue John *128*
blue schists 74, 129, 175
body waves see seismic waves
boehmite 141
bomb, volcanic *72*
bony fishes 202, **266–9**, *267*
borax 140
bornhardts 48, 81
bornite **124**, 144
boron 33, 112, 138, **140**
bosses 68–9, *70*
boudinage 67
boulder clay see till
bournonite 147
Bowen, N. **298–9**
brachiopods 191, 192, 194, 197, 198, 199, 200, 201, 202, 205, 206, 212, 213, 219, 223, **255–7**, 261
Brachiosaurus 217, 275
Branchiopoda 262
branchiosaurs 269
braunite 143
breccias 17, **170–1**
bromine 147, 159
Brongniart, A. **294**
Brontosaurus 217
brontotheres see titanotheres

bronze 125
bronzite **124**, 127, 130, 135
bronzite-chondrites 124
brucite 141
bryozoans 192, 194, 198, 200, **255**
Buch, C. von 294, **299**
Buckland, W. 295, 296, **299**, 305
Buffon, G. 292, **299**
building stone *158*, **159**, *170*, *174*, 177, 207, 253
Burgess Shale 192
buttes *51*, *81*, *84*, *216*
bytownite 134

cadmium 137, **146–7**
caesium **147**
Calamites 201, 251
calcarenites 225
calcareous rocks *168*, *169*, **171–2**, *171*, *172*; see also chalks
calcite 47, 53, 59, 61, *113*, 118, 119, *120*, 122, *124*, **124–5**, 127, 129, 131, *134*, 137, 139, 140, 142, *161*, 167, 169, *171*, *172*, *174*, 247, 254
calcium 10, 34, 122, 127, 128, **142**, 161, 162, 164, 165, 166, 171, 179; bicarbonate *83*; carbonate 59, 81, *82*, *83*, 99, 168, 169, 173; see also aragonite, calcite
calcrete 84
calcsilicates 174
calderas 96
Caledonian-Appalachian orogeny 192, 196, 197, 203
caliches 203
Cambrian 187, **190–3**, 194, 196, 197, 213, 247, 253, 254, 295, 305, 307; faunal provinces 193; faunas 190–2, *191*; floras *191*; geography *191*, 192–3
Canadian Shield 81–2, 188, 198, 207, 208
cap shells see monoplacophora
captorhinomorphs 270, 272
carbon 9, 33, 35, 122, **127**, *129*, **140**, 143, 152, 153, 168, 176, 181–4, 187; dioxide 13, 20, 35, 47, 59, 72, 81, 82, 160, 166, 173, 174, 181; monoxide 20, 35
carbonaceous chondrites 10, 25, 26
carbonates 35, 125, 164, 172
carbonatite 125, **164**
carbon-14 dating 181, **184**, 233
Carboniferous 40, 155, 190, 201, **204–9**, *206*, 208, 215, 240, 243, 250, 251, 257, 263, 269, 270, 295; climate 204; fauna, 205, 206, 213; flora 205, 211; geography 204, 205–9; formation of coal in 156, 206, 207; Mississippian **207**, 208; Pennsylvanian *39*, **207–8**; stratigraphic correlation 204–5
carbuncle 129
carnallite 142
carnivores 229, **280**, 283
carnosaurs 275
carnotite 143, 149
carpolestids 283
cartilagenous fishes **266**
cassiterite 117, 125, 139, 147,

148
Catastrophism *185*, 187, 296, 299, 300
catena 179
Cathaysia 211, 212, 215
caves 50, 82, *83*; phreatic 82; vadose 81, *108*
caviomorphs 283
Caytoniales 251
celestine 117, 138, 146
celsian 147
Cenozoic **225–34**, 243–4; see also Tertiary, Quaternary
Cephalaspis 265
cephalopods 192, 204
cerargyrite 146
ceratites 259
ceratopians 276
cerium 149, 164
cerussite 125, 128, 149
cesium see caesium
chabazite 139
chalcedony *120*, 121, **125**, *131*, 135, 139
chalcopyrite 30, 116, 117, **125**, *134*, 142, 144
chalicotheres 282
chalk 125, *168*, *169*, 221, *224*, 225
chalybite see siderite
Chamberlin, T. **299–300**
chamosite **125**, 173
charnockites 124, 130
chelicerates 261, 262, **263**
chelonians see tortoises, turtles
chernozems 179
cherts *112*, 125, 135, 172, **173**, 188, 225, 236, 248, 254
chiastolite 122
china clay see kaolinite
chloanthite 145
chlorapatite 122
chlorine 17, 35, 72, 147, 159, 164
chlorite 59, **125–6**, 138, 176, 177, 188
chondrites 12, 25, 26, 127, 130, 160, 236
chondrosteans 223, *240*, **268**
chondrules 25
chromite 121, **126**, 137, 143
chromium 98, 112, 126, 127, 129, 130, 136, **143**, 189
chrysoprase 125, *131*
chrysotile *123*, 137
cidaroids 219
cinnabar **126**, 148
cinnamon stone 129
cirques 50, *56*
cirripedes 262
citrine 122, 136
clastic rocks 135, 169, 171
clay minerals 47, 122, **126**, 130, 131, 168, 169, 170
clays *106*, 108, 126, 160, **168**, 174, 176
cleavage: mineral 119, *120*, 124, 127; rock 175, *176*
climatic zones **36–7**, *37*, 75, 168, 178, 179, 197, 232, 244
clinopyroxene 34
Cloos, H. 297, **300**
clubmosses (lycopsids) 243, **250–1**,
clymenids 204
coal 41, 59, *110*, 140, 143, 153–7, *153*, *154*, 155, 168, 173, 206, 207, *208*, 209, 217, 221; limnic 155; paralic

154–5; rank of 155–6; world resources *154, 156–7; see also* Carboniferous, Jurassic
Coal Measures 205
coastal plains 89–90, *90*
coastlines 93–4, *93, 94*
cobalt 17, *113*, **143**
cobaltite 116, *143*
coccolithophores, coccoliths 53, 125, *169*, 172, 201, 219, 220, 223, 224, 225, 249, **250**
cockpits 83
coelacanths 202, 266, 268
coelenterates 190, 243, 254
Coelophysis 217
coelurosaurs 275
colemanite 140
collembolids 263
collophane 122, **126**
color index 162
colubroids 272
compaction 58, 168, 169
concretion *58*
conductivity (electrical) of rocks 13, **30–1**
cone-sheets 68
conglomerates 169, 170, **171**
coniferophytes **252**
conifers 212, 215, 219, 222, 223, 252
connate water 107
conodonts 192, 197, 200, 201, 204, 205, **263**
consequents 75
continental drift **39–41**, *39*, 45, 63, 297, 298, 301, 302, 309
Cooksonia 250
Cope, E. 296, **300**, 306
Copepoda 262
copper 30, *109*, *111*, 112, *113*, 116, 119, 123, *124*, 125, 127, 132, 138, **144**, *144–5*, 148, 159, 164, 189
copper pyrites *see* chalcopyrite
Corallina 249
corals (anthozoans) 53, 94, *102*, 172, 192, 197, 198, 199, 200, 201, 204, 205, 206, 214, 222, 247, 248, 254–5; rugose 194, 202, **254–5**; scleractinian 214, 219, 255; tabulate 194, 202, 254; *see also* atolls, reefs
Cordaitales 212, 252; *Cordaites* 212
cordierite **126**, 176
core 10, 11, **12**, 27, 29, **34**, 35, 36, 41, 160
cornelian 125
corundum *118*, 120, **126**, 136, 141, *173*
cosets *55*
cotylosaurs 212, **270**, 272
craters, impact: lunar 15, *16*, 21, 27; Martian 20, **21**, *22*, 23, 27; Mercurian 19, 21, 27; terrestrial 24, *25*, **27**, 127
craters, volcanic *95*
creodonts 226, 229, 280
crestal mountains 98, 99, 101, 102
Cretaceous 39, 90, *168, 169*, *171*, 172, 209, 217, **221–5**, 243, 248, 249, 250, 253, 260, *267*, 295; climate 222, 276–7; fauna 222, 223–4, 278; flora 222–3, 252; geography 221–222, 225
crinoids 192, 194, 198, 206, 219, 223, *256*, **257**
cristobalite 17, **126–7**, 133–4
crocidolite (blue asbestos) *123*, **127**, 136
crocodiles 217, 221, 240, 243, 269, 270, 273, **277**
Cro-Magnon Man 286
cross-bedding *55*, 167, 174
cross-lamination *55*
crossopterygians 266, **268**
cross profiles **77–8**
crust 10, **11**, 12–3, 27, 127, 160, 162, 174; continental 11, 13, 40, 41, 43, 74, 163; oceanic 11, 13, 40, 41, 43, 74,

178; formation 188, 189
crustaceans 194, 206, 223, **262**
cryolite **127**
crystal habit 113, 120
crystallography **113–8**, *113*, 303; X-ray 116
crystal systems **113–8**, *115*; cubic *114*, 115, **116**, *117*, *120*, 127, 128, *129*; hexagonal 115, **118**; monoclinic 115, **117–8**, *118*, 122; orthorhombic 115, **117**, *122, 138*; tetragonal 115, **116–7**, *117*; triclinic 115, **118**; trigonal 115, **118**, *121, 124, 135*
cubic system *see* crystal systems
cuprite **127**, 144
Cuvier, G. 241, 294, 295, 296, 297, **300**, 305
cycadophytes **251**, 252
cycads 212, 215, 219, 222, 223, 251
cynodonts 212, 278
cystoids **257**

dacites 96, **164**, 165, 188, 216
Daly, R. **300**
dams *103, 104*, **105–6**, *107*
Dana, J. 294, 297, **300**, 303
Dapedius 265
Darwin, C. 102, 164, 190, 237, *293*, 294, 296, **300–1**, 304, 305, 307
Darwin, G. **301**
dating **180–4**, *187*
Davis, W. 297, **301**, 302
deformability **103–4**
de Geer, G. **301**
Deimos 23–4
Deinosuchus 277
Deinotherium 231
De la Beche, H. 294, 296, **301**, 305
deltas 53, **90–3**, *91, 92*, 173; arcuate *91, 92*; elongate (birdfoot) 91–3; lobate 91–3
dendrochronology (tree-ring dating) 184
dendroids 264
deposition **52–6**, *53, 54, 55*, 59, 76, 90, 94, 168, 172, 185
desert roses *85*, 123
deserts *see* arid landscapes
Devonian 190, **201–4**, *203*, 206, 208, 243, 248, 263, 265, 266, 268, 295, 305, 307; climate 201, 204; fauna 201–3, *202*, 204; flora 201, *202*, 211; geography 201, *202*
diachronism 187
diagenesis **58–9**, *58, 59*, 167, 172
diamonds *111*, *119*, 120, *126*, **127**, 140, 166, *167*
diaspore 141
diatomites 249
diatoms 99, 201, 225, **249**
dicotyledons 252
Dictyonema 264
dicynodonts 270, 277–8
dikes 68, 69, 70, 72, *163*; swarms 68–69, 187, 189
dilatancy hardening **64–5**
Dimetrodon 212, 277
dinocephalians 277
dinoflagellates 219, 223, 224, **249–50**
dinosaurs 239–40, 243, 270, 273, *274*, **275–7**, *275*, 305; Triassic 214; Jurassic 217, 221; Cretaceous 223, 225
diopside 34, **127**, 135
diorites 123, 124, 130, 162, 163, **164**
Diplodocus 217, 275
dipnoans *see* lungfish
DNA 113, 237–8
docodonts 278
"dog-tooth" calcite *113, 124*
Dokuchaiev, V. 178, 179
dolerite 68, 123, *163*
dolines 81, *82*
dolomite: mineral 47, 59, 118,

119, 122, **127**, 141, 142, *161*, *171, 172*, 173, 174; rock 81, 126, 127, 128, 142, *172*, 174, 216
double refraction (birefringence) *120*, 125
drainage basins 75
drape folds 66
drumlins 88
dunes *55*, **84–5**, 86, 90, 94, 100, 203, 247
dunite 17, 133
duricrusts 85
du Toit, A. 40, **301**

Earth *passim, et*: age of 9, 14, 35, **180–4**, *182–3*, 187, 291, 297, 299–300, 303, 309 (*see also* dating); atmosphere **34–5**, 47, 72, 76, 160, 181, 189, 191, 236; chemical differentiation 10, **34**; composition **10–3**, *33*, 33–4; gravitational field **29–30**, *31*; heat flow within 70–1, 88, 160; internal temperatures **12–3**, 30, 174; magnetic field 18, **31–2**, 41–2, *42*, 65–6, 187, 224, 233; magnetic poles 32; Man's ideas of **288–97**; natural vibrations of 29; obliquity of 40; obliquity of axis 36; origin of **9–10**, 14, 35, 180, 292, 299–300; origin of life on 72, **235–7**, 298; physical measurements **10–3**; structure **10–3**, 29, *291*; length of year 242; *see also* core, crust, mantle
earth pillars 49, 81, *91*
earthquakes 28, 40, 61, **62–6**, *63*, 65, 75, 100, 101, 106; associated phenomena 65–6; prediction and control 64–5; Mercalli intensity 64; Richter magnitude 64; in relation to plate movements 43, 44, 62, 63–4, 71, 78
echinoderms 192, 219, 223, *256*, **257–8**
echinoids 53, 194, 219, 223, 257, **258**
eclogite 11, 131, **167**, 175
Edaphosaurus 212, 277
edentates 229, **282**, 300
edrioasteroids 192
elapids 272
elaterite 119
electromagnetic geophysical studies **30–1**, *31*; resistivity methods 105
electromagnetic prospecting (EM) techniques 31
electron microprobe analyzer 34
elements 35, **139–49**, *140*; cosmic abundances of *33*; periodic table of *139*, 140
elements, Aristotelian **289**, 292, 297, 298
elephants 231, 232, 280, **281**
Elie de Beaumont, L. 294, 296, **301–2**
embolomeres 269
emerald 118, 119, 122, 124, **127**
emery 126
engineering geology **103–7**, *103, 104, 105, 106*, 110
enstatite 10, 34, **127**, 135
entelodonts 282
Eocene 39, 225, 226, **227–9**, *227, 228*, 250, 279–80
Eocoelia 199–200
eocrinoids 192
Eohippus 282
epidiorite 175
epidote 117, **127**, 178
epsomite 115
Equisetum 251
ergs 85
erosion **48–52**, *49, 50, 51*, 54, 59, 68, 70, 74, 75, 76, 77, 78, 80, *81*, 83, 84, 94, 95, 100, *160*, 163, 168, *171*, 173, 185,

188, 302; by ice **49–50**, *50*, 58, 75, 86, 89; by water **48–9**, **51–2**, 89–90, *93*, 106, *177*; by wind **50–1**, *51*, 84; lunar 15; Martian 21, 23; Mercurian 19; Venusian 20; cycle of 301; *see also* soil erosion
erratics, glacial *87*, 88
Eryops 269
Eskola, P. **302**
Etna *71, 72*
eugeosynclines 72, 196
Euparkeria 245, 273
europium 17
eurypterids 263
evaporites 53–4, 61, 112, 121, 122, 129, 134, 138, 140, 141, 142, *157*, 168, 171, 172, **173**, 193, 204, 206, 213, 215, 231
evolution 39, 180, **237–40**, *238*, 240–1, 242, 247–8, 294, 297, 300, 301, 304, 305, 306
exfoliation 46
expanding Earth hypothesis 42
extrusive rocks *see* igneous rocks

fairy crosses (fairy stones) 137
fault, transform 64
faulting **60–1**, *60*, 64, 67, 74, 98, 104, *110*, *151*, 164, 177, *309*
fault-plane solutions 64
fayalite 119, 133
feldspars 11, 12, 17, 18, 34, 47, 69, *118*, *121*, 124, 126, **128**, 130, 134, 142, 162, *163*, 164, *165*, 167, 168, 169, 170, 174, *175*, *177*
feldspathoids **128**, 133, 136, 162, 164
ferns 201, 212, 215, 219, 243, *244*, 250, 251; *see also* pterophytes
ferricretes 85
fishes **240–1**, **264–9**, *265, 267*; Cambrian 192; Ordovician 194; Silurian 198–9; Devonian 201–4, *243*; Carboniferous 206; Jurassic 218; Cretaceous 223; Tertiary 231
fissility 169
fissipeds 280
fjords 94
flint 125, 135, *173*, 224, 225
Flood, the 185, 241, 291, *294*, 299, 309
floodplains 77, **77–8**, 100
flowering plants *see* angiosperms
fluorapatite 122, 142
fluorescence 121, **128**
fluorine 122, 138, 147, 164
fluorite (fluorspar) *113*, 116, 120, 127, **128**, 142
flutes 88, 170
flysch 73, *228*
folding 61, **66–7**, *66, 67*, 72, 73, 76, *189*, 216, *309*
foliation *see* gneissose banding
fool's gold *see* pyrite
foraminifera (forams) 45, 125, 205, 212, 219, 222, 223, 226, 232, 242, 244, **253**, 254, 261
forsterite 119, 133
fossil record 187, **240–4**, 245, 295
fossils 124, 125, 159, *161*, 173, 180, 187, *190*, 209, 240, *243*, 294, 295, 302, 307; debates over nature of 290–1, 293, 298; preservation of 58, 59, 174, 241; *see also* stratigraphic correlation
Fox, Sidney 236, 237
fractional crystallization 162, 164, 165, 174
fracture 120
Frasch process 142
frost shattering 47, 78, 130, *176*
fuller's earth 168
fumaroles 96
fusulinids 212, 213, 253

gabbro 11, 39, 123, *161, 162*, **165**, 189; tholeiitic 130
galena 30, 116, 119, 125, 127, **128**, *129*, 137, 142, 149
gallium 137, **144**, 145
gamma-ray spectrometry 111
gangue *111, 119*, 135
garnet 11, 34, 116, **128–9**, *167*, 175, *177*, 178
garnierite 143
gas *see* natural gas
gastropods 192, 194, 221, 223, *244*, 258, **260–1**
geanticlines 72
Geikie, A. 294, **302**, 306
gemstones **118–9**
geochemistry 33–5, *33*, *34–5*, 36, *111*, 112, 302
geodes 127
geophone 29, *30*
geophysics **28–33**, *29, 30, 31*, *32*, 105, 108, *111*, 152
geos 94
geosynclines 72, 74, 129, *170*, 189, 196, 204, 215–6, 303
geothermal energy 109
germanium **144–5**
Gesner, C. 291, **302**
geysers 96
Giant's Causeway 46, 47
gibbsite 141
Gigantopteris 211, 212
Gilbert, G. 91, 294, **302**
ginkgos 212, 215, 219, 252
glacial theory 297, 299
glaciation, glaciers **56–8**, *56, 57*, *86–8*, *86, 87*, 130, 171, 209, 211, 227, 296, 297; erosion by 48, **49–50**, *50*, 75, 89; *see also* ice, ice ages, ice sheets
glauconite *120*, **129**, 169, 225
glaucophane 122, **129**, 176
Globigerina 230; Globigerinacea 253; Globigerina ooze 97, 99, 225; globigerinids 223
Glossopteridales 251; *Glossopteridophyta* 212; *Glossopteris* 39, 212, 215
gneisses 68, 106, 126, 129, 135, 137, *161*, 177, 178, 181, 188, 189; gray 189
gneissose banding (foliation) 175
goethite (limonite) 20, 125, **129**, 143, 169
gold 30, 61, *111*, 112, *113*, 116, *119*, *120*, 122, 125, 134, **148**, 189, 289
Goldschmidt, V. **302**
Gondwanaland 39, 58, 192, 195, 200, 201, 204, **209**, 210, 211, 212, 213, 215, **217**, 221–2
goniatites 204, 205, 259
gorgonopsids 277
gossans 112, 129
gours 82
Grabau, A. 297, **302**
graben *see* rift valleys
graded beds 54, *91*, 170
grain support 171
Grand Canyon 49, 186, *211*
granite 47, 74, 121, 122, 124, 125, 127, 138, *161, 162*, **163**, 164, 165, 181, 208, 216
granodiorite 11, *163*, **164**
granulites 124, 130, 175, 188, 189, *190*
grapestone 172
graphite 30, 127, **129**, 140, 146, 155, 176
graptolites 192, *194–195*, 197, 198, 199–200, 201, **263–4**, 297, 304
gravimeter 29–30
gravity studies **29–30**, *31*, 108
gravity tectonics 53, 67
graywackes 54, 169, **170**, 188
Great Auk 279
greenalite *112*
greenockite 146
greensands 129, 169, 225
greenschists 125, 175
greenstone belts 188–9, *190*

greenstones *see* prasinites
Grenville belt 189
grossularite 128, 129
groundwater 30, 47, 72, *95*, 104, 105, 106, *107*, **108**, 109, 173
Gryphaea 219
Guettard, J-E. **302**
Gussow's Principle 153
Gutenberg, B. **302**
guyots 98, 101–2
gymnosperms 219, 222, **251**–2
gypsum 47, 54, 61, 84, *85*, *114*, 117, *118*, 120, 121, 122, **129**, 142, 173

Hadley Cell 36
hadrosaurs 276
hafnium 149
halite 53, 116, *117*, **129**–30, 138, 141, **157**–9, 173; *see also* salt
Hall, J. 294, **302**–3
Hall, Sir J. 294, 295, **302**
Hall-Héroult process *141*
halloysite 126, **130**
hardness, mineral 120
Haug, G. 297, **303**
Haüy, R-J. **113**–4, 116, 124, **303**
Hawaii 21, 71, 102, 166
heliodor 124
heliotrope 125
helium 9, 10, *33*, 35, 152, 180
hematite 17, *112*, 118, 129, **130**, 143, 169, 176, 201
Hercynian mountains 176, 196, 216
Hess, H. 42, 43, **303**
heulandite 139
hexagonal system *see* crystal systems
hiddenite 137
Himalayas 39, *44*, 73, 74, 79, 189, 232
Hipparion 231
Holmes, A. 40, **303**
Holocene 234
holosteans 218, 223, *240*, **268**
holosymmetric class 116, *117*
holothuroids 257
hominids 231, 232, 234, 244, **284**–7, *285*, 300, 305
Homo see hominids
Hooke, R. 241, 292, 293, **303**
hornblende 11, 117, 121, 122, **130**, 162, 163, 164, 175, 176, 178, 188, 189
Horne, J. 297, **303**, 306
hornfels 126, *161*, **177**
horn silver 146
horses 231, 281, **282**, 304, 305
horsetails 211, 212, 215, 219, 243, 250, **251**
horsts 80, *81*
hotspots 101–2
hot springs *95*, 96, 124, 126, 130
Humboldt, Baron F.H.A. von 294, **303**, *304*
humus *178*–9
Hutton, J. 186, 292, 293, 294, 295, 296, 298, 302, **303**–4, 305, 306
Huxley, T. 294, 296, **304**, 305, 306
hydrocarbons 140, 152, *169*; *see also* natural gas, petroleum
hydrogen 8, 9, 10, *33*, 35, 236; sulfide 35
hydrogen-burning 33
hydrogeology, hydrology 83, **107**–9, *107,108*
hydrological gradient 82
hydrologic cycle **107**–8, *107*, 109
hydrothermal deposits *112*, 142
hydroxyapatite 122
Hyenia 251
hyolithids 192
hypersthene 123, 127, **130**, 135
Hypsilophodon 275, 276
hyraces 280, 281
hystrichospheres 249

ice 9, 10, **49**–**50**, 56, **56–8**, **130**
ice ages (glaciations) **58**, 78, 297, 298; Precambrian 58, 191, 193, 243; Paleozoic 58; Tertiary 231–2; Pleistocene 40, **57**–8, 85, 89, 90, 232, 244
Iceland spar *120*, 125
ichthyosaurs 217, 223, 239, 245, *270*, **271**
Ichthyostega 269
idocrase (vesuvianite) 117, **130**
igneous intrusion 45, **68**–70, *68, 69, 70*, 72, 124, 138, 143, *163*, 166, 174, 177
igneous rocks 34, 35, 123, 125–6, 128, 130, 134, 143, **160**–7, *160, 161*, 170, 173, 174, 176, 177, 178, 180, 300, 302; extrusive (volcanic) 11, **160**; intrusive (plutonic) 17, *160*, **160**–1, 187; *see also* lavas
ignimbrites 95, 164
Iguanodon 223, 275–6
illite 47, 59, 126, **130**
ilmenite 17, **130**, 143, 146
imbricate stacking *55*
imbricate structure 60
induced polarization (IP) techniques 30–1
inert gases (noble gases) 9, 10
infiltration 107, 108
inoceramids 261
insectivores **280**
insects 262, **263**; Carboniferous 206, 209; Cretaceous 223, 253; development of flight 201
inselbergs 81, 85
insequents 75–6
intrusive rocks; *see* igneous intrusion, igneous rocks
invertebrates **253**–**64**
iodine 26, **147**
iolite 126
iridescence *121*, *124*
iridium 137, 148
Irish Deer 234, 282
iron 10, 17, 30, *33*, 34, 35, 47, *48*, 59, *112*, 119, *120*, 121, *123*, 125, 126, *130*, 132, 137, **143**, 161, 162, 164, 165, 166, 168, 169, *172*, 173, 179, 289
iron-nickel 10, 12, 17, 24, 27, 160
irons 10, 24, 25, 26, 27, 143
ironstones *112*, 125, *130, 137*, 143, **172**–3, *172*, 220, 221; banded ironstone formations 172–3, 188, 189; oolitic 172, 173; clay 172, 173
island arcs 43, 73, *100*, **100**–1, 164, 165, 188
isostasy 40, 58, 61, 297

jade 121, 130, 133
jadeite **130**–1, 135
jasper 125, **131**, 135, 173
jawless fishes *see* Agnatha
jointing 46–7, **61**, 104
Jupiter 9, 10, 26
Jurassic 39, 143, *155*, 186, *216*, **217**–**21**, *221*, 222, 243, 258, 266, 295; fauna 217–9, *218*, 221; flora 211, *218*, 219; geography 217, *218*; formation of coal in *156*, 221; stratigraphic correlation 218–9
juvenile water 107

kamacite 24, 27
kamenitzas *82*
kaolin *135*
kaolinite (china clay) 47, 59, 125, 126, 130, **131**, 168
karren 48, 81
Karroo 40, 79, 221, *245*, 277
karst **81**–3, *82*, 83, 84, *108*, 125
Kelvin, Lord 296, 297, 299–300, 301, 303, **304**
Kenyapithecus 284
kerogen 150, *169*
kettleholes 88
kidney iron ore *130*

Kilauea 71–2, 96
kimberlite 33, 124, 127, *165*, **166–7**, *167*
kinchpoints *77*, 78
komatiite 188
kopjes 81
Krakatoa 70, 71, 72, 96, *295*
kratons 188, 195
krypton 35
kunzite *see* spodumene
kyanite 118, 121, **131**, 137

labradorite 134, 165
labyrinthodonts 268, **269**
laccoliths 68, *70*
lagomorphs **283**
Lamarck, J. 237, 292, 300, **304**, 305
lamellibranchs 214
lampreys 265
lapis lazuli **131**, 137; *see also* lazurite
Lapworth, C. 190, 194, 198, 294, 297, **304**
laterites 48, *112*, 125, 144, **173**, 178
lateritic soils 179
laterization 48
Latimeria 268
lattices, crystal 114, 115
Laurasia 204, 206–9, 210, 213, 215, 216–7
lavas 16, 18, 42, 72, 80, 95, 127, 131, 160, 164, 166, 173, 187; aa 96; pahoehoe 96; pillow 96, *98*, 188; *see also* igneous rocks (extrusive)
lazurite 128, **131**
leaching 47, 84, *112*, 178; *see also* weathering (chemical)
lead 26, 30, 35, 61, *113*, 121, 128, 144, **149**, 159, 184, 289; *see also* tetraethyl lead
Leakey L. 284, **305**
Le Gros Clark, Sir W. **305**
Lehmann, J. 293, 294, **305**
lemuroids 284
Leonardo da Vinci 291, **305**
Lepidodendron 250
lepidolite *120*, 140
lepospondyls **269**, 270
leucite 128, **131**, 142, 162
levorotation 150
life, origin of 72, **235**–7, 298
lignite 155
limestones 59, **81**–3, *82, 83*, 104, 108, *124*, 125, 126, 127, 128, 142, 152, 160, *161*, 164, 168, *171*, **171**–2, *172*, 173, *174*, 176, 179, 192, 201, *203*, 206, *207*, 208, 209, *211*, 216, 218, *219*, 220, 247, 253, 258, 263
limonite *see* goethite
linnaeite 143
Linnaeus, C. 237
Lithophaga 219
lithium 33, 137, 138, **140**, 147
lizardite 137
lizards 270, **272**–3
Logan, Sir W. **305**
long-profile 76, **76**–7
lopoliths 68, *70*
lunettes 84
lungfish (dipnoans) 202, **266**–8
lycopods 201, 211, 212, 250; *see also* clubmosses
Lyell, Sir C. 185, *186*, 225, 227, 229, 231, 294, 295, 296, 298, 300, **305**, 306

magma 34, 68, 71, 74, 95, 97, 112, 113, 122, 124, 126, 160, 162, *165*, 166, 174, 180
magmatic deposits *112*
magnesite **131**–2, 141
magnesium 9, 11, *33*, 34, 124, 127, *131*, 138, **141**, 160, 161, 162, 164, 165, 166, 171, 172, 188
magnetic fields: lunar 18; terrestrial 18, **31**–2, 41–2, 65–6, 187, 224, 233 (reversals in polarity 41, *42*); Venusian 19; *see also* paleomagnetism

magnetite 30, 31, *112*, 116, 121, 126, 129, 130, **132**, 137, 143, 189, 236
magnetometer 32, 41–2
malachite 123, 127, **132**, 144
mammal-like reptiles 212, 214–5, 239–40, *245*, 270, **277**–8, 279
mammals *239*, 243, 244, 270, 276–7, 278, **279**–87; Permian 212; Triassic 215, 278; Jurassic 217–8; Cretaceous 223; Tertiary 229, 231, 232
mammals, primitive **279**
mammoths 143, *281*
manganese 121, 129, 136, **143**; nodules *112*, *113*, 143
manganite 143
Mantell, G. **305**
mantle 10, **11**–2, 13, 27, 29, **34**, 35, 36, 43, 74, 127, 129, 133, 137, 160, 162, 166, *167*, 174; convection currents in 40, 45, 101, 188, 190
marble 125, *174*, 201, 236
marcasite **132**, 134
maria 15, 17, 199
Marsh, O. 296, **305**
marsh gas 153
marsupials *239*, 247, **279**–80
Massif Central 80, 155, 176, 220, 225
mass numbers 35, 180
mass spectrometry 35, 36, 181, 184
mastodons 231, 232, 237, 295
matrix support 171
Matthews, D. 42, 303
Mauna Loa 95, 96
meanders 76–7, 81
median valleys 98, 99
megachiropterans 280
Megaloceros 282
Megalosaurus 275, 299
Megatherium, 282, *300*
Megistotherium 280
Mendeléev, D. 139
Mercalli intensity scale *see* earthquakes
Mercury 9, 14, 15, **18**–9, *19*, 27
mercury 126, **148**
Merychyppus 231
mesas *51*, *81*
Mesosaurus 39, 212
mesosiderites 24
Mesozoic **213**–**25**, 243–4, 248, 250, 270; *see also* Triassic, Jurassic, Cretaceous
metabasites 175, 176
metamorphic rocks 34, 58, 59, 61, 67, 123, 160, *161*, 170, **174**–8, *174*, *175*, *176*, *177*, 180, 302
metamorphism 74, **174**, 189, 216, 241; prograde 174, *177*; retrograde 174
meteoric water 107, 112
meteorites 10, 12, 15, 18, 21, **24**–7, *25*, 26, 27, *33*, 124, 127, 130, 132, 133, 134, 135, 143, 160, 184, 188, 236
meteoroids 21, *25*
meteors 24, *25*
methane 9, 35, 59, 152, 153, 236
micas 10, 17, 34, 67, *120*, 124, 126, 127, 129, 130, **132**, 142, 162, 167, 169, 174, 175, *176*, *177*
micaschists 124, 133
microchiropterans 280
microcline 128, **132**, 134, 136, 162, 163
microcrinoids 257
micropaleontology 241–2
mid-Atlantic ridge 41, 42
mid-ocean canyons 100
mid-ocean ridges 32, 34, **41**–2, *42, 45*, 64, *75*, 78, **97–9**, *98*, 101, 102, 188, 191, 201

migmatites 177
millerite 143
Millstone Grit 205
mimetite *143*
mineral facies **175**
minerals 111, 113, 116, **119**–**39**, 161, 180
mines, mining geology 104, **109**–**12**, *109, 110, 111*
Miocene 225, 227, **229**–**31**, *230*, 232, 284
miogeosynclines 72, 196
miogypsinids 229, 230
Mississippian *see* Carboniferous
mitochondria 238–9
mixosaurs 271
Moho, Mohorovičić Discontinuity 11
Mohs' hardness scale **120**
mollusks 122, 192, 198, 204, 206, 214, 218, 229, 231, *247*, **258**–**61**, 271
molybdenite 146
molybdenum 143, **146**
monadnocks 81
monazite **132**, 142, 146, 149
monoclinic system *see* crystal systems
monocotyledons 252
monoplacophora (cap shells) 192
monotremes **279**
montmorillonite 47, 59, 126, **132**, 224
Moon **14**–**8**, 19, 27, 29, 36, 46, 135, 184, 187, 188, 189, *301*; asthenosphere *18*; structure *18*; moonquakes 18, 66
moonstone 134
moraines 56; lateral *57*, 86; medial 86; terminal *56*, 88
morganite 124
Morrison Formation 217, 221
mosasaurs 223, 273
moss agate, *121*, 125
Moulton, F. 300
mountain-building *see* orogeny
mountain chains, mountain ranges 39, 40, *45*, 46, 66, 72, **73**–**5**, *73, 74, 75*, 81, 164, 166, 174, 176, 185, 189, 190, 215, 302, 303
Moythomasia 202, *240*
Muchocephalus 245
mudstones 54, **168**
multituberculates 218, 226, 229, 279
Murchison, Sir R. 190, 194, 197–8, 199, 201, 209, 294, 295, 296, 304, **305**, 306, 307
Murray, J. 294, 297, **305**–6
muscovite *120*, 124, 132, **133**, 174, 176, *177*
Muscovy glass 133
mushroom fold 67
mylonites **177**
myomorphs 283
mysticetes 283

nappes 66
natrolite 139
natural bridges *51*
natural gas 28, 129, 142, *150*, 152, **152**–3, 169, 221; biogenic 152–3
natural selection 237, 301
nautiloids 194, 202, **244**, 258, **259**–**60**; *Nautilus* 259, 260
Neanderthal Man 286
nebular hypothesis 9–*10*, 300
neodymium 142
Neogene 225
neon 33, 35
nepheline 118, 126, 128, **133**, 162, 164
nephrite 121, 122, **133**, 138
Neptune 9, 10
Neptunism 185
niccolite 143
nickel 17, 34, *113*, **143**–4, 189; *see also* iron-nickel
niobium 164
nitrogen 9, 35, 152, 181
noble gases *see* inert gases
nodules *112*, *113*, 134

norite 17
nothosaurs 270, **271**, 272
notoungulates 281
nuées ardentes 164
nudibranchs 260
Nummulites 226, 227, 229, 230
nummulitids 253

obsequents 75
obsidian 119, 127, **167**
ocean floor 45, 71, 79, 93,
97–102, 97, 98, 100, 101,
112, 162, 165, 167, 187, 188,
306
ocean trenches **42–3**, 73, 74,
94, 100, **100–1**
Octocorallia 254
odontocetes 283
oil 28, 29, 124, 168, 169, 221;
crude 142; *see also* petroleum
Old Red Sandstone 200–1, 203,
204, 305, 307
Oligocene 121, 225, 226, **229**,
263, 284
oligoclase 134
Oligopithecus 284
olivenites *133*
olivine 11, 12, 13, 17, 18, 24,
27, 34, 117, 119, **133**, 137,
141, 143, *161*, 162, *165*, 166,
167, 175
omphacite 167
omphalosaurs 271
oncolites 219
onyx 121, 125
oolites 172, 173; ooliths 53,
112, 125, *171*, *172*
oozes 97, 99, 171, 172, 225,
249
opal 48, 119, 127, **133–4**, 173
ophiacodonts 277
ophiolites 166
ophiuroids 257
Orbigny, A. d' **306**
orbitolines 253
Ordovician 190, 192, **194–7**,
195, *197*, 198, *199*, 262, 294,
304; climate 196–7; fauna
194–5, *196–7*, 200; flora *194*;
geography *194*, 196;
stratigraphic correlation 195;
volcanic activity 72
ore deposits 28, 29, 30, 61, 69,
109, **110**, **111**, **112–3**, 122,
123, 189
ornithischians **275–6**
ornithomimids 275, 276
Ornithosuchus 273
orogeny (mountain-building)
11, 18, 27, 74, 104, 160, 164,
170, 185, 187, 192, 215
Orohippus 305
orpiment **136**, 145
orthoclase 117, 120, 124, 128,
132, *133*, **134**, 136, 162, 163,
170
orthorhombic system *see*
crystal systems
Osborn, H. 296, **306**
ostracoderms 201, 202, 265
ostracods 198, 200, 219, 241–2,
262
Ostrea 219
Owen, Sir R. **306**, *307*
oxbow lakes 76, 81, *90*
oxygen 8, 9, *33*, 35, 47, 59,
122, 126, 135, **140–1**, 142,
173, 181, 191, 247; isotopes
of in seawater 232, 245–7
ozokerite 150

pachycephalosaurs 276
Pachydiscus 223
pahoehoe 96
Paleocene **225–6**, *226*
paleoecology 242, **245–8**, 302
Paleogene 225
paleogeography 187
paleomagnetism **31–2**, 40,
41–2, 143, 187, 196, 200, 204
paleontology 300; *see also*
fossil record, fossils,
micropaleontology,
paleoecology
Paleozoic **190–213**, 243, 244,

247–8, 269; *see also*
Cambrian, Ordovician,
Silurian, Devonian,
Carboniferous, Permian
paliguanids 272
palladium 148
pallasites 24, 26, 134
Pangaea 39, 40, 192, 209, 210,
215, 217, 275, 309
panning *119*
pantotheres 279
paralic coals 154–5
paramomyids 283–4
parapithecids 284
parapsids 270, 271
Parasemionotus 240
pareiasaurs 270
patronite 143
Peach, B. 297, 303, **306**
peacock ore *see* bornite
pearlspar *see* dolomite
peat 153–7, *155*, *179*, 206, 207
pebble-phosphate 173
Pecora 282
pedimentation 49
pediments 85
pediplain *see* peneplain
pedalfers 179
pedocals 179
pegmatites 68, 69, 122, 124,
127, 147
pelagic sediments 98, 99,
100
Pelée, Mont 96
pelites 122, 131, 137, 174, 175,
176
Pelomyxa 239
pelycosaurs 212, **277**
Penck, A. 297, **306**
peneplain (pediplain) 52, 80,
81, 85, 301
Pennsylvanian *see*
Carboniferous
pentlandite 143
percolation 108
periclase 12, 141
peridot 133, **134**
peridotite 11, 17, 39, 123, 137,
161, 162, **166**, 167, 175
Periodic Table **139**, 140
perissodactyls 229, **281–2**
permafrost 37, 88, 89
permeability 103, **104**, **108**,
169
Permian 156, 190, 192,
209–13, 214, 239, 243, 270;
climate 210–1; fauna 210,
212–3; flora 210, 211–2;
geography 210; stratigraphic
correlation 209–10
perthite 132, 134, 162, 163
petalite 140
petroleum 59, 66, 100, 111,
124, 129, **149–52**, *152*, 156,
157, 159, 168, 248, 294; traps
151, 159
phacoliths 68, *70*
phenocrysts 164
phlogopite *120*, 124, 132, 167
Phobos *23–4*
phonolite **164**
phosphates *139*, 142
phosphorites 173
phosphorus 33, 122, 126, **142**,
144, 164, 173
photosynthesis 35, 140, 191
phreatic caves 82
phreatic surface 108
phyllites **176**, *189*
phyllocarids 194
phytosaurs 273
picrodontids 283
piezometric surface 108
pigeonite 130
pigeon's blood 136
Piltdown Man 305
pinnipeds 280
pipes 68, 127, **166–7**
pisolites 172
pitchblende *see* uraninite
pitchstone 119
Pithecanthropus erectus 286
placentals 226, 279, **281–7**
placers 112
placoderms 201–2, **266**

placodonts 270, **271–2**
plagioclase 17, 24, 34, *118*,
121, 128, **134**, 162, 163, 164,
165, 167, 175, 178
planetesimal hypothesis 300
plankton 172, 173, 232
plants 201, **248–53**
plateaux 50, *51*, 72, 73, **79–81**,
80, 81, 85, 95, 102, 164, 166,
217
plate margins 70–1, 100–1;
constructive 71, 97–9, *101*;
destructive (subduction
zones) 35, **42–3**, *43*, 44, 45,
46, 69, 71, 72, 129, 131, 165,
176, 178, 192, 200, 201, 216,
221
plate tectonics 10, 11, 27, 28,
35, 39, 40, **41–6**, *43*, 44, 58,
61, 67, 69, 73, 93–4, 97,
100–1, 143, 168, 176, 187,
190, 196, 201, 210, 239, 242,
297, 303, 309; earthquakes
and 43, 44, *62*, **63–4**, 71, 72,
78; paleomagnetic evidence
for 31, *32*, 40, **41–2**;
paleontological evidence for
39, 40; rift valleys and 78–9;
vulcanicity and 41, 43, *45*,
70–1, *71*, 72; *see also*
continental drift, plate
margins, sea-floor spreading
platinum 17, **148**
playas 85
Playfair, J. 294, 295, **306**
Pleistocene 225, 231, **234**
pleochroism 137
plesiadapids 283–4
plesiosaurs 217, 223, 270, **271**
Pliocene 225, **231–2**, *231*, 250
plugs, volcanic 68, *70*, 95
Pluto 9
plutonic rocks *see* igneous rocks
Plutonism 185
plutons 68, 69
podzol *178*, 179
polar wandering *32*
poljes 82, 83
pollen 187; analysis 184, 241–2
pollucite 147
polyhalite **134**
polymorphism 119, 122, 126,
131, 137
pongids 284
porosity **108**, 169
porphyroblasts 137, 175, *177*
porphyry deposits 111, 112,
144, 146
potassium 10, 11, 12, 17, 26,
34, 35, 45, 112, 128, 129,
142, 147, 159, *160*, 180, 181,
187, 190, 304
potassium-argon dating 129,
181, 233
potholes 49, 81
Powell, J. W. 294, 302, **306**
prase 125
prasinites 175
Precambrian **187–90**, *188*, *189*,
190, 195, 196, 235, 254, 258,
264; fossils *190*; geography
188; ice ages in 58, 191, 193,
232; ironstones *112*, 125,
130, 143, 172–3, 188, 189;
rocks 18, 180, 184, *189*, *190*,
192
pressure waves *see* seismic
waves
Primary rocks 293
primates 226, **283–4**
primitive mammals **279**
procolophonids *238*, 270
Proconsul 284, 305
Propliopithecus 284
prosauropods 275
prosimians 226, 229, 283
proterochampsids 273
Proterosuchus 273
prototheres 239
Proterozoic 188, **189–90**
proto-Atlantic 192, **196**, 197,
198, 200
Protosuchus 277
proustite 146
psammites 174
Pseudobeaconia 240

pseudomorphs *215*
pseudotachylites 177
psilomelane 143
psilopsids 201, **250**
Pteranodon 223, 273
Pterodactylus 273
pterophytes 250, **251**, 252; *see
also* ferns
pterosaurs 217, 223, 270, **273**
Purgatorius 283
P-waves *see* seismic waves
Pygaster 219
pyrargyrite 146, 147
Pyrenees 227
pyrite (fool's gold) 30, 59, *112*,
116, 119, *120*, 122, 125, 129,
131, 132, **134**, 142, 143, 169
pyroclastic rocks 72, 160
pyrolusite 143
pyrope 128, 129
pyroxenes 11, 12, 17, 18, 24,
27, 34, 122, 123, 126, 127,
130, **135**, 162, 163, 164, *165*,
166, 167, 174, 175, 177, 178
pyroxenoids 135
pyrrhotite **135**, 143

quartz 11, 47, 59, 61, 69, *118*,
120, 121, 122, *123*, 124, 125,
126, 127, 129, 132, *135*,
135–6, 142, 160, 162, *163*,
164, *165*, 168, 169, 170, 174,
175, 176, *177*, 178, 189
quartz sandstones *see*
sandstones (quartz)
quartzites 59, 106, 135, 174,
175
Quaternary 39, 187, 225,
232–4; climate 233, 233;
fauna 233–4; flora 232, 233;
geography 232; *see also* ice
ages (Pleistocene)
Quetzalcoatlus 273

radiolaria 45, 99, 173, 201,
219, 223, 224, 225, 253, **254**;
radiolarites 254
radiometric dating
(radioisotope dating) **180–1**,
184, 187, 190
radium 138, **149**
radon 111
rainwater 47, 48–9, 59, 81, *82*,
112, 125, 173; *see also*
meteoric water
raised beaches *234*
Ramapithecus 284
rank (of coals) **155–6**
rare earth elements 17, **142–3**,
149, 159, 164
realgar **136**, 145
Recent rocks 293
reefs 94, 102, 170, 172, 249,
255, *293*, 301
reptiles 204, 239, 240, **270–8**,
270, *274*; Carboniferous 206,
209, 269; Permian 212, 243;
Triassic 214, *272*; Jurassic
217, 221; Cretaceous 222,
223, 225
rhamphorhynchoids 273
rhinoceroses 281–2
rhipidistians 266, 269
rhodium 148
rhodochrosite **136**, 143
rhodonite 135, **136**, 143
rhourds 84
rhynchocephalians 272
Rhyniophytales 250
rhyodacites **164**, 165
rhyolite 95, 96, 127, *161*, **164**,
165, 216
rias 94
Richter, C. 302
Richter magnitude *see*
earthquakes
riebeckite 122, 123, 127, **136**
rift valleys **78–9**, 80, *81*, 100;
East African (Great) 41, 61,
78, 79, 164; Rhinegraben
61, 78
Ring of Fire 74, 165
ripple drift 54–*55*; ripple marks
167, 172, 174; ripples 100,
247
river valleys **75–8**, *76*, 77, 94,

301
roches moutonnées 50, 86
rock crystal *135–6*
rock roses *see* desert roses
rocks 119, **160–79**, *293*; *see also*
igneous rocks, metamorphic
rocks, sedimentary rocks
rocks, oldest known
terrestrial 10, **184**, *781*, 236
rock salt *see* halite
Rocky Mountains 79, *193*,
204, 207, 211, 220, 224
rodents 229, 232, **283**
Rosenbusch, K. 299, **306**, 307,
309
rubidium 17, 26, 35, 164, 180,
181, 187
rubidium-strontium dating
181, 184, 190
ruby 119, 126, 136, 137
rudaceous rocks 168, 169,
170–1, 174
rudistids 261
rusting 47, *48*
ruthenium 148
rutile 117, 136, 143, 146

saber-toothed tigers 234, 239,
280
sabkhas 157, 173
Sahara Desert *84*
Saint-Hilaire, G. 300
salt 119, 141, *151*, **156–7**,
157–9, *215*; *see also* halite
saltation 50, 167
salt domes 129, *151*, 159
salt pans *157*
salt weathering 47, 84
San Andreas Fault 61, **63–4**,
65, 94
sands 105, *106*, 119, 135; as
carriers for petroleum 151
sandstones 59, 67, 135, *160*,
167, 169, *170*, 200, 206, 207,
208, 216, 220, *223*, 225;
quartz *161*, **169–70**, 174; *see
also* arenaceous rocks
sanidine 128, 132, 134, **136**,
162
sapphire 126, **136**
saprolite 48
sard 125
sardonyx 121
satin-spar 129
Saturn 9, 27, 289
saurischians **275**
sauropods 223, 275
Scandinavian Shield 220
scandium 142
scapolite 128, **136**
scheelite 148
schistosity 175, **176–7**
schists 74, 121, 124, 126, 127,
128–9, 130, 131, 137, *175*,
176–7, *176*, 177, 236; blue
74, 175; green 125, 175; *see
also* micaschists, peliticschists
schriebersite 27
scintillometers 111
sciuromorphs 283
Scleractinia *see* corals
scree 47, 86, 170
Scrope, G. 294, 296, **306–7**
sea-floor spreading 11, **41–2**,
45, 63, 64, *71*, 74, 78, **97–9**,
100, 102, 217, 303
seamounts 98, 101–2
seawater 192
Secondary rocks 293
Sedgwick, A. 190, 194, **197–8**,
201, 295, 296, 300, 304, 305,
306, **307**
sedimentary deposits 112
sedimentary rocks 34, 35, 52,
58–9, 160, *161*, 165, **167–73**,
168, 169, *170*, 171, *172*, *173*,
174, 180, 298, 302
sedimentation 27; *see also*
diagenesis
seed ferns 201, 212
seiches 65
seifs 84
seismic waves 10–2, 41, *62*,
62–3, 64, 65; body waves
28–9, 62 (pressure waves (P-

waves) 11, 29, *62*; shear waves (S-waves) *62*); surface waves 28–9, **62–3** (Love waves (L-waves) *62*; Rayleigh waves (R-waves) *62*–3)
seismograms 29, 62
seismology, seismic studies 10, 18, **28–9**, *29*, 30, 43, 105, 109, 111
seismometers 29, *62*
selenite 129
serpentine 10, *123*, **137**, 138, 143, 167, 175
serpentinites 137, 166, 175, 176
Seymouria 269; seymouria-morphs 269
shales 54, 67, 105, 152, **168–9**, *169*, 170, 173, *175*, *176*, 195, 199, 206, 207, *208*, 216, 218, 220, 263, 264
sharks 201–2, 266; cartilagenous 223; spiny *see* acanthodians
siderite (chalybite) 59, *112*, 122, 125, 127, **137**, 140, 143, 173
Sierra Nevada 176, *233*, 298
silcretes 85
silica 34, 125, 126–7, 128, 140, *160*, *161*, 162, 164, 165, 166, 168, 179, 254
silicate minerals, silicates 10, 11, 12, 13, 34, 119, 122, 128, 141, 142, 160, 164
silicon 9, 11, 12, 18, *33*, 34, 122, 135, **142**, 145, 160
sillimanite 122, 131, **137**, 175
sills 68, 72
silt *169*
siltstones 264
Silurian 190, 192, 194, **197–201**, *199*, 202, 265, 294, 295, 305, 306; climate 200–1; fauna *198*–9; flora *198*; geography *198*, 200–1; stratigraphic correlation 199–200
silver 61, 113, *116*, 128, **146**, *147*, 148, 189
sinkholes 81, *82*
slates 125, *148*, *175*, **176**, 177
smaltite 143, 145
Smith, W. **185–6**, 241, 294–5, **307**
smithsonite 144
snakes 270, **272**
Snider-Pellegrini, A. 39
sodalite 128, *139*, 162, 164
sodium 17, 33, 34, 128, **141**, 159, *160*, 166
soil erosion 86
soil mechanics 104
soil profile 179
soils 48, 59, 83, 89, 104, 160, **178–9**, *178*, *179*, 203, 207; ahumic 89
Solar System 15, 36; composition of 18; origin and formation of 9–*10*, 26, 33, 34, 184; scale of *12*–3
Solfatara 96
solid solution series 119
Sorby, H. 297, **307**, 309
space groups 115–6
spectrometry, spectros-copy 25; atomic absorption 34; atomic emission 33–4; gamma-ray 111; X-ray fluorescence 34; *see also* X-ray spectrometry
sperylite 148
spessartine 128, 129
sphalerite (zinc blende) *112*, 116, 128, **137**, 142, 144
sphenacodonts 277
sphene (titanite) **137**, 143, 178
sphenoid 116–7
sphenopsids *see* horsetails
spinels 11, 13, 116, *117*, **137**, 141
spodumene 135, **137**, 140
sponges 173, 192, 202, 219, 223, 225, **254**, 255
spores 187, 200, 205, 219
stacks 94
stalactites 82, *83*; stalagmites 82, *83*
stannite 147
staurolite **137**
Stegosaurus 217, 276
Steno, N. 185, 241, 292, 293, **307–8**, 309
stereographic projection *116*
stibnite 117, 136, **137–8**, *138*, 147
stilbite 139
stishovite 12
stocks 68–9, *70*
stony-irons 24, 25, 26, 27
stony meteorites (stones) 10, 12, 24, *26*, 27
Strabo 253, 290
Strachey, J. 185
strain meters 29
strandflat 89
stratigraphic correlation **180**, **186**
stratigraphy **184–7**, 307
stratovolcanoes 95
strength (of rocks) 103
striations, glacial 50, *87*, 171
Stricklandia 199–200
stromatolites *190*, 203, 219, 235–6, 249
stromatoporoids 201, 202, 203, 204, 248, 254, **255**
strontianite **138**, 146
strontium 17, 26, 35, *112*, 125, **146**
subduction zones *see* plate margins (destructive)
submarine canyons 93, *98*, 99, 100, 300
submarine mountains 41; *see also* crestal mountains, guyots, seamounts
subsequents 75
subungulates 229, **280–1**
Suess, E. **308–9**
Suina 282
sulfur 12, 17, 30, 33, 34, *117*, 119, 121, 122, 134, **142**, *150*, 152, 159; dioxide 35, 72, 111
sulfur bacteria 134
Sun 9, 14, 15, 289; composition of 10, 25, 33; radiation from 36, 58
superphosphate 142
Superposition, Law of 185–6
surface waves *see* seismic waves
Surtsey 34, 71, 95, 101
syenites 124, 126, 161, 162, **164**
syenodiorites 162
sylvanite 148

sylvite **138**, 142
symmetry (in crystals) **114–6**
synapsids 270, 277
synecology 245

taenite 24
taiga 88
talc 120, **138**
Tapinocephalus 212
tapirs 281
tarsioids 284
tectonic dating 187
tektites *25*, **27–8**, *28*
Telanthropus 286
teleosts 218, 223, *240*, **268–9**
tellurides 148
temnospondyls 269, 270
tennantite 145
tephra 72, 95, 96
Tertiary 39, *163*, *169*, **225–32**, 250, 295, 305; climate 226, 227, 231–2; fauna *226*, 227–8, 229, 230–1; flora *226*, 230, 231; geography 225, *226*, 227, *229*, 230, *231*; formation of coal 156; volcanic activity 72; *see also* Paleocene, Eocene, Oligocene, Miocene, Pliocene
Tethys 212, 215, 216, 217, 221, 222, 225, 226, 227, *228*, 271
Tetraethyl lead 141, 149
tetragonal system *see* crystal systems
tetrahedrite 147
texture: of rocks 167, 174, 176–7; of soils 178
thallium 137
thecodonts 217, *245*, 270, **273**, 275, 277
therapsids **277–8**
therocephalians 277
Thoatherium 239
tholeiitic rock 130
thorium 10, 11, 17, 35, 45, **149**, 164, 180, 181, 190, 304
tidal waves *see* tsunamis
tiger-eye *123*, 125, 136
tile ore 127
till 40, 57, 58, 87, 88, 90, 171
tillites 87, **171**, 200
tilloids 87
tin 122, 125, 142, **147**, *148*
titanite *see* sphene
titanium 17, 130, 136, **143**
titanotheres 229, 282
tombolos 94
topaz 117, 120, 125, **138**, 147
torbernite 149
tors 48, 81
tortoises 270–1
tourmaline 47, 118, 125, 127, **138**, 140, 147, *163*
trace fossils 192, 220, 245, 258, 262, **264**
trachytes 131, **164**
tranquillityite *17*
transform faults 64, *98*, 99
Transitional rocks 293
transpiration *107*
tree-ring dating *see* dendrochronology
trematosaurs 269
tremolite 121, 122, 123, 133, **138**
Triadobatrachus 270

Triassic 209, **213–7**, *216*, 243; climate 213, *215*, 217; fauna 213, *214*–5; flora 211, 213–*214*, 215; geography 213, *214*, 215–7
Triceratops 223, 276
triclinic system *see* crystal systems
triconodonts 218, 279
trigonal system *see* crystal systems
trilobites 190–1, 192, 193, 194, 197, 198, 199, 200, 203, 261–**262**
Trimerophytales 250
tritylodonts 278
troilite 10, 24, 27, 135
trona 141
tsunamis (tidal waves) 65
tuffs 123, 160
tundra 57–8, **88–9**, *88*, 89
tungsten 122, 139, **148**
turbidites 54, 56, 98, *100*
turbidity currents 54, 93, 99, 100, 101, 170, 195, 300
turquoise **138**
turtles 217, 221, 223, **270–1**
Tuttle, O. 298–9
twinning *114*, *124*, *128*, 134
tylopods 282
Tyrannosaurus 223, 275

ulexite 140
Ulmannia 212
ultramarine 131
unconformities *151*, *186*, 190, 297
Uniformitarianism 185, 186, 187, 295, 296, 297, 300, 301, 304, 305, 307
unit cell 113
Uralian geosyncline 210, 212, 213; Uralian Sea 212, 226, 229
Ural Mountains 204, 209
uraninite (pitchblende) **138**, 149
uranium 10, 11, 17, 26, 35, 45, 111, 121, 138, 142, **149**, 152, 180, 181, 190, 304
uranium-lead dating 181, 184
Uranus 9, 10
U-shaped valleys *50*, *57*, 75, 77, *86*, 87
Ussher, J. **309**
uvarovite 128, 129

vadose caves 81
vadose water *107*–*108*
valence (valency) 139
vanadinite *143*
vanadium **143**
varves 169, 301
vascular plants 201, **250–1**
vauclusian springs *82*
veins 61, 69, 122, 125, 128, 135
Vening Meinesz, F. 43, 303
Venus 9, **19–20**, *20*; greenhouse effect 35
vermiculite 126, **138–9**
Verneuil process 136
vertebrates 212, 239, **264–87**
vesuvianite *see* idocrase
Vesuvius 96, 130, 131
Vine, F. 42, 303
viperids 272
vitrite 156
volcanic bombs *72*

volcanic rocks *see* igneous rocks
Volcanic rocks 293
volcanoes *see* vulcanicity
Voltzia 215
vulcanicity (volcanism) 35, 41, **70–2**, *71*, *72*, 73, 75, **95–6**, *95*, 96, 101–2, 109, 164, 166, 187, *230*, 235, 236–7, 292, 298, 306–7; lunar 15, 18; Martian 21–*23*; *see also* plate tectonics
Vulcano 96

wadis 85
Walchia 211
Walther, J. **309**
water 10, 13, 30, 35, 59, 82, 106, **107–9**, *107*, 112, 122, 139, 160, 166, 168, 169, 173, 174; connate 107; juvenile 107; meteoric 107, 112 (*see also* rainwater); in chemical weathering 47, *48*; in erosion 48–*49*, 51–*52*, 106, 177; in physical weathering *47*; *see also* groundwater, ice, pore waters, rivers
waterfalls 77
water parting 75
watershed 75
water table 104, 108, 112, 170
wavellite *139*
weathering **46–8**, 59, 78, 80, 83–4, 104, 111, 159, *160*, 169, 170, 171; chemical 47–48, 52, 53, 83, 124, 168, 171, 173, 174; physical *46*–47, 130, 174, *176*; Mars 20; meteorites 27
Wegener, A. **39–40**, 41, 297, 301, **309**
Werner, A. 293, 294, 299, 305, **309**
whalebacks 67
whales 229, 271, **283**
Widmanstätten pattern 24, *26*
witherite **139**, 147
wolframite 125, **139**, 148
wollastonite 135, **139**
Woodward, J. 292, **309**
wulfenite 146

xenoliths 11, 13, *165*, 167
xenotime 142
X-ray crystallography 116
X-ray diffraction *see* X-ray spectrometry
X-ray diffraction patterns 116
X-ray fluorescence spectro-scopy 34
X-ray spectrometry 18, *34*–5

yardangs 50; Martian 23
year, length of 242
yttrium 142

zeolites 121, **139**
zinc 17, 61, 113, 121, 128, 137, **144**, 146, 159
zinc blende *see* sphalerite
Zinjanthropus 286
zinnwaldite 140
zircon 47, 117, **139**, 146
zirconium 17, 139, **146**, 149
Zirkel, F. 299, 306, 307, **309**
zoantharia 254
zooxanthellae 250, 254
Zosterophyllales 250

Acknowledgments

Unless otherwise stated, all the illustrations on a given page are credited to the same source.

Allard Design Group Ltd., St. Albans 12, 182, 183
Mike Andrews, Bristol 109, 111, 145 (bottom)
Associated Press Ltd., London 20
F. B. Atkins, Oxford 26, 34, 114 (top), 120 (bottom left), 121 (top left, and bottom), 127 (bottom), 129, 135 (bottom left)
H. Bäcker, Hannover 113 (top right)
J. D. Bell, Oxford 71 (bottom right), 95 (left), 163 (bottom left), 165 (top right)
Bodleian Library, Oxford 302
G. S. Boulton, Norwich 87 (lower center right), 89
Paul Brierley, Harlow 113 (top left), 117 (top, and bottom left), 118 (top right), 120 (top, center left, center right, and bottom right), 121 (top right), 122, 123 (bottom left, bottom center, and bottom right), 124 (top), 125, 128 (right), 131 (top), 132 (top), 133 (top), 135 (top left), 136, 137, 138, 139 (top), 141 (bottom), 148 (left), 149 (top), 163 (bottom right), 165 (top left), 170 (top left), 177 (center)
British Museum (Natural History), London 308
British Museum, London 184
G. M. Brown, Durham 17; by courtesy of the G. P. Slide Co., Houston 16 (bottom)
A. J. Charig, London 243
J. D. Collinson, Keele 50 (bottom), 55 (bottom right), 59 (bottom), 82 (top), 176 (top), 189, 197, 207, 208, 219; by courtesy of the Greenland Geological Survey 57, 91 (top), 234
K. G. Cox, Oxford 96, 101, 230 (top)
Diagram Visual Information Ltd., London 69, 70 (bottom), 78, 114 (bottom), 240
Elsevier Archives, Amsterdam 38, 92, 161, 178, 179, 186 (top), 220, 238, 242, 278, 285
A. J. Erlank, Capetown 167
Robert Estall, London 144 (bottom)
Mary Evans Picture Library, London 291 (bottom), 292, 300, 307
Fotolink, London 105
S. W. Fox, Miami 236
D. G. Fraser, Oxford 36
J. Fuller, Cambridge 249, 250, 251, 252, 253 (bottom), 254, 255, 257, 258, 259 (bottom), 260, 261 (top), 262, 263, 264, 265 (top), 266, 268, 269, 270 (bottom), 271, 272 (bottom), 273, 275, 276, 277, 279, 280, 281 (bottom), 282, 283, 284, 286
Roger Gorringe, London 10, 25 (bottom), 42, 45 (top), 51 (bottom), 55 (left), 60 (top, and bottom), 66, 67 (bottom), 76 (bottom), 81 (left), 82 (bottom), 91 (center, and bottom), 94, 95 (top right), 106, 107 (bottom), 110 (top), 151, 156, 239 (bottom), 301 (bottom)
A. Hallam, Oxford 203
Robert Harding Associates, London 46, 47, 48 (top), 49, 51 (top), 52, 54, 72, 73, 74, 75, 77 (bottom), 84 (bottom), 85, 86, 87 (left, top right, and center right), 90, 93, 95 (bottom right), 103, 104, 110 (bottom), 119, 123 (top), 126, 131 (bottom), 139 (bottom), 141 (top left), 143, 144 (top), 145 (top right) 146, 147, 150, 152, 153, 157, 158, 168, 169 (bottom), 175, 177 (bottom), 193
Bryon Harvey, London 18, 37, 45 (bottom), 98 (top), 100, 102
J. R. Heirtzler, Woods Hole 98 (bottom)
Alan Hutchison Library, London 141 (top right), 145 (top left), 148 (right)
Imitor, Bromley 272 (top)
Institute of Geological Sciences, London 28 (top), 124 (bottom left, and bottom right), 128 (left), 130, 132 (bottom), 133 (bottom), 134, 172 (bottom), 173, 177 (top), 259 (top), 265 (bottom), back jacket
Institute of Oceanographic Sciences, Godalming 97
H. L. James, Port Townsend; by courtesy of *Economic Geology* 112 (right)
W. J. Kennedy, Oxford 55 (center right), 70 (top), 81 (right), 169 (top), 171 (top), 244 (top and center), 247, 253 (top), 256, 261 (bottom)
P. D. Lane, Keele 163 (top), 165 (bottom), 195, 199
Andrew Lawson, London 48 (bottom)
Lovell Johns, Oxford 22, 27, 28 (bottom), 31 (top), 32, 33, 39, 43 (bottom), 44 (top), 62 (bottom right), 71 (bottom left), 84 (top), 140, 154 (bottom), 188, 191 (bottom), 194 (bottom), 198 (bottom), 202 (bottom), 205 (bottom), 210 (bottom), 214 (bottom), 218 (bottom), 222 (bottom), 226 (bottom), 227, 229, 230 (bottom), 231, 232, 239 (top)
P. J. McCabe 211, 216, 221, 228
S. Moorbath, Oxford 181
P. Morris, Esher 267, 270 (top), 274
Tony Morris, Towcester 191 (top), 194 (top), 198 (top), 202 (top), 205 (top), 210 (top), 214 (top), 218 (top), 222 (top), 226 (top)
D. G. Murchison, Newcastle upon Tyne 154 (top), 155
T. A. Mutch, Rhode Island 23
NASA, Washington 21
Natural Science Photos, Watford 53, 67 (top), 79
O. W. Nicolls, London, by courtesy of the Selection Exploration Trust, Canada 31 (bottom)
Maurice Nimmo, Haverfordwest 61, 87 (bottom right), 135 (right), 162, 171 (bottom), 176 (bottom), 224
Novosti Picture Agency, London 281 (top)
Oxford Illustrators, Oxford 30, 58, 62 (top, and bottom left), 113 (bottom), 115, 116, 117 (top, and bottom right), 118 (top left, bottom left, bottom center, and bottom right), 160, 245, 310, 311, 312, 313, 314
Picturepoint Ltd., London 63, 287
N. J. Price, London 60 (center)
E. Roedder, Bethesda 112 (left)
Ronan Picture Library, Cambridge 291 (top), 293, 295, 296, 298, 299, 301 (top), 304, 309; by courtesy of E. P. Goldschmid and Co., Ltd. 289, 290; by courtesy of the Royal Astronomical Society 288
Servizio Editoriale Fotografico, Turin 107 (top), 223
Shell U.K. Ltd., London 149 (bottom)
Space Frontiers Ltd., Havant, frontispiece, 14, 16 (top), 18, 19, 24, 25 (top), 29, 43 (top), 44 (bottom), 233; Official Naval Observatory Photograph 8
Spectrum Colour Library, London 65, 68, 71 (top), 80, 127 (top), 142, 166, 170 (top right), 235, 244 (bottom)
Tate Gallery, London 185, 294
A. C. Waltham, Nottingham, front jacket, 50 (top), 76 (top), 77 (top), 83, 108, 170 (bottom), 186 (bottom), 206, 215
G. J. Waugham, Durham 88
R. C. L. Wilson, Milton Keynes 55 (top right), 56, 59 (top), 87 (upper center right), 99, 172 (top), 190, 248
A. E. Wright, Birmingham 174
ZEFA, London 246

The Publishers have attempted to observe the legal requirements with respect to the rights of the suppliers of photographic materials. Nevertheless, persons who have claims are invited to apply to the Publishers.